Chromatography and its Varied Approaches

Chromatography and its Varied Approaches

Edited by **Carlos Dayton**

NY RESEARCH PRESS
New York

Published by NY Research Press,
23 West, 55th Street, Suite 816,
New York, NY 10019, USA
www.nyresearchpress.com

Chromatography and its Varied Approaches
Edited by Carlos Dayton

International Standard Book Number: 978-1-63238-082-1 (Hardback)

Printed in China.

Contents

Preface

This book aims to highlight the current researches and provides a platform to further the scope of innovations in this area. This book is a product of the combined efforts of many researchers and scientists, after going through thorough studies and analysis from different parts of the world. The objective of this book is to provide the readers with the latest information of the field.

A number of chromatographic techniques have been developed for the purpose of separation of mixed substances. Chromatography is the most functional and widely used method in chemical testing and experiments. It plays an important role in the development of chemistry, biology and medicine and more such related areas. Its elementary simplicity and simple operation make it versatile to apply along with other detection methods such as electrolysis, photometric methods etc. This book overviews latest developments in the techniques of chromatography which would prove to be valuable to the readers involved in separation science. It covers three major areas i.e. theories, its applications and sample preparations. This book targets not just students but also practitioners and professionals engaged in the field of chemistry, pharmacology, biotechnology and related genres.

I would like to express my sincere thanks to the authors for their dedicated efforts in the completion of this book. I acknowledge the efforts of the publisher for providing constant support. Lastly, I would like to thank my family for their support in all academic endeavors.

Editor

Use of Chromatography in Animal Ecology

Emma B. Casanave, M. Soledad Araujo and Gustavo H. López

Additional information is available at the end of the chapter

1. Introduction

There are several applications of chromatographic science in the field of animal ecology, biology and physiology. One of the most spread uses is identifying the bile acid profiles either by Thin Layer Chromatography, Gas chromatography or High Performance Liquid Chromatography in wild collected feces.

The analysis of wild collected feces is a fundamental ecological tool, applied in studies about population size, diet analysis and to identify the presence of a species in a certain area. Moreover, it is useful when it is necessary to monitor those species which are elusive, difficult to observe, threatened, nocturnal or sympatric, or exist in low densities. The identification can be done by external physical characteristics such as size, shape, odor and color, or through specific signals associated with the deposition of feces, for example tracks and scrapes [1,2]. However, this technique is sometimes useless because of the difficulties that exist in the correct identification of feces. Often, this sort of evidence is not present mainly because many of these external characteristics are sensitive to environmental conditions such as heat, desiccation or fast decomposition in humid and rainy regions, and can be affected by another type of factors: health, diet, size and age of the individual [3, 4].

Because of these reasons is that other techniques become necessary. The use of chromatographic techniques to identify or confirm the identity of wild collected feces is of great importance for biologists, because invasive procedures such as capture and manipulation are avoided. During the last years, the chromatographic determination of fecal bile acids has become a more precise method to identify unknown feces from the wild. The comparison of the whole pattern of fecal bile acids between field-collected scats and scats with known origin allows identifying the species. It has been demonstrated in several studies that fecal bile acids and their relative concentration follow patterns that are species-specific, particularly for mammals [5-7], including our recent studies in Xenarthra species [8].

In this chapter we will discuss the general chromatographic methods commonly used for the analysis of bile acids in biological samples. Moreover, we will emphasize the TLC and HPLC methodologies we used at our laboratory and the most relevant results we obtained, applied to the identification of mammal species, particularly of the Magnaorden Xenarthra (Mammalia).

2. Bile acids

Bile acids are the main components of bile and are among the first products isolated in a pure form, generally from gallbladder bile where it is present in high concentrations. They are acidic steroids produced during cholesterol metabolism in the liver and are secreted in the gallbladder or in the intestine.

Bile acids are produced by all vertebrates and show a great structural diversity among different species [9, 10]. Indeed, no other class of small molecules shows such striking variety across vertebrates. The diversity in bile salt chemical structures originates from differences in the two basic structural components of bile salt molecules: the 19-carbon (C_{19}) steroid nucleus and a side-chain. In all bile salts characterized to date, the four-ring cyclopentanophenanthrene ('steroid') nucleus (rings labelled A, B, C and D) is fully saturated. The A/B ring juncture is *cis* in most bile salts but *trans* in some species, a shift that greatly influences the overall shape of the steroid nucleus. A/B *trans* (5α) bile salts have an extended, planar orientation of the steroid rings, while A/B cis 5β) bile salts have a 'bent' orientation of the A ring relative to the other three rings [10].

The structural variation in the C_{19} of the steroid nucleus would be stereochemistry of the A/B ring juncture, sites of the oxo or hydroxy groups and orientation of hydroxyl groups (α or β). The structural variation in the side chain includes the length, the presence and orientation of OH groups, the presence of unsaturation, the stereochemistry of the C_{25} carbon atom and the site of carboxyl group in bile acids and OH group in bile alcohols [9, 10]. Other than length, further structural variation in the side-chain includes the presence and orientation of hydroxyl groups, the presence of unsaturation in the side-chain, and above all, the substituent on the terminal carbon atom, which are a hydroxyl group in bile alcohols and a carboxyl group in bile acids. Side-chain length and the state of oxidation at C_{27} is used to assign bile salts to three broad classes: C_{27} bile alcohols, C_{27} bile acids and C_{24} bile acids [10]. The structure of the side chain determines the class of compound; bile acids have a carboxyl group at the end of the chain and bile alcohols have a primary alcohol group [11] (Fig. 1).

Unlike the majority of biological small molecules, whose structures have remained constant since the formation of prokaryotic and eukaryotic cells, the molecular structure of bile acids show a distinctive evolution which parallels that of the vertebrate species which formed them (from C_{27} bile alcohols, C_{27} bile acids to C_{24} bile acids). The progressive nature of bile acid evolution is detectable between different genera, between members of different families and members of different orders [10].

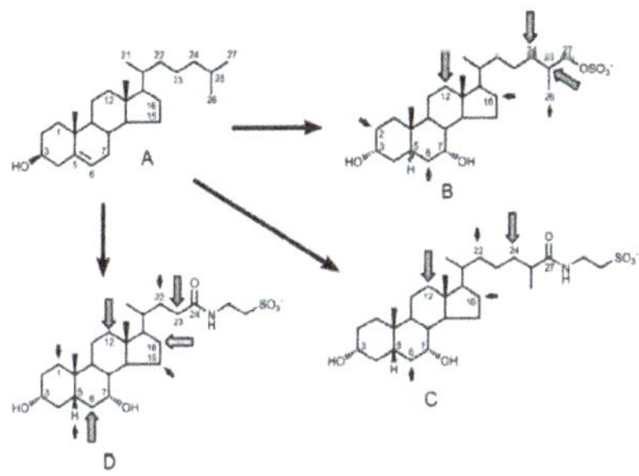

Figure 1. Sites of hydroxylation of the three main classes of bile salts. A: cholesterol; B: C_{27} bile alcohols (sulfates); C: C_{27} bile acids (as tauroamidated) and D: C_{24} bile acids (also as tauroamidated) (from [11]).

Conjugation of bile acids with the amino acids glycine and taurine occurs in the liver, before storage in the gallbladder and subsequent secretion into the duodenum via the bile duct. Within the intestinal lumen, bile acids interact with lipases and assist the lipolysis and absorption of fats, including fat-soluble vitamins, by the formation of mixed micelles. During enterohepatic circulation, the primary bile acids, CA and CDCA which are both synthesized in the liver may be modified by intestinal bacteria to form secondary bile acids, mainly DCA, LCA and UDCA (Fig. 2). In the colon of animals with a cecum, anaerobic bacteria remove the hydroxyl group at C_7 to form 7-deoxy bile acids. In the side chain, bile acids suffer deconjugation [11, 12-14].

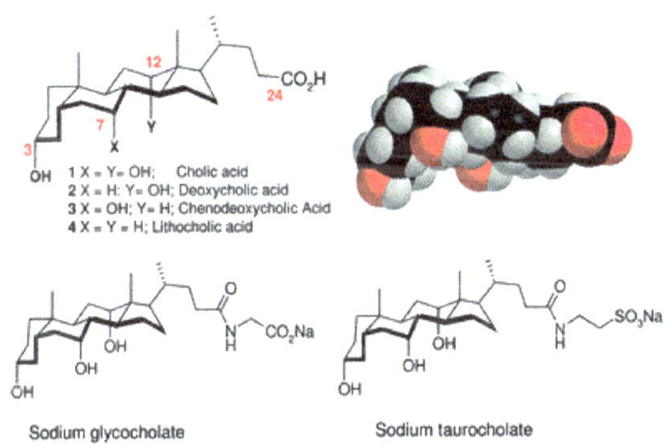

Figure 2. Structure of the main primary and secondary bile acids in mammals (from [15]).

Fecal bile acid patterns are complex due to bacterial metabolism during intestinal transit, which gives mono-, di- and/or –trioxo compounds, and also iso-(3β-hydroxy), urso-(7β-hydroxy) and lago-(12β-hydroxy) bile acids [16-18]. The colon bacteria deconjugate bile acids. The majority of the unconjugated bile acids are 7α-dehydroxylated, being LCA and DCA the predominant secondary fecal bile acids. They are reabsorbed from the colon, modified by hepatic enzymes and circulate in the enterohepatic circulation. Bile acids in the jejunum remain conjugated because of the absence of bacteria in the small intestine [19]. Moreover, in feces, certain bile acids are firmly bound to bacteria [12].

3. Bile acid analysis

The simultaneous separation and quantification of bile acids and their conjugates are challenging and have always presented technical difficulties due to marked differences in physicochemical properties, the presence of isomeric forms, their structural similarity or their relatively low concentrations in some biological samples [14, 20].

The most used techniques for the analysis of bile acid patterns in different biological samples include TLC coupled to densitometry, colorimetry, fluorimetry [5, 7, 21, 22], GC [16, 23, 24] and HPLC [25-27]. Moreover, non-chromatographic methods have been developed non chromatographic methods such as enzymatic, immunological and electrochemical ones. However, several of those methods are either not sensitive or are non-specific and have been displaced by GC, HPLC and radioimmunoassay. Radioimmunoassay, although it is highly sensitive, is not quite specific and overlaps of different bile acids occur [28]. On the other hand, both GC and HPLC are highly specific and sensitive to the extent of measuring even a few picomoles of bile acids [19].

Enzymatic methods using bile acid hydroxysteroid dehydrogenase are used for routine analysis of physiological fluids due to their simplicity. However, one of the main disadvantages is that differentiation of individual bile acids is not possible because total bile acids, rather than individual bile acids, are quantified [14].

Immunoassays of bile acids in different matrixes have provided relatively simple and sensitive methods for determining the concentration of selected sub-groups of bile acids. However, these assays do not provide full data for each of the individual bile acids.

Bile acids are haptens, nonantigenic low molecular weight, therefore complexes of bile acids covalently bound to large immunogenic molecules, for example bovine serum albumin, are required to obtain specific antibodies. In early experiments, [^{14}C]- or [^{3}H]-labelled bile acids were used as tracers for analysis of serum bile acids, whereas [^{125}I]-labelled tracers were used in later publications. RIA can detect bile acids at the picomol level and are useful for high-throughput analysis. RIA for different bile acids in serum was described early for CA [29] and for CDCA, UDCA and LCA [30].

EIA use a hapten-enzyme conjugate as a tracer instead of radioisotope-labelled antigens. The construction of an effective EIA system requires the preparation of an enzyme-labelled antigen that possesses a high specific activity without decreasing the enzyme activity or immunoreativity. The labelling enzymes that have been employed for these assays include alkaline phosphatase and peroxidase [31]. Therefore, a quality antibody, preferably a monoclonal one, is very important for reliable analysis of bile acids in biological fluids.

The drawback of most RIAs and EIAs remains the limited specificity of the antibody. Usually, antibodies demonstrate high structural specificity for bile acids but often there is a low level of cross-reactivity with bile acids having similar structure to the target bile acid.

4. Sample collection

The collection of samples for analysis is a critical step and constitutes a potential source of variation. For this reason, the procedure must be standardized. To determine the fecal bile acid pattern, it is necessary to collect feces from a known origin of the species under study. These reference feces can be collected in zoos, reserves or any place where one can assure the procedence of the fecal material. Moreover, feces can be collected during the capture and handling of a wild individual.

Feces are put in paper bags with silica gel so as to avoid humidity. They are correctly identified with: number, collector, date and place of collection, GPS coordinates, a microhabitat description, fecal external aspect (fresh, old, very old) and any information that it is considered relevant. Unknown feces from the study area, should match the external physical characteristics of the species under study, shape, size, colour and odour.

In our study, we analyzed the fecal bile acid patterns of different species from the Magnaorden Xenarthra (Mammalia). Our aim was to identify wild collected feces through their fecal bile acid patterns, and find in chromatography, a simple and rapid ecological tool.

For this purpose, samples were collected in zoos, reserves from the wild in different areas of Argentina. The studied species were: pichi (*Zaedyus pichiy*), screaming hairy armadillo or crying armadillo (*Chaetophractus vellerosus*), big hairy armadillo (*Chaetophractus villosus*), southern long-nosed armadillo (*Dasypus hybridus*), giant armadillo (*Priodontes maximus*), southern tamandua (*Tamandua tetradactyla*), giant anteater (*Myrmecophaga tridactyla*), southern three-banded armadillo (*Tolypeutes matacus*) and six-banded armadillo (*Euphractus sexcinctus*).

5. Sample pre-treatment and solvent extraction

The method of extraction will depend on the physicochemical properties of the biological material to be analyzed. Solid materials, such as feces, present particular problems due to the need of disrupting the matrix before extraction, either by homogenization, saponification, enzymatic digestion or lyophilization. Bile acids are hydrophobic steroids and hence, capable of extraction from aqueous media by organic solvents. The ease of extraction of bile acids from biological samples depends on four factors: the extent and type of conjugation, the number of hydroxyl groups, the state of ionization of the carboxyl group and the degree of binding to proteins. An increase in the number of hydroxyls or the presence of sulfates, glucuronides and to a lesser extent, glycine and taurine conjugates, will greatly increase the relative polarity (water solubility) of the bile acid. The state of ionization primarily depends upon the pK of the terminal group of the side chain, which varies from around 5.8 for the carboxyl group in the free bile acids to 4.3 for the glycine-conjugates and 1.9 for the sulfonic acid group of the taurine-conjugates. Although it would be expected that a decreased pH would favor extraction of bile acids because of the suppression of ionization, in fact bile acid extraction is, in many instances, favored at alkaline pH, presumably because of the reduced protein binding that occurs [12].

Although there are a wide variety of methods available for the extraction of bile acids from biological materials, there is no general method that covers all eventualities. For that reason, the researcher must establish the solvent system which best fits to the aim of each particular study.

Previously to the extraction step, feces are dried in oven at 30°C for one day to eliminate humidity and are stored in hermetic flasks in a dry and dark place. Feces are crushed with the use of a mortar and pestle and sieved with a fine mesh. They are macroscopically observed and rests of preys (plants, seeds, insects, bones, feathers) are eliminated. Fecal extracts must be filtered and concentrated to an adequate final volume, preferably under nitrogen stream.

For the extraction step, the ratio solvent/sample should be kept high, 20:1 or 10:1 (v/w). In our study, we extracted 1 gram of the fecal powder with a mixture of benzene:methanol (1:1 v/v). With the use of this solvent system, we could extract from feces the highest number of compounds in a single step. In addition, we also tested another extraction solvent system composed of dichloromethane:methanol (1:1 v/v). We replaced benzene with dichloromethane because it has the same extractive properties as benzene but is less toxic and thus easier to handle [3]. For HPLC, dried fecal extracts were resuspended in methanol and filtered with a polytetrafluorethylene filter of 0.45 μm before injection.

6. Thin-layer chromatography

TLC of bile acids has been used to differentiate feces of several mammal species, mainly carnivores such as the lesser grisson (*Galictis cuja*), guiña cat (*Leopardus guigna*), red fox (*Lycalopex culpaeus*), grey fox (*Lycalopex griseus*), pampas's fox (*Lycalopex gymnocercus*), puma (*Puma concolor*), jaguar(*Panthera onca*), snow leopard(*Panthera pardus ciscaucasica*), pandas and different species of bears [6, 7, 32-37]. It had also been applied to a wide variety of other species: manatees [38], sperm whales [32], storks and herons [39]. Our study was the first one which reported the use of TLC to differentiate Xenarthra species, through their fecal bile acid patterns [8].

TLC is a type of planar chromatography in which the stationary phase is a solid adsorbent of fine particles, and the mobile phase is liquid. Aluminum or glass plates are used and they are covered with a fine and uniform layer of the stationary phase which is generally, silica. In some cases, it is chemically modified so as to provide suitable resolution conditions depending on the analyte. The correct choice of the mobile or stationary phase is essential to obtain an efficient separation of compounds to be analyzed [40-42]. Mobile phase generally consists of a solvent system composed of 2, 3 or 4 components which vary in their polarity and selectivity and they can include water, organic solvents and buffers [43, 44]. Silica gel is the most common adsorbent used in TLC as stationary phase, especially for the identification and separation of steroids such as bile acids in several biological samples, due to its adsorptive properties, great active surface and high pore size [45].

TLC separation mechanism consists of the differential migration of the mixture components through the chromatographic plate, dragged by the moving solvent; the distance travelled by each compound depends on its chemical structure and its affinity for both phases. Several mechanisms are involved in the separation process and predominant forces depend on the phases and solutes properties [43, 44].

Although various separation techniques are commonly applied for the determination of bile acids in biological samples, TLC offers practical advantages, mainly its simplicity, economical equipment needed, ease of operation, short analysis time and high efficiency in analysing simultaneously a large number of samples [43, 46]. It enables reliable separation and analysis of a wide variety of compounds from different types of biological samples. Moreover, this technique is versatile because it can be modified using different types of mobile and stationary phases and visualizing methods [43, 45].

The choice of the correct stationary and mobile phases is of great importance and it depends upon the type of sample and the objectives of the study. Thus, the greatest problem in using TLC in specific cases, such as bile acid analysis, is the selection of suitable mobile phases [47].

6.1. Experimental protocol

Since there are a great variety of TLC methodologies to separate bile acids, in this section we report the technique used at our laboratory, which allows us to resolve the highest number of compounds in a single chromatographic run. The best results were obtained using HPTLC silicagel $60F_{254}$ plates with aluminium base of 20 x 20 cm, with a bed thickness of 0.2 mm. We follow the protocol cited in [5]. Each sample extract and standards for the most common bile acids for mammals are spotted on the plates with the use of a capillary tube: LCA, TCA, GCA, CA, CDCA, DCA, DHCA, GCDCA and CHOL. Bile acid standard stock solutions are prepared in methanol at a concentration of 0.1 %. We tested the use of different sample (75, 90, 105, 120, 150 and 180 μl) and standard (7.5, 15, 22.5 and 30 μl) quantities being spotted on the plates, so as to standardize the optimal concentrations for a better visualization.

As development solvent we use a mixture of toluene:aceticacid:water in a proportion of 5:5:1.5 v/v. We first saturate the mobile phase with water vapour and then we extract the rest of the liquid water. This gives more reproducibility to the analysis since variations in ambient humidity and temperature can alter the results.

Bile acid spots are visualized by spraying the plates with a revealing solution of anysaldehide:glacial acetic acid:sulphuric acid in a concentration of 0.5:50:1 v/v; plates are heated in oven at 150 °C for 15 minutes.

The bile acid pattern of each species is determined by the comparison of R_f values (relation between distance travelled by the compound and distance travelled by the eluent) and colour of the compounds with those of standard solutions. The relative intensity (concentration) of sample spots helps in the identification of the fecal bile acid pattern of each species.

6.2. Results and discussion

In our study, we could differentiate all the Xenarthra species through their fecal bile acid patterns. We detected 15 compounds in feces; seven bile acids, seven unidentified compounds and cholesterol.

Moreover, we showed the resolution of TLC because some pairs of bile acids with very similar R_f values, CDCA-DCA and CA-GCA, were discriminated. Although the R_f value is the main parameter to identify a compound, several other characteristics of the bands had to be carefully analysed in the chromatographic plates to determine bile acid patterns.

Some standard R_f values partially overlapped; however, we established a R_f range for each compound and, together with their colour, the spots were correctly identified. The concentration of the compounds was proportional to the relative intensity of each band in the chromatographic plate, as determined visually. Specific colours obtained through the oxidation by anisaldehyde for each bile acid allowed the identification of the spots in the plates. DHCA differed from the other standards because it showed two and, in some cases, three bands with a distinctively orange colour. Standards which showed more intense colours were LCA (dark green), TCA (blue-grey) and CHOL (dark pink); the rest of the bile acids had less intense colours (Fig. 3).

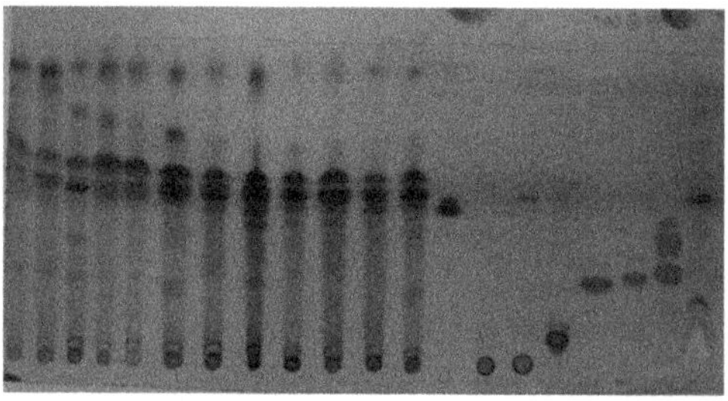

Figure 3. A chromatographic plate revealed with anisaldehyde:glacial acetic acid:sulphuric acid (0.5:50:1 v/v), shows bands of different colours. Left: 12 sample lines. Right: 8 standard bile acids.

Therefore, we stress the importance of spotting standard solutions together with the samples in each chromatographic plate to correctly identify compounds.

Even if TLC of fecal bile acids offers practical advantages such as simplicity and ease of operation [43, 46], like any laboratory technique it needs much of practical knowledge and skills [7] and it requires a careful and detailed analysis from the researcher.

In our study, TLC was useful to identify feces from Xenarthra species as it allowed the extraction, visualization and identification of fecal bile acids from individuals of all the studied species.

For several years, TLC has been used to identify wild-collected feces, mainly for species with low contents of vegetal material in their diets, such as carnivores [5, 35, 48, 49]. Nevertheless, in [34] they were not able to differentiate feces from *Lycalopex culpaeus* and

L. griseus in Chile; they argued that there was too much variability in the spot pattern even among feces from the same individual. In spite of that, in [3] they could discriminate feces from both species. Moreover, in [6] authors demonstrated that bile acid patterns were specific for some threatened carnivore species in Chile.

Several factors should be considered when analysing TLC results. One of them is the concentration of the sample; in the case of our study, samples from captive animals showed more intense bands while samples from wild animals had little concentration of some bile acids, making the correct identification sometimes difficult. However, we could find the optimal concentration, i.e., the concentration that allowed the detection of the compounds in the chromatographic plates, for wild and captive animals, being higher for wild ones.

A second factor that should be considered, especially when working with wild animals, is the effect of the type of diet, which has been reported in other investigations. It is possible that the presence of chemical substances in the feces, as products of the diet, has an effect on the detectability of bile acids, masking the spots [34] as it was observed previously, for example in coyotes [50].

Xenarthra species are omnivores and carnivores-omnivores [51], including invertebrates, small vertebrates, carrion, plant roots, tubers and seeds in their diets [52-55]. During plate running, other colour bands were observed; they corresponded to plant pigments which were yellow or orange for captive animals and green for wild ones. Differences in extract colours among species may reflect variations in the diet composition, mainly due to vegetal pigments which are naturally colored contrary to bile acid which are uncolored. Although we found some plant material in the feces, they had only small amounts and there was no evidence that prey items interfere with bile acid identification.

Age of wild-collected scats and thus, weathering, plays an important role in the analysis of feces by TLC [7, 36]. Some authors [7, 35, 48, 56] suggested that feces weathering can lower the concentration of bile acids in scats, leading to their erroneous identification. In our study, it was demonstrated that bile acids were clearly identified even in two-year old feces.

Further, small sample size can worsen TLC performance [36, 56, 57]; however, this was not the case of our study since we used sufficient quantity of dry pulverized fecal material to allow a good detection of the compounds [5].

The experience of the analyst in applying TLC is also important. For example, the spraying with the visualizing agent is not always uniform and some areas of the chromatographic plate may be uncoloured, resulting in an incorrect interpretation of the bands.

Several authors reported different TLC methods for the analysis and separation of bile acids. In [58] they reported a solvent system which allows group separation; free bile acids are separated with the first solvent system acetic acid:carbón tetrachloride:di-isopropyl ether:iso-amyl acetate:n-propanol:benzene (5:20:30:40:10:10 v/v), then with the second solvent system propionic acid:iso-amyl acetate:water:n-propano (15:20:5:10 v/v) they separated taurine and glycineconjugated standard bile acids. In [21], authors reported 15 solvent systems to separate bile acids, being the most efficient the acidic one.

In [59] they reported two solvent systems. The first one (isopropanol:glacial acetic acid, 93:7 v/v) allows separation of free bile acids, glycine conjugates and taurine conjugates from one another, and a second solvent system consisting of hexane:methylethylketone:glacial acetic acid, 56:36:8 v/v (Petcoff's solution), for the separation of the four free bile acids found in human bile. Authors in [48] used a TLC methodology to identify the fecal bile acid patterns in several carnivore species; plates were developed in a paper-lined, equilibrated bath containing Petcoff´s solution. After air drying, plates were visualized by spraying with a solution of acetic acid:sulphuric acid:anisaldehyde 50:1:0.5 (v/v) as revealing agent, and placing in a 120°C oven [60].

On the other hand, as cited in [33], they developed plates in Petcoff´s solution and then visualized with a 50:1:0.5 v/v solution of acetic acid:sulfuric acid:p-anisaldehyde to identify bile acids from grizzly and black bear feces. In [35] they used used a developing bath of Petcoff's solution to run the plates and they visualized bile acid bands by spraying the plates with 50% v/v sulfuric acid, which allowed them to observe spot colors in cougar and jaguar feces [3]. Sulfuric acid was also present in our visualizing agent. [61] reported a method to separate standard bile acids consisting of a reversed thin-layer chromatography using a mixture of metanol:water as the solvent system, in the concentration range 50–75% (v/v).

In [6] they reported a TLC method to determine the fecal bile acid patterns in carnivore feces. It consisted of an alkaline extraction to purify the bile acids from other lipophilic steroidal compounds; plates were eluted with Petcoff's solution and developed with a mixture of glacial acetic acid, water, sulfuric acid and 3,4-dimethoxybenzaldehyde.

From our results, and because TLC of fecal bile acids has proved to offer robust data to establish habitat use and to study food habits of some sympatric mammal species [6, 36], we assume that this technique would also be useful for future ecological studies in Xenarthra and in other species. Considering the scarcity of available information about some ecological and biological aspects of Xenarthra, these results, the first ones on the application of TLC for the identification of their feces, could be very important for future studies about the conservation, distribution and eco-physiology of this group.

7. Gas chromatography

GC and GC-MS have been widely used for quantitative and qualitative analysis of bile acids in different biological samples [10, 14, 37, 62-65]. In broad terms, GC works on the same principle as HPLC, however, in GC compounds are in gaseous form and an inert gas, like nitrogen or helium, is used for elution. The various compounds are resolved on the basis of their retention/elution behavior on the stationary phase of the column. There are two basic requirements for GC analysis, an appropriate column and derivatization of the compounds, to cause vaporization under the chromatographic conditions used [19].

Then, those techniques are limited by complex sample preparation, derivatization and most importantly, because requires the hydrolysis of conjugated bile acids into their free forms, prior to their analysis [66]; although GC-MS provides high sensitivity and resolution of isomeric bile acids, it is necessary to isolate and purify bile acids before analysis [14, 19].

Bile acids are present in unconjugated and/or conjugated form in biological fluids. They range from 1-2 μg/ml in plasma and urine to significant amounts in the intestinal content and as much as 10 mg/ml in the gallbladder bile. Bile acids are present in association with proteins, sterols and their esters, free or esterified fatty acids, bile pigments and water-soluble small molecules, which must be either removed before chromatography or the chromatographic conditions should be such that these compounds do not interfere in bile acid analysis. Since bile acids are present mainly as glycine or taurine conjugate in plasma and bile, unconjugated in feces and unconjugated as well as conjugated with glycine, taurine, glucuronic acid and sulfuric acid in the urine, the methods for isolation of bile acids from these sources need to be appropriately modified [19].

However, a major difficulty in quantitative analysis of fecal bile acids is their strong binding with the bacterial debris in the stool, and quantitative extraction is difficult. Stool contains bile acids, neutral sterols, cholesterol and its bacterial metabolites, plant sterols and their bacterial metabolites and fatty acids, which need to be removed before GC. Since fatty acids are less strongly retained on the capillary columns, bile acids can usually be quantitated in their presence.

For GC fecal bile acid analysis, several methods have been reported and most of them are quite complex [23, 62, 67].

In [68] they used continuous soxhlet extraction of aliquots of stool using chloroform and methanol. Extracts were derivatized and the methyl esters obtained were subjected to preparative TLC, respective bands were eluted and prepared for GC. On the other hand, in [23] they did a similar procedure but with tedious and longer steps. In alternate methods, bile acids from feces have been extracted with ammoniacal alcohol, methanol-hydrochloric acid, acetic acid-toluene, after removal of neutral sterols [23, 67].

In [19] they digested fecal samples with internal standard, extracted neutral sterols by repeated extractions, extracted bile acids with ethyl acetate, remove mineral acid and the residue was subjected to methyl ester formation; part of the aliquot was used for GC after trimethylsilylation. However, this method was not applicable to samples which contained conjugated bile acids.

Since the basic requirement for GC is that the compounds are in gaseous form at the column temperature, it is necessary to derivatize the polar functional groups. The carboxyl group in bile acids is often converted into the methyl ester. In addition, other bile acid esters have been used for GC, such as ethyl, n-propyl, isobutyl and n-butyl esters. Moreover, other derivatizations are common: of the hydroxyl group or the oxo group.

8. High performance liquid chromatography

The first report of the application of HPLC for bile acid separation was in 1976 [68 in 12]; and although its usefulness has been demonstrated for bile acid analysis in a wide variety of biological samples, up to the present it has never been used for the identification of species through their fecal bile acid pattern.

During the last years, HPLC has been one of the most used techniques for the analysis and identification of different lipid classes [44] and, particularly, for the separation and quantification of bile acids and their derivatives in different biological materials, due to its possibility to be coupled to a great variety of detectors [14, 27, 69-74]. HPLC presents several advantages in relation to other techniques such as high resolution, high sensitivity and specificity [44, 75].

A chromatographic system consists of four main components: a device for sample introduction, a mobile phase, a stationary phase (column) and a detector. The injector is simply required to allow introduction of the analytes into a flowing liquid stream without introducing any discrimination effects, being manually or automatically operated. The two components which are associated with the separation that occurs in a chromatographic system are the mobile and stationary phases. In HPLC the mobile phase is a liquid delivered under high pressure to ensure a constant flow rate, and reproducible retention times. The stationary phase is packed into a cylindrical column with a particulate support to which is bonded the stationary phase and it is capable of withstanding the high pressures which are necessary.

Detectors and columns vary depending on the objectives of the study and the type of sample. A small quantity of liquid sample is injected into the mobile phase which is flowing through the column, so the individual components of the mixture are separated [76, 77].

To take advantage of an HPLC column, it is necessary to use integrated chromatographic systems; all the components are specifically designed and coupled to get a maximum efficiency.

A chromatographic separation occurs if the components of a mixture interact to different extents with the mobile and/or stationary phases (Fig. 3), and therefore take different times to move from the position of sample introduction to the position at which they are detected. So, there are two extremes, as follows: a) all analytes have total affinity for the mobile phase and do not interact with the stationary phase; all analytes move at the same rate as the mobile phase, they reach the detector very quickly and are not separated, and b) all analytes have total affinity for the stationary phase and do not interact with the mobile phase; all analytes are retained on the column and do not reach the detector. The role of the analyst is therefore, based on a knowledge of the sample, to manipulate the properties of the stationary and/or mobile phases to move between these extremes and reach the desired separation [77].

Reversed-phase HPLC is the most used type of liquid chromatography for compound separation; it has been reported that almost 75% of all HPLC separations are done in the reverse mode. It is one of the most reliable and powerful methods for a direct and no destructive analysis of those compounds which are less volatile than others, polar and particularly, steroids and their conjugates [27]. Particularly, the resolution of compounds with similar polarities to bile acids results efficient if reversed-phase HPLC is used [78, 79].

The main characteristic of reversed-phase HPLC is that the mobile phase is more polar than the stationary phase, and therefore the components of the sample elute from the column in order of their hydrophobicity, number, position and configuration of hydroxyl groups in the steroid nucleus [44, 46, 80].

The retention behavior of a certain compound in reversed-phase HPLC depends on its solubility in and the partition between mobile and stationary phases. The polarity of a molecule controls the solubility in the mobile phase while the hydrophobic surface plays a key role in the interaction with the stationary phase. Bile acids have a hydrophobic region in the β face of the steroid nucleus, an area which is constant among common bile acids. Therefore, when increasing the number of OH groups in the α face of the molecule, increases the polarity and improve the solubility in an aqueous mobile phase. On the other hand, the presence of a β OH or carbonyl group, reduces the hydrophobic area of β face, resulting in a diminished of retention time [79]. The most polar bile acids are trihydroxilated tauroconjugated ones and the least polar are free monohydroxylated bile acids [74, 79].

The choice of the mobile phase will depend not only on the characteristics of the analyte, but also on the detection mode. Almost all methods use an elution gradient with a binary or tertiary solvent system, which are combined depending on their polarities; so, a satisfactory resolution of different lipid classes is achieved [44].

8.1. HPLC detectors

In different biological samples, bile acids are present as a complex mixture with a wide variety of polarities due to the presence of conjugated and free forms. For that reason is that the choice of the detection system will be largely governed by the structures of the analytes [14, 27].

There are several detectors that can be coupled to an HPLC system. However, at the present it is not available a simple, sensitive and direct method for the simultaneous analysis and quantification of individual bile acids in different tissues and fluids [20].

UV detectors have been widely used for the determination of bile acids in biological samples [25, 38, 81-83], mainly due to their relative low costs [44] and also because it allows the direct analysis of conjugated bile acids without prior derivatization [27, 71] and the reduction of the analysis time. However, HPLC-UV does not provide the required sensitivity and selectivity to detect the trace amounts of some bile acids or bile acid-sulfates in biological matrices [14, 20, 66, 73, 84], for example free bile acids [27, 80]. The simultaneous analysis of individual bile acids from a mixture presents several technical difficulties due to their similar and complex chemical structures, to their low UV absorbance, their low volatility, the presence of isomeric forms and the low concentration in certain type of samples [20, 27], particularly feces [46, 73, 84].

In several studies, it is necessary to pretreat samples which involve extraction, purification and derivatization so as to increase sensitivity and specificity [40].

Moreover, HPLC has been coupled to other types of detectors, for example RID, FLD and Electrochemical Detector.

The use of RID is limited by poor sensitivity and unstableness of base-line in gradient elution conditions, which is necessary for the separation of a complicated mixture with relatively short analysis time. Furthermore, the use of FLD or electrochemical detection does not detect non-chromophoric bile acids without a suitable pre-column derivatization [27]. Methods for sample derivatization are complex, with fluorescing chemicals [71, 85] or other complex derivatizations [27], which can introduce contaminants and produce secondary incomplete reactions, involving more complexity and longer analysis times [14, 27, 44, 73]. That is why derivatization methods are not the choice for a comprehensive analysis of fecal bile acid patterns [84].

The majority of lipids show a maximum absorbance in the range of wavelengths from 190 to 210 nm [44]; particularly, bile acids show different capabilities of absorption to UV light depending on their structure. UV detection can be used at 200–210 nm with moderate sensitivity for glycine- and taurine-amidated bile acids, which can be analyzed directly without previous derivatization steps, using conventional UV detectors [27, 71]. Unconjugated bile acids and their sulfated and glycosylated conjugates have a markedly low absorbance in this range [27, 80].

For that reason, in certain cases, depending on the aim of the study, is necessary to pretreat samples. This involves steps of extraction, purification and derivatization to increase sensitivity and specificity [40]. Although a great variety of pre-column derivatizations have been used to increase detection of bile acids to UV light, the complete conversion of compounds it is not assured and in general, required long steps that finally led to the loss of sample [14, 27].

In recent years, another detector used is ELSD. Theoretically, this detector responds to all solutes that are not volatile and the response is proportional to the mass of the solute present. The eluent is atomized in a stream of nitrogen and the finely divided spray passes down a heated chamber during which time the solvent is evaporated. The removal of the solvent produces a stream of particles which pass through a collimated beam of light. The scattered light at an angle in relation to the incident light, is focused onto a photomultiplier tube and the output is processed in an appropriate manner electronically. For a linear response, the droplet size must be carefully controlled. The device is fairly compact and relatively simple to operate. Detector sensitivity is in the range of 10-20 ng of solute. However, the magnitude of the response varies widely between different substances.

8.2. Liquid chromatography-mass spectrometry

Methods using FAB-MS and ESI-MS are applied in bile acid analysis. FAB-MS, however, has inferior quantitative capabilities and yields less intact ions (more in-source fragmentation) compared to ESI. Therefore, ESI-LC–MS remains a powerful technique for direct quantitative analysis of bile acids in biological matrices. Several methods have been developed and used to quantify bile acids in biological matrices using ESI-LC–MS [73, 86, 87].

The persistent need for rapid and sensitive methods has motivated efforts to exploit the high sensitivity, specificity and the minimal sample preparation requirements of HPLC–MS/MS for bile acid analysis in biological fluids, including the complex profile of secondary bile acids in feces resulting from colonic bacterial metabolism [88].

In conventional ESI the sample eluting from HPLC is pumped through a thin capillary (internal diameter approximately 0.1 mm) which is raised to a high potential (4 kV). Small charged droplets are sprayed from the ESI capillary into a stream of inert gas, generally nitrogen, at atmospheric pressure and travel down towards an orifice in the mass-spectrometer high-vacuum system. As the droplets traverse this path they become desolvated and reduced in size to such an extent that surface-coulombic forces overcome surface-tension forces and the droplets break up into smaller droplets. This process continues until they reach a point in which either an ion desorbs from a droplet or solvent is completely removed. This mechanism results in a beam of ions, which are directed to the mass spectrometer.

In a quadrupole mass analyzer MS/MS instrument, a precursor ion is mass-selected by first mass analyzer and focused into a collision cell preceding a second mass analyzer. The mass analyzers are arranged in series. Inert gas is generally introduced into the collision region and collisions occur between the precursor ion and inert gas molecules. In these collisions part of the precursor ions translational energy can be converted into internal energy, and as a result of single or multiple collisions an unstable excited precursor ions decompose to product ions. Product ions are mass-analyzed by the second mass analyzer [89].

MS/MS of steroids and bile acids has been an area of considerable study [90, 91]. When analyzed by negative-ion FAB or ESI-MS, bile acids, steroid sulphates and steroid glucuronides give abundant [M-H]–ions and few fragment ions. To gain structural information, MS/MS spectra are recorded. Any modification of the steroid skeleton or side chain would result in a change in the pattern of fragment ions [91].

LC-MS/MS is a simple, sensitive and rapid technique for the analysis of bile acids in complex matrix as feces [46].

8.3. Experimental protocol

In this section, we will describe the HPLC methods used to identify fecal bile acids in Xenarthra species.

We used a HPLC Thermo Finnigan made up of a gradient quaternary pump, on-line degassifier, a thermostatic module for the column and an UV-Vis detector with double wavelength, set at 200 and 210 nm. Analyses were performed on a reversed-phase C-18 column and a similar pre-column. We adapted the HPLC experimental protocol from [25] to our objectives.

Standard bile acids were CA, DCA, DHCA, CDCA, LCA, UDCA, GCA, sodium glicocholate, GDCA, GCDCA, TCA, TDCA, TLCA and CME, and CHOL. All were prepared in methanol at a concentration of 0.1%; and, when it was necessary, at 0.2% and 0.4%, and filtered through a 0.45 μm syringe driven HPLC filter.

As mobile phase we used a solution of 0.3% ammonium carbonate in water/acetonitrile. The analyses were performed under a linear gradient constituted as follows: 73:27 v/v for 10 minutes, 68:32 v/v for 10 minutes, 50:50 v/v for 10 minutes and from 50 to 100% acetonitrile until 60 minutes. The flow rate was 0.8 ml/min. Column temperature was set at 25-27°C, and the sample volume injected was 20 μl. For the most retained compounds, solvent gradient was modified, gradually increasing from the minute 30 to the minute 60 from 50% to 100% acetonitrile.

We run a mixture of all standards before each series of samples, so as to compensate any possible variation in retention times due to column efficiency loss or environmental conditions. Sample peaks were tentatively assigned comparing their retention times with those of standard solutions, previously injected in the same conditions.

For purposes of quantification, calibration curves for all standards in methanol were constructed. Regression equations were obtained through linear regression analysis and applied to the peak area of each bile acid as a function of concentration.

In each chromatogram we measured the retention time (R_t) expressed in minutes which is defined as the time that each compound spends in eluting from the column, the peak area (A) expressed in absorbance units (mAU), which determines the quantity of each compound present in the sample and the peak width measured at the base line (W).

8.4. Results and discussion

HPLC has been applied in a great variety of studies with different aims; however it has never been used as an ecological tool for differentiating species through fecal bile acids. In our work, we could identify the fecal bile acid pattern for all Xenarthra studied species, allowing the differentiation of wild-collected feces. No differences were observed between males and females, or between captive and wild individuals of the same species. All the species presented DHCA, TCA, TLCA and LCA, besides from CHOL, and three unidentified peaks.

When we calculated the percentage composition of each bile acid in samples from different individuals of the same species, we could observe that the compounds were the same;

however, their quantities vary between them. This can be due to intra and/or interindividual variations.

Variations in the composition of bile acids were reported for other biological samples. In [71] authors found differences in the hepatic composition of bile acids among different individuals of rats. Moreover, in [92] they recently reported diurnal variations, between morning and afternoon, in liver and plasma bile acids in laboratory mice. Bile acid circadian rhythm has been recognized for several years by various authors in different biological materials such as plasma, liver, gallbladder and intestinal contents. Our results coincide with those studies in which they did not report variations in the bile acid pattern of the species but only in the quantity of some of them [92].

Individual retention times for all 15 standard bile acids and CHOL, were obtained. A typical chromatogram of the mixture of standards shown in Fig. 4. Free bile acids, taurine-conjugated and glycine-conjugated bile acids were resolved in less than 32 minutes, achieving also the resolution of DCA and GDCA acids which could not be separated in previous works [25, 80].

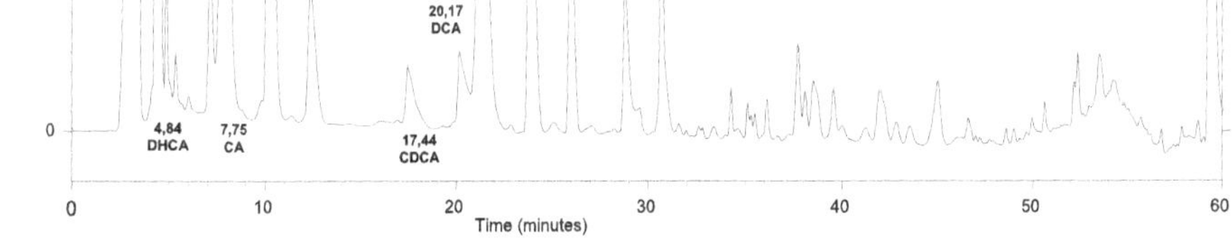

Figure 4. A typical HPLC chromatogram of the mixture of all the standards used. CA, GCA, TCA, CME, DHCA, DCA, CDCA, GDCA, GCDCA, UDCA, TDCA, TCDCA, LCA, TLCA, CHOL.

For several years, a great variety of analytical methods for quantitative determination of bile acids in various biological materials have been described [19, 27, 73, 74], including HPLC–UV assays [25, 38, 81, 83]. The different classes of bile acids have different absorption intensities to UV light, showing some limitations in their detection [27, 80]. The main disadvantages of these methods are the limited sensitivity and specificity of UV detection, especially in complex biological matrices, such as tissues and feces, due to the lack of a chromophore in the bile acid molecule [14, 66, 73, 84].

In our study, compounds showed greater absorbance at 200 nm, demonstrated by their larger areas. As it was demonstrated before by other authors [27, 69]; free bile acids were harder to detect than conjugated ones. Tauro-conjugated bile acids showed greater absorbance values than their corresponding glyco-conjugated ones.

We could demonstrate not only the great resolution power of HPLC even with a UV detector, but also we achieved the resolution of the majority of the identified compounds in a relatively short time. In previous works, although they reported separation of different bile acid types, running times were longer [71].

The C-18 column was the most appropriated for the resolution of the majority of the compounds. Although when doing HPLC, there are several parameters that should be taken into account to achieve an efficient separation of all bile acids; among the most important ones, are the composition and strength of the mobile phase. The strength is involved in the peaks symmetry control and in the bile acid elution order [40, 93]; in this work, for most retained compounds we increased the mobile phase strength increasing the proportion of acetonitrile, so for example, we could elute cholesterol. Under these conditions, the analysis reproducibility in terms of retention time and areas between different runs, even among long periods of inactivity, was satisfactory, allowing a precise identification of the peaks. Moreover, the high efficiency of this chromatographic system was demonstrated and this is reflected in the peak sharpness.

As we found unidentified compounds which did not coincide with any of the standards used, we are in process of identifying them and also confirming the identity of fecal bile acids found by HPLC-UV, through ESI-MS/MS.

Finally, we were able to differentiate all Xenarthra species through their fecal bile acid patterns, by HPLC. This study is of great relevance because is the first one in reporting HPLC as an ecological tool for the identification of wild-collected mammal feces. Moreover, it has proved to be a relatively simple method, without large preparation and derivatization steps, achieving resolution and identification of most of the compounds in a short time.

9. Conclusion

As a non-invasive method, the analysis of feces is a fundamental tool for field work in ecological studies, not only to identify the presence of certain species in a particular area [94], but also for studying threatened species or animals difficult to observe and trap. For the identification, the original fecal shape must be maintained; however, as several factors can corrode it through time, visual identification is not always reliable. Particularly, feces from Xenarthra are sometimes difficult to identify in the wild because they are, commonly, total or partially mixed with the substrate.

The chromatographic determination of fecal bile acids has become a more precise method to identify unknown feces from the wild. The comparison of the whole pattern of fecal bile acids between field-collected scats and scats with known origin allows identifying the species from fecal material, avoiding capture and manipulation of animals.

We were able to establish the fecal bile acid patterns for Xenarthra species, which were different for all of them. Moreover, these patterns were consistent among different individuals of the same species. As it was reported before for other mammal species [3, 5-7, 25, 35, 36], we confirm that chromatographic determination of fecal bile acids is a precise technique to identify unknown wild-collected feces.

Chromatographic techniques are the method of choice for a detailed analysis of bile acid profiles in biological samples. However, there is no a single satisfactory method for the analysis of all bile acids in biological fluids. All techniques present limitations in their specificity, analysis times or simplicity. Some types of samples, such as urine or feces, can contain complex mixtures of bile acids; other samples, such as tissues and cells, can contain small quantities of bile acids, being, then, easier to analyze. Thus, the choice of the analytical method will depend on the particular aim of the study and the type of sample. Certainly, in our case, the use of multiple analytical techniques (TLC, HPLC, HPLC-ESI-MS/MS) allows a precise resolution and confirmation of complex bile acid patterns.

HPLC-UV analysis has been widely used for the determination of bile acids in several biological fluids [68]. The main disadvantage is its limited sensitivity and specificity to UV detection in complex biological samples, such as tissues and feces [66].

In our study, both techniques, TLC and HPLC, presented advantages and disadvantages in the analysis of Xenarthra feces. Although TLC offers advantages such as relative simplicity, short analysis times, ease of operation and simultaneous analysis of a big number of samples, as reported previously [43, 47, 93], it can be affected by external factors such as environmental conditions, humidity and temperature, and by the operability of the researcher. Moreover, it has lower resolution power and reproducibility than HPLC [44].

TLC separation selectivity allowed resolving and visualizing CHOL, free and amidated bile acids in a single run, as it was reported before [95]. TLC could also resolve pairs of bile acids with very similar Rf values, for example CA-GCA and DCA-CDCA. Previous works have reported the performance of both methodologies for bile acid analysis in gallbladder bile or liver. In [95] they reported that TLC produces quick, precise and reproducible results, with shorter analysis times and low costs.

On the other hand, HPLC most important advantages were precision, higher resolution power than TLC, high sensitivity and specificity, as it was observed by other authors before [44, 75]. However, one disadvantage is longer analysis times due to the injection of only one sample at the time.

Diet of animals, and particularly in the case of Xenarthra species, can have an effect on the detectability of some bile acids. In our study, in feces with high contents of vegetal material, pigments appeared as colored bands in the TLC chromatographic plates and unidentified peaks in HPLC. However, they did not interfere with bile acid detection and we were able to correctly identify compounds.

With the application of HPLC we corroborate TLC results, not only confirming the presence of the compounds found by TLC but also identifying new compounds which were not resolved by TLC, in Xenarthra feces.

Bile acid detection constituted an essential step in both techniques. The choice of the detection system is governed mainly by bile acid structures. In TLC, spraying the plates with the revealing reagent is the critical step, in which it is important the ability and experience of the operator. Spraying must be uniform to avoid uncolored areas that later interfere with the correct identification of spots. In HPLC, different classes of bile acids have different absorption capabilities to UV light; this is a critical factor especially for free bile acids, showing limitations for their detection. However, in our study we could detect free bile acids.

In relation to the components of the chromatographic system, the choice of the mobile and stationary phase is very important, as they depend on the type of sample and the objectives of the study. In TLC we used silica gel as stationary phase, which is the most used phase for lipid analysis [47]. The solvent system composed of toluene:acetic acid:water (5:5:1.5 v/v) proposed by [5] was the most adequate for our aims, achieving a good resolution of the great majority of the compounds, except for some tauro-conjugated bile acids such as TDCA and TLCA. In HPLC we used a reversed-phase column with a ODS stationary phase, and the mobile phase composed of ammonium carbonate and acetonitrile; under these conditions we were able to separate almost all bile acids, including tauro-conjugated bile acids not resolved by TLC.

Thus, our work demonstrates that both techniques, TLC and HPLC, are complementary, and they should be used together to take advantage of the positive aspects of each one.

The analysis of fecal bile acids is a useful tool not only for ecological and biological studies, but also is of great clinical interest. These determinations can be helpful in the evaluation of intestinal, biliary and hepatic functions, and in the diagnosis and treatment of some related diseases such as colon cancer [46, 96, 97].

Chromatographic techniques have been widely used to identify bile acids in different biological materials, mainly in gallbladder bile of mammal species. However, our study is the first one to report the use of TLC and HPLC to differentiate Xenarthra species through their fecal bile acid patterns.

Finally, our work established the validity of fecal bile acids to differentiate close related species, being useful to assess habitat use and to study food habits of sympatric species. The determination of these species-specific patterns offers robust data for the elaboration of conservation strategies in the long term.

Author details

Emma B. Casanave
Cátedra de Fisiología Animal, Dpto. de Biología, Bioquímica y Farmacia, Universidad Nacional del Sur, Bahía Blanca, Buenos Aires, Argentina
Consejo Nacional de Investigaciones Científicas y Técnicas, (CONICET) Argentina

M. Soledad Araujo
Cátedra de Fisiología Animal, Dpto. de Biología, Bioquímica y Farmacia, Universidad Nacional del Sur, Bahía Blanca, Buenos Aires, Argentina

Gustavo H. López
Cátedra de Bioanalítica II, Dpto. de Biología, Bioquímica y Farmacia, Universidad Nacional del Sur, Bahía Blanca, Buenos Aires, Argentina

Abbreviation list

CA: Cholic acid
CDCA: Chenodeoxycholic acid
CHOL: Cholesterol
CME: Cholic-methyl-esther
DCA: Deoxycholic acid
DHCA: Dehydrocholic acid
EIA: Enzyme immunoassays
ELSD: Evaporative Light Scattering Detector
ESI: Electrospray Ionization
FAB: Fast Atom Bombardment
FLD: Fluorescence Detector
GC: Gas Chromatography
GCA: Glycocholic acid
GCDCA: Glycochenodeoxycholic acid
GDCA: Glycodeoxycholic acid
HPLC: High Performance Liquid Chromatography
HPTLC: High Performance Thin Layer Chromatography
LC: Liquid Chromatography
LCA: Lithocholic acid
MS: Mass Spectrometry
ODS: octadecylsilane
RIA: Radioimmunoanalysis
RID: Refractive Index Detector
TCA: Taurocholic acid
TCDCA: Taurochenodeoxycholic acid
TDCA: Taurodeoxycholic acid
TLC: Thin layer Chromatography

TLCA: Taurolithocholic acid
UDCA: Ursodeoxycholicacid
UV: Ultraviolet

Acknowledgement

This work was supported by Secretaría General de Ciencia y Tecnología, Universidad Nacional del Sur, PGI 24/B152.

10. References

[1] Soler L.; Lucherini, M.; Manfredi, M.C.; Ciuccio, M.; Birochio D. & Casanave E. B. (2009). Characteristics of defecation sites in the Geoffroy's Cat Leopardus geoffroyi. Mastozoología Neotropical. 16(2): 485-489.

[2] Castillo, D.F.; Birochio D.E.; Lucherini, M. & Casanave, E.B. (2011). Diet of adults and cubs of *Lycalopexgymnocercus*in Pampas grassland: a validation of the Optimal Foraging Theory?. Annales Zoologici Fennici 48: 251-256.

[3] Capurro, A.F., Novaro, A.J.; Travaini, A. & Romero, M. (1997). Improved bile acid thin layer chromatography to identify feces of Neotropical carnivores. Journal of Wildlife Management 61: 1424-1427.

[4] Chame, M. (2003). Terrestrial Mammal Feces: a Morphometric Summary and description. Memórias do Instituto Oswaldo Cruz 98: 71-94.

[5] Cazón, A.V. & Sühring, S.S. (1999). A technique for extraction and thin layer chromatography visualization of fecal bile acids applied to Neotropical felid scats. Revista de Biología Tropical 47: 245-249.

[6] Guerrero, C.; Espinoza, L.;Niemeyer, H.M. & Simonetti, J.A. (2006). Empleo del perfil de ácidos biliares fecales para evaluar uso del hábitat por carnívoros amenazados en el bosque maulino de Chile central. Revista Chilena de Historia Natural 79: 89-95.

[7] Khorozyan, I.G.; Cazón, A.V.; Malkhasyan, A.G. & Abramov, A.V. (2007). Using Thin-Layer Chromatography of Fecal Bile Acids to Study the Leopard (Panthera parduscis caucasica) Population.Biology Bulletin 34: 361–366.

[8] Araujo, M.S.; Ciuccio, M.; Cazón, A.V. & Casanave, E.B. (2010). Differentiation of Xenarthra (Mammalia) species through the identification of their fecal bile acid patterns: an ecological tool. 2010. Revista Chilena de Historia Natural 83: 557-566.

[9] Hofmann, A.F.; Hagey, L.R.; Krasowski, M.D. (2010). Bile salts of vertebrates: structural variation and possible evolutionary significance. Journal of Lipid Research 51: 226-246.

[10] Hagey, L.R.; Vidal, N.; Hofmann, A.F. & Krasowski, M.D. (2010a) Evolutionary diversity of bile salts in reptiles and mammals, including analysis of ancient human and extinct giant ground sloth coprolites. BMC Evolutionary Biology 10: 133.

[11] Hofmann, A.F. & Hagey, L.R. (2008). Bile Acids: Chemistry, Pathochemistry, Biology, Pathobiology and Therapautics. Cellular and Molecular Life Sciences 65: 2461-2483.

[12] Street, J.M., Trafford, D.J.H. & Makin, H.L.J. (1983). The quantitative estimation of bile acids and their conjugates in human biological fluids. Journal of Lipid Research 24: 491-511.

[13] Russell, D.W. (2003). The enzymes, regulation, and genetics of bile acid synthesis. Annual Reviews in Biochemistry 72: 137-174.
[14] Alnouti, Y.; Csanaky, I.L. & Klaassen, C.D. (2008). Quantitative-profiling of bile acids and their conjugates in mouse liver, bile, plasma, and urine using LC-MS/MS. Journal of Chromatography B 873: 209-217.
[15] Mukhopadhyay, S. & Maitra, U. (2004). Chemistry and biology of bile acids. Current Science 87 (12): 1666-1683.
[16] Eneroth, P.B.; Gordon, B.;Ryhage, R.R. & Sjövall, J. (1966a). Identification of mono and dihydroxy bile acids in human feces by gas-liquid chromatography and mass spectrometry. Journal of Lipid Research 7: 511-523.
[17] Eneroth, P.B.; Gordon, R. & Sjövall, J. (1966b). Characterization of trisubstituted cholanoic acids in human feces. Journal of Lipid Research 7: 524-530.
[18] Sjövall, J. (2004). Fifty years with bile acids and steroids in health and disease. Lipids 39(8): 703-722.
[19] Batta, A.K. & Salen, G. (1999). Gas chromatography of bile acids. Journal of Chromatography B. Biomedical Sciences and Applications 723(1-2): 1-16.
[20] [20] Huang, J.; Bathena, S.P.R.;Csanaky, I.L. & Alnouti, Y. (2011). Simultaneous characterization of bile acids and their sulfate metabolites in in mouse liver, plasma, bile, and urine using LC-MS/MS. Journal of Pharmaceutical and Biomedical Analysis.Journal of Pharmaceutical and Biomedical Analysis 55 (5): 1111–1119.
[21] Eneroth, P. (1963). Thin-layer chromatography of bile acids. Journal of Lipid Research 4(1): 11-16.
[22] Zarzycki, P.K.; Wierzbowska, M. & Lamparczyk, H. (1999). Retention and separation studies of cholesterol and bile acids using thermostated thin-layer chromatography. Journal of Chromatography A 857: 255-262.
[23] Grundy, S.M.; Ahrens, E.H. & Miettinen, T.A. (1965). Quantitative isolation and gas-liquid chromatographic analysis of total fecal bile acids.Journal of Lipid Research 6: 397-410.
[24] Batta, A.K.;Salen, G.; Batta, P.; Tint, G.S.;Alberts, D. & Earnest, D.L. (2002). Simultaneous quantitation of fatty acids, sterols and bile acids in human stool by capillary gas–liquid chromatography. Journal of Chromatography B 775: 153-161.
[25] Yeh, Y.H. & Hwang, D.F. (2001). High-Performance Liquid Chromatographic determination for bile components in fish, chicken and duck. Journal of Chromatography B 751: 1-8.
[26] Zhang, C.; Brown, S.B. & Hara, T.J. (2001). Biochemical and physiological evidence that bile acids produced and released by lake char (Salvelinusnamaycush) function as chemical signals. Journal of Comparative Physiology B 171: 161-171.
[27] Kakiyama, G.; Hosoda, A.; Iida, T.; Fujimoto, Y.; Goto, T.; Mano, T.; Goto, J. & Nambara, T. (2006). A direct method for the separation and quantification of bile acid acyl glycosides by High-Performance Liquid Chromatography with an Evaporative Light Scattering Detector. Journal of Chromatography A 1125: 112-116.
[28] Beckett, G.J.; Hunter, W.M. & Percy-Robb, I.W. (1978). Investigations into the choice of immunogen, ligand, antiserum and assay conditions for the radioimmunoassay of conjugated cholic acid.ClinicaChimicaActa 88 (2): 257–266.

[29] Simmonds, W.J.; Korman, M.G.; Go, V.L. &Hofmann, A.F. (1973). Radioimmunoassay of conjugated cholyl bile acids in serum. Gastroenterology 65(5): 705-11.
[30] Demers L.M. &Hepner G. (1976). Radioimmunoassay of bile acids in serum. Clinical Chemistry 22(5): 602-606.
[31] Baqir, Y.A.;Murison, J.; Ross, P.E. & Bouchier, I.A. (1979). Radioimmunoassay of primary bile salts in serum. Journal of Clinical Pathology 32:560-564. doi:10.1136/jcp.32.6.560.
[32] Hagey, L.R.; Crombie, D.L.; Espinosa, E.; Carey, M.C.; Igimi, H. & Hofmann, A.F. (1993). Ursodeoxycholic acid in the Ursidae: biliary bile acids of bears, pandas, and related carnivores. Journal of Lipid Research 34: 1911-1917.
[33] Picton, H.D. & Kendall, K.C. (1994). Chromatographic (TLC) differentiation of grizzly bear and black bear scats. International Conference Bear Research and Management 9: 497-501.
[34] Jiménez, J.E.;Yáñez, J.L. & Jaksic, F.M. (1996). Inability of Thin layer chromatography to distinguish feces from congeneric foxes by their bile acid contents.Acta Theriologica 41: 211-215.
[35] Fernandez, G.J.;Corley, J.C. & Capurro, A.F. (1997). Identification of cougar and jaguar feces through bile acid chromatography. Journal of Wildlife Management 61: 506-510.
[36] Taber, A.B.; Novaro, A.J.; Neris, N. & Colman, F.H. (1997). The food habits of sympatric jaguar and puma in the Paraguayan Chaco. Biotropica 29: 204-213.
[37] Solá, S.; Garshelis, D.L.; Amaral, J.D.; Noyce, K.V.; Coy, P.L.; Steer, C.J.; Iaizzo, P.A. & Rodrigues, C.M.P. (2006). Plasma levels of ursodeoxycholic acid in black bears, Ursusamericanus: seasonal changes. Comparative Biochemistry and Physiology, Part C 143: 204-208.
[38] Kuroki, S.; Schteingart, C.D.; Hagey, L.R.; Cohen, B.I.; Mosbach, E.H.; Rossi, S.S.; Hofmann, A.F.; Matoba, M.; Une, M. & Hoshita, T. (1988). Bile salts of the West Indian manatee, Trichechusmanatus latirostris: novel bile alcohol sulfates and absence of bile acids. Journal of Lipid Research 29: 509-522.
[39] Hagey, L.R.; Schteingart, C.D.; Ton-Nu, H.T. & Hofmann, A.F. (2002). A novel primary bile acid in the Shoebill stork and herons and its phylogenetic significance.Journal of Lipid Research 43: 685-690.
[40] Scalia, S. (1995). Bile acid separation. Journal of Chromatography B 671: 299-317.
[41] Pyka, A. &Dolowy, M. (2004). Separation of selected bile acids by TLC. III. Separation on Various Stationary Phases. Journal of Liquid Chromatography & Related Technologies 27(16): 2613-2623.
[42] Pyka, A. & Dolowy, M. (2005). Separation of selected bile acids by TLC. VII. Separation by reversed partition HPTLC. Journal of Liquid Chromatography & Related Technologies 28: 1573-1581.
[43] Sherma, J. & Fried, B. (2005). Thin layer chromatographic analysis of biological samples. A review. Journal of Liquid Chromatography & Related Technologies 28: 2297-2314.
[44] Peterson, B.L. & Cummings, B.S. (2006). A review of chromatographic methods for the assessment of phospholipids in biological samples. Biomedical Chromatography 20: 227-243.

[45] Dolowy, M. (2007). Separation of selected bile acids by TLC. IX. Separation on Silica gel 60 and on silica gel 60F254 aluminum plates impregnated with Cu (II), Ni (II), Fe (II) and Mn (II) cations. Journal of Liquid Chromatography and Related Technologies 30: 405-418.
[46] Perwaiz, S.; Tuchweber, B.; Mignault, D.; Gilat, T. & Yousef, I.M. (2001). Determination of bile acids in biological fluids by liquid chromatography–electrospray tandem mass spectrometry. Journal of Lipid Research 42: 114-119.
[47] Palamarev, C.E.; Kandinska, M.I. & Palamareva, M.D. (2004). Automatic Selection of Mobile Phases. IV. Thin-Layer Chromatography on Silica of Simple and Complex Organic Compounds. Journal of Liquid Chromatography & Related Technologies 27: 2003–2018.
[48] Major, M.; Johnson, M.K.; Davis, W.S. &Kellog, T.F. (1980). Identifying scats by recovery of bile acids. Journal of Wildlife Management 44: 290-293.
[49] Johnson, M.K.; Belden, R. &Aldred, D.R. (1984). Differentiating mountain lion and bobcat scats. Journal of Wildlife Management 48: 239-244.
[50] Quinn, T. & Jackman, W.R. (1994). Influence of diet on detection of fecal bile acids by thin-layer chromatography. Journal of Wildlife Management 58: 295-299.
[51] Redford, K.H. & Wetzel, R.M. (1985). "Euphractus sexcinctus". Mammalian Species 252: 1-4.
[52] Casanave, E.B.; Manfredi, C. & Luengos Vidal, E.M. (2003). Ecología comportamental de los armadillos en un pastizal serrano, In: A*ctas de las II Jornadas Interdisciplinarias del Sudoeste Bonaerense*, María del C. Vaquero y Mabel N. Cernadas de Bulnes (eds), pp. 115-125. EdiUNS, ISBN 987-9281-76-4, Bahía Blanca, Argentina.
[53] Dalponte, J.C. & Tavares-Filho, J.A. (2004). Diet of the yellow armadillo, Euphractus sexcinctus, in south-central Brazil. Edentata 6: 37-41.
[54] Soibelzon, E.; Daniele, G.; Negrete, J.; Carlini, A.A. & Plischuk, S. (2007). Annual diet of the little hairy armadillo, Chaetophractusvellerosus (Mammalia, Dasypodidae) in Buenos Aires Province, Argentina. Journal of Mammalogy 88: 1319-1324.
[55] McDonough, C.M. & Loughry, W.J. (2008). Behavioral ecology of armadillos, in: Vizcaíno, S.F. & W.J. Loughry (eds.), The biology of the Xenarthra, Univ. Press of Florida, Gainesville, pp. 281-293.
[56] Ray, J.C. (1996). Resource use patterns among mongooses and other carnivores in a central African rainforest. Ph.D. dissertation, University of Florida, Gainesville.
[57] Ray, J.C. & Sunquist, M.E. (2001). Trophic relations in a community of African rainforest carnivores. Oecologia 127: 395-408.
[58] Hofmann, A.F. (1961). Thin-layer adsorption chromatography of free and conjugated bile acids on silicic acid. Journal of Lipid Research Notes on Methodology 3: 127-128.
[59] Chavez, M.N. & Krone, C.L. (1976). Silicic acid Thin-Layer Chromatography of conjugated and free bile acids. Journal of Lipid Research 17: 545-547.
[60] Kritchevsky, D.;Martak, D.S. & Rothblat, G.H. (1963). Detection of bile acids in thin-layer chromatography. Analytical Biochemistry 5(5): 388-392.
[61] Sarbu, C.; Kuhajda, K. & Kevresan, S. (2001). Evaluation of the lipophilicity of bile acids and their derivatives by thin-layer chromatography and principal component analysis. Journal of Chromatography A 917: 361-366.

[62] Setchell, K.D.R.; Schwarz, M.; O'Connell, N.C.; Lund, E.G.; Davis, D.L.; Lathe, R.; Thompson, H.R.; Tyson, W.R.; Sokol, R.J. & Russell, D.W. (1998). Identification of a New Inborn Error in Bile Acid Synthesis: Mutation of the Oxysterol 7 α-Hydroxylase Gene Causes Severe Neonatal Liver Disease. Journal of Clinical Investigations 102: 1690-1703.

[63] Keller, S. & Jahreis, G. (2004). Determination of underivatised sterols and bile acid trimethylsilyl ether methyl esters by gas chromatography–mass spectrometry–single ion monitoring in faeces. Journal of Chromatography B 813: 199–207.

[64] Hagey, L.R.; Vidal, N.; Hofmann, A.F. & Krasowski, M.D. (2010b). Complex evolution of bile salts in birds. Auk. 127(4): 820–831. doi:10.1525/auk.2010.09155.

[65] Hagey, L.R.; Møller, P.R.; Hofmann, A.F. & Krasowski, M.D. (2010c). Diversity of bile salts in fish and amphibians: evolution of a complex biochemical pathway. Physiological and Biochemical Zoology 83: 308–321.

[66] Roda, A.; Piazza, F. & Baraldini, M. (1998). Separation techniques for bile salts analysis. Journal of Chromatography B 717: 263-278.

[67] Setchell, K.D.; Lawson, A.M.; Tanida, N. & Sjövall, J. (1983). General methods for the analysis of metabolic profiles of bile acids and related compounds in feces. Journal of Lipid Research 24: 1085-1100.

[68] Eneroth, P., Hellstrom, K. & Sjövall, J. (1968). A method for quantitative determination of bile acids in human feces. Acta Chemica Scandinavica 22: 1729-1744.

[69] Shaw, R. & Elliott, W.H. (1978). Bile acids. LVII. Analysis of bile acids by high-pressure liquid chromatography and mass spectrometry. Lipids 13: 971-975.

[70] Rossi, S.S.; Converse, J.L. & Hofmann, A.F. (1987). High pressure liquid chromatography analysis of conjugated bile acids in human bile: simultaneous resolution of sulfated and unsulfated lithocholyl amidates and the common conjugated bile acids. Journal of Lipid Research 28: 589-595.

[71] Prydz, K.; Kase, B.F. & Pedersen, J.I. (1988). High pressure liquid chromatography solvent systems for studies of bile acid biosynthesis. Journal of Lipid Research 29: 532-537.

[72] Sakakura, H.; Kimura N.; Takeda, H.; Komatsu, H.; Ishizaki, K. & Nagata, S. (1998). Simultaneous determination of bile acids in rat liver tissue by high-performance liquid chromatography. Journal of Chromatography B 178: 33-40.

[73] Lin, D-L.; Chang, H-C. & Chen, C-Y. (2000). Identification and quantitation of bile acids in bear bile by HPLC. Journal of Food and Drug Analysis 8(4): 283-288.

[74] Burkard, I.; Von eckardstein, A. & Rentsch, K.M. (2005). Differentiated quantification of human bile acids in serum by high-performance liquid chromatography-tandem mass spectrometry. Journal of Chromatography B 826: 147-159.

[75] Persson, E.; Lofgren, L.; Hansson, G.; Abrahamsson, B.; Lennernas, H. & Nilsson, R. (2007). Simultaneous assessment of lipid classes and bile acids in human intestinal fluid by solid-phase extraction and HPLC methods. Journal of Lipid Research 48: 242-251.

[76] Lima, E.S. & Abadía, D.S.P. (2002). High-Performance Liquid Chromatography of fatty acids in biological samples. Analytica Chimica Acta 465(1-2): 81-91.

[77] Robards, K.; Haddad, P.R. & Jackson, P.E. (1994). Principles and Practice of Modern Chromatographic Methods, Academic Press, London, San Diego, pp. 495.

[78] Ardrey, R.E. (2003). Liquid Chromatography, in: John Wiley & Sons, Ltd. (eds.), Liquid Chromatography-Mass Spectrometry: An introduction. The Atrium, Southern Gate, Chichester, West Sussex, England, pp. 7-31.

[79] Parris, N.A. (1977). Liquid chromatography separation of bile acids. Journal of Chromatography 133(2): 273-279.

[80] Sjövall, J.; Griffiths, W.J.; Setchell, K.D.R.; Mano, N. & Goto, J. (2010), Analysis of Bile Acids, in: Makin, H.L.J. & Gower, D.B. (eds.). Steroid Analysis, 2nd edition, Springer, pp. 837-966.

[81] [80] Torchia, E.C.;Labonté, E.D. & Angellon, L.B. (2001). Separation and quantitation of bile acids using an isocratic solvent system for High Performance Liquid Chromatography coupled to an Evaporative Light Scattering Detector. Analytical Biochemistry 298: 293-298.

[82] Carducci, C.N.; Matejka, M.; Ergüven, H.; Labonia, N. & Mamianetti, A. (1985). High-performance liquid chromatographic analysis of bile acids in hamster bile. Journal of Chromatography 337(1): 91-97.

[83] Mamianetti, A.; Garrido, D.; Carducci, C.N. & Vescina, M.C. (1999). Fecal bileacidexcretionprofile in gallstonepatients. Medicina 59 (3): 269-273.

[84] Natalini, B.; Sardella, R.; Camaioni, E.; Natalini, S. & Pellicciari, R. (2006). Dominant factors affecting the chromatographic behaviour of bile acids. Chromatographia 64(5): 343-349.

[85] Griffiths, W.J. & Sjövall, J. (2010). Bile acids: analysis in biological fluids and tissues. Journal of Lipid Research 51: 23-41.

[86] Kamada, S.M.; Maeda, M. & Tsuji, A. (1983). Fluorescence high-performance liquid chromatographic determination of free and conjugated bile acids in serum and bile using 1-bromoacetylpyrene as a pre-labeling reagent. Journal of Chromatography 272: 29-41.

[87] Roda A.; Gioacchini, A.M.; Cerrè, C. & Baraldini, M. (1995). High-performance liquid chromatographic-electrospray mass spectrometric analysis of bile acids in biological fluids.Journal of Chromatography B: Biomedical Sciences and Applications 665 (2): 281–294.

[88] Bootsma, A.H.; Overmars, H.; Van rooij, A.; Van lint, A.E.M.; Wanders, R.J.A.; Van Gennip, A.H. & Vreken, P. (1999). Rapid analysis of conjugated bile acids in plasma using electrospray tandem mass spectrometry: Application for selective screening of peroxisomal disorders. Journal of Inherited Metabolic Disease 22 (3): 307-310.

[89] Hong, Y.J.; Turowski, M.; Lin, J.T. & Yokoyama, W.H. (2007). Simultaneous characterization of bile acid, sterols, and determination of acylglycerides in feces from soluble cellulose-fed hamsters using HPLC with Evaporative Light-Scattering Detection and APCI-MS. Journal of Agricultural and Food Chemistry 55(24): 9750-9757.

[90] Griffiths, W.J.; Jonsson, A.P.; Liu, S.; Rai, D.K. & Wang, Y. (2001). Electrospray and tandem mass spectrometry in biochemistry. Biochemestry Journal 355: 545-561.

[91] Tomer, K.B.; Jensen, N.J.; Gross, M.L. & Whitney, J. (1986). Fast atom bombardment with tandem mass spectrometry for determination of bile salts and their conjugates. Biomedical and Environmental Mass Spectrometry 13: 265-272.

[92] Chatman, K.; Hollenbeck, T.; Hagey, L.; Vallee, M.; Purdy, R.; Weiss, F. & Siuzdak, G. (1999). Nanoelectrospray Mass Spectrometry and Precursor Ion Monitoring for Quantitative Steroid Analysis and Attomole Sensitivity. Analytical Chemistry 71 (13): 2358–2363.

[93] Zhang, Y.K.J.; Guo, G.L. & Klaassen, C.D. (2011). Diurnal Variations of Mouse Plasma and Hepatic Bile Acid Concentrations as well as Expression of Biosynthetic Enzymes and Transporters. PLoS ONE 6(2): e16683. doi:10.1371/journal.pone.0016683.

[94] Zarzycki, P.K.; Kulhanek, K.M.; Smith, R.; Bartoszuk, M.A. & Lamparczyk, H. (2005). Planar Chromatography versus Column Chromatography: A Performance Comparison. LCGC North America 23(3): 286-300.

[95] Ciuccio, M.; Araujo, M.S. & Casanave, E.B. (2007). Estudio ecológico de las especies de armadillos presentes en cercanías de Bahía Blanca. In *Ambiente natural, campo y ciudad: Estrategias de uso y conservación en el Sudoeste Bonaerense.* N. Cazzaniga y María del C. Vaquero, pp. 183-190, EdiUNS, ISBN 978-987-23429-1-3, Bahía Blanca, Argentina.

[96] Müllner, S., Hofmann, R., Saar, K. & Karbe-Thönges, B. (1992). Economic assay for the evaluation of bile acid sequestrants using ox bile and quantitative TLC analysis.Journal of planar chromatography, modern TLC 5(6): 408-416.

[97] Kasbo, J.; Saleem, M.; Perwaiz, S.; Mignault, D.; Lamireau, T.; Tuchweber, B. & Yousef, I. (2002). Biliary, fecal and plasma deoxycholic acid in rabbit, hamster, guinea pig, and rat: Comparative study and implication in colon cancer. Biological and Pharmaceutical Bulletin 25(10): 1381-1384.

[98] Meance, S.; Boutron-Ruault, M.C.; Myara, A.; Gerhardt, M.F.; Marteau, P.; Lavergne, A.; Franchisseur, C. & Bouley, C. (2003). Fecal primary bile acids and serum cholesterol are associated with colorectal adenomas. Digestive Diseases and Sciences 48(9): 1751-1757.

Chapter 2

Purification of Phospholipases A_2 from American Snake Venoms

Rodrigo G. Stábeli, Rodrigo Simões-Silva, Anderson M. Kayano, Gizeli S. Gimenez, Andrea A. Moura, Cleópatra A. S. Caldeira, Antonio Coutinho-Neto, Kayena D. Zaqueo, Juliana P. Zuliani, Leonardo A. Calderon and Andreimar M. Soares

Additional information is available at the end of the chapter

1. Introduction

Snake venoms are a complex mixture of compounds with a wide range of biological and pharmacological activities, which more than 90% of their dry weight is composed by proteins, comprising a variety of enzymes, such as proteases (metalo and serine), phospholipases A_2, L-aminoacid oxidases, esterases, and others [1-5]. A great number of proteins were purified and characterized from snake venoms [1, 2]. Some of these proteins exhibit enzymatic activity, while many others are non-enzymatic proteins and peptides. Based on their structures, they can be grouped into a small number of super-families based on remarkable similarities in their primary, secondary and tertiary structures, however showing distinct pharmacologic effects [3].

One of the most important protein super-families present in snake venoms are the phospholipases A_2 (PLA_2, E.C. 3.1.1.4), a class of heat-stable and highly homologous enzymes, which catalyse the hydrolysis of the 2-acyl bond of cell membrane phospholipids releasing arachidonic acid and lysophospholipids (Figure 1). These proteins are found in a wide range of cells, tissues and biological fluids, such as macrophages, platelets, spleen, smooth muscle, placenta, synovial fluid, inflammatory exudate and animal venoms. There is a high medical and scientific interest in these enzymes due to their involvement in a variety of inflammatory diseases and accidents caused by venomous animals. Since the first PLA_2 activity was observed in Naja snake venom, PLA_2s were characterized as the major component of snake venoms, being responsible for several pathophysiological effects caused by snake envenomation, such as neurotoxic, cardiotoxic, myotoxic, cytotoxic, hypotensive and anti-coagulant activities [1-10].

Phospholipases constitute a diverse subgroup of lipolytic enzymes that share the ability to hydrolyse one or more ester linkages in phospholipids, with phosphodiesterase as well as acyl hydrolase activity. The amphipathic nature of phospholipids creates obstacles for the enzymes, as the substrates are assembled into bilayers or micelles and are not present in significant amounts as a single soluble substrate [11]. According to Waite [12], all phospholipases target phospholipids as substrates, they vary in the site of action on the phospholipid molecule, their function and mode of action, and their regulation. Phospholipases function in various roles, ranging from the digestion of nutrients to the formation of bioactive molecules. This diversity of function suggests that phospholipases are relevant for life; the continuous remodelling of cell membranes requires the action of one or more phospholipases. The most common phospholipids in mammalian cells are phosphatidylcholine (PC), phosphatidylserine (PS), phosphatidylinositol (PI) and phosphatidylethanolamine (PE). The plasma membrane of most eukaryotic cells contains predominantly PC and sphingomyelin in the outer leaflet, and PI, PE and PS in the inner leaflet [11].

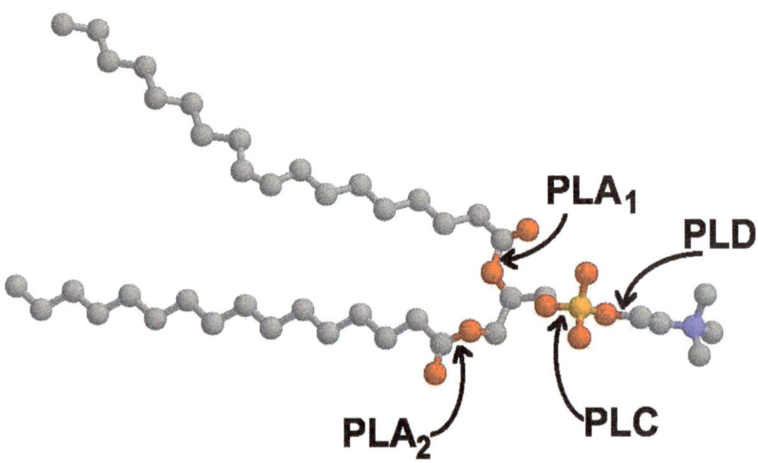

Figure 1. Phospholipase hydrolysis specificity sites in a 1,2-diacylglycerolphospholipid molecule (structure design from the ACD/l Lab. via Chem. Sketch – Freeware Version 1994 – 2009 software).

Phospholipases are classified according to their site of action in the phospholipid molecule. Thus, a phospholipase A_1 (PLA_1) hydrolyzes the 1-acyl group of a phospholipid, the bond between the fatty acid and the glycerine residue at the 1-position of the phospholipid. A phospholipase A_2 (PLA_2) hydrolyzes the 2-acyl, or central acyl, group and phospholipases C (PLC) and D (PLD), which are also known as phosphodiesterases, cleave on different sides of the phosphodiester linkage (Figure 1). The hydrolysis of a phospholipid by a PLA_1 or a PLA_2 results in the production of a lysophospholipid. The phospholipase metabolites are involved in diverse cellular processes including signal transduction, host defense (including antibacterial effects), formation of platelet activating cofactor, membrane remodeling and general lipid metabolism [12-14].

According to the latest classification [6], these proteins constitute a superfamily of different enzymes belonging to 15 groups and their subgroups including five distinct types of enzymes: the ones called secreted PLA_2 ($sPLA_2$), the cytosolic ($cPLA_2$), the Ca^{2+} independent ($iPLA_2$), the acetyl-hydrolases from platelet activating factors (PAF-AH) and the liposomal. The classification system groups these enzymes considering characteristics such as their origin, aminoacid sequence and catalytic mechanisms, among others.

The $sPLA_2$s have a Mr. varying from 13,000 to 18,000, usually containing from 5 to 8 disulphide bond. They are enzymes that have a histidine in the active site and require the presence of the Ca^{2+} ion for the catalysis. Phospholipases A_2 from the IA, IB, IIA, IIB, IIC, IID, IIE, IIF, III, V, IX, X, XIA, XIB, XII, XIII, XIV groups are representative of the $sPLA_2$s. The $cPLA_2$s are proteins with Mr between 61,000 to 114,000 that also use a serine in the catalytic site (groups IVA, IVB, IVC, IVD, IVE, IVF). The $iPLA_2$s are enzymes which also use a serine for catalysis (groups VIA-1, VIA-2, VIB, VIC, VID, VIE, VIF). The PAF-AH are phospholipases A_2 with serine in the catalytic site that hydrolyze the acetyl group from the *sn*-2 position of the platelet activating factors (PAF), whose representative groups are VIIA, VIIB, VIIIA, VIIB. The liposomal PLA_2s are assembled in group XV and are enzymes with an optimum pH close to 4.5 that have preserved histidine and aspartate residues, suggesting the presence of the catalytic triad Ser/His/Asp and also a supposed sequence N-terminal sign and N-bond glycosylation sites [6].

With the discovery of a great variety of phospholipase A_2 in the last decade and the present expansion of the research in the area, more PLA_2s should be discovered yet. Phospholipase A_2 found in snake venoms ($svPLA_2$s) are classified into groups I and II. The phospholipase A_2 from group I have two to three amino acids inserted in the 52-65 regions, called "elapid loop", being isolated from the snake venoms of the Elapidae family (subfamily: Elapinae and Hydrophiinae). The ones from group II are characterized by the lack of the Cys11-Cys77 bond which is substituted by a disulphide bond between the Cys51-Cys133, and besides that had five to seven amino acids extending the C-terminal regions, being bound in snake venoms of the Viperidae family (subfamily Viperinae and Crotalinae) [15,16].

The myotoxic PLA_2s of the IIA class have been subdivided in two main groups: The Asp49, catalytically active; and the Lys49, catalytically inactive. The essential co-factor for the phospholipase A_2 catalysis Ca2+. The phospholipase A_2 Asp49 require calcium to stabilize the catalytic conformation, presenting a calcium bond site that is constituted by the β-carboxylic group of Asp49 and the C=O carbonylic groups of the Tyr28, Gly30 and Gly32. The presence of two water molecules structurally preserved complete the coordination sphere of Ca^{2+} forming a pentagonal pyramid [9,15].

The catalytic mechanism of the PLA_2-phospholipid involves the nucleophilic attack of a water molecule to the *sn*-2 bond of the phospholipid substrate (Figure 2). In the proposed model, the proton from position 3 of the imidazole ring of the His48 residue involved in a strong interaction with the carboxylate group of the Asp49 prevents the imidazole ring rotation to occur and keeps the nitrogen at position 1 of this ring, in an appropriate special position. A water molecule then promotes the nucleophilic attack to the carbon of the ester group of the substrate and, at this moment, the imidazole ring of the His48 receives a proton from the water molecule, favoring the reaction. Subsequently to the acyl-ester bond hydrolysis at the sn-2 position of the phospholipid, this proton is donated by the imidazole ring to the oxygen, which then forms the alcohol group of the lysophospholipid to be released together with the fatty acid [15,17].

The Ca^{2+} ion, coordinated by the Asp49 residue, a water molecule and the oxygen atoms from the Gly30, Trp31 and Gly32 (not shown), are responsible for the stabilization of the reactive intermediary [15].

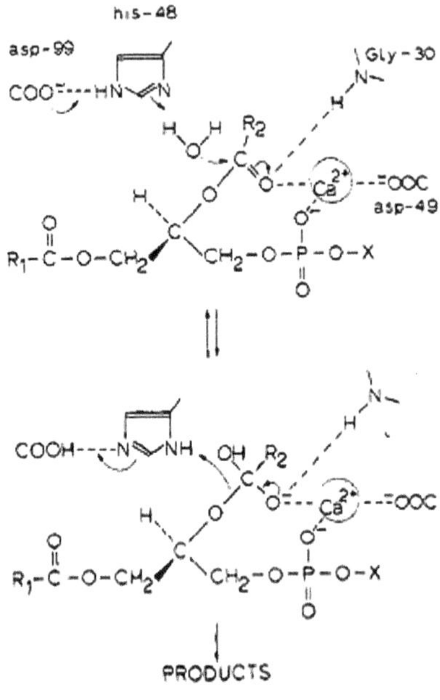

Figure 2. Schematic representation of the catalysis mechanism proposed for the PLA_2s. Interaction of the residues from the catalytic site of $sPLA_2s$ and the calcium ion with the transition state of the catalytic reaction in which a water molecule polarized by the His48 and Asp99 residues binds to the carbonyl group of the substrate [18].

The substitution of the Asp49 residue by the Lys49 significantly alters the binding site of Ca^{2+} in the phospholipase A_2, preventing its binding and resulting in low or inexistent catalytic activity. Thus, the Asp49 residue is of fundamental importance for the catalytic mechanism of the phospholipase A_2. It is likely that this occurs due to its capability of binding and orienting the calcium ion, however, there is no relevant difference between Asp49 and Lys49 in relation to the structural conformation stability of these enzymes [9,15,19].

The absence of catalytic activity does not affect myotoxicity. Most snake PLA_2s from the *Bothrops* genus already described are basic proteins, with isoelectric point between 7 to 10, showing the presence or absence of catalytic, myotoxic, edematogenic and anticoagulating activities [9,20].

On the other hand, acid PLA_2s present in Bothrops snake venoms were not studied as well as basic PLA_2s, resulting in little knowledge regarding the action mechanism of these enzymes [21-25].

PLA_2s catalytic activity represents a key role in envenomation pathophysiology, however, recent studies have shown that some effects are independent of PLA_2s catalytic activity, such as myotoxicity [19,26]. The absence of a tight correlation between PLA_2 catalytic and non-catalytic activities, together with the diversity of biological effects produced by these proteins increases the scientific interest in the understanding of the structural basis of PLA_2 mechanisms of action.

Evidences suggest that these activities can be mediated by interactions between PLA_2s and endogen acceptors on the target cell membrane [27-29].

2. PLA_2 purification

Snake venom components, obtained with high degree of purity, could be used for the understanding of the role of these components in the physiopathological processes resulted from poisoning, as well as biotechnological/nanotechnological applications. Hence, many purified PLA_2s from snake venoms, as well as epitopes of these molecules, are being mapped in order to identify determinants responsible for the deleterious actions seen, as well as possible applications in biotechnological models.

New advances in materials and equipments have contributed with protein purification processes, allowing the obtaining of samples with high degree of purity and quantity. These advances have allowed process optimization, providing reduction of steps, reagents use and thus avoiding the unnecessary exposure to agents that may, in some way, alter the sample's functionality or physical-chemical stability.

Thus, the selection of adequate techniques and chromatographic methods oriented by physical chemical properties and biological/functional characteristics, are of fundamental importance to obtain satisfactory results. The information pertinent to protein structure, such as the homology to others already purified, should be taken into consideration and could make the purification processes easier.

Ion exchange chromatography was introduced in 1930 [30] and still one of the main techniques used for protein purification. It has been extensively used in single step processes as well as associated to other chromatographic techniques. Ion exchange chromatography allows the separation of proteins based on their charge due to amino acid composition that are ionized as a function of pH.

Proteins with positive net charge, in a certain pH (bellow their isoelectric point), can be separated with the use of a cation exchange resin and on the other hand, proteins with negative net charge in a pH value above their isoelectric point, can be separated with an anion exchange resin.

Scientific publications have shown that the use of cation-exchange resins is a very efficient method to obtain PLA_2s from bothropic venoms, particularly those with alkaline pH (Table 1). The versatility of this technique can be observed in the work done by Andriao-Escarso et al. [21] who compared the fractioning of many bothropic venoms. In this work, the venoms were fractioned in a column containing CM-Sepharose® (2 x 20 cm), equilibrated with ammonium bicarbonate 50 mM pH 8.0 and eluted with a saline gradient of 50 to 500 mM of the same reagent. Under these conditions, MjTX-I and MjTX-II from *B. moojeni* snake venom were co-purified (isoforms of PLA_2 with pIs of 8.1 and 8.2 values, respectively). The same occurs with *B. jararacussu* venom, where the BthTX-I and BthTX-II were purified. However, the most expressive result was observed with *B. pirajai* venom, from which 3 isoforms of myotoxins, called as PrTX-I (pI 8.50), PrTX-II (pI 9.03) and PrTX-III (pI 9.16) were purified. In the above cases, it is important to note that the protein elution occurs always following pIs increasing value. In our lab we used this technique routinely in order to isolate myotoxins from bothropic venoms, which can be observed in the chromatograms shown in Figure 3.

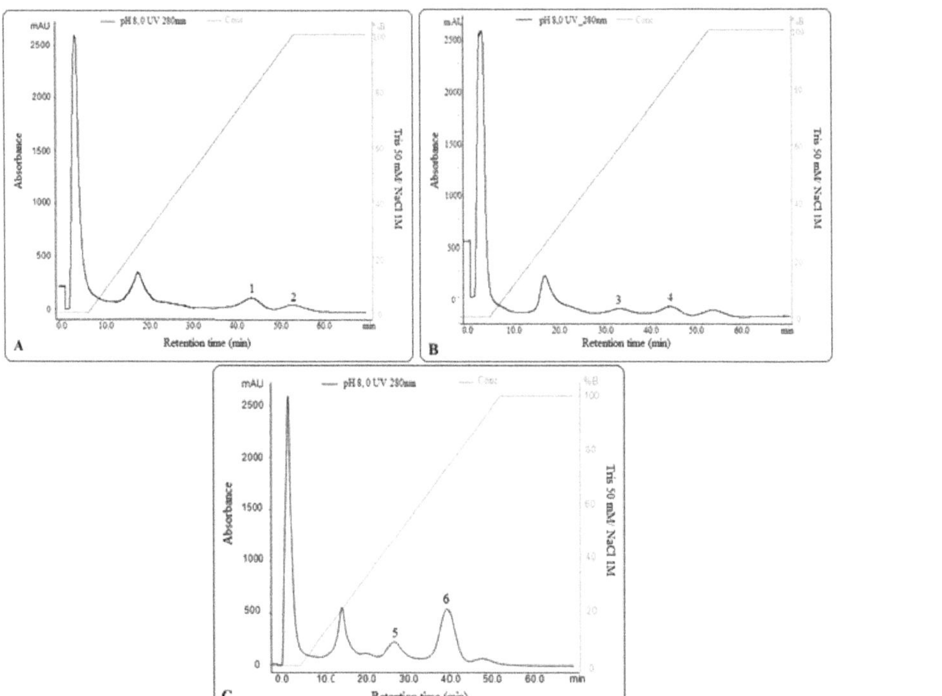

Figure 3. Chromatographic profile using CM-sepharose® Column 1ml (*Hitrap*) equilibrated with Tris 50 mM buffer (buffer A) and eluted with a linear gradient of Tris 50 mM/NaCl 1 M (buffer B) in pH 8.0. **A.** Chromatography of the crude venom from *Bothrops brazili* **B.** Chromatography of the crude venom from *Bothrops moojeni* **C.** Chromatography of the crude venom from *Bothrops jararacussu*. Absorbance read at 280 nm. ,2,3,4,5 and 6 marks indicate the fractions corresponding to the PLA_2s of each venom.

Species	PLA_2	PLA_2 Activity	MW (kDa)	pI	Access Number (Uniprot)	Purification strategy	Ref.
Agkistrodon bilineatus	PLA_2	Absence	14.0	10.2	Q9PSF9	Gel filtration chromatography on Sephadex G-75® and then submitted to ion-exchange on CM-Cellulose® column.	[78]
Agkistrodon contortrix contortrix	PLA_2	Presence	14.0			Ion-exchange chromatography on DEAE-Cellulose® column, followed by affinity chromatography with immobilized BSA and then submitted to gel filtration on Cellulofine GCL-2000® column.	[79]
Agkistrodon contortrix laticinctus	MT1	Absence	14.0	9.0	49121	Anion-exchange chromatography on Waters DEAE-5PW® column and then submitted to cation-exchange on Protein Pak SP-SPW® column.	[80]
Agkistrodon contortrix laticinctus	ACL-I	Presence	14.0			Gel filtration chromatography on Superdex-200® column and then submitted to ion-exchange on CM-Sepharose FF® column.	[81]
Atropoides nummifer	Myotoxin IH	Absence	16.0			Cation-exchange chromatography on CM-Sephadex C-25® column.	[82]
Atropoides nummifer	Myotoxin I	Absence	16.0			Cation-exchange chromatography on CM-Sephadex C-25® column.	[83]
Atropoides nummifer	Myotoxin II	Absence	13.7	8.7	P82950	Cation-exchange chromatography on CM-Sephadex C-25® column.	[84]
Bothriechis (Bothrops) schlegelii	Miotoxina II	Presence	15.0	>9.5	P80963	Ion-exchange chromatography on CM-Sephadex® column.	[85]
Bothrocophias hyoprora	PhTX-I	Presence	14.2			Reverse Phase chromatography on Bondapack® C-18 column.	[86]
Bothropoides insularis	SIII-SPVI	Presence	15.0			Gel filtration chromatography on Sephadex G-150® column and then submitted to SP-Sephadex C25® column.	[87]
Bothropoides insularis	BinTX-I BinTX-II	Presence Presence	13.9 13.7	5.0 4.4	Q8QG87 P84397	Reverse Phase chromatography on Vydac® C18 column.	[88]
Bothropoides insularis	Bj PLA_2	Presence	13.9	8.6		Gel filtration chromatography on Superdex 75® column and then submitted to cation-exchange on Protein pack SP-5PW® column and Reverse Phase chromatography on μ-Bondapack® C18 column.	[89]
Bothropoides jararaca	BjPLA_2	Presence	14.0		P81243	Ion-exchange chromatography on DEAE Sephacel® column and then submitted to Reverse Phase chromatography on Ultrapore RPRC-C3® column.	[90]
Bothropoides jararaca	PLA_2	Presence	14.2	4.5	Q9PRZ0	Gel filtration on Sephacryl S-200® column and then submitted to reverse phase on Pep-RPC HR 5/5® column.	[91]
Bothropoides pauloensis	BpPLA_2	Presence	15.8	4.3	D0UGJ0	Cation-exchange chromatography on CM-Sepharose® column followed by Phenyl-Sepharose CL-4B® column and then submitted to reverse phase chromatography on C8 column.	[23]
Bothropoides pauloensis	BnSP-7	Absence	13.7	8.9	Q9IAT9	Cation-exchange chromatography on CM-Sepharose® column or heparin agarose® column.	[26]
Bothrops alternatus	BA SpII RP4	Presence	14.1	4.8	P86456	Gel filtration chromatography Sephadex G-75® column followed by reverse phase chromatography on C18 column.	[92]
Bothrops alternatus	PLA_2	Presence	15.0	5.0		Gel filtration chromatography on Sephadex G-50® column followed by ion-exchange on SP Sephadex C-50® column and then submitted to gel filtration chromatography on Sephadex G-75® column.	[93]
Bothrops alternatus	BaTX	Absence	13.8	8.6	P86453	Gel filtration chromatography on Superdex 75® column followed by reverse phase chromatography on μ-Bondapack® C18 column.	[94]
Bothrops asper	MTX-I MTX-II MTX-III MTX-IV Basp-I-PLA_2	Presence Absence Presence Absence Presence	14.1 14.2 14.2 Nd 14.2	8.1- 8.3 8,1- 8.3 8.1- 8.3 8.1- 8.3 4.6		Ion-exchange chromatography on CM-Sepharose® column followed by hydrophobic interaction chromatography on Phenyl-Sepharose® column.	[95]
Bothrops asper	Myotoxin I	Presence	10.7	nd		Ion-exchange chromatography on CM-Sephadex C-25® column followed by gel filtration chromatography on Sephadex G-75® column.	[7]
Bothrops asper	Myotoxin II	Absence	13.3	nd	P24605	Ion-exchange chromatography on CM-Sephadex C-25® column.	[96]
Bothrops asper	Myotoxin III	Presence	13.9	>9.5	P20472	Ion-exchange chromatography on CM-Sephadex C-25® column.	[97]
Bothrops asper	Myotoxic PLA_2	Presence	14.1	nd		Gel filtration chromatography on Sephadex G-75® followed by ion-exchange chromatography on CM-cellulose® column.	[98]
Bothrops asper	BaspPLA_2-II	Presence	14.2	4.9	P86389	Ion-exchange on CM-Sephadex C-25® followed by chromatography on DEAE Sepharose® column, active fractions subjected to reverse phase chromatography on C8 column and finally chromatography with CM-	[22]

Bothrops atrox	$BaPLA_2I$ $BaPLA_2$ III	Presence Presence	15.0 15.0	9.1 6.9		Gel filtration chromatography on Sephacryl S-100 HR® column followed by reverse phase on C4 column.	[99]
Bothrops atrox	Basic Myotoxin	Presence	13.5			Ion-exchange chromatography on CM-Sephadex C-25® column and then re-chromatographed on the same column and same conditions.	[32]
Bothrops atrox	Myotoxin I	Absence	13.8	8.9	Q6JK69	Ion-exchange chromatography on Carboximetil-Sephadex C-25® followed by reverse phase chromatography on C8 column.	[100]
Bothrops brazili	MTX-I MTX-II	Presence Absence	14.0 14.0			Ion-exchange chromatography on CM-Sepharose® column.	[101]
Bothrops brazili	BbTX-II BbTX-III	Absence Presence	13.9 13.6	8.7 8.4		Reverse phase chromatography on C18 column.	[102]
Bothrops erythromelas	BE-I-PLA_2	Presence	13.6			Gel filtration chromatography on Superdex 75® followed by chromatography on monoQ® column, fractions being subjected to reverse phase chromatography on C4 column afterwards.	[103]
Bothrops jararacussu	BthTX-I BthTX-II	Absence Presence	13.0 13.0	8.2	Q90249 P45881	Gel filtration chromatography on Sephadex G-75®, followed by cation-exchange chromatography on SP-Sephadex C-25® column.	[34]
Bothrops jararacussu	BJ IV	Presence	15.0		P0CAR8	Ion-exchange chromatography on Protein Pack SP 5PW® column followed by reverse phase chromatography on μ-Bondapack® C18 column.	[104]
Bothrops jararacussu	BthA-I-PLA_2	Presence	13.7	4.5	Q8AXY1	Ion-exchange chromatography on CM-Sepharose® column, followed by reverse phase chromatography on C18 column.	[58]
Bothrops jararacussu	SIIISPIIA SIIISPIIB SIIISPIIIA SIIISPIIIB	Presence Presence Presence Presence	15.0 15.0 15.0 15.0	5.3 5.3 5.3 5.3		Gel filtration chromatography on Sephadex G-75® column, followed by ion-exchange chromatography on SP-Sephadex C-25® column, and finally HPLC on C18 column.	[105]
Bothrops lanceolatus	PLA_2-1 PLA_2-2 PLA_2-3	Presence Presence Presence	15.0 13.0 18.0	5.3 5.3 5.3		Reverse phase chromatography on Lichrosfera RP100® C18 column.	[106]
Bothrops leucurus	BLK-PLA_2 BLD-PLA_2	Absence Presence	14.0 14.0		P86974 P86975	Gel filtration chromatography on Sephacryl S-200®, followed by ion-exchange on Q-Sepharose and then submitted to reverse phase chromatography on HPLC Vydac® C4.	[38]
Bothrops leucurus	Bl-PLA2	Presence	15.0	5.4	P0DJ62	Ion-exchange chromatography on CM- Sepharose® column, followed by hydrophobic interaction chromatography on Phenyl-Sepharose® column.	[43]
Bothrops marajoensis	BmjeTX-I BmjeTX-II	Presence Presence	13.8 13.8		P86803 P86804	Ion-exchange chromatography on Protein Pack SP 5PW®, followed by reverse phase chromatography on μ-Bondapack® C18 column.	[107]
Bothrops marajoensis	BmarPLA_2	Absence	14.0	nd	P0DI92	Ion-exchange chromatography on Protein Pack SP 5PW®, followed by reverse phase chromatography.	[108]
Bothrops marajoensis	Bmaj-9	Presence	13.7	8.5	B3A0N3	Reverse phase chromatography on μ-Bondapack® C18 column.	[109]
Bothrops moojeni	BthA-I	Presence	13.6	5.2	G3DT18	Ion-exchange on CM-Sepharose® column, followed by hydrophobic interaction chromatography on Phenyl-Sepharose® column.	[24]
Bothrops moojeni	MjTX-III MjTX-IV	Absence Absence	14.6 14.6			Gel filtration chromatography on Superdex -75XK® column, followed by reverse phase chromatography on C18 column.	[110]
Bothrops moojeni	MjTX-I ou Miotoxina-I	Absence	13.4	8.2	P82114	Ion-exchange chromatography on CM-Sepharose® column.	[26]
Bothrops moojeni	MjTX-II	Absence	14.0	8.2	Q9I834	Ion-exchange chromatography on CM-Sepharose® column.	[111]
Bothrops moojeni	BmooTX-I	Presence	15.0	4.2		Ion-exchange on DEAE-Sepharose®, gel filtration on Sephadex G-75® column and hydrophobic interaction chromatography on Phenyl-Sepharose®.	[42]
Bothrops moojeni	BmTX-I	Presence	14.2	7.8	P0C8M1	Reverse phase chromatography on μ-Bondapack® C18 column.	[112]
Bothrops moojeni	BmooMtx	Absence	16.5			Ion-exchange chromatography on DEAE-Sephacel® column and then submitted to gel filtration on Sephadex G-75® column.	[113]
Bothrops pirajai	Piratoxin-I	Absence	13.8	8.3	P58399	Gel filtration chromatography on Sephadex G-75® column, followed by ion-exchange chromatography on Sephadex C25® column.	[114]
Bothrops pirajai	Piratoxin-III ou MPIII 4R	Presence	13.8		P58464	Ion-exchange chromatography on semi-preparative u-Bondapack® column, followed by ion-exchange chromatography on Protein Pack SP 5PW® column.	[115]
Bothrops pirajai	Bpir-I PLA2	Presence	14.5		C9DPL5	Ion-exchange chromatography on CM- Sepharose FF® column, followed by reverse phase chromatography on C18 column.	[25]
Bothrops pirajai	Piratoxin -II	Absence	13.7	9.0	P82287	Gel filtration chromatography on Sephadex G-75® column and ion-exchange chromatography on Sephadex C25® column.	[116]

Cerrophidion goodmani	Myotoxin I Myotoxin II	Presence Absence	14.3 13.4	8.2 8.9		Ion-exchange chromatography on CM-Sephadex® column.	[117]
Cerrophidion goodmani	GodMT-II	Absence	13.7			Ion-exchange chromatography on CM-Sephadex® column.	[118]
Cerrophidion goodmani	Pgo K49	Absence	13.8			Gel filtration chromatography on Sephadex G-75 HR® column, followed by reverse phase chromatography on Vydac® C8 column.	[119]
Crotalus atrox	PLA_2- 1 PLA_2- 2	Absence Presence	15.3 15.5	4.6 8.6		Gel filtration chromatography on DEAE-cellulose® column.	[119]
Crotalus atrox	Cax-K49	Absence	13.8		Q81VZ7	Gel filtration chromatography on DEAE-cellulose® column.	[119]
Crotalus durissus cascavella	PLA_2	Presence	15.0			Gel filtration chromatography (pharmacia), followed by reverse phase on µ-Bondapak® C-18 column.	[120]
Crotalus durissus collilineatus	F6a	Presence	14.9	5.8	P0CAS2	Reverse phase chromatography on µ-Bondapack® C18 column.	[121]
Crotalus durissus cumanensis	Cdc-9 Cdc-10	Presence Presence	14.1 14.2	8.25 8.4	P86805 P86806	Reverse phase chromatography on µ-Bondapack® C18 column.	[122]
Crotalus durissus cumanensis	Cdcum6	Presence	14.3	Nd	P0CAS1	Gel filtration chromatography followed by reverse phase chromatography.	[123]
Crotalus durissus ruruima	PLA2A	Presence	14.2		P86169	Gel filtration chromatography followed by reverse phase chromatography.	[124]
Crotalus durissus ruruima	Cdr-12 Cdr-13	Presence Presence	14.3 14.2	8.1 8.1	P0CAS3 P0CAS4	Reverse phase chromatography on µ-Bondapack® C18 column.	[121]
Crotalus durissus terrificus	CdtF16	Presence	14.8		P0CAS6	Gel filtration chromatography on Superdex 75® column, followed by reverse phase chromatography on µ-Bondapack® C18.	[125]
Crotalus durissus terrificus	Crotoxin B	Presence	14.5	5.1		Gel filtration chromatography on Sephadex G75® column, followed by chromatography on Mono-Q® and finally ion-exchange chromatography followed by DEAE-cellulose® column.	[126]
Crotalus durissus terrificus	CdtF17	Presence	14.6	8.15	P0CAS7	Reverse phase chromatography on µ-Bondapack® C-18 column.	[127]
Crotalus durissus terrificus	CdtF15	Presence	14.5	8.8	P0CAS5	Gel filtration chromatography on Superdex 75® column followed by reverse phase chromatography on µ-Bondapack® C-18 column.	[128]
Crotalus scutulatus scutulatus	MTX-a MTX-b		14.5 14.4	9.2 7.4	P18998 P62023	Reverse phase on Vydac® C8 column.	[129]
Lachesis muta	LmTX-I LmTX-II	Presence Presence	14.2 14.1	8.7 8.6	P0C942 P0C943	Gel filtration chromatography on Superdex 75® column, followed by reverse phase chromatography on µ-Bondapack® C-18 column and finally reverse phase chromatography on C8 column.	[130]
Lachesis muta	LM-PLA_2-I LM-PLA_2-II	Presence Presence	17.0 18.0	4.7 5.4	P0C932 P0C933	Gel filtration chromatography on Sephacryl S-200® column, followed by reverse phase chromatography on C2 column and finally reverse phase chromatography on C18 column.	[131]
Lachesis stenophys	LSPA-1	Presence	13.8	nd	P84651	Gel filtration chromatography on Sephacryl S-200® column followed by ion-exchange chromatography using MonoQ HR 5/5® column and finally reverse chromatography on Sephasil® C-18 column.	[132]
Porthidium nasutum	$PnPLA_2$	Presence	15.8	4.6		Reverse phase chromatography on C18 column.	[133]
Micrurus tener tener	MitTx-beta	Presence	16.7		G9I930	Reverse phase chromatography on C18 Vydac® column followed by reverse phase on Vydac® C18 column.	[134]
Micrurus tener microgalbineus	PLA_2-1	Presence			P25072	Gel filtration chromatography on Sephadex G-50® followed by ion-exchange chromatography on CM-cellulose column.	[135]
Micrurus pyrrhocryptus	PLA_2 A1 PLA_2 B1 PLA_2 D5 PLA_2 D6	Presence Presence Presence Presence			P0CAS8 P0CAS9 P0CAT0 P0CAT1	Gel filtration chromatography on Superdex G 75 HR® followed by reverse phase chromatography on Vydac® C18 column.	[136]
Micrurus nigrocinctus	Nigroxin A Nigroxin B	Presence Presence			P81166 P81167	Ion-exchange chromatography on Mono Q FF® column followed by reverse phase chromatography on Vydac® C4 column.	[137]
Micrurus nigrocinctus	PLA_2-1 PLA_2-2 PLA_2-3	Presence Presence Presence			P21790 P21791 P21792	Gel filtration chromatography on Sephadex G-75® column followed by ion-exchange chromatography on CM-cellulose® column.	[138]
Micrurus dumerilli carinicauda	MiDCA1	Presence	15.5	8.0		Reverse phase chromatography on Sephasil Peptide® C18 column followed by reverse phase chromatography on µ-Bondapak® C18 column.	[139]

Table 1. PLA_2s isolated from American snake venoms and respective chromatographic methods used.

Some authors have proposed changes to the methodology described above. Spencer et al. [31] described the purification of BthTX-I with the use of Resourse S® (methyl-sulphonate functional group), equilibrated in pH 7.8 (phosphate buffer 25 mM). Sample elution was done in increasing ionic strength conditions (NaCl 0 to 2 M), under 2.5 ml/min flow. In this model, the BthTX-I was eluted in NaCl 0.42M with a high degree of purity. However, the chromatographic profile in the conditions tested differs significantly from the observed in other works that describe the fractioning of this venom. This difference is due to the resin composition. This is corroborated with data obtained in experiments performed in our lab, where the effect of pH in the separation of myotoxin isoforms from *B. jararacussu* venom was used, as shown in Figures 4. SDS-PAGE showed that fractions corresponding to myotoxins showed protein bands with apparent molecular mass compatible with $PLA_{2}s$ class II (Figure 5).

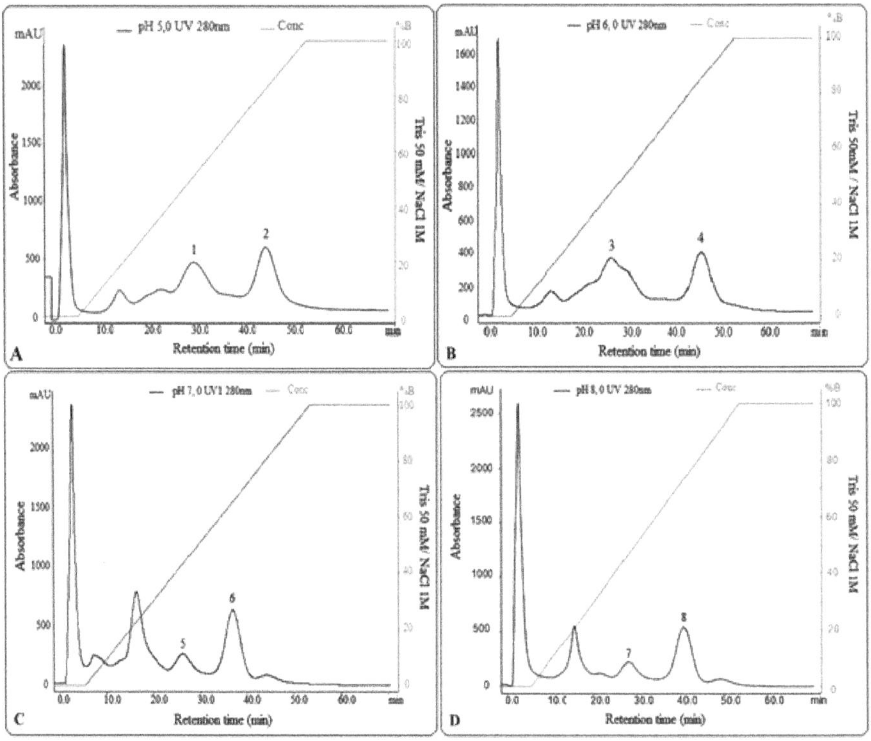

Figure 4. Chromatographic profile of the *B.jararacussu* venom in CM-sepharose® column 1 ml (*Hitrap*) equilibrated with Tris 50 mM buffer (buffer A) and eluted with a linear gradient of Tris 50 mM/NaCl 1M (buffer B) in different pH conditions. **A.** pH 5.0 **B.** pH 6.0 **C.** pH 7.0 **D.** pH 8.0. Absorbance was read at 280 nm. Fractions numbered (1 to 8) indicate the fractions selected for SDS-PAGE analysis in order to confirm the presence of PLA2s (BthTx I e BthTx II).

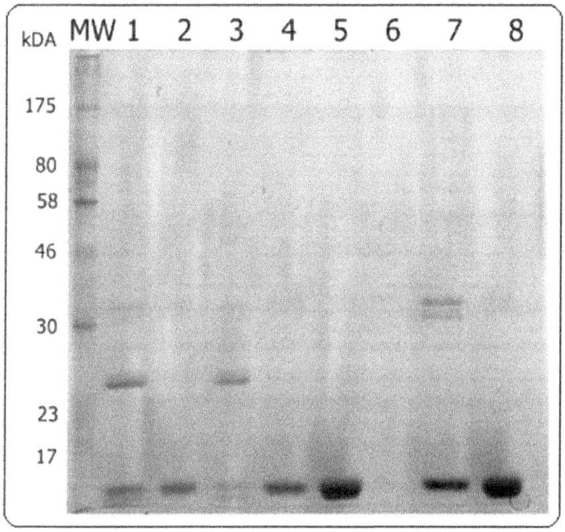

Figure 5. SDS Page analysis. Lines 1 and 2 (pH 5.0); 3 and 4 (pH 6.0); 5 and 6 (pH 7.0); 7 and 8 (pH 8.0). BthTx I was obtained in high degree of purity with pIs 5.0, 6.0, and 8.0. BthTx II was obtained with pH 7.0.

Resolution differences were also observed by other authors. As performed by Lomonte et al. [26], the isolation of two basic myotoxins, MjTX-I e MjTX-II, from the *B. moojeni* venom was obtained using CM-Sephadex C-25 equilibrated with Tris-HCl 50 mM pH 7.0 and eluted in saline gradient up to 0.75 M of Tris-HCl. Also, Soares et al. [33] described the isolation of MjTX-II with high purity using the combination of CM-Sepharose resin and ammonium bicarbonate buffer. According to the authors, the increase of pH to 8.0 has favored the elution of several fractions, allowing MjTX-II to be eluted separately with ionic strength equal to 0.35 M of ammonium bicarbonate. Moreover, the use of CM-Sepharose® seems to have also contributed a lot in the increasing of resolution for this chromatographic separation.

The combination of chromatographic techniques has also been used to purify these toxins. The association of the Ion-exchange chromatography and molecular exclusion has been one of the most recurrent in isolation and purification of phospholipases from bothropic venoms. Gel filtration chromatography is a technique based in particle size to obtain the separation. In this type of separation there is no physical or chemical interaction between the molecules of the analyte and the stationary phase, being currently used for separation of molecules with high molecular mass. The sample is introduced in a column, filled with a matrix constituted by small sized silica particles (5 to 10 μm) or a polymer containing a uniform net pores of which solvent and solute molecules diffuse. The retention time in the column depends on the effective size of the analyte molecules, the higher sized being the first ones to be eluted. Different from the higher molecules, the smaller penetrate the pores being retained and eluted later. Between the higher and lower molecules, there are the intermediary sized molecules, whose penetration capacity in the pores depends on their diameter. In addition to that, this technique has also some very important characteristics, such as operational simplicity, physical chemical stability, inertia (absence of reactivity and adsorptive properties) and versatility, since it allows the separation of small molecules with mass under 100 Da as well as extremely big molecules with various kDa.

The work performed by Homsi-Brandeburgo et al. [34] is a example of combination of different chromatographic techniques for the isolation of myotoxins with PLA_2 structure. It describes for the first time the BthTX-I purification using the combination of molecular exclusion chromatography in Sephadex G-75® resin followed by Ionic exchange chromatography in SP-Sephadex C-25®. In the first step, four fractions were obtained, called S_I, S_{II}, S_{III} and S_{IV}. The Functional analysis of these fractions showed that the proteolytic activity over casein and fibrinogen was detected on fraction S_I, while the phospholipase activity was concentrated in fraction S_{III}. The apparent molecular mass profile of this fraction showed that it was composed by proteins between 12,900 and 28,800 Da, compatible with the mass profile of the class II PLA_2s.

On the second step, S_{III} fraction was submitted to ionic exchange chromatography and five fractions were obtained, identified as $S_{III}SP_I$ to $S_{III}SP_{IV}$. The pIs and apparent molecular mass evaluation showed the following profile: $S_{III}P_I$ (pI 4.2 and 22.400 Da), $S_{III}P_{II}$ (pI 4.8 and 15.500 Da), $S_{III}P_{III}$ (pI 6.9 and dimeric structure, each monomer with a molecular mass of 13.900 Da), $S_{III}P_{IV}$ (pI 7.7 and 13.200 Da) e $S_{III}P_V$ called BthTX-I that presented pI 8,2 and 12.880 Da. Pereira et al. [35] obtained the complete sequence of BthTX-II, a myotoxin homologous to the BthTX-I, which corresponds to the $S_{III}SP_{IV}$ fraction described by Homsi-Brandeburgo et al. [34].

Another chromatographic technique regularly used in PLA_2s purification procedures is the Reverse-phase associated with High performance liquid chromatography (RP- HPLC). This technique is characterized by its high resolution capacity and is normally used in a more refined step of the purification process, being very useful in separating isoforms. The retention principle of reverse-phase chromatography is based in hydrophobicity and is mainly due to the interactions between hydrophobic domains of the proteins and the stationary phase. This technique has many advantages, such as: use of less toxic mobile phases together with lower costs, such as methanol and water; stable stationary phases; fast column equilibrium after mobile phase change; easy to use gradient elution; faster analysis and good reproducibility.

Rodrigues et al. [36] described the isolation of two PLA_2s isoforms from the *B. neuwiedi pauloensis* venom using the combination of ion (cation) exchange chromatography and molecular exclusion setting up a preparative phase. Subsequently, a reverse-phase chromatography was used for the analytical phase of the procedure. Initially, the venom was fractioned in a column containing CM-Sepharose® equilibrated with ammonium acetate solution 0.05 M, pH 5.5 and eluted in linear gradient up to 1 M of the same buffer, resulting in six fractions. The pH, more acid than the ones used in the work previously mentioned, has increased the surface residual charge, intensifying the interaction force between the protein and the resin, thus altering the elution profile when compared to the performed by Rodrigues et al. [37]. Proceeding with purification, the sample with phospholipase activity (S-5) was submitted to a new fractioning in a Sephadex G-50® column yielding 3 fractions, of which the denominated S-5-SG-2 showed catalytic activity. It was then submitted to RP- HPLC in C18 column to obtain toxins with high purity degree.

Also, with the use of a multiple step procedure [38] successfully isolated two isoforms of PLA_2s from *B. leucurus* venom. After a first molecular exclusion chromatography using Sephacryl S-200®, 7 fractions were obtained, from which the named "P6" showed to be composed by proteins with apparent molecular mass bellow 30 kDa, and a major fraction of approximately 14 kDa concentrated the phospholipase activity. This fraction was re-chromatographed in a Q-Sepharose® resin (ion exchange) and equilibrated with Tris-HCl 20 mM pH 8.0, yielding 6 fractions. The fraction corresponding to the negatively charged fraction was eluted without significant interaction with the resin, hence with a positive residual charge (basic pI) was selected, showing to be a homogeneous fraction of 14 kDa and presenting phospholipase activity. This fraction was submitted to a RP- HPLC in C4 column, yielding as result two major fractions with close hydrophobicity (eluted with 33% and 36% acetonitrile) and apparent molecular mass of 14 kDa.

Myotoxins with PLA_2s structure from bothropic venoms that have acid pI have being more difficult to isolate. Different from cation exchange resins (CM Sepharose®, Resource S® and CM Sephadex®), anion exchange resins have not been so efficient in the separation of components from bothropic venoms, which requires, complementary steps to obtain these toxins with a satisfactory purity degree, as shown in Table 1.

Daniele et al. [32] described the fractioning of the *B. neuwiedii* venom using a combination of double molecular exclusion chromatography followed by anion exchange chromatography. The first step of the molecular exclusion chromatography was done using Sephadex G-50® where a single fraction with PLA_2s activity was eluted. This fraction was re-chromatographed in Sephacryl S-200® resin, yielding 2 active fractions. The first fraction was re-chromatographed in Mono Q® column (functional group quaternary ammonium) yielding a PLA_2s named P-3. From the second fraction, submitted to the same chromatographic procedure, two other PLA_2s isoforms were isolated, named P-1 and P-2. Although showing different behavior over the molecular exclusion resin, the three isoforms showed very close apparent molecular mass (15 kDa) when assayed by SDS-PAGE. This difference could be resulted from differential interactions of aromatic residues located on the protein surface with the stationary phase [40, 41] and can be also verified in other acid PLA_2s, like the one obtained from *B. jararacussu* venom by Homsi-Brandeburgo et al. [34].

Other procedures used hydrophobic interaction chromatography to isolate these PLA_2s. This is a method that separates the proteins by means of their hydrophobicity: the hydrophobic domains of the proteins bind to the hydrophobic functional groups (phenyl and aryl) of the stationary phase. Proteins should be submitted to the presence of a high saline concentration, which stabilize then and increases water entropy, thus amplifying hydrophobic interactions. In the presence of high salt concentrations, the matrix functional groups interact and retain the proteins that have surface hydrophobic domains. Hence, elution and protein separations can be controlled altering the salt or reducing its concentration.

Santos-Filho et al. [42], working with *B. moojeni* venom, applied three sequential steps to obtain BmooTX-I, a PLA_2 with apparent molecular mass of 15 kDa and pl 4.2. In this work, the crude venom was chromatographed in DEAE-Sepharose® (Dietylaminoetyl) resin, equilibrated with ammonium bicarbonate 50mM, pH 7.8 and brought to a saline gradient of 0.3M of the same salt. A fraction named E3 showed phospholipase activity, being then submitted to molecular exclusion chromatography in Sephadex G-75® resin. Three fractions were obtained, from which one named S2G3 was submitted to hydrophobic interaction chromatography in Phenyl-Sepharose® resin, the BmooTX-I being eluted in the end of the process.

In a work published in 2011, Nunes et al. [43] described the isolation of an acid phospholipase named BL-PLA_2, obtained from *Bothrops leucurus* through two sequential chromatographic steps. On the first step, the acid proteins were separated from the others with the use of a cation exchange column (CM-Sepharose®) equilibrated with ammonium bicarbonate, pH 7.8. The acid fraction (eluted without interaction with the resin) was lyophilized and applied to a Phenyl-Sepharose CL-4B® column (1 x 10 cm), previously equilibrated with a Tris-HCl 10mM buffer, NaCl 4M, pH 8.5. The elution occurred under decreasing NaCl gradient in a buffered environment (Tris-HCl 10 mM, pH 8.5), concluding the process in an electrolyte free environment. An enzymatically active fraction (BL-PLA_2), (with pI 5.4 and apparent molecular mass of approximately 15 kDa) was obtained at the end of the process.

The bioaffinity chromatography differs from others chromatographic methods because it is based in biological or functional interactions between the protein and the ligand. The nature of these interactions varies, being the most used those which are based on the interactions between: enzymes and substrate analogous and inhibitors; antigens and antibodies; lectins and glycoconjugates; metals and proteins fused with histamine tails. The high selectivity, the easiness of performance together with the diversity of ligands that can be immobilized in a chromatographic matrix make this method a useful tool for the purification of phospholipases. Based on the neutralization of myotoxic effects of the venom from *B. jararacussu* by heparin [44-46], the use of a column containing Agarose-heparin® could be used for the purification of myotoxins. They also ratify the interactions between heparin and myotoxin through the reduction of many biological effects, such as: edema induction, myotoxicity (*in vivo*) and cytotoxicity over mice myoblasts culture (L.6 – ATCC CRL 14581) and endothelial cells.

Following this strategy, Soares et al. [26] described the purification of BnSP-7, a myotoxin Lys-49 from *B. neuwiedi pauloensis*, with the use of chromatographic process based in this heparin functionality, which corroborates previous results obtained by Lomonte et al. [46], that showed the efficient inhibitory activity of heparin against myotoxicity and edema induced by myotoxin II, a lysine 49 phospholipase A_2 from *Bothrops asper*. Also in this study, it was possible to infer the participation of the C-terminal region of the protein in the damaging effects on the cytoplasmic membrane.

Snake venom components share many similar antigenic epitopes that can induce to a crossed recognition by antibodies produces against a determined toxin. In this context, Stabeli et al. [47] showed that antibodies that recognize a peptide (Ile1-Hse11) from Bm-LAAO present crossed immunoreactivity with components not related to the LAAOs group present in venoms from *Bothrops, Crotalus, Micrurus* e *Lachesis* snake venoms. Also, Beghini et al. [48] showed that the serum produced against crotoxin and phospholipase A_2 from *Crotalus durissus cascavella* was able to neutralize the neurotoxic activity produced by *B. jararacussu* venom and BthTX-I.

Based on this information, pertinent to the crossed immunoreactivity existent between venom components, Gomes et al. [49] described the co-purification of a lectin (BJcuL) and a phospholipase A_2 (BthTX-1) using a immunoaffinity resin containing antibodies produced against the crotoxin. 20 mg of crotoxin was solubilized in coupling buffer (sodium bicarbonate 100 mM, NaCl mM, pH 8.3) and incubated overnight at 4 °C with 1 g of Sepharose® activated by cyanogen bromide (CNBr). After washing with the same buffer, the resin was blocked with Tris-HCl 100 mM buffer. This resin was packed and thoroughly washed with saline phosphate buffer (PBS) pH 7.4. Crotalic counter-venom hiperimune horse plasma (20 mg) was applied over the resin at a flow of 10 mL/hr and re-circulated overnight through the column. Then, it was washed until the absorbance went back to basal levels, showing that the material was retained (IgG anti-Ctx), then eluted with glycin-HCl 100 mM pH 2.8. The IgG anti-Ctx was then immobilized in CNBr activated Sepharose® resin through a procedure analogous to the above cited, generating a new resin called Sepharose-Bound Anti-CtxIgG. 20 mg of the crude venom from *B. jararacussu* was applied over this resin, yielding two fractions: the first, composed by proteins that were not recognized by the immobilized antibodies and a second fraction composed by components of venom from *B. jar*aracussu that reacted crosswise with the Anti-Ctx antibodies, called Bj-F. A posterior analysis of this fraction, done by mass spectrometry, amino-terminal sequencing by Edman degradation and search by homology in the NCBI *protein data bank*, showed that it was composed by lectin and BthTX-I.

Different authors used substrate analogous or reversible inhibitors coupled to the chromatographic resin. Rock and Snyder [50] were the first ones to use phospholipid analogous to build a bioaffinity matrix [Rac-1-(9-carboxy)-nonil-2-exadecilglycero-3-phosphocoline]. In addition to them, Dijkman [51] described the synthesis of an analogous of acylamino phospholipid[(R)-1-deoxy-1-thio-(ω-carboxy-undecyl)-2-deoxy- (n-decanoylamino)-3-glycerophosphocholine] which was coupled to a Sepharose 6B® resin containing a spacer arm. With the use of this resin it was possible to purify phospholipases from horse pancreas, and venoms from *Naja melanonleuca* and *Crotallus adamanteus*.

3. Characterization

Venomic can be defined as an analysis in large scale of the components present in the venom of a certain species. In this context, the proteomic approach has allowed a better understanding of the venom components, through the application of many instruments that enables the analysis of their expression, structure, pos-traductional modifications and classification by homology or function. An approach developed by Calvete [52] for the analysis of snake venomic consists in an initial fractioning step of the crude venom using RP - HPLC, followed by characterization of each fraction by a combination of amino-terminal sequencing, SDS-PAGE, IEF or 2DE and mass spectrometry to determine molecular mass and cysteine content. Additionally, the modern venomic analysis use techniques such as Peptide Mass Fingerprint and the search for sequence similarity in data banks.

SDS-PAGE is a method related to the migration of charged particles in a medium under the influence of a continuous electric field [53]. From the electrophoretic point of view, the most important properties of the proteins are molar mass, charge and conformation. Mono dimensional polyacrylamide gel electrophoresis permit the analysis of the protein in its native or denatured form. In the first case, there are no alterations in conformation, biological activity and between protein subunits. This system is called non-dissociating or native, which proteins are separated based on their charge, using the isoelectric focusing method (IEF), or else, in vertical gel without SDS. During the IEF, a pH gradient is formed and the charged species move through the gel until they reach a specific pH. In this pH, the proteins have no effective charge (known as protein pI). The IEF shows high resolution, being able to separate macromolecules with pI differences of just 0.001 pH units [54, 55]. In dissociating or denaturing systems, the proteins are solubilized in buffer containing the reagent used to promote protein denaturation. SDS-PAGE, originally described by Laemmli [56], is an electrophoresis technique in polyacrylamide gel (PAGE) that used SDS as a denaturing agent, with interacts with the proteins giving them negative charges, allowing them to migrate, through a polyacrylamide gel towards a positive electrode a be separated by the differences related to their mass

Teixeira et al [25] described the purification of an acid phospholipase from *B. pirajai* (BpirPLA_2I). As a biochemical characterization step, polyacrylamide gel electrophoresis in denaturing conditions (SDS-PAGE) was done. Using this approach, carried out in reducing and non-reducing conditions, the author could infer that the purified protein had the form of a monomer with apparent molecular mass of 14 kDa, both in reducing conditions as well as in non-reducing conditions the proteins presented the same mass, being confirmed afterwards by mass spectrometry.

Moreover, Torres [57] fractionated B. marajoensis venom using a cationic ion exchange column followed by an analytical phase in RP- HPLC, obtaining a phospholipase BmarPLA_2 that was submitted to SDS-PAGE in reducing conditions showing apparent molecular mass of 14 kDa. However, in non-reducing conditions, the author observed the appearance of a single band at 28 kDa, concluding that BmarPLA2 was a dimeric structured protein joined by disulphide bridges. Thus, the above-cited examples demonstrate the importance of this procedure (SDS-PAGE) as a protein characterization step.

The determination of the isoelectric point is another important biochemical characterization of phospholipases A_2. Previous studies involving phospholipases from snake venoms have shown that the acid phospholipases are catalytically more active than their basic isoforms

[22, 42, 58]. Therefore, many authors have included, as a biochemical characterization parameter, the determination of the isoelectric point of the by isoelectric focusing. Due to pI determination importance, Teixeira [25] used the methodology proposed by Vesterberg and Eriksson [59] to evaluate pI of BpirPLA2-1. In order to obtain the pI value, a 7% polyacrylamide gel was prepared and polymerized over a glass plate of 12 x 14 cm using a U shaped rubber as support. A millimeter plate was previously greased with glycerin for better refrigeration of the gel. Two strips of Pharmacia Biotech were used to connect the gel and the platinum electrodes. The cathode was in contact with NaOH 1 M solution and the anode was in phosphoric acid 1 M. The platinum electrodes were centered over the paper strips and the system was then closed. The high voltage source was adjusted to the maximum values of 500 V, 10 mA, 3 watts and 30 minutes for a pre-run. Following, the samples were applied always in the intersection of two blue lines, exactly over the more central line of the gel. Then the source was programed for 1500 V, 15 mA, 10 watts and 5 h. The end of the run was determined when the source showed a high voltage and low amperage (around 1 mA). After isoelectric focusing, about 1 cm width (lengthwise) were sliced from each extremity of the gel and placed in test tubes containing 200 μL of distilled water for the pH reading after 2 hours of rest. Next, the pH gradient determination graph was plotted. The remaining gel containing the samples was fixed in solution of trichloroacetic acid for 30 minutes, followed by silver staining.

Another important technique as a step to characterize components from snake venoms is the bidimensional electrophoresis (2D). This one was initially developed by O'Farrell [60]. The original methodology consisted of the preparation of polyacrylamide cylindrical gels, in which a pH gradient was established through a pre-run with specific amphoterics (also called ampholytes), that present high buffering capability in pHs close to their isoelectric points (pIs). The proteins were then submitted to an isoelectric focusing (IEF) and subsequently to an electrophoresis in the presence of SDS by a conventional system described by Laemmli [56]. Then, proteins were separated in the first dimension according to their pIs (IEF) and in the second dimension based on their molecular mass (SDS-PAGE).

Bidimensional electrophoresis is laborious, time consuming and difficult to be reproduced in different laboratories and depended on the ability of the researcher to obtain consistent results. Nowadays, many of these problems were solved with the development of new technologies. An important advance which has contributed to the increase of the 2D electrophoresis reproducibility was developed by Gorg [61] of the strip form gels with immobilized pH gradient (IPG - *immobilized pH gel*). The strips are made by the copolymerization of acrylamide with the *Immobiline®* (*Amershan Biosciences/GE Heathcare*) reagent, which contains acid and alkaline buffering groups. Another important technological progress was the improvement of the protein samples preparation methods, together with the discovery of new non-ionic detergents, such as CHAPS surfactants and SB 3-10, used with reducing agents adequate for IEF, like Dithiothreitol (DTT) and Tributyl Phosphine (TBP). Studies performed by Herbert [62] demonstrated that these advances had strongly contributed to the solubilization of a greater number of proteins to be analyzed in bidimensional electrophoresis.

The proteomic analysis of snake components has made use of the 2D electrophoresis as a tool, due to its high-resolution capability that allows, in a single process, the determination of apparent molecular mass and isoelectric point of the venom constituents. Fernandez et al. [22] described the determination of the isoelectric point and apparent molecular mass of Basp-PLA_2-II using this technique. In order to do it, the protein was focused in IPG Immobiline® Dry Strip of 7 cm and pH 3-10, under a 200 V tension for 1 min, followed by a second stage of 3500 V for 120 minutes. The second dimension was done in SDS-PAGE 12% and then subsequently dyed with Coomassie blue. It was demonstrated that Basp-PLA_2 –II had a pI of 4.9, which is close to the theoretical isoelectric point value (pI 5.05) defined by the primary sequence, evaluated using the Compute pI/MW tool (www.expasy.ch/tools) software and apparent molecular weight between 15 and 16 kDa, consistent with the molecular weight (MW 14,212±6 Da) obtained by ESI/MS (Electrospray Ionization/Mass Spectrometry).

The advantage of this technique is the high resolution. Alape-Giron [63] working with *B. asper* venom, performed an ontogenic analysis and an analysis based on the snake's capture location in different regions of Costa Rica. Using tryptic digestion, MALDI-TOF mass fingerprinting analysis and aminoacid sequencing by MALDI-TOF submitted to similarity search by BLAST, the author showed the intra-specific variability in venom composition. It was hence evidenced that among the venoms obtained from adult species collected in the Caribe area and the Pacific area, there are around 30 proteins that are found in a snake group from a place which find no correspondents in the other.

In our lab, this technique has been used as follows: The proteins are separated by the isoelectric point in 13 cm strips with pH values varying between 3 and 10 in a nonlinear form. These strips contain polyacrylamide gel, where the gradient pH is formed by the presence of ampholytes. To re-hydrate the strips, 250 μL of sample [400 μg of proteins plus re-hydration solution (7 M of urea, 2 M of Thiourea, 2% of Triton X-100 (v/v/), 1% of IPG *Buffer*® (v/v) and DTT)] is applied in a channel of the apparatus over which the strips are set. The strip's gel is re-hydrated at room temperature for about 12 hours. After this period, the strips are taken to the focusing system in the following conditions: (1) 500 V step until accumulates 500 Vh; (2) 500 to 1000 V gradient until it accumulates 800 Vh; (3) 1000 to 8000 V gradient until it accumulates 11300 Vh and (4) 8000 V step until it accumulates 3000 Vh. In average, the program run during 5.5 hours, but the time of the final step can be lengthened, if the sample does not reach to the end of the strip during the running according the initial program, it could be confirmed by a bromophenol blue line. At the end of focusing, the strips are equilibrated in two steps. On the first, 10 mL of the solution containing 6 M of urea, 2% of SDS (m/v), 30% of glycerol (v/v), 75 mM of Tris-HCl (pH 8,8), 0,002% of bromophenol blue and 1% DTT (m/v) for each strip is used. In the second, the same solution is used, but DTT is replaced by 2.5% of iodoacetamide (m/v). Each strip equilibrium step run during 15 minutes, under light stirring. Following that, the strips are applied on 10 % polyacrylamide gels previously prepared on 180 X 160 X 1.0 mm plates. After each strip and the standard stay appropriately accommodated in the polyacrylamide gel, a 0,5% agarose (m/v) heated (40 °C) solution is added. The agarose polymerization, provides an effective contact between the strip and the gel, thus avoiding the appearance of air bubbles. Protein

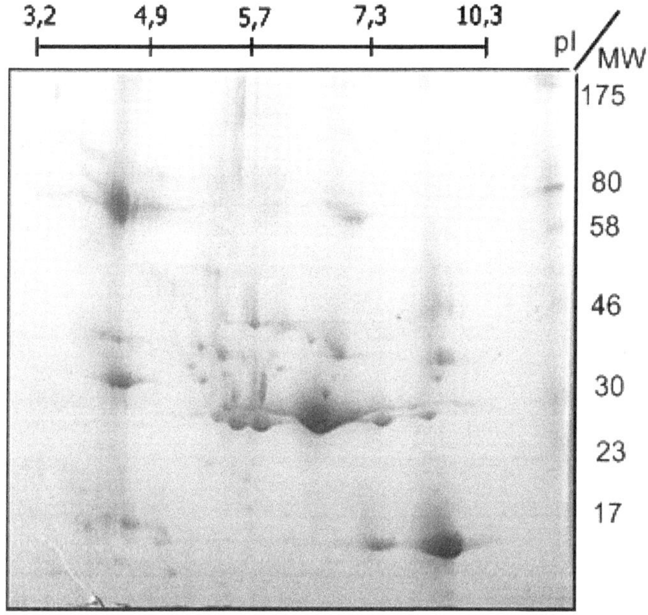

Figure 6. Electrophoretic profile in 2D-PAGE 10%, 13 cm strip pH 3-10 non-linear of proteins from crude venom from *Bothrops moojeni*. Molecular weight (MW) –*Color Plus Prestained Protein Marker – Broad Range* (7-175 kDa) (P7709S). Coomassie G-250.

separation, according to molecular mass, is done by applying 25 mA per gel and 100 W during approximately 5.5 hours. After this period, the gel is washed with deionized water. Then, the proteins are fixed using a solution containing acetic acid 10% (v/v) and ethanol 40% (v/v) during one hour. Then, the fixing solution is removed and the gel is washed again with deionized water 3 times during 10 minutes. The proteins present in the gel are exposed using traditional methods for protein coloring, such as Coomassie blue or Silver nitrate. An example of the practical application of this methodology can be seen in Figure 6.

4. Functional characterization

Many biological activities are related to myotoxins with PLA_2 structure obtained from snake venoms. In bothropic snake bite accidents and in experimental models with the use of these venoms, the noxious activity induced by these toxins on the striated muscles is striking [64]. The detection of the myotoxic activity associated to the phospholipase activity detection (in the case of Snake venom PLA_2 Asp49) is used as an important auxiliary biological marker in the purification procedures, monitoring its presence.

The myotoxic activity assay can be done in two ways: *in vivo* and *in vitro*. The analysis can be done through the quantification of the released intracellular enzymes activity to the periphery blood or to the supernatant of the culture medium of cellular lineages. There are two main enzymes used to this end:

Creatine Kinase (EC 2.7.3.2): is a dimeric protein formed by the combination of subunits (B and M) and in its cytosolic form is found in many tissues, especially in skeletal muscle tissue (CK-MM), cardiac (CK-MB) and in the brain (CK-BB).

Lactate dehydrogenase (EC 1.1.1.27): is an enzyme widely distributed in many tissues and organisms. It is presented in the form of homo or hetero tetramers of subunits M and H, being present in muscular tissue in the homotetrameric form of subunit M.

In vivo, the CK activity quantification in murine models has been the most used to assay the presence of myotoxic PLA_2, especially due to their low cost, ease of performance and high specificity as skeletal muscular tissue lesion markers when exposed to myotoxins.

As for the *In vitro* assays, myoblast lineages C2C12 (ATCC CRL-1772), differentiated until the formation of myotubules, have been used as models to assay the cellular toxicity, through the quantification of LDH levels in the supernatant of cell cultures exposed to toxins.

Regarding the phospholipase activity detections, it can be done by direct and indirect methods. Directly, it is possible to detect the presence of PLA_2s with the use of chromogenic substrates, such as 4N3OBA(4-nitro-3-octanoyloxybenzoic acid) that induce the formation of detectable product at 425 nm [65] and fluorescent substrates (NBD) coupled to phospholipids that are used to quantitively and qualitatively survey the PLA_2s activity isolated from snake venom [23].

Indirectly, the approach used consists in the potentiometric assay of the fatty acids released after the enzymatic hydrolysis of the phospholipids, through the quantification with standard alkaline solution [66]. Moreover, fatty acids released by the enzymatic degradation can be quantified through the alteration of the optical density of the pH indicator solution, such as phenol red [67], brilliant yellow [68] and bromothymol blue [69]. Another indirect method to assay PLA_2 activity present in samples consists in the detection of hemolysis induced by lysophospholipids derived from phospholipids submitted to enzymatic digestion. This can be done through the quantification of hemoglobin present in solution or through the visualization of hemolytic halo in agarose matrix with immobilized red blood cells.

4.1. *In vivo* assay of the myotoxic activity

Mice is used for the *in vivo* assay of the myotoxic activity. Swiss males weighing between 18 g and 22 g, kept in controlled environment (12 h in the light and 12 h in the dark), with food and water *ad libitum* up to the moment of use. PBS solubilized sample and control (PBS) are filtered through 44 μm pores immediately prior to use. Reagents for CK activity dosage are prepared and used according to manufacturer's instructions.

A Sample (50 μL) or control (50 μL) will be injected in mice gastrocnemic muscle using adequate device in order to guarantee a precise volume control. After a time lap (3 and/or 6 h), blood sample is collected in heparinized tubes and centrifuged to separate plasma. CK concentration is determined according to manufacturer's instructions and expressed in U/L, where one unit corresponds to the production of 1 mmol of NADH per minute [26,70-72].

4.2. *In vitro* myotoxic activity assay

In order to assay myotoxic *in vitro* activity, myoblast lineage cells are used, such as murine skeletal muscle C2C12 myoblasts (ATCC CRL-1772) as described by Lomonte et al. [73], cultivated in modified Dubelco Eagle medium, supplemented with 1% bovine fetal serum. PBS solubilized sample, negative control (PBS) and positive control (Triton X-100) should be filtered through 22 μm pore filters immediately prior to use. Reagents for LDH activity dosage are prepared and used according to manufacturer's instructions.

In 96 well plate, $2X10^5$ cells/150 μL are set, sample and/or control (50 μL) are incubated in humid atmosphere at 37 °C and 5% CO_2 for a 3 hour period. Afterwards, collect supernatant aliquot and quantify LDH activity released by cells with cytoplasmic membrane integrity compromised, according to manufacturer's instructions and expressed in U/I, where one unit corresponds to the production of 1 mmol of lactate per minute.

5. Phospholipasic activity

5.1. 4N3OBA Substrate enzymatic hydrolysis

Phospholipase A_2 activity can be measured according to the technique described by Holzer and Mackessy [65], modified for 96 wells plate [74].

Prepare aliquots of 100 μL of 4N3OBA 0.1% solution in acetonitrile and lyophilize. Keep the aliquots at -20° C until ready to use. The color reagent is prepared solubilizing the contents of one aliquot of 4N3OBA in 1ml of reagent containing Tris 10 mM, $CaCl_2$ 10 mM, NaCl 100 mM, and pH 8.0. For the test in micro plates, add 180 μL of color reagent and 20 μL of sample or water (blank), incubate the mixture at 37 °C for 5 minutes, measuring the optical density at 425 nm and 600 nm (to correct sample turbidity) at 30 second intervals. The activity will be expressed according to the equation (1) where 1 unit of phospholipase activity corresponds to the production of 1 μmol of 4-nitro-3-hydroxy-benzoic acid per minute.

$$\begin{gathered}\text{PLA}_2\text{ activity }\left[\mu\text{mol} \bullet \text{min}-1\bullet \mu\text{g}-1\right] = \\ \left[\text{OD425nm} - \text{OD 600nm / min}\right] \text{ x } 0.07862 \left[\text{umol / OD425nm}\right] \text{ / } 1\text{/protein }\left(1/\mu\text{g}\right)\end{gathered} \tag{1}$$

5.2. Enzymatic hydrolysis of fluorescent substrates (NBD)

The phospholipase activity can also be assayed with the used of chromogenic substrates, using acyl-NBD reagents: NBD-PC (Phosphatidylcholine), NBDPG (phosphatidylglycerol), NBD-PE (phosphatidylethanolamine) or NBD-PA (phosphatidic acid). A solution of fluorescent lipids should be previously prepared in a 1 mg/ml concentration in chloroform. 100 μL aliquots are distributed and then dried under nitrogen flow. The dried lipid will be solubilized in 1 ml of NaCl 0.15 M and sonicated until the obtention of a limpid solution. For the test, the lipids should be diluted in a solution containing Tris-HCl 50 mM, $CaCl_2$ 1 mM pH 7.5. Initially, incubate the solution at 37 °C and, after 2 minutes, make an initial reading, configuring the equipment for excitation at 460 nm and emission at 534 nm. Following, apply the sample and make a second reading after 12 minutes. The change in fluorescence intensity is converted to nanomoles of product per minute (nmoles/min) using a calibration curve, prepared by hydrolyzing completely a substrate solution through sodium hydroxide treatment. The fluorescence intensity unit was converted to nmoles/min [33].

5.3. Potentiometric titration of fatty acids

The phospholipase activity can be assayed by potentiometric titration as described by de Haas [75], using as substrate an egg yolk emulsion in the presence of sodium deoxycholate 0.03 M and $CaCl_2$ 0.6 M. Fatty acids released enzymatically are titrated with a standard solution of NaOH 0.1 N at pH 8.0 at room temperature. The phospholipase activity is generally done with different concentrations of toxin, and calculated per amount of microequivalents of alkali consumed per minute, by mg of protein. One unit of phospholipase activity can be defined as the quantity of enzyme that releases 1 μmol of fatty acid per minute, in the reaction conditions.

5.4. Phenol red

The spectrophotometric detections of phenol red solution, induced by the increase of free fatty acids concentration can also be used to assay the phospholipase activity in samples, as described by Radvanyi [67].

In order to use this technique, prepare the reagent solution containing Phosphatidylcholine 0.25% (w/v) TritonX-100 0.4% (v/v), phenol chloride 32 mM. In a thermostatic cuvette at 37 °C, add 1mL of reagent solution and 10 μL of sample. After stabilization for 20 seconds, determine the optical density measuring at 558 nm for 3 minutes, in kinetic intervals of 15 seconds. One unit of phospholipase can be defined as the quantity of enzyme necessary to convert 0.001 UA 558 nm per minute.

5.5. Indirect hemolysis

In this test, phospholipids (from egg yolk, soy lecithin or other sources) are used as substrates, with the production of fatty acid and corresponding lysophospholipids. These lysophospholipids have membrane activity over red blood cells, producing hemolysis that can be detected through the quantification of hemoglobin present in solution or through a hemolysis halo present in agarose gel containing intact red blood cells [76].

For the test, collect blood in a heparinized tube, wash the red blood cells with PBS, centrifuging at 800 xg for 5 minutes and prepare the suspension at 3%. Prepare solution containing phosphate buffer 20 mM, sodium chloride 100 mM and $CaCl_2$ 10 mM, erythrocyte suspension 3% (1:30 v/v) and egg yolk solution 0.1% (1:30 v/v). Add 10 μL of sample or PBS (control 0%) or Triton X-100 0.1% (control 100%) and incubate at 37 °C for 30 minutes. Then, centrifuge, collect the supernatant and determine the optical density at 405 nm, using PBS as blank. The results will be expressed in % of hemolysis compared with the positive control.

Hemoglobin dosage present in solution with the use of the Drabkin reagent (potassium ferrocyanide in buffered environment) [77] can be done by comparing the optical density of the samples with the standard curve made with the hemoglobin cyanide solution, according to manufacturer instructions.

To assay the hemolytic activity in agarose gel, carefully heat the suspension containing agarose 2% in PBS until complete fusion. After partial cooling (45 °C), add an equal volume of PBS containing $CaCl_2$ 0.02 M; egg yolk suspension (1:30 m/v), erythrocytes washed in PBS (1:30 m/v), pouring over Petri plate until the formation of a layer approximately 2 mm thick. After solidification of the gel, orifices of uniform diameter (0.2 cm diameter) to apply the sample are made. The gel is incubated for 12 hours, at 37 °C and humid environment. The formation of a translucid halo around the gel application point is indicative of phospholipase activity, contrasting with the rest of the gel which remains with a reddish tone due to the presence of integral red blood cells retained in the gel net.

Abbreviations

PLA_2s: Phospholipases A_2
$sPLA_2$: secreted PLA_2
$cPLA_2$: cytosolic PLA_2
$iPLA_2$: Ca^{2+} independent PLA_2
PAF-AH: acetyl-hydrolases from platelet activating factors and liposomal
Mr: relative mass
PAF: platelet activating factors
$svPLA_2s$: phospholipase A_2 found in snake venoms
SDS-PAGE: Sodium dodecyl sulfate – PolyAcrylamide Gel Electrophoresis
MW: Molecular weight
RP-HPLC: reverse-phase chromatography
DEAE-Sepharose: Dietylaminoetyl Sepharose
tEnd cells: endothelial cells
CNBr: cyanogen bromide
PBS: phosphate buffer saline
IEF: isoelectric focusing
2DE: bi-dimensional electrophoresis
IPG: immobilized pH gel
pIs: isoelectric points
DTT: Dithiothreitol
TBP: Tributyl phosphine
ESI: Electrospray Ionization

MS: Mass Spectrometry
CK-MM: Creatine Kinase - skeletal muscle tissue
CK-MB: Creatine Kinase – cardiac
CK-BB: Creatine Kinase - brain
CK: Creatine Kinase
LDH: lactate dehydrogenase
NBD: N-4-Nitrobenzo-2-Oxa-1,3-Diazole
NBD-PC: N-4-Nitrobenzo-2-Oxa-1,3-Diazole Phosphatidylcholine
NBD-PG: N-4-Nitrobenzo-2-Oxa-1,3-Diazole
NBD-PE: N-4-Nitrobenzo-2-Oxa-1,3-Diazole phosphatidylglycerol
NBD-PA: N-4-Nitrobenzo-2-Oxa-1,3-Diazole Phosphatidic acid

Author details

Leonardo de A. Calderon, Juliana P. Zuliani and Rodrigo G. Stábeli
Center of Applied Biomolecular Studies in Health (CEBio), Fiocruz Rondônia, Oswaldo Cruz Foundation - FIOCRUZ and Medicine Department, Federal University of Rondônia - UNIR, Porto Velho, RO, Brazil

Andreimar M. Soares
Center of Applied Biomolecular Studies in Health (CEBio), Fiocruz Rondônia, Oswaldo Cruz Foundation - FIOCRUZ, Porto Velho, RO, Brazil

Anderson M. Kayano, Antonio C. Neto, Andrea A. de Moura and Gizeli S. Gimenez
Experimental Biology, Federal University of Rondônia - UNIR, Porto Velho, RO, Brazil

Kayena D. Zaqueo and Rodrigo S. Silva
Biodiversity and Biotechnology, Bionorte Web, Porto Velho, RO, Brazil

Experimental Biology, Federal University of Rondônia - UNIR, Porto Velho, RO, Brazil

Cleópatra A. S. Caldeira
Biological Sciences, Federal University of Rondônia - UNIR, Porto Velho, RO, Brazil

Acknowledgement

The authors are grateful to Professor Luiz Hildebrando Pereira da Silva, George A. Oliveira, Marjorie J. M. Nascimento and Rafaela D. Souza by the assistance and to the Ministry of Science and Technology (MCT), Conselho Nacional de Desenvolvimento Científico e Tecnológico (CNPq/MCT), Financiadora de Estudos e Projetos (FINEP/MCT), Fundação de Tecnologia do Acre/Fundo de Desenvolvimento Científico e Tecnológico (Funtac/FDCT), Secretary of Development of the Rondônia State (PRONEX/CNPq), Instituto Nacional para Pesquisa Translacional em Saúde e Ambiente na Região Amazônica (INCTINPeTAm/CNPq/MCT), and Rede de Biodiversidade e Biotecnologia da Amazônia Legal (Rede Bionorte/MCT) for the financial support.

6. References

[1] Kordis D, Gubensek F (2000). Adaptive evolution of animal toxin multigene families. Gene 1: 43-52.
[2] Menez A (1998). Functional architectures of animal toxins: a clue to drug design Toxicon 11: 1557-1572.
[3] Kang T.S, Georgieva D, Genov N, Murakami M.T, Sinha M, Kumar R.P, Kaur P, Kumar S, Dey S, Sharma S, Vrielink A, Betzel C, Takeda S, Arni R. K, Singh T.P, Kini R.M (2011). Enzymatic toxins from snake venom: structural characterization and mechanism of catalysis. FEBS J. 23: 4544-4576.
[4] Calvete J.J, Juarez P, Sanz L (2007). Snake venomics. Strategy and applications. J. Mass Spectrom. 11: 1405-1414.
[5] Mebs D (2001). Toxicity in animals. Trends in evolution. Toxicon 1: 87-96
[6] Dennis E.A, Cao J, Hsu Y.H, Magrioti V, Kokotos G. (2011). Phospholipase A_2 enzymes: physical structure, biological function, disease implication, chemical inhibition, and therapeutic intervention. Chem. Rev. 10: 6130-6185.
[7] Gutierrez J.M, Ownby C.L, Odell G.V (1984). Isolation of a Myotoxin from *Bothrops asper* Venom - Partial Characterization and Action on Skeletal-Muscle. Toxicon 1: 115-128.
[8] de Paula R.C, Castro H.C, Rodrigues C.R, Melo P.A, Fuly A.L (2009). Structural and pharmacological features of phospholipases A_2 from snake venoms. Protein Pept. Lett. 8: 899-907.
[9] Soares A.M, Sestito W.P, Marcussi S, Stabeli R.G, Andriao-Escarso S. H, Cunha O.A, Vieira C.A, Giglio J.R (2004). Alkylation of myotoxic phospholipases A_2 in *Bothrops moojeni* venom: a promising approach to an enhanced antivenom production. Int. J. Biochem. Cell. Biol. 2: 258-270.
[10] Rodrigues V.M, Marcussi S, Cambraia R.S, de Araujo A.L, Malta-Neto N.R, Hamaguchi A, Ferro E.A, Homsi-Brandeburgo M.I, Giglio J.R, Soares A.M (2004). Bactericidal and neurotoxic activities of two myotoxic phospholipases A_2 from *Bothrops neuwiedi pauloensis* snake venom. Toxicon 3: 305-314.
[11] Taghrid S, Istivan, T.S, Coloe, P. J (2006). Phospholipase A in Gram-negative bacteria and its role in pathogenesis. Microbiology 152: 1263-1274.
[12] Waite M (1996). Phospholipases. In Biochemistry of Lipids, Lipoproteins and Membranes, pp. 211-236. Edited by D. E. Vance & J. Vance. Amsterdam: Elsevier.
[13] Dennis E.A (1994) Diversity of Group Types, Regulation, and Function of Phospholipase A_2. J. Biol. Chem. 269: 13057-13060.
[14] Dennis E.A (1997). The growing phospholipase A_2 superfamily of signal transduction enzymes. Trends Biochem. Sci. 22: 1-2.
[15] Arni R.K, Ward R.J (1996). Phospholipase A_2 a structural review. Toxicon 8: 827-841.
[16] Burke J.E, Dennis E.A (2009). Phospholipase A_2 biochemistry. Cardiovasc. Drugs Ther. 1: 49-59.
[17] de Oliveira T.C, de Amorim H.L.N, Guimaraes J.A (2007). Interfacial activation of snake venom phospholipases A_2 svPLA2 probed by molecular dynamics simulations. J. Mol. Struct. Theochem. 1-3: 31-41.

[18] Verheij H.M, Volwerk J.J, Jansen E.H.J.M, Puyk W.C, Dijkstra B.W, Drenth J, Dehaas G.H (1980). Methylation of Histidine-48 in Pancreatic Phospholipase A_2 - Role of Histidine and Calcium-Ion in the Catalytic Mechanism. Biochemistry 4: 743-750.

[19] dos Santos J.I, Fernandes C.A, Magro A.J, Fontes M.R (2009). The intriguing phospholipases A_2 homologues: relevant structural features on myotoxicity and catalytic inactivity. Protein Pept. Lett. 8: 887-893.

[20] Lomonte B, Rey-Suarez P, Tsai W.C., Angulo, Y, Sasa M, Gutierrez J.M, Calvete J.J (2012). Snake venomics of the pit vipers *Porthidium nasutum, Porthidium ophryomegas,* and *Cerrophidion godmani* from Costa Rica: toxicological and taxonomical insights. J. Proteomics 5: 1675-1689.

[21] Andriao-Escarso S.H, Soares A.M, Rodrigues V.M, Angulo Y, Diaz C, Lomonte B, Gutierrez J.M, Giglio J.R (2000). Myotoxic phospholipases A_2 in Bothrops snake venoms: effect of chemical modifications on the enzymatic and pharmacological properties of bothrops toxins from *Bothrops jararacussu*. Biochimie 8: 755-763.

[22] Fernandez J, Gutierrez J.M, Angulo Y, Sanz L, Juarez P, Calvete J.J, Lomonte B (2010). Isolation of an acidic phospholipase A_2 from the venom of the snake *Bothrops asper* of Costa Rica: biochemical and toxicological characterization. Biochimie 3: 273-283.

[23] Rodrigues R.S, Izidoro L.F, Teixeira S.S, Silveira L.B, Hamaguchi A, Homsi-Brandeburgo M.I, Selistre-de-Araujo H.S, Giglio J.R, Fuly A.L, Soares A.M, Rodrigues V.M (2007). Isolation and functional characterization of a new myotoxic acidic phospholipase A_2 from *Bothrops pauloensis* snake venom. Toxicon 1: 153-165.

[24] Silveira L.B, Marchi-Salvador D.P, Santos-Filho N. A, Silva F.P.Jr, Marcussi S, Fuly A. L, Nomizo A, da Silva S.L, Stabeli R.G, Arantes E.C, Soares A.M (2012). Isolation and expression of a hypotensive and anti-platelet acidic phospholipase A_2 from *Bothrops moojeni* snake venom. J. Pharm. Biomed. Anal. In Press

[25] Teixeira S.S, Silveira L.B, da Silva F.M, Marchi-Salvador D.P, Silva F.P. Jr, Izidoro L. F, Fuly A.L, Juliano M.A, dos Santos C.R, Murakami M.T, Sampaio S.V, da Silva S.L, Soares A.M (2011). Molecular characterization of an acidic phospholipase A_2 from *Bothrops pirajai* snake venom: synthetic C-terminal peptide identifies its antiplatelet region. Arch. Toxicol 10: 1219-1233.

[26] Soares A.M, Andriao-Escarso S.H, Angulo Y, Lomonte B, Gutierrez J.M, Marangoni S, Toyama M.H, Arni R.K, Giglio J.R (2000). Structural and functional characterization of myotoxin I, a Lys49 phospholipase A(2) homologue from *Bothrops moojeni* (Caissaca) snake venom. Arch. Biochem. Biophys 1: 7-15.

[27] Hanasaki K, Arita H (1999). Biological and pathological functions of phospholipase A_2 receptor. Arch. Biochem. Biophys 2: 215-223.

[28] Lambeau G, Lazdunski M (1999). Receptors for a growing family of secreted phospholipases A_2. Trends Pharmacol. Sci. 4: 162-170.

[29] Valentin E, Lambeau G (2000). What can venom phospholipases A_2 tell us about the functional diversity of mammalian secreted phospholipases A_2. Biochimie 9-10: 815-831.

[30] Wixom R.L, (2001). The beginnings of chromatography — The pioneers (1900–1960), In: Gehrke C.W, Wixom R.L, and Bayer E, Ed(s). Journal of Chromatography Library, Elsevier 64: 1-38.

[31] Spencer P.J, Aird S.D, Boni-Mitake M, Nascimento N, Rogero J.R (1998). A single-step purification of bothropstoxin-1. Braz. J. Med. Biol. Res. 9: 1125-1127.
[32] Lomonte B, Gutierrez J.M, Furtado M.F, Otero R, Rosso J.P, Vargas O, Carmona E, Rovira M.E (1990). Isolation of basic myotoxins from *Bothrops moojeni* and *Bothrops atrox* snake venoms. Toxicon 10: 1137-1146.
[33] Soares A.M, Rodrigues V.M, Borges M.H, Andriao-Escarso S.H, Cunha O.A, Homsi-Brandeburgo M.I, Giglio J.R (1997). Inhibition of proteases, myotoxins and phospholipases A_2 from *Bothrops* venoms by the heteromeric protein complex of *Didelphis albiventris* opossum serum. Biochem. Mol. Biol. Int. 5: 1091-1099.
[34] Homsi-Brandeburgo, M.I, Queiroz L.S, Santo-Neto H, Rodrigues-Simioni L, Giglio J.R (1988). Fractionation of *Bothrops jararacussu* snake venom: partial chemical characterization and biological activity of bothropstoxin. Toxicon 7: 615-627.
[35] Pereira M.F, Novello J.C, Cintra A.C, Giglio J.R, Landucci E.T, Oliveira B, Marangoni S (1998). The amino acid sequence of bothropstoxin-II, an Asp-49 myotoxin from *Bothrops jararacussu* (Jararacucu) venom with low phospholipase A_2 activity. J. Protein Chem. 4: 381-386.
[36] Rodrigues V.M, Marcussi S, Cambraia R.S, de Araujo A.L, Malta-Neto N.R, Hamaguchi A, Ferro E.A, Homsi-Brandeburgo M.I, Giglio J.R,Soares A.M (2004). Bactericidal and neurotoxic activities of two myotoxic phospholipases A_2 from *Bothrops neuwiedi pauloensis* snake venom. Toxicon 3, 305-314.
[37] Rodrigues V.M, Soares A.M, Mancin A.C, Fontes M.R, Homsi-Brandeburgo M.I, Giglio J.R (1998). Geographic variations in the composition of myotoxins from *Bothrops neuwiedi* snake venoms: biochemical characterization and biological activity. Comp. Biochem. Physiol. A Mol. Integr. Physiol. 3: 215-222.
[38] Higuchi D.A, Barbosa C.M, Bincoletto C, Chagas J.R, Magalhaes A, Richardson M, Sanchez E.F, Pesquero J.B, Araujo R.C, Pesquero J.L (2007). Purification and partial characterization of two phospholipases A_2 from *Bothrops leucurus* (white-tailed-jararaca) snake venom. Biochimie 3: 319-328.
[39] Daniele J.J, Bianco I.D, Delgado C, Carrillo D.B, Fidelio G.D (1997). A new phospholipase A_2 isoform isolated from *Bothrops neuwiedii* (Yarara chica) venom with novel kinetic and chromatographic properties. Toxicon 8: 1205-1215.
[40] Janson J.C (1967). Adsorption phenomena on Sephadex. J. Chromatogr. 1: 12-20.
[41] Williams K.W (1972). Solute-gel interactions in gel filtration. Lab. Pract. 9: 667-670.
[42] Santos-Filho N.A, Silveira L.B, Oliveira C.Z, Bernardes C.P, Menaldo D.L, Fuly A.L, Arantes E.C, Sampaio S.V, Mamede C.C, Beletti M.E, de Oliveira F, Soares A.M (2008). A new acidic myotoxic, anti-platelet and prostaglandin I2 inductor phospholipase A_2 isolated from *Bothrops moojeni* snake venom. Toxicon 8: 908-917.
[43] Nunes D.C, Rodrigues R.S, Lucena M.N, Cologna C.T, Oliveira A.C, Hamaguchi A, Homsi-Brandeburgo M.I, Arantes E.C, Teixeira D.N, Ueira-Vieira C, Rodrigues V.M (2011). Isolation and functional characterization of proinflammatory acidic phospholipase A_2 from *Bothrops leucurus* snake venom. Comp. Biochem. Physiol. C. Toxicol. Pharmacol. 3: 226-233.
[44] Melo P.A, Suarez-Kurtz G (1988). Release of sarcoplasmic enzymes from skeletal muscle by *Bothrops jararacussu* venom: antagonism by heparin and by the serum of South American marsupials. Toxicon 1: 87-95.

[45] Melo P.A, Homsi-Brandeburgo M.I, Giglio J.R, Suarez-Kurtz, G (1993). Antagonism of the myotoxic effects of *Bothrops jararacussu* venom and bothropstoxin by polyanions. Toxicon 3: 285-291.

[46] Lomonte B, Moreno E, Tarkowski A, Hanson, L.A, Maccarana, M (1994). Neutralizing interaction between heparins and myotoxin II, a lysine 49 phospholipase A_2 from *Bothrops asper* snake venom. Identification of a heparin-binding and cytolytic toxin region by the use of synthetic peptides and molecular modeling. J. Biol. Chem. 47: 29867-29873.

[47] Stabeli R.G, Magalhaes L.M, Selistre-de-Araujo H.S, Oliveira E.B (2005). Antibodies to a fragment of the *Bothrops moojeni*amino acid oxidase cross-react with snake venom components unrelated to the parent protein. Toxicon 3: 308-317.

[48] Beghini D.G, da Cruz-Hofling M.A, Randazzo-Moura P, Rodrigues-Simioni L, Novello J.C, Hyslop S, Marangoni S (2005). Cross-neutralization of the neurotoxicity of *Crotalus durissus terrificus* and *Bothrops jararacussu* venoms by antisera against crotoxin and phospholipase A_2 from *Crotalus durissus cascavella* venom. Toxicon 6: 604-611.

[49] Gomes P.C, Machado de Avila R.A, Selena Maria W, Richardson M, Fortes-Dias C.L, Chavez-Olortegui C (2007). The co-purification of a lectin (BJcuL) with phospholipases A_2 from *Bothrops jararacussu* snake venom by immunoaffinity chromatography with antibodies to crotoxin. Toxicon 8: 1099-1108.

[50] Rock C.O, Snyder F (1975). Metabolic interrelations of hydroxy-substituted ether-linked glycerolipids in the pink portion of the rabbit harderian gland. Arch. Biochem. Biophys 2: 631-636.

[51] Dijkman R, Beiboer S.H, Verheij H.M (1997). An affinity column for phospholipase A_2 based on immobilised acylaminophospholipid analogues. Biochim. Biophys Acta 1: 1-8.

[52] Calvete J.J, Sanz L, Angulo Y, Lomonte B, Gutierrez J.M (2009b). Venoms, venomics, antivenomics. FEBS Lett. 11: 1736-1743.

[53] Gomez-Ariza, J.L.T, Garcia-Barrera (2005). Analytical Characterization of bioactive metal species in the cellular domain (Metallomics) to simplify environmental and biological proteomics. Int. J. Environ. Anal. Chem. 85: 255-266.

[54] Gao Y.X, Chen C.Y, Zhang P.Q, Chai Z.F, He W, Huang Y.Y (2003). Detection of metalloproteins in human liver cytosol by synchrotron radiation X-ray fluorescence after sodium dodecyl sulphate polyacrylamide gel electrophoresis. Anal. Chim. Acta 1: 131-137.

[55] Szpunar J. (2004). Metallomics: a new frontier in analytical chemistry. Anal. Bioanal Chem. 1: 54-56.

[56] Laemmli U.K (1970). Cleavage of Structural Proteins during Assembly of Head of Bacteriophage-T4.Nature 5259: 680.

[57] Torres A.F.C, Dantas R.T, Toyama M.H, Diz E, Zara F.J, de Queiroz M.G.R, Nogueira N.A.P, de Oliveira M.R, Toyama D.D, Monteiro H.S.A, Martins A.M.C (2010). Antibacterial and antiparasitic effects of *Bothrops marajoensis* venom and its fractions: Phospholipase A_2 and L-amino acid oxidase. Toxicon 4: 795-804.

[58] Andriao-Escarso S.H, Soares A.M, Fontes M.R, Fuly A.L, Correa F.M, Rosa J.C, Greene L.J, Giglio J.R (2002). Structural and functional characterization of an acidic platelet aggregation inhibitor and hypotensive phospholipase A_2 from *Bothrops jararacussu* snake venom. Biochem. Pharmacol. 4: 723-732.

[59] Vesterberg O, Eriksson, R (1972). Detection of fibrinolytic activity by a zymogram technique after isoelectric focusing. Biochim. Biophys. Acta 2: 393-397.
[60] O'Farrell P.H (1975). High-resolution two-dimensional electrophoresis of proteins. J. Biol. Chem. 10: 4007-4021.
[61] Gorg A, Postel W, Weser J, Patutschnick W, Cleve H (1985). Improved resolution of PI (alpha 1-antitrypsin) phenotypes by a large-scale immobilized pH gradient. Am. J. Hum. Genet. 5: 922-930.
[62] Herbert B (1999). Advances in protein solubilization for two-dimensional electrophoresis. Electrophoresis 4-5: 660-663
[63] Alape-Giron A, Flores-Diaz M, Sanz L, Madrigal M, Escolano J, Sasa M, Calvete J.J (2009). Studies on the venom proteome of *Bothrops asper*: perspectives and applications. Toxicon 7: 938-948.
[64] Gutierrez J.M, Lomonte B (1995). Phospholipase A2 myotoxins from *Bothrops* snake venoms. Toxicon 11: 1405-1424.
[65] Holzer M, Mackessy S.P (1996). An aqueous endpoint assay of snake venom phospholipase A2. Toxicon 10: 1149-1155.
[66] Martin-Mounot N, Rochat H (1979). Isolation and characterization of a toxic phospholipase A2 in the spitting cobra (*Naja mossambicamossambica*) venom. Toxicon 17: 127-36.
[67] Radvanyi F, Saliou B, Bon C, Strong P.N (1987). The interaction between the presynaptic phospholipase neurotoxins beta-bungarotoxin and crotoxin and mixed detergent-phosphatidylcholine micelles. A comparison with non-neurotoxic snake venom phospholipases A2. J. Biol. Chem. 19: 8966-8974.
[68] Yao Y, Wang M.H, Zhao K.Y, Wang C.C (1998). Assay for enzyme activity by following the absorbance change of pH-indicators. J. Biochem. Biophys. Methods 2-3: 119-130.
[69] Price J.A (2007). A colorimetric assay for measuring phospholipase A2 degradation of phosphatidylcholine at physiological pH. J. Biochem. Biophys Methods 3: 441-444.
[70] Soares A.M, Guerra-Sá R, Borja-Oliveira C, Rodrigues V, Rodrigues-Simioni L, Rodrigues V, Fontes M.R.M, Giglio J.R (2000). Molecular cloning and functional characterization of BnSP-7, a myotoxin Lys-49 phospholipase A2 homologue, from *Bothrops neuwiedi pauloensis* venom. Arch. Biochem. Biophys. 378: 201–209.
[71] Soares A.M, Mancin A.C, Cecchini A.L, Arantes E.C, Franca S.C, Gutiérrez J.M, Giglio J.R (2001). Effects of chemical modifications of crotoxin B, the phospholipase A2 subunit of crotoxin from *Crotalus durissus terrificus* snake venom, on its enzymatic and pharmacological activities. Int. J. Biochem. Cell Biol. 33: 877–888.
[72] Soares A.M, Andriao-Escarso S.H, Bortoleto R.K, Rodrigues-Simioni L, Arni R.K, Ward R.J, Gutiérrez J.M, Giglio J.R (2001). Dissociation of enzymatic and pharmacological properties of piratoxins-I and-III, two myotoxic phospholipases A2 from *Bothrops pirajai* snake venom. Arch. Biochem. Biophys. 387: 188–19.
[73] Lomonte B, Angulo Y, Rufini S, Cho W, Giglio J.R, Ohno M, Daniele J.J, Geoghegan, P, Gutierrez J.M (1999). Comparative study of the cytolytic activity of myotoxic phospholipases A2 on mouse endothelial (tEnd) and skeletal muscle (C2C12) cells in vitro. Toxicon 37: 145–158.

[74] Ponce-Soto L.A, Bonfim V.L, Rodrigues-Simioni L, Novello J.C, Marangoni S (2006). Determination of primary structure of two isoforms 6-1 and 6-2 PLA_2 D49 from *Bothrops jararacussu* snake venom and neurotoxic characterization using in vitro neuromuscular preparation. Protein J. 25: 47-55.

[75] de Haas G.H, Postema M.M, Niewenhuizen W, Van Deenen L.L (1968). Purification and properties of phospholipase A and its zymogen from porcine pancreas. Bull. Soc. Chim. Biol. (Paris) 9: 1383.

[76] Gutierrez J. M, Avila C, Rojas E, Cerdas L (1988). An alternative in vitro method for testing the potency of the polyvalent antivenom produced in Costa Rica. Toxicon 4: 411-413.

[77] Assendelft O.W.Van (1970). Spectrophotometry of haemoglobin derivatives. Royal Vangorcum Ltd. Assen. The Netherlands and Charles C. Thomas Springfield. III: 152.

[78] Nikai T, Komori Y, Yagihashi S, Ohara A, Ohizumi Y, Sugihara H (1994). Isolation and Characterization of Phospholipase-A_2 from *Agkistrodon bilineatus* (Common Cantil) Venom. Int. J. Biochem. 26: 879-884.

[79] Takagi J, Sekiya F, Kasahara K, Inada Y, Saito Y (1988). Venom from southern copperhead snake *Agkistrodon contortrix contortrix* II. A unique phospholipase A_2 that induces platelet aggregation. Toxicon 2: 199-206.

[80] Johnson E.K, Ownby C.L (1993). Isolation of a myotoxin from the venom of *Agkistrodon contortrix laticinctus* (broad-banded copperhead) and pathogenesis of myonecrosis induced by it in mice. Toxicon 3: 243-255.

[81] Leite R.S, Franco W, Ownby C L, Selistre-de-Araujo H.S (2004). Effects of ACL myotoxin, a Lys49 phospholipase A_2 from *Agkistrodon contortrix laticinctus* snake venom, on water transport in the isolated toad urinary bladder. Toxicon 1: 77-83.

[82] Rojas E, Saravia P, Angulom Y, Arce V, Lomonte B, Chavez J.J, Velasquez R, Thelestam M, Gutierrez J.M (2001). Venom of the crotaline snake *Atropoides nummifer* (jumping viper) from Guatemala and Honduras: comparative toxicological characterization, isolation of a myotoxic phospholipase A_2 homologue and neutralization by two antivenoms. Comp. Biochem. Physiol. C. Toxicol. Pharmacol. 2: 151-162.

[83] Gutierrez J.M, Lomonte B, Cerdas L (1986). Isolation and partial characterization of a myotoxin from the venom of the snake *Bothrops nummifer*. Toxicon 9: 885-894.

[84] Angulo Y, Olamendi-Portugal T, Alape-Giron A, Possani L.D, Lomonte B (2002). Structural characterization and phylogenetic relationships of myotoxin II from *Atropoides (Bothrops) nummifer* snake venom, a Lys49 phospholipase A_2 homologue. Int. J. Biochem. Cell. Biol. 10: 1268-1278.

[85] Angulo Y, Chaves E, Alape A, Rucavado A, Gutierrez J.M, Lomonte B (1997). Isolation and characterization of a myotoxic phospholipase A_2 from the venom of the arboreal snake *Bothriechis (Bothrops) schlegelii* from Costa Rica. Arch. Biochem. Biophys 2: 260-266.

[86] Huancahuire-Veja S, Ponce-Soto L.A, Martins-de-Souza D, Marangoni S (2011). Biochemical and pharmacological characterization of PhTX-I a new myotoxic phospholipase A_2 isolated from *Porthidium hyoprora* snake venom. Comp. Biochem. Physiol. C. Toxicol. Pharmacol. 2: 108-119.

[87] Selistre H. S, Queiroz L.S, Cunha O.A, De Souza G.E, Giglio J.R (1990). Isolation and characterization of hemorrhagic, myonecrotic and edema-inducing toxins from *Bothrops insularis* (jararaca ilhoa) snake venom. Toxicon 3: 261-273.

[88] Cogo J.C, Lilla S, Souza G.H, Hyslop S, de Nucci G (2006). Purification, sequencing and structural analysis of two acidic phospholipases A_2 from the venom of *Bothrops insularis* (jararaca ilhoa). Biochimie 12: 1947-1959.

[89] Machado Braga M.D, Costa Martins A.M, Alves C.D, de Menezes D.B, Martins R.D, Ferreira Barbosa P.S, de Sousa Oliveira I.M, Toyama M.H, Toyama D.O, Dos Santos Diz Filho E.B, Ramos Fagundes F.H, Fonteles M.C, Azul Monteiro H.S (2008). Purification and renal effects of phospholipase A_2 isolated from *Bothrops insularis* venom. Toxicon 2: 181-190.

[90] Serrano S.M, Reichl A.P, Mentele R, Auerswald, E.A, Santoro M.L, Sampaio C.A, Camargo A.C, Assakura M.T (1999). A novel phospholipase A_2, BJ-PLA_2, from the venom of the snake *Bothrops jararaca:* purification, primary structure analysis, and its characterization as a platelet-aggregation-inhibiting factor. Arch. Biochem. Biophys. 1: 26-32.

[91] Machado O.L, Oliveira-Carvalho A.L, Zingali R.B, Carlini C.R (1993). Purification physicochemical characterization and N-terminal-amino acid sequence of a phospholipase A_2 from *Bothrops jararaca* venom. Braz. J. Med. Biol. Res. 2: 163-166.

[92] Garcia Denegri M.E, Acosta O.C, Huancahuire-Veja S, Martins-de-Souza D, Marangoni S, Marunak S.L, Teibler G.P, Leiva L.C, Ponce-Soto L.A (2010). Isolation and functional characterization of a new acidic PLA_2 Ba SpII RP4 of the *Bothrops alternatus* snake venom from Argentina. Toxicon 1: 64-74.

[93] Nisenbom H.E, Seki C, Vidal J.C (1986). Phospholipase A_2 from *Bothrops alternatus* (vibora de la cruz) venom. Purification and some characteristic properties. Toxicon 3: 259-272.

[94] Ponce-Soto L.A, Lomonte B, Gutierrez J.M, Rodrigues-Simioni L, Novello J.C, Marangoni S (2007a). Structural and functional properties of BaTX, a new Lys49 phospholipase A_2 homologue isolated from the venom of the snake *Bothrops alternatus*. Biochim. Biophys. Acta 4: 585-593.

[95] Quintero A, Soares A.M (2010). Functional and structural characterization of phospholipases A_2 isolated from *Bothrops asper* snake venom in Panama. J. Venom. Anim. Toxins incl. Trop. Dis. 4: 664-664.

[96] Lomonte B, Gutierrez J.M (1989) A new muscle damaging toxin, myotoxin II, from the venom of the snake Bothrops asper (terciopelo). Toxicon 27: 725-733.

[97] Kaiser I. I, Gutierrez, J. M, Plummer D, Aird S. D, Odell G. V (1990). The Amino-Acid-Sequence of a Myotoxic Phospholipase from the Venom of *Bothrops asper*. Arch. Biochem. Biophys 2: 319-325.

[98] Mebs D, Samejima Y (1986). Isolation and characterization of myotoxic phospholipases A_2 from crotalid venoms. Toxicon 2: 161-168.

[99] Kanashiro M.M, de Cassia M. E.R, Petretski J.H, Prates M.V, Alves, E. W, Machado O.L, da Silva W.D, Kipnis T.L (2002). Biochemical and biological properties of phospholipases A_2 from *Bothrops atrox* snake venom. Biochem. Pharmacol. 7: 1179-1186.

[100] Nunez V, Arce V, Gutierrez J.M,Lomonte B (2004). Structural and functional characterization of myotoxin I, a Lys49 phospholipase A_2 homologue from the venom of the snake *Bothrops atrox*. Toxicon 1: 91-101.

[101] Costa T.R, Menaldo D.L, Oliveira C.Z, Santos-Filho N.A, Teixeira S.S, Nomizo A, Fuly A.L, Monteiro M. C, de Souza B M, Palma M.S, Stabeli R.G, Sampaio S.V,Soares A.M

(2008). Myotoxic phospholipases A_2 isolated from *Bothrops brazili* snake venom and synthetic peptides derived from their C-terminal region: cytotoxic effect on microorganism and tumor cells. Peptides 10: 1645-1656

[102] Huancahuire-Vega S, Ponce-Soto L.A, Martins-de-Souza D, Marangoni S (2009). Structural and functional characterization of brazilitoxins II and III (BbTX-II and -III), two myotoxins from the venom of *Bothrops brazili* snake. Toxicon 6: 818-827.

[103] de Albuquerque Modesto J.C, Spencer P.J, Fritzen M, Valenca R.C, Oliva M.L, da Silva M.B, Chudzinski-Tavassi A.M, Guarnieri M.C (2006). BE-I-PLA_2, a novel acidic phospholipase A_2 from *Bothrops erythromelas* venom: isolation, cloning and characterization as potent anti-platelet and inductor of prostaglandin I2 release by endothelial cells. Biochem. Pharmacol. 3: 377-384.

[104] Bonfim V.L, Toyama M.H, Novello J.C, Hyslop S, Oliveira C.R, Rodrigues-Simioni L, Marangoni S (2001). Isolation and enzymatic characterization of a basic phospholipase A_2 from *Bothrops jararacussu* snake venom. J Protein Chem 3: 239-245.

[105] Ketelhut D.F, de Mello M.H, Veronese E.L, Esmeraldino L.E, Murakami M.T, Arni R. K, Giglio J.R, Cintra A.C, Sampaio S.V (2003). Isolation, characterization and biological activity of acidic phospholipase A_2 isoforms from *Bothrops jararacussu* snake venom. Biochimie 10: 983-991.

[106] Gutierrez J.M, Sanz L, Escolano J, Fernandez J, Lomonte B, Angulo Y, Rucavado A, Warrell D.A, Calvete J.J. (2008). Snake venomics of the Lesser Antillean pit vipers *Bothrops caribbaeus* and *Bothrops lanceolatus*: correlation with toxicological activities and immunoreactivity of a heterologous antivenom. J. Proteome Res. 10: 4396-4408.

[107] Ponce-Soto L.A, Martins-de-Souza D, Marangoni S (2010). Neurotoxic, myotoxic and cytolytic activities of the new basic PLA_2 isoforms BmjeTX-I and BmjeTX-II isolated from the *Bothrops marajoensis* (Marajo Lancehead) snake venom. Protein J. 2: 103-113.

[108] Costa Torres A.F, Dantas R.T, Toyama M.H, Diz Filho E, Zara F.J, Rodrigues de Queiroz M, G Pinto Nogueira N.A, Rosa de Oliveira M, de Oliveira Toyama D, Monteiro H. S, Martins A.M (2010). Antibacterial and antiparasitic effects of *Bothrops marajoensis* venom and its fractions: Phospholipase A_2 and L-amino acid oxidase. Toxicon 4: 795-804.

[109] Galbiatti C, Rocha T,Randazzo-Moura P, Ponce-Soto L.A, Marangoni S, Cruz-Höfling M.A, Rodrigues-Simioni, L (2012). Pharmacological and partial biochemical characterization of Bmaj-9 isolated from *Bothrops marajoensis* snake venom. J. Venom. Anim. Toxins incl. Trop. Dis 18: 62-72.

[110] Perchuc A.M, Menin L, Favreau P, Buhler B, Bulet P, Schoni R, Wilmer M, Ernst B, Stocklin R (2010). Isolation and characterization of two new Lys49 PLA_2s with heparin neutralizing properties from *Bothrops moojeni* snake venom. Toxicon 6: 1080-1092.

[111] Soares A.M, Rodrigues V.M, Homsi-Brandeburgo M.I, Toyama M.H, Lombardi F.R, Arni R.K, Giglio J.R (1998). A rapid procedure for the isolation of the Lys-49 myotoxin II from *Bothrops moojeni* (caissaca) venom: biochemical characterization, crystallization, myotoxic and edematogenic activity. Toxicon 3: 503-514.

[112] Calgarotto A.K, Damico D.C, Ponce-Soto L.A, Baldasso P. A, Da Silva S.L, Souza G. H, Eberlin M.N, Marangoni S (2008). Biological and biochemical characterization of new basic phospholipase A(2) BmTX-I isolated from *Bothrops moojeni* snake venom. Toxicon 8: 1509-1519.

[113] Queiroz M.R, Mamede C.C, Fonseca K.C, Canabrava L.C.M.N, Franca L.V, Silva M. C, Stanziola L, Beletti M.E, Canabrava H.A.N, Oliveira, F (2011). Biological characterization of a myotoxin phosphoplipase A2 homologue purified from the venom of the snake *Bothrops moojeni*. J. Venom. Anim. Toxins incl. Trop. Dis.1: 49-58.
[114] Mancuso L C, Correa M.M, Vieira C.A, Cunha O.A, Lachat J, de Araujo H.S, Ownby C. L, Giglio J.R (1995). Fractionation of *Bothrops pirajai* snake venom: isolation and characterization of piratoxin-I, a new myotoxic protein. Toxicon 5: 615-626.
[115] Toyama M.H, Costa P.D, Novello J.C, de Oliveira B, Giglio J.R, da Cruz-Hofling M.A, Marangon S (1999). Purification and amino acid sequence of MP-III 4R D49 phospholipase A2 from *Bothrops pirajai* snake venom, a toxin with moderate PLA2 and anticoagulant activities and high myotoxic activity. J. Protein Chem. 3: 371-378
[116] Toyama M.H, Mancuso L.C, Giglio J.R, Novello J.C, Oliveira B, Marangoni S (1995). A quick procedure for the isolation of dimeric piratoxins-I and II, two myotoxins from *Bothrops pirajai* snake venom.N-terminal sequencing. Biochem. Mol. Biol. Int. 6: 1047-1055.
[117] Diaz C, Gutierrez J.M, Lomonte B (1992). Isolation and Characterization of Basic Myotoxic Phospholipases-A2 from *Bothrops Godmani* (Godman Pit Viper) Snake-Venom. Arch. Biochem. Biophys 1: 135-142.
[118] de Sousa M.V, Morhy L, Arni R.K, Ward R.J, Diaz C, Gutierrez J.M (1998). Amino acid sequence of a myotoxic Lys49-phospholipase A2 homologue from the venom of *Cerrophidion (Bothrops) godmani*. Biochim. Biophys Acta 2: 204-208.
[119] Tsai I.H, Chen Y.H, Wang Y.M, Tu M.C, Tu A.T (2001). Purification, sequencing, and phylogenetic analyses of novel Lys-49 phospholipases A2 from the venoms of rattlesnakes and other pit vipers. Arch. Biochem. Biophys 2: 236-244.
[120] Fonseca F.V, Antunes E, Morganti R.P, Monteiro H.S, Martins A.M, Toyama D.O, Marangoni S, Toyama M.H (2006). Characterization of a new platelet aggregating factor from crotoxin *Crotalus durissus cascavella* venom. Protein J. 3: 183-192.
[121] Ponce-Soto L.A, Baldasso P.A, Romero-Vargas F.F, Winck F.V, Novello J.C, Marangoni, S (2007). Biochemical pharmacological and structural characterization of two PLA2 isoforms Cdr-12 and Cdr-13 from *Crotalus durissus ruruima* snake venom. Protein J. 1: 39-49.
[122] Romero-Vargas F.F, Ponce-Soto L.A, Martins-de-Souza D, Marangoni S (2010). Biological and biochemical characterization of two new PLA2 isoforms Cdc-9 and Cdc-10 from *Crotalus durissus cumanensis* snake venom. Comp. Biochem. Physiol. C. Toxicol Pharmacol. 1: 66-74.
[123] Pereanez J.A, Nunez V, Huancahuire-Vega S, Marangoni S, Ponce-Soto L.A (2009). Biochemical and biological characterization of a PLA2 from crotoxin complex of *Crotalus durissus cumanensis*. Toxicon 5: 534-542.
[124] Diz Filho E.B, Marangoni S, Toyama D.O, Fagundes F.H, Oliveira S.C, Fonseca F.V, Calgarotto A.K, Joazeiro P.P,Toyama M.H (2009). Enzymatic and structural characterization of new PLA2 isoform isolated from white venom of *Crotalus durissus ruruima*. Toxicon 1: 104-114.
[125] Hernandez-Oliveira S, Toyama M.H, Toyama D.O, Marangoni S, Hyslop S, Rodrigues-Simioni L (2005). Biochemical, pharmacological and structural characterization of a new

PLA2 from *Crotalus durissus terrificus* (South American rattlesnake) venom. Protein J. 4: 233-242.

[126] Faure G, Bon C (1987). Several isoforms of crotoxin are present in individual venoms from the South American rattlesnake *Crotalus durissus terrificus*. Toxicon 2: 229-234

[127] Oliveira D.G, Toyama M.H, Novello J.C, Beriam, L.O, Marangoni, S (2002). Structural and functional characterization of basic PLA2 isolated from *Crotalus durissus terrificus* venom. J. Protein Chem. 3: 161-168.

[128] Toyama, M.H, de Oliveira D.G, Beriam L.O, Novello J.C, Rodrigues-Simioni L, Marangoni S (2003). Structural, enzymatic and biological properties of new PLA2 isoform from *Crotalus durissus terrificus* venom. Toxicon 8: 1033-1038.

[129] Zepeda H, Rael E.D, Knight R.A (1985). Isolation of two phospholipases A2 from Mojave rattlesnake (*Crotalus scutulatus scutulatus*) venom and variation of immunologically related venom proteins in different populations. Comp. Biochem. Physiol. B. 2: 319-324.

[130] Damico D.C, Bueno L.G, Rodrigues-Simioni L, Marangoni S, da Cruz-Hofling M.A, Novello J.C (2006). Functional characterization of a basic D49 phospholipase A2 (LmTX-I) from the venom of the snake *Lachesis muta muta* (bushmaster). Toxicon 7: 759-765.

[131] Fuly A.L, de Miranda A.L, Zingali R.B, Guimaraes J.A (2002). Purification and characterization of a phospholipase A2 isoenzyme isolated from *Lachesis muta* snake venom. Biochem. Pharmacol. 9: 1589-1597.

[132] de Assis E.B, Estevao-Costa M.I, do Carmo Valentim A, Silva-Neto A, Agostini Cotta G, Alvarenga Mudado M, Richardson M, Fortes-Dias C.L (2008). Purification and complete primary structure of the first PLA2 from *Lachesis stenophrys* (the Central American Bushmaster) snake venom. Protein J. 5: 327-333.

[133] Vargas L.J, Londono M, Quintana J.C, Rua C, Segura C, Lomonte B, Nunez V (2012). An acidic phospholipase A2 with antibacterial activity from *Porthidium nasutum* snake venom. *Comp* Biochem. Physiol. B. Biochem. Mol. Biol. 4: 341-347.

[134] Bohlen C.J, Chesler A.T, Sharif-Naeini R, Medzihradszky K.F, Zhou S, King, Sanchez E.E, Burlingame A.L, Basbaum A.I, Julius D (2011). A heteromeric Texas coral snake toxin targets acid-sensing ion channels to produce pain. Nature 7373: 410-414.

[135] Possani L.D, Alagon A.C, Fletcher P.L, Jr, Varela M.J, Julia J.Z (1979). Purification and characterization of a phospholipase A2 from the venom of the coral snake, *Micrurus fulvius microgalbineus* (Brown and Smith). Biochem J. 3: 603-606.

[136] Dokmetjian J.C, Del Canto S, Vinzon S, de Jimenez Bonino M.B (2009). Biochemical characterization of the *Micrurus pyrrhocryptus* venom. Toxicon 3: 375-382.

[137] Alape-Giron A, Persson B, Cederlund E, Flores-Diaz M, Gutierrez J.M, Thelestam M, Bergman T, Jornvall H (1999). Elapid venom toxins: multiple recruitments of ancient scaffolds. Eur. J. Biochem. 1-2: 225-234.

[138] Mochca-Morales J, Martin B.M, Possani,L.D (1990). Isolation and characterization of helothermine, a novel toxin from *Heloderma horridum horridum* (Mexican beaded lizard) venom. Toxicon 3: 299-309.

[139] Dal Belo C.A, Toyama M.H, Toyama D, Marangoni S, Moreno F.B, Cavada B.S, Fontana M.D, Hyslop S, Carneiro E.M, Boschero A.C (2005). Determination of the amino acid sequence of a new phospholipase A(2) (MIDCA1) isolated *from Micrurus dumerilii carinicauda* venom. Protein J. 3: 147-153.

Chapter 3

2D-NanoLC-ESI-MS/MS for Separation and Identification of Mouse Brain Membrane Proteins

Phan Van Chi and Nguyen Tien Dung

Additional information is available at the end of the chapter

1. Introduction

Comprehensive proteomics analysis has the potential to provide new knowledge on cellular responses in development, aging, drug action, environmental stress, and disease pathogenesis (carcinogenesis, cardiovascular disease, etc). However, the separation and identification of proteomes/proteins is a challenging task due to their heterogeneous constituents or complex structures and closely related physico-chemical behaviors. It is clear that the combination of many analytical techniques is necessary to fulfill this complex task. At the start of proteomics research, two-dimensional electrophoresis (2DE) was routinely used to separate complex proteomic sample because of its high resolving power. In this technique, proteins are separated in a two-step process (two dimensions) based on their different physical properties. The first dimension is isoelectrofocusing in which proteins are separated based on their isoelectric points (pI, the pH where a *protein's net charge* is zero) using immobilized pH-gradient strips. Proteins then are separated according to their mass using sodium dodecylsulfate-polyacrylamide gel electrophoresis (SDS-PAGE) in the second dimension. With 2DE, thousands of proteins can be detected in a single experiment depending on the used staining techniques (Coomassie blue, silver, fluorescent dyes staining) [11]. Mass spectrometry (MS), using either electrospray ionization (ESI) or matrix-assisted laser desorption/ionization (MALDI), is the key technology for the identification of protein spots including membrane proteins, for which differential expression has been demonstrated [16, 30].

2DE, however, has some major drawbacks/disadvantages. It is time-consuming, difficult to reproduce and automation is hard to achieve. Furthermore, 2DE faces with many difficulties in analyzing several groups of proteins, such as low-abundance proteins, hydrophobic proteins (membrane proteins/membrane-bound and membrane-associated proteins), very large as well as very small proteins and proteins with extreme pI values. Unfortunately, these proteins have high proportion in comparison to total cellular proteins and are usually the most promising targets for drug development or disease diagnostics. About 30% of the mammalian genome encodes integral membrane proteins [27]. However, the comprehensive proteomic analysis of these proteins by mass spectrometry is difficult due to the amphipathic (containing regions that are hydrophobic and hydrophilic) nature in integral membrane proteins and their general low abundance levels [23]. Since the analysis of membrane proteins remains a significant challenge in proteomics, other techniques need to be established to address these problems. There have been many strategies developed for enriching, isolating and separating membrane proteins for proteomic analysis that have moved this field forward.

In recent years, two-dimensional liquid chromatography (2D-LC) has been employed as a complementary or alternative separation technique to 2DE. The combination of liquid chromatography as a separation tool for proteins and peptides with tandem mass spectrometry as an identification tool referred to as LC-MS/MS has generated a powerful and broadly used technique in the field of proteomics [6, 9, 10, 21, 22], particularly in the analysis of membrane proteomes [18, 19]. With the development of new quantitative strategies and bioinformatics tools to cope with the analysis of the large amounts of data generated in proteomics experiments, the resolution and sensitivity state-of-the art LC-MS/MS systems has reached dimensions allowing not only the analysis of individual proteins but also investigations on the level of complete proteomes [8]. This approach is usually based on the injection of the digested protein sample onto a strong cation-exchange (SCX) column as a first-dimension separation. Peptides bound in SCX column are eluted and separated from the column as fractions by an injecting salt plugs/salt step gradient of increasing salt concentration. Each fraction is subsequently separated on a reversed-phase (RP) column as the second orthogonal separation dimension before being presented to mass spectrometry analysis. Different stationary phases in chromatography columns provide variable levels of resolution. Reversed-phase chromatography is highly compatible with subsequent mass spectrometric analysis due to the lack of salts in the buffers and provides relatively high-resolution separation. Most reversed-phase stationary phases for LC-MS analysis consist of silica beads of 3–5 μm in diameter with alkyl chains of either eight or eighteen carbons in length (C8 or C18) attached. Using column switching, the entire procedure is on-line and fully automated. In order to improve sensitivity the reversed phase separation is usually performed in the nanoflow scale and mass spectrometry is used as the final detection method.

In this chapter a strategy for enrichment, isolation, separation, identification and characterization of mouse brain membrane proteins with the basic *setup* of two-dimensional nano liquid chromatography (2D-nanoLC) system (UltiMate™/FAMOS/Switchos™, LC Parking, Dionex, The Netherlands) coupled online with QSTAR®XL MS/MS mass spectrometer (Appllied Biosystems/MDS SCIEX, Ontario, Canada) is presented.

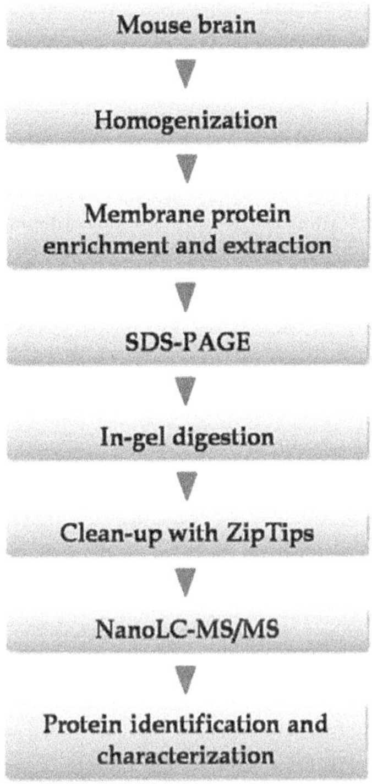

Figure 1. A scheme illustrating the necessary steps, including enrichment and extraction, separation, identification and characterization for proteomic analyses of mouse brain membrane proteins using gel-based approach in combination with comprehensive two-dimensional nano liquid chromatography (2D-nanoLC) coupled online with tandem mass spectrometry.

2. Membrane protein enrichment and extraction

Swiss mouse brains were collected as soon as possible after the animals were killed. The samples (3-5 g) were excised into approximately 5 mm wide pieces using scissors and washed with 10 ml of ice cold PBS buffer (0.2 g KCl, 8 g NaCl, 1.44 Na_2HPO_4, 0.24 g KH_2PO_4) and then resuspended in 3 volumes of the homogenization medium (0.25 M sucrose in 5 mM Tris-HCl pH 7.4 with 1 mM tetrasodium EGTA, 1 mM sodium orthovanadate (Na_3VO_4) and 2 mM sodium fluoride in deionized filter-sterilised MilliQ water) containing protease inhibitors (Calbiochem Protease Inhibitor Cocktail Set 111, catalog number 39134, contains AEBSF, aprotinin, bestatin, E-64, leupeptin, pepstatin A). After the medium has been drained off, new medium was replaced and drained off again. 10 ml of homogenisation medium (containing inhibitors) was added and the sample was homogenised using a Polytron in a Potter homogeniser with motor driven teflon pestle at approximately 1,000 rpm. Completely homogenized samples were centrifuged at 10,000 rpm for 15 min at 4°C to sediment large organelles. The obtained supernatant was used for recentrifugation again at 10,000 rpm for 15 min at 4°C. The supernatant was collected and centrifuged at 40,000 rpm at 4°C for 1 hr. After discarding the clear supernatant, the membrane pellets were retained and washed by resuspending in ice-cold 0.1 M Na_2CO_3 containing protease inhibitors for 1 hr. The mouse brain membrane protein fractions were obtained by centrifugation again at 40,000 rpm for 1 h at 4°C. The sample was divided and stored at −80°C until use. The protein concentration of the extracted membrane fractions was assessed using a Quick Start™ Bradford Protein Assay Kit (Bio-Rad, Hercules, CA 94547 USA).

3. Protein quantification

Protein concentration of the extracted membrane fractions was determined using Bio-Rad's Quick Start™ Bradford Protein Assay [5]. The assay is based on the observation that the maximum absorbance for an acidic solution of Coomassie Brilliant Blue G-250 shifts from 465nm to 595 nm when binding to protein occurs. Both hydrophobic and ionic interactions stabilize the anionic form of the dye, causing a visible colour change. For the standard curve, bovine serum albumin over a wide range of concentrations (0.1 - 20 μg/μl) was used. The low concentration range assay was used in the test tube format. 2 μl of standard or sample was added to 798 μl of MilliQ water. 200 μL of Bio-Rad reagent was added, mixed, and incubated for 10 min at room temperature. The absorbance at the wavelength of 595 nm was measured in a spectrophotometer. Glass or polystyrene (cheap) cuvettes have been used, however the color reagent stained both. Disposable cuvettes were recommended.

4. Sodium dodecyl sulfate polyacrylamide gel electrophoresis (SDS-PAGE) and in-gel digestion

4.1. SDS-PAGE

All chemicals including Precision Plus Protein Unstained Standards (catalog number: 161-0363EDU), Coomassie Brilliant Blue G-250 Staining Solution Kit (catalog number: 161-0435EDU) using for SDS-PAGE, were purchased from Bio-Rad (Bio-Rad Laboratories, Inc., CA 94547, USA). The procedure was carried out according to Laemmli [14].

The following stock solutions were prepared: (i) 1.5 M Tris–HCl, pH 8.8; (ii) 0.5 M Tris–HCl, pH 6.8; (iii) 30% acrylamide/bisacrylamide solution (37.5:1); (iv) *N,N,N,N* tetramethylethylenediamine (TEMED); (v) freshly prepared 10% ammonium persulphate (APS) solution; (vi) 10% sodium dodecyl sulphate (SDS) solution; (vii) 10X SDS gel running buffer (30 g Tris-base, 144 g glycine, 10 g SDS, dissolved in MilliQ water and adjusted to a volume of one liter); (viii) 5X sample buffer (10% SDS, 50% glycerol, 300 mM Tris–HCl (pH 6.8), 0.05% bromphenol blue. Dithiothreitol (DTT) was added to a final concentration of 100 mM prior to use).

The membrane fraction was solubilized in lysis buffer containing 3% SDS. Equal volumes containing approximately 25 μg/lane of MP were separated by 12% SDS-PAGE and were visualized by staining with Coomassie Brilliant Blue G-250.

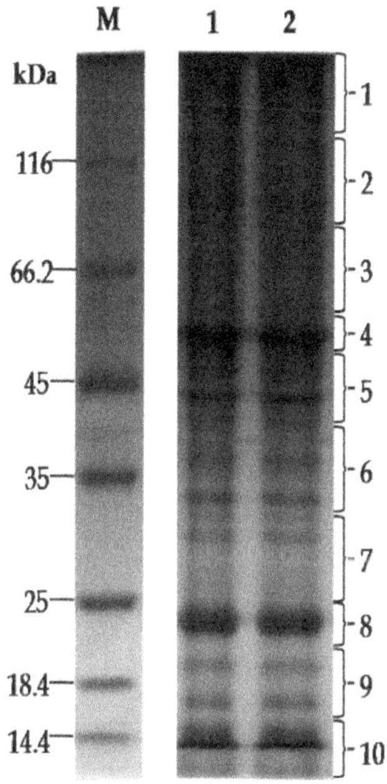

Figure 2. The separation of membrane proteins (MPs) by SDS-PAGE. The gel was cut into 10 slices that covered known apparent mass ranges. Lane M, protein standard markers; lane 1 & lane 2: membrane protein fractions isolated from mouse brain; 1-10: slices to be cut for trypsin in-gel digestion, separation and analyses by nanoLC-MS/MS.

4.2. In-gel digestion

In-gel digestion of proteins isolated by gel electrophoresis was carried out according to the protocol published by Shevchenko *et al* [25] with some modifications described in our previous study [3, 28, 29]. All chemicals including DTT, iodoacetamide (IAA), ammonium bicarbonate, ammonium acetate, trypsin (proteomics sequencing grade), sodium bicarbonate and Triton X-100 were purchased from Sigma-Aldrich (St. Louis, MO, USA) prepared using deionized filter-sterilised MilliQ water.

Upon electrophoresis, proteins were fixed within a polyacrylamide matrix by incubating the entire gel in 5% (vol/vol) acetic acid in 1:1 (vol/vol) water:methanol. Coomassie blue-stained protein bands were excised from gels and placed into 1.5 ml eppendorf tubes, destained with 50% ACN in 25 mM NH_4HCO pH 8.0 at room temperature with occasional vortexing, until gel pieces became white and shrank, and then acetonitrile was removed. The gel pieces were then reduced by incubating with 5 mM DTT solution at 56°C for 45 min and alkylated for 1 hr with 20 mM IAA solution in darkness at room temperature. The membrane proteins were digested by adding trypsin buffer (0.03 μg/μl in 10 mM ammonium bicarbonate containing 10% (vol/vol) acetonitrile) and incubating overnight at 37°C. Check if all solution was absorbed and add more trypsin buffer, if necessary. Gel pieces should be completely covered with trypsin buffer (typically, 50 μl or more).

4.3. Sample cleanup with C-18 ZipTips

The resulting peptide digestion products were extracted by adding 100 μl of extraction buffer (1:2 (vol/vol) 5% formic acid/acetonitrile) to each tube and incubated for 15 min at 37 °C in a shaker. All extracts were saved and dried and re-dissolved in 10–20 μl of 0.1% FA, incubated for 2–5 min in the sonication bath and centrifuged for 15 min at 10,000 rpm at the bench-top centrifuge. The obtained supernatant was applied for binding the samples onto micro pipette tips (μC18), catalog number ZTC18S096 (ZipTip®, Millipore Co., Billerica, MA 01821 USA), equilibrated by being aspirated and dispensed with 100% acetonitrile, 40% acetonitrile/0.1% FA, 0.1% FA solutions. The samples were washed (4 times by aspirating and dispensing) with 15 μl of 0.1% FA), then eluted with 10 μl of 40% acetonitrile/0.1% FA. Appropriate aliquots were withdrawn for LC-MS/MS analysis or store at −20°C as contingency.

5. Two-dimensional nano liquid chromatography (2D-nanoLC)

The basic setup of an online two-dimensional nano liquid chromatography (2D-nanoLC) system (LC Parking, Dionex, The Netherlands) was developed for improved separation and hydrophobic peptide recovery, especially for complex peptides made from enzymatic digests of selected proteomes. The system works with the principle of elution of the digested peptides from the first dimension SCX column with injected salt solution plugs of increasing concentration. The eluted peptides are again trapped and introduced into the nanoflow path for separation and analysis by second dimension RP column and tandem mass spectrometry. The great advantage of the system is a robust and fully automated separation. The methods are easy to set up and composed of identical runs differing only in the concentration of injected salt plugs.

For the mentioned above online 2D-nanoLC system, the following columns were used: (i) strong cation exchange (SCX) column (500 μm i.d.×1.5 cm) packed with BioX-SCX, 300 Å, 5 μm, (LC Parking, Dionex, P/N 161395); (ii) Trap column: 300 μm i.d.× 0.5 cm, packed with PepMap™ C18, 100 Å, 5 μm, (LC Parking, Dionex, P/N 160454); (iii) Reversed phase (RP) column: 75 μm i.d.× 15 cm, packed with C18 PepMap100, 100 Å, 3 μm, (LC Parking, Dionex, P/N 160321). The column physico-chemical properties, functions, and the mobile phase, loading/eluted solvents for the flow diagram in online 2D-nanoLC system (UltiMate™/FAMOS/Switchos™, LC Packings, Dionex) with the 10-port valve automatic swiching configuration are shown in details in table 1 and figure 3.

1	2	3	4
Type of column	1st Dimension - Exchange Ion Chromatography- SCX column	PepMap nano RP Trapping column	2nd Dimension- Reversed Phase C18 column
Function	Separation of a protein/peptide on an *ion exchange*	Pre-concentrating sample	Separation of a protein/peptide on a reversed phase
Physical properties of column	500 μm i.d. x 1.5 cm, 300 Å, 5 μm (LC Parking, Dionex, P/N 161395)	300 μm i.d. x 0.5 cm, PepMap™ C18, 5 μm, 100Å (LC Parking, Dionex, P/N 160454)	75 μm i.d. x 15 cm, packed with PepMap™C18, 100 Å, 3 μm, (LC Parking, Dionex, P/N 160321)
Injected volumn	20 μl		
Flow rate	200 nl/min		
Loading flow		30 μl/min	
Loading solvent		0.1% FA, pH2.9	
Eluted solvent (Elute positively charged peptides on SCX)	Ammonium acetate solutions: 10 mM, 20 mM, 40mM, 60mM, 80mM, 100 mM, 200 mM, 500mM, 1M, 2M		
Mobile phase			A (0.1% FA in LC-MS grade water); B (0.1% FA in 85% LC-MS grade ACN)

Table 1. The type of columns with their physico-chemical properties, functions and the mobile phase, loading/eluted solvents that were used for basic experimental setup of an online two-dimensional nano liquid chromatography system (2D-nanoLC, UltiMate™/FAMOS/Switchos™, LC Parking, Dionex, The Netherlands).

For in-gel digest samples, as the first step, the peptide mixture was re-dissolved in 30 μl of 0.1% FA and directly loaded onto a strong cation exchange (SCX) column (500 μm i.d.×1.5 cm, 5 μm, 300 Å) at a flow rate of 30 μl/min. Bound peptides were eluted by following ammonium acetate gradients from 10 mM to 2 M: 10 mM, 20 mM, 40mM, 60mM, 80mM, 100 mM, 200 mM, 500mM, 1M and 2M and then desalted and concentrated independently on a C18 trap column (300 μm i.d. ×0.5 cm, 5 μm, 100 Å). The eluted peptides were further separated onto a reversed phase C18 column (75 μm i.d.×15 cm, 5 μm, 100 Å), for the second dimension. The flow rate was maintained at 200 nl/min with solvent A (0.1% FA in LC-MS grade water). With 10 different concentrations of ammonium acetate (plugs), there should be 10 identical runs.

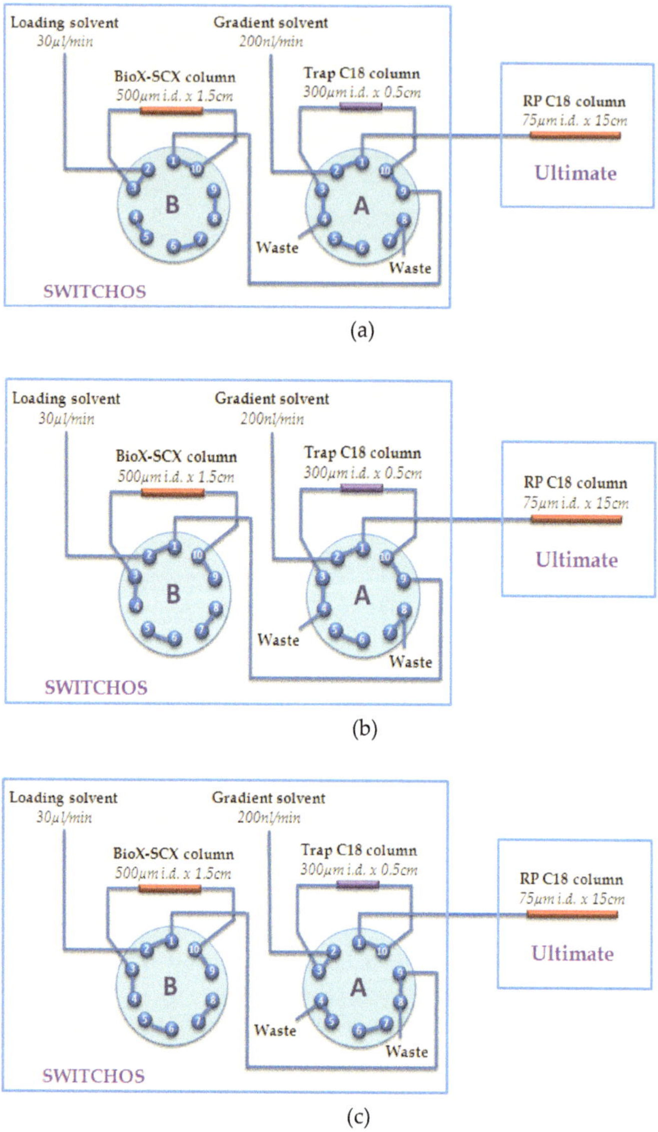

Figure 3. Illustration of the flow diagram in online 2D-nanoLC system (UltiMate™/FAMOS/Switchos™, LC Packings, Dionex) with the 10-port valve automatic swiching configuration and localization of BioX-SCX column, Trap C18 column and RP C18 column in: (a) loading mode, (b) clean-up mode and (c) analysis mode.

After washing (~12 min), peptides were eluted from a reversed phase C18 column using the solvent B (0.1% FA in 85% LC-MS grade ACN) gradients: from 5 to 20% of solvent B in 25 min, 20 to 70% in 28 min, 70 to 100% in 10 min and maintaining 100% B in 10 min, and back to 5% B in 5 min.

6. Integrating NanoLC system and tandem mass spectrometer

In our example, samples were delivered into the instrument by an automated in-line (integrated LC Parking's System, 5 mm C18 nano-precolumn and 75 μm i.d. × 15 cm column, packed with C18 PepMap100, 100 Å, 3 μm, (LC Parking, Dionex, P/N 160321) via a nanoelectrospray source head and 10 μm inner diameter PicoTip (New Objective, Massachusetts, USA) (Figure 4).

According to the workflow, after 2D-nanoLC separation, peptides were independently analyzed by a QSTAR®XL MS/MS mass spectrometer (Appllied Biosystems/MDS SCIEX, Ontario, Canada) equipped with a nanoESI source. MS and MS/MS spectra were recorded and processed in IDA mode (Information Dependent Acquisition) controlled by Analyst QS software. Typical settings are chosen to select multiply charged ions for MS/MS that produce at least 45-50 ion counts/s in a 0.5 s survey scan. The range of the MS full scan was from 400 to 1200 amu followed by MS/MS fragmentation of the three most intense precursor peptide ions for 1 s each.

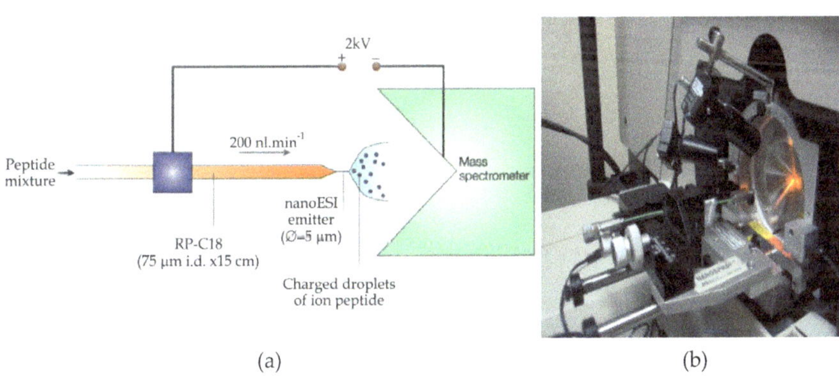

Figure 4. Setup and demonstration of nanoLC-MS interface, link between nanolC with nanoelectrospray ionization source and tandem mass spectrometry: (a) Schematic diagram of a nanoLC-MS interface; (b) Interface of nanoLC (LC Parkings, Dionex, Netherlands) with QSTAR®XL MS/MS mass spectrometer (Appllied Biosystems/MDS SCIEX, Ontario, Canada). By using this system, complex sample can be injected, desalted, separated and analyzed in complete automatization.

7. Protein identification and validation

There are a number of different methods for identifying the proteins in the sample, and the most frequently used is the searching of the uninterpreted MS/MS data. The FASTA formatted protein sequences from National Center for BioTechnology Information (NCBI) and UniProtKB/Swiss-Prot databases are collected for proteins identified or identification by each MS experiment. Searching uninterpreted MS/MS data from a single peptide or from a complete nanoLC-MS/MS run was automatically analyzed with a non-redundant protein database by the program SEQUEST, which allows the correlation of experimental data with theoretical spectra generated from known protein sequences [7].

The precursor mass is used as a filter to find a list of candidate peptide sequences from the theoretical digest of the database. A variety of different systems are used to score the experimental MS/MS spectrum against spectra predicted from the candidate peptide sequences. For protein identification, experimental data were searched against the NCBInr and Swiss-Prot mouse protein database using Mascot v1.8 software in which the criteria were based on the manufacturer's definitions (Matrix Science Ltd, London, UK) [20]. The parameters were set as follows: enzymatic cleavage with trypsin; 1 potential missed cleavage; a peptide and fragment mass tolerance of ± 0.25 Da, and fixed modification of carbamidomethyl (cysteine); variable modification of oxidation (methionine); 1^+, 2^+, and 3^+ peptide charge. Protein identifications were performed using a Mowse scoring algorithm with a confidence level of 95% and at least two matched peptides, showing a high score [12].

For further verification, proteins might be validated by MSQuant software [1, 4, 24] available at http://msquant.sourceforge.net. The MSQuant software is used as a validation and quantitation tool that produces the Mascot peptide identifications (HTLM files) and allows manual verification against the raw MS data (QSTAR XL raw files). The MSQuant software will pick up significant and verified hits from the Mascot output file and export information of identified proteins into an .xls file, including the GI (genInfo identifier) number and molecular-mass values.

8. Prediction of transmembrane domains (TMDs)

The identified proteins were categorized based on their cellular locations and biological processes according to Gene Ontology (GO) information obtained at http://www.ebi.ac.uk/pub/databases/GO/goa/mouse and ftp://ftp.geneontology.org/pub/go/ [2]. The TMHMM (www.cbs.dtu.dk/services/TMHMM/) algorithm was used to predict transmembrane domains (TMDs) [15, 26].

The average hydrophobic values and transmembrane domains of the identified proteins were calculated using the SOSUI system that is available at http://bp.nuap.nagoya-u.ac.jp/sosui/ [11]. The proteins exhibiting positive GRAVY values were recognized as hydrophobic and those with negative values were hydrophilic [13].

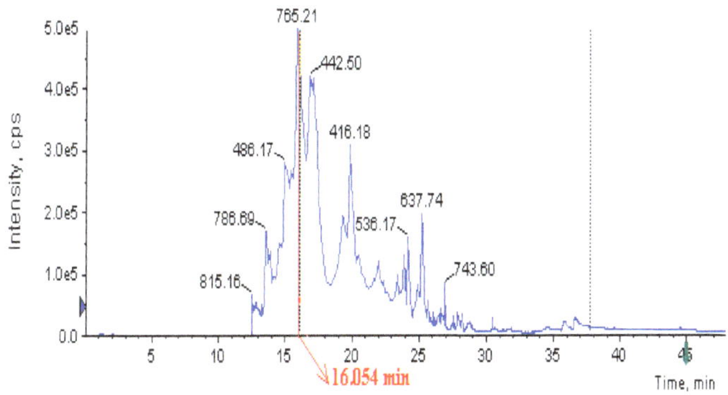

(a)

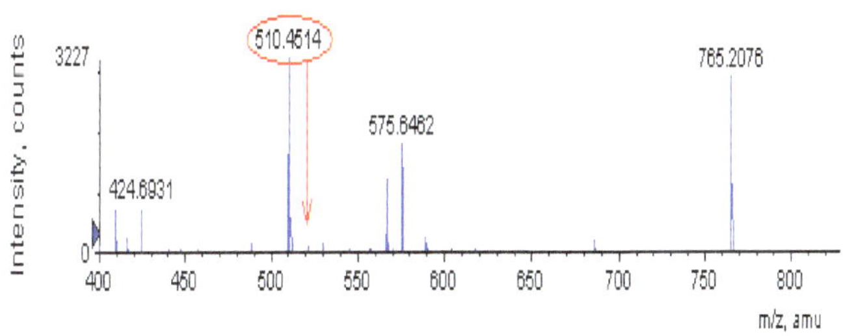

(b)

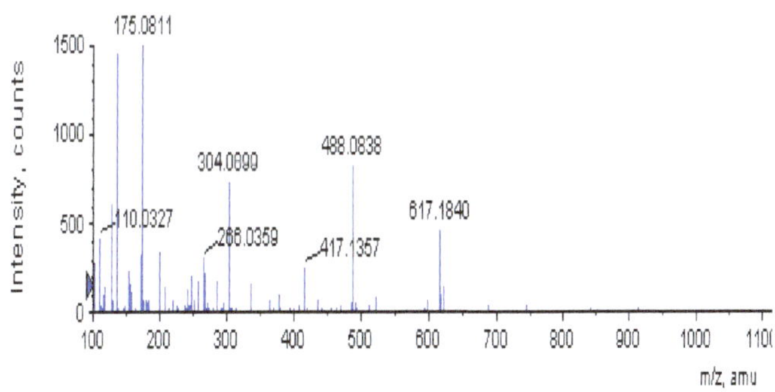

(c)

Figure 5. Illustration of 2D-nanoLC-ESI-MS/MS spectra: (a) The total ion current (TIC) profile tryptic digest of membrane proteins (band 6) at the concentration of 100 mM ammonium acetate for run time 0-50 min; (b) TOF-MS spectrum at 16.054 min; (c) TOF product spectrum of a peptide ion with m/z = 510.45.

Peptide Summary Report (Mou ×

D:\PROJECTS\NASFOTED\MASCOT RESULT\B6

{MATRIX SCIENCE} Mascot Search Results

User	: **Nguyen Tien Dung**
Email	: **nguyentiendung_bk@yahoo.com**
Search title	: **Mouse membrane - Band 6**
MS data file	: **CM6D_CID.pkl**
Database	: **SwissProt 2011 (533049 sequences; 189064225 residues)**
Taxonomy	: **Mus musculus (house mouse) (16480 sequences)**
Timestamp	: **15 Dec 2011 at 10:25:38 GMT**
Protein hits	: ADT2_MOUSE ADP/ATP translocase 2 OS=Mus musculus GN=Slc25a5 PE=1 SV=3
	MOG_MOUSE Myelin-oligodendrocyte glycoprotein OS=Mus musculus GN=Mog PE=1 SV=1
	PHB_MOUSE Prohibitin OS=Mus musculus GN=Phb PE=1 SV=1
	NDUS3_MOUSE NADH dehydrogenase [ubiquinone] iron-sulfur protein 3, mitochondrial
	ADT1_MOUSE ADP/ATP translocase 1 OS=Mus musculus GN=Slc25a4 PE=1 SV=4
	VDAC1_MOUSE Voltage-dependent anion-selective channel protein 1 OS=Mus musculus
	TBA1A_MOUSE Tubulin alpha-1A chain OS=Mus musculus GN=Tuba1a PE=1 SV=1
	VDAC2_MOUSE Voltage-dependent anion-selective channel protein 2 OS=Mus musculus
	TBB2A_MOUSE Tubulin beta-2A chain OS=Mus musculus GN=Tubb2a PE=1 SV=1
	M2OM_MOUSE Mitochondrial 2-oxoglutarate/malate carrier protein OS=Mus musculus
	MPCP_MOUSE Phosphate carrier protein, mitochondrial OS=Mus musculus GN=Slc25a3
	TBB5_MOUSE Tubulin beta-5 chain OS=Mus musculus GN=Tubb5 PE=1 SV=1
	TBB4B_MOUSE Tubulin beta-4B chain OS=Mus musculus GN=Tubb4b PE=1 SV=1
	TBB3_MOUSE Tubulin beta-3 chain OS=Mus musculus GN=Tubb3 PE=1 SV=1
	ATPG_MOUSE ATP synthase subunit gamma, mitochondrial OS=Mus musculus GN=Atp5c1
	CY1_MOUSE Cytochrome c1, heme protein, mitochondrial OS=Mus musculus GN=Cyc1
	TBB6_MOUSE Tubulin beta-6 chain OS=Mus musculus GN=Tubb6 PE=1 SV=1
	TBA4A_MOUSE Tubulin alpha-4A chain OS=Mus musculus GN=Tuba4a PE=1 SV=1
	NSF_MOUSE Vesicle-fusing ATPase OS=Mus musculus GN=Nsf PE=1 SV=2
	VDAC3_MOUSE Voltage-dependent anion-selective channel protein 3 OS=Mus musculus
	H12_MOUSE Histone H1.2 OS=Mus musculus GN=Hist1h1c PE=1 SV=2
	CN37_MOUSE 2',3'-cyclic-nucleotide 3'-phosphodiesterase OS=Mus musculus GN=Cnp
	1433E_MOUSE 14-3-3 protein epsilon OS=Mus musculus GN=Ywhae PE=1 SV=1
	KCC2A_MOUSE Calcium/calmodulin-dependent protein kinase type II subunit alpha
	ETFB_MOUSE Electron transfer flavoprotein subunit beta OS=Mus musculus GN=Etfb
	1433G_MOUSE 14-3-3 protein gamma OS=Mus musculus GN=Ywhag PE=1 SV=2
	AINX_MOUSE Alpha-internexin OS=Mus musculus GN=Ina PE=1 SV=2
	RAB3A_MOUSE Ras-related protein Rab-3A OS=Mus musculus GN=Rab3a PE=1 SV=1
	MBP_MOUSE Myelin basic protein OS=Mus musculus GN=Mbp PE=1 SV=2
	1433T_MOUSE 14-3-3 protein theta OS=Mus musculus GN=Ywhaq PE=1 SV=1
	GSTM1_MOUSE Glutathione S-transferase Mu 1 OS=Mus musculus GN=Gstm1 PE=1 SV=2
	PHB2_MOUSE Prohibitin-2 OS=Mus musculus GN=Phb2 PE=1 SV=1
	NFL_MOUSE Neurofilament light polypeptide OS=Mus musculus GN=Nefl PE=1 SV=5
	GFAP_MOUSE Glial fibrillary acidic protein OS=Mus musculus GN=Gfap PE=1 SV=4
	1433Z_MOUSE 14-3-3 protein zeta/delta OS=Mus musculus GN=Ywhaz PE=1 SV=1
	MYPR_MOUSE Myelin proteolipid protein OS=Mus musculus GN=Plp1 PE=1 SV=2
	H14_MOUSE Histone H1.4 OS=Mus musculus GN=Hist1h1e PE=1 SV=2
	NFH_MOUSE Neurofilament heavy polypeptide OS=Mus musculus GN=Nefh PE=1 SV=3

Figure 6. An example of Mascot search result shows list of the identified mouse brain membrane proteins isolated from band 6 (see figure 2) and their accession numbers, using SwissProt database (533049 sequences, 189064225 residues)

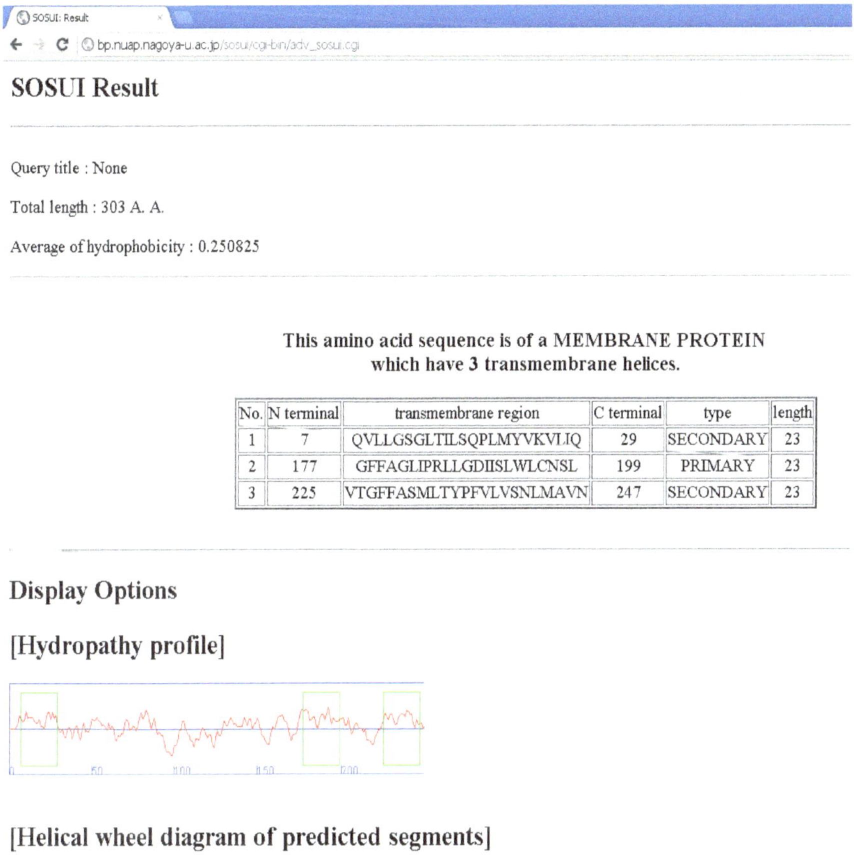

SOSUI: Result

bp.nuap.nagoya-u.ac.jp/sosui/cgi-bin/adv_sosui.cgi

SOSUI Result

Query title : None

Total length : 303 A. A.

Average of hydrophobicity : 0.250825

This amino acid sequence is of a MEMBRANE PROTEIN which have 3 transmembrane helices.

No.	N terminal	transmembrane region	C terminal	type	length
1	7	QVLLGSGLTILSQPLMYVKVLIQ	29	SECONDARY	23
2	177	GFFAGLIPRLLGDIISLWLCNSL	199	PRIMARY	23
3	225	VTGFFASMLTYPFVLVSNLMAVN	247	SECONDARY	23

Display Options

[Hydropathy profile]

[Helical wheel diagram of predicted segments]

Figure 7. An example of hydropathy profile and transmembrane regions/domains of an identified mouse brain membrane protein calculated using the SOSUI system that is available at http://bp.nuap.nagoya-u.ac.jp/sosui/ [11].

9. Conclusion

Identification and characterization of membrane proteins is a crucial challenge in proteomics research. Thus, we have designed a strategy of gel-based approach in combination with comprehensive two-dimensional nano liquid chromatography (2D-nanoLC) that is robust and offers high separation capacity and high analysis throughput for mouse brain membrane proteins. By using this system, mixtures of in-gel trypsin-digested mouse brain membrane proteins were injected, desalted, separated and analyzed in complete automatization. The workflow started by the extraction and purification of the membrane fractions, then the SDS-PAGE was carried out as a useful preparative separation step. After staining, the gel slides with protein bands were cut, reduced, alkylated and trypsin-digested. The peptide mixtures extracted from each gel slice were fractionated by 2D-nanoLC coupled online with tandem mass spectrometry analysis (nanoESI-Q-TOF-MS/MS). The proteins were identified by MASCOT search against mouse protein database using a peptide and fragment mass tolerance of ±0.25 Da. Protein identification was carried out using a MOWSE scoring algorithm with a confidence level of 95% and processed by MSQuant software for further validation. In total, 298 identified membrane proteins from mouse brain tissues were verified by UniProt database, SOSUI and TMHMM prediction algorithms. Of these, 129 (43.3%) proteins have at least one transmembrane domain according to SOSUI and TMHMM. Furthermore, the function, subcellular location, molecular weight, post-translational modifications, transmembrane domains (TMD) and average of hydrophobicity of the identified membrane proteins might be categorized and analysed.

Author details

Phan Van Chi and Nguyen Tien Dung
Department of Protein Biochemistry, Institute of Biotechnology (IBT),
Vietnam Academy of Science & Technology (VAST), Hanoi, Vietnam

Acknowledgement

The present study was carried out at the National Key Laboratory of Gene Technology (NKLGT) and was approved by the Ethics Committee of the Institute of Biotechnology (IBT), Vietnam Academy of Science and Technology (VAST)

Funding

The work was funded by National Foundation for Science and Technology Development (NAFOSTED), Research Project Code Number: 106.05-2010.35.

10. References

[1] Andersen JS, Wilkinson CJ, Mayor T, Mortensen P, Nigg EA and Mann M (2003) Proteomic characterization of the human centrosome by protein correlation profiling. Nature 426(6966): 570-4.

[2] Ashburner M, Ball CA, Blake JA, Botstein D, Butler H, Cherry JM, Davis AP, Dolinski K, Dwight SS, Eppig JT, Harris MA, Hill DP, Issel-Tarver L, Kasarskis A, Lewis S, Matese JC, Richardson JE, Ringwald M, Rubin GM, Sherlock G (2000) Gene ontology: tool for the unification of biology. The Gene Ontology Consortium, Nat. Genet. 25(1): 25-9.

[3] Binh DV, Thanh TT and Chi PV (2010) Proteomic characterization of the thermostable toxins from Naja naja venom. J. Venom. Anim. Toxins incl. Trop. Dis. 16(4): 631-8.

[4] Blagoev B, Ong SE, Kratchmarova I and Mann M (2004) Temporal analysis of phosphotyrosine-dependent signaling networks by quantitative proteomics. Nat. Biotechnol. 22(9): 1139–45.

[5] Bradford MM (1976) A rapid and sensitive for the quantitation of microgram quantitites of protein utilizing the principle of protein-dye binding. Anal. Biochem. 72: 248-54.
[6] Cutillas PR and Timms JF (2010) LC-MS/MS in Proteomics. In: John M. Walker Ed. Methods in Molecular Biology, Volume 658, Humana Press, Springer New York Dordrecht Heidelberg London, DOI 10.1007/978-1-60761-780-8.
[7] Eng JK, McCormack AL and Yates III JR (1994). An approach to correlate tandem mass spectral data of peptides with amino acid sequences in a protein database. J. Am. Soc. Mass Spectrom. 5(11): 976–89.
[8] Fröhlich T and Arnold GJ (2006) Proteome research based on modern liquid chromatography-tandem mass spectrometry: separation, identification and quantification. J. Neural. Transm. 113(8): 973-94.
[9] Fournier ML, Gilmore JM, Martin-Brown SA, Washburn MP (2007) Multidimensional separations-based shotgun proteomics. Chem.Rev. 107:3654–86
[10] Gilmore J and Washburn M (2010) Advances in shotgun proteomics and the analysis of membrane proteomes. J. Proteomics 73(11): 2078-91.
[11] Hirokawa T, Boon-Chieng S and Mitaku S (1988) SOSUI: classification and secondary structure prediction system for membrane proteins. Bioinformatics 14(4): 378-9.
[12] Issaq H and Veenstra T (2008) Two-dimensional polyacrylamide gel electrophoresis (2D- PAGE): advances and perspectives. Biotechniques 44(5): 697-8.
[13] Kyte J and Doolittle RF (1982) A simple method for displaying the hydropathic character of a protein J. Mol. Biol. 157(1): 105-32.
[14] Laemmli UK (1970) Cleavage of structural proteins during the assembly of the head of bacteriophage T4. Nature 227(5259): 680-85.
[15] Melen K, Krogh A and von Heijne G (2003) Reliability measures for membrane protein topology prediction algorithms. J. Mol. Biol. 327(3): 735-44.
[16] Mirza SP, Halligan BD, Greene AS and Olivier M (2007) Improved method for the analysis of membrane proteins by mass spectrometry. Physiol. Genomics 30(1): 89–94.
[17] Pappin DJ, Hojrup P, Bleasby AJ (1993) Rapid identification of proteins by peptide-mass fingerprinting. Curr. Biol. 3(6): 327-32.
[18] Peng L, Kapp E, Fenyö D, Kwon MS, Jiang P, Wu S, Jiang Y, Aguilar MI, Ahmed N, Baker MS, Cai Z, Chen YJ, Chi PV, Chung MCM, He F, Len ACL, Liao PC, Nakamura K, Ngai SM, Paik YK, Pan TL, Poon TCW, Salekdeh H, Simpson RJ, Sirdeshmukh R, Srisomsap C, Svasti J, Tyan YC, Dreyer FS, McLauchlan D, Rawson P and Jordan TW (2010) The Asia Oceania Human Proteome Organisation Membrane Proteomics Initiative. Preparation and characterization of the carbonate-washed membrane standard. *Proteomics* 10:1-7.
[19] Peng L, Kapp E, McLauchlan D and Jordan T (2011) Characterization of the Asia Oceania Human Proteome Organisation MembraneProteomics Initiative Standard using SDS-PAGE shotgun proteomics. Proteomics 11(22): 4376-84.
[20] Perkins DN, Pappin DJ, Creasy DM and Cottrell JS (1999) Probability-based protein identification by searching sequence databases using mass spectrometry data. Electrophoresis 20(18): 3551–67.

[21] Rabilloud T, Chevallet M, Luche S, Lelong C (2010). Tow-dimensional gel electrophoresis in proteomics: past, present and future. J. Proteomics 73(11):2064-77.
[22] Rabilloud T, Lelong C (2011) Two-dimensional gel electrophoresis in proteomics: a tutorial. J. Proteomics. 74(10):1829-41
[23] Santoni V, Malloy M and Rabilloud T (2000) Membrane proteins and proteomics: un amour impossible? Electrophoresis 21(6): 1054-70.
[24] Schulze WX and Mann M (2004) A novel proteomic screen for peptide-protein interactions. J. Biol. Chem. 279: 10756-64.
[25] Shevchenko A, Tomas H, Havlis J, Olsen JV, Mann M (2007) In-gel digestion for mass spectrometric characterization of proteins and proteomes. Nat. Protocols 1(6): 2856-60.
[26] Sonnhammer EL, von Heijne G, Krogh, A (1998) A hidden Markov model for predicting transmembrane helices in protein sequences. Proc. Int. Conf. Intell. Syst. Mol. Biol. 6: 175-182.
[27] Stevens TJ and Arkin IT (2000) Do more complex organisms have a greater proportion of membrane proteins in their genomes? Proteins 39(4): 417-20
[28] Thanh TT, Phuong NT, Nhi NT and Chi PV (2008) Changes of serum glycoproteins patients with lung cancer. J. Proteomics Bioinform. 1(1):15-22.
[29] Thanh TT and Chi PV (2010) Separation and identification of mouse liver membrane proteins using a gel-based approach in combination with 2DnanoLC-Q-TOF-MS/MS. Adv. Nat. Sci.: Nanosc. Nanotechnology I(2010)015015, doi:10.1088/2043-6254/1/1/015015.
[30] Wu CC and Yates JR (2003) The application of mass spectrometry to membrane proteomics. Nat. Biotechnol. 21(3): 262–7.

Chapter 4

Current Trends in Sample Treatment Techniques for Environmental and Food Analysis

Paolo Lucci, Deborah Pacetti, Oscar Núñez and Natale G. Frega

Additional information is available at the end of the chapter

1. Introduction

Nowadays, there is a growing need for applications in food and environmental areas able to cope with the analysis of a large number of analytes in very complex matrices [1]. The new analytical procedures demand sensitivity, robustness, effectiveness and high resolution with reduced analysis time. Many of these requirements may be met to a certain extent by the total or partial automation of the conventional analytical methods, including sample preparation or sample pre-treatment coupled on-line to the analytical system. Furthermore, the recent use of ultra-high-performance liquid chromatography (UHPLC) for environmental and food chemical analysis has increased the overall sample throughput and laboratory efficiency without loss (and even with an improvement) of resolution obtained by conventional HPLC systems.

Nonetheless, despite the advances in chromatographic separations and mass spectrometry techniques, sample treatment is still one of the most important parts of the analytical process and effective sample preparation is essential for achieving good analytical results [1]. Ideal sample preparation methods should be fast, accurate, precise and must keep sample integrity. Therefore, over the last years, considerable efforts have been made to develop modern approaches in sample treatment techniques that enable the reduction of the analysis time without compromising the integrity of the extraction process. The use of on-line solid-phase extraction (SPE), which minimizes sample manipulation and provides both high pre-concentration factors and recoveries [2-5], is an increasingly powerful and rapid technique used to improve the sample throughput and overcome many of the limitations associated with the classical off-line SPE procedure. However, in most of the cases, matrix related compounds may also be co-extracted and could interfere in the analysis. Consequently, in order to minimize the effect of all these possible interferences a selective clean-up step may be required. Higher specificity and selectivity together with satisfactory extraction efficiency can be obtained using sorbents based on molecularly imprinted polymers (MIPs) [6-8].

Solid-phase extraction (SPE) based on MIPs is a highly attractive and promising approach for matrix clean-up, enrichment and selective extraction of analytes in such kind of complex samples. Another modern trend in sample preparation for multi-residue applications is the use of the QuEChERS (*Quick, Easy, Cheap, Effective, Rugged and Safe*) method. The QuEChERS method is a recent and fascinating alternative procedure that has become particularly popular for the multi-residue analysis of pesticides in various food matrices [9,10], although this methodology is also being successfully employed for the extraction of other families of compounds [11,12]. Recently, the use of turbulent-flow chromatography (TFC) has also been reported for direct analysis of complex matrices such as honey, milk and animal tissues with reduced or without any sample manipulation [13-15].

The aim of this chapter is to discuss new trends in sample preparation techniques applied into food and environmental analysis. It includes a selection of the most interesting and promising sample treatment procedures such as on-line SPE methods, MIPs, QuEChERS, and turbulent flow chromatography. The applicability of each technique in food and environmental analysis will be discussed through the analysis of the most relevant papers recently published.

2. Solid-phase extraction (SPE)

Solid-phase extraction (SPE) is the most popular sample preparation technique for environmental and food samples. Due to its high versatility, the SPE procedure is used for many purposes, such as purification, trace enrichment, desalting, derivatization and class fractionation. The principle of SPE is similar to that of liquid-liquid extraction (LLE). It involves partitioning between a liquid (sample matrix or solvent with analytes) and a solid sorbent phase. Anyway, many of the problems associated with LLE, such as incomplete phase separations (emulsion), less-than-quantitative recoveries, use of expensive, breakable specialty glassware, disposal of large quantities of organic solvents, can be prevented by using SPE procedure. In addition, SPE resulted more efficient than LLE because yields quantitative extractions that are easy to perform, is rapid, and can be automated [16,17].

The general SPE procedure has to provide sample extracts that are free of interfering matrix components and concentrated enough for detection. The SPE process basically consists in four different steps: conditioning, sample addition, washing and elution (**Figure 1**).

First, the most suitable solid sorbent will be selected and conditioned using an appropriate solvent. During the conditioning the functional groups of the sorbent bed are solvated in order to make them able to interact with the sample. The sample addition consists in the percolation of the samples through the solid sorbent. During this step, the analytes as well as some matrix components are retained and thus concentrated on the SPE packing material. Successively, the analytes and interferents separation could be realized by the three following ways: selective extraction, selective washing or selective elution. Selective extraction is performed when the SPE procedure is used to remove the interfering components (trace enrichment). In this way, only selected components are retained. Selective washing is accomplished when the target analytes and the impurities are retained on the sorbent bed: the impurities will be rinsed through with wash solutions that are strong enough to remove them, but weak enough to leave the analytes behind. Differently, selective elution consists in the elution of the adsorbed compounds of interest by a solvent that leaves the strongly retained impurities behind. The elution of target analytes could require different solvents, when SPE is applied in order to perform the class compound fractionation.

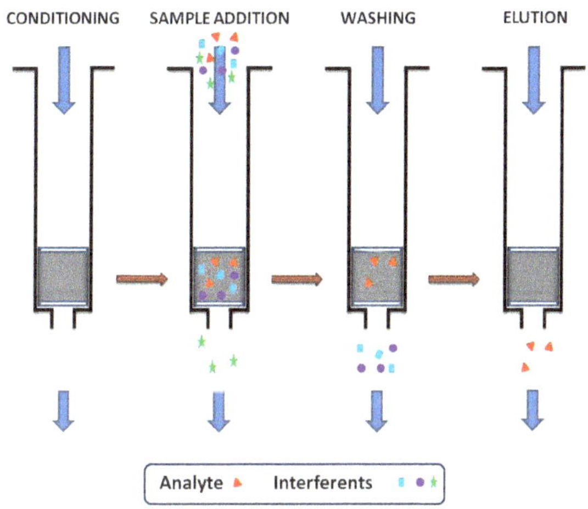

Figure 1. Schematic representation of SPE clean-up procedure.

To achieve optimal SPE extraction conditions, the choice of sorbent is a key factor because this can control parameters of primary importance such as selectivity, affinity and capacity [18]. This choice depends strongly on the nature of the analytes and their physical and chemical properties, which should define the interaction with the chosen sorbent. However, results can also depend heavily on the sample matrix and its interactions with both sorbent and analytes.

After the sorbent choice, the eluotropic strength of adsorption on silica and the polarity index will be helpful in order to select a suitable solvent. The polarity index is an accurate measure of solvent's ability to interact as proton donator, proton acceptor or dipole whereas eluotropic series arranges solvents in order of decreasing elution strength for solutes from a particular sorbent.

2.1. SPE sorbents

The sorbent selectivity depends on the attractive forces between the analytes and the functional groups on the sorbent surface. The sorbent can interact with analytes by hydrophobic (non polar-non polar, van der Waals), hydrophilic (polar-polar, hydrogen bonding, dipole-dipole, dipole-induced dipole), cationic-anionic and selective antigen-antibody interactions.

Each sorbent offers a unique mix of these types of interactions.

The sorbent widely used for SPE packing can be classified into polar phases (normal phase), non-polar phases (reversed phase), ion exchange and immunoaffinity adsorbents.

Polar phases are used under normal phase chromatography conditions. These phases include polar adsorption media (LC Florisil, ENVI-Florisil, and LC-Alumina) and polar-functionalized bonded silica materials. The retention mechanism of an analyte is primarily due to interactions between polar functional groups of the analyte and polar groups on the sorbent surface. These include hydrogen bonding, π-π interactions, dipole-dipole interactions, and dipole-induced dipole interactions, among others. A compound adsorbed by these mechanisms is eluted by passing a solvent that disrupts the binding mechanism; usually a solvent that is more polar than the sample's original matrix.

Polar adsorption media comprises underivatized silica material (SPE-Si), magnesium silicate (SPE- Florisil) and aluminum oxide materials (SPE-alumina).

SPE-Si is suitable to adsorb polar compounds from non polar matrices. All samples used with this material must be relatively water-free since the functional group involved in the adsorption of compounds are the free hydroxyl groups on the surface of silica particles.

Polar-functionalized bonded silica sorbent consists of a silica material modified by bonding functional groups, such as ciano (SPE-CN), aminopropyl (SPE-NH_2), diol (SPE-Diol) to the surface of the SPE material. These phases are less retentive than SPE-Si toward very polar analytes and therefore permit extractions impossible to achieve with unmodified silica gel [19]. They result useful to adsorb and selectively elute compounds of very similar structure (e.g. isomers), or complex mixtures or classes of compounds such as drugs and lipids. Moreover, SPE-CN, can be used also under reversed phase conditions (with aqueous samples) to extract moderately polar compounds. The SPE-NH_2 can also be applied under ion exchange conditions in order to separate charged compounds.

Non polar phases are used under reversed phase chromatography conditions. These sorbents comprise alkyl silica and polymer based materials. Alkyl silica sorbents are manufactured by bonding alkyl or aryl functional groups, such as cyano (SPE-CN), octyl (SPE-8), octadecyl (SPE-18) and phenyl (SPE-Ph) to the silica surface. These phases are suitable for the extraction of hydrophobic or polar organic analytes from aqueous matrices. The retention of analytes is due primarily to the non polar-non polar attractive forces between the carbon-hydrogen bonds in the analytes and the functional groups on the silica surface. The elution of adsorbed compounds is generally made by using a non polar solvent to disrupt the forces that bind the compound to the packing. Since all silica based bonded phases contain not-uncapped silanols, which can cause the strongly binding (sometime irreversibly, i.e tetracyclines) of some group of compounds, the addition of a more polar solvent may be often necessary. The main drawback of alkyl silica sorbent, especially of SPE-8 and SPE-18, is their poor water wettability. These cartridges require an initial conditioning step with a water-miscible organic solvent. When the internal surface of sorbent fails to be wetted because of the omission of the conditioning step or if the sorbent runs dry, the accessibility of sorbent surface for adsorbing analytes is severely reduced. For instance, low recovery of analytes can be observed when SPE-18 sorbent is accidentally dried down before sample application.

Nevertheless, the narrow pH stability range of all modified silica reversed phase must be taken into account when SPE is carried in extremely acidic or basic media. For this purpose, a reversed phase polymerically bonded, such as copolymers of styrene-divinylbenzene (SPE-PS-DVB) resulted more resistant to pH extremes, and thus is more suitable for environmental applications for trapping organic compounds from acidified aqueous samples. Moreover, PS-DVB resin copolymer is a hydrophobic resin which has greater analyte retention, mainly for polar compounds, than their hydrophobic surface containing a relatively large number of active aromatic sites which allow π-π interactions with unsaturated analytes [20]. The higher potential of PS-DVB over SPE-18 for trapping aromatic compound, especially phenols, is largely demonstrated [21,22]. Anyway, PS-DVB has some drawbacks, such as lack of selectivity and low breakthrough volumes for highly polar compounds, which leads to their incomplete extraction from predominantly aqueous matrices. Over the years, the performance of SPE-PS-DVB has been enhanced by attaching polar groups (i.e. acetyl, hydroxmethyl, benzoyl, o-carboxybenzoil, sulfonate, trimethylammonium) to the aromatic ring on the polymer DVB [23] or by changing the copolymer composition. The SPE-DVB phase modified with *o*-carboxybenzoyl was useful applied for the determination of pesticides and phenolic compounds in environmental waters. The HLB sorbent, a macroporous copolymer prepared from a balance ratio of two monomers the lipophilic divinylbenzene and the hydrophilic N-vinylpyrrolidone has been formulated. It can absorb a wide range of polar and no-polar compounds [24] and its performance is unaffected by sorbent dry. It represents the most common hydrophilic sorbent used in the herbicides extraction [25,26].

Ion Exchange phases are comprised of positively (aliphatic quaternary amine, aminopropyl) or negatively (aliphatic sulfonic acid, aliphatic carboxylic acid) charged groups that are bonded to the silica surface. These sorbents are really suitable for extraction of charged analytes, such as acidic and basic compounds, from aqueous or non-polar organic samples. They exert a retention mechanism based mainly on the electrostatic attraction of the charged functional group of the analytes to the charged groups that are bonded to the silica surface. In order to retain a compound by ion exchange from an aqueous solution, the pH of the sample matrix must be one at which both the compound of interest and the functional group on the bonded silica are charged. Also, there should be few, if any, other species of the same charge as the compound in the matrix that may interfere with the adsorption of the compound of interest. A solution having a pH that neutralizes either the compound's functional group or the functional group on the sorbent surface is used to elute the compound of interest. When one of these functional groups is neutralized, the electrostatic force that binds the two together is disrupted and the compound is eluted. Alternatively, a solution that has a high ionic strength, or that contains ionic species that displaces the adsorbed compound, is used to elute the compound.

Positively charged compounds are isolated under cation exchange conditions by using SPE sorbent containing silica linked with aliphatic sulfonic acid (SPE-SCX) or aliphatic carboxylic acid (SPE-WCX). The sulfonic acid group is strongly acidic and attracts or exchanges cationic species in a contacting solution. It is charged over the whole pH range, and therefore can be used to isolate strong cationic (very high pKa >14) or weak cationic (moderately high pKa <12) compounds, as long as the pH of the solution is one at which the compound of interest is charged. Anyway, SPE-SCX cartridges should be used to isolate strong cations only when their recovery or elution is not desired. Weak cations can be isolated and eluted from SPE-SCX; elution is done with a solution at 2 pH units above the cation's pKa (neutralizing the analytes), or by adding a different cation that displaces the analytes. If recovery of a strongly cationic species is desired, SPE-WCX is more suitable. The carboxylic acid group, present in SPE-WCX material, is a weak anion, and is thus considered a weak cation exchanger (WCX). It has a pKa of about 4.8, will be negatively charged in solutions of at least 2 pH units above this value, and will isolate cations if the pH is one at which they are both charged. SPE-WCX can be used to isolate and recover both strong and weak cations because the carboxylic acid functional group on the silica surface can be neutralized (2 pH units below its pKa) in order to elute the strong or weak cations. Weak cations also can be eluted from LC-WCX with a solution that neutralizes the adsorbed cations (2 pH units above its pKa), or by adding a different cation that displaces the analytes.

Negatively charged compounds can be isolated under anion exchange condition by using SPE sorbent containing silica functionalized with positively charged groups, such SPE-SAX and SPE- NH_2.

SPE-SAX material presents an aliphatic quaternary amine as functional group. This is a strong base that exchanges or attracts anionic species in the contacting solution. Its pKa is very high (greater than 14), which makes the bonded functional group charged at all pHs in aqueous solution. As a result, LC-SAX is used to isolate strong anionic (very low, pKa <1) or weak anionic (moderately low, pKa >2) compounds, as long as the pH of the sample is one at which the compound of interest is charged. For an acidic compound of interest, the pH of the matrix must be 2 pH units above its pKa for it to be charged. In most cases, the compounds of interest are strong or weak acids. Because it binds so strongly, LC-SAX is used to extract strong anions only when recovery or elution of the strong anion is not desired (the compound is isolated and discarded). Weak anions can be isolated and eluted from LC-SAX because they can be either displaced by an alternative anion or eluted with an acidic solution at a pH that neutralizes the weak anion (2 pH units below its pKa). If recovery of a strongly anionic species is desired, the use SPE-NH_2 is recommended. Generally, SPE-NH_2 is used for normal phase separations but it is also considered to be a weak anion exchanger (WAX) when used with aqueous solutions. It has an aliphatic aminopropyl group bonded to the silica surface. The pKa of this primary amine is around 9.8. For it to be used as an anion exchanger, the sample must be applied at a pH at least 2 units below 9.8. SPE-NH_2 is used to recover both strong and weak anions because the amine group can be neutralized (2 pH units above its pKa) in order to elute the strong or weak anions.

Immunoaffinity SPE phases, also called immunosorbents (ISs), are very interesting materials because of their high selectivity. ISs cartridges are filled with antibody materials bonded onto silica gel support. They allow extraction, concentration and clean up from complex matrices in a single step, and from large sample volumes. The retention mechanism of these sorbents involves reversible and selective antigen-antibody interactions.

Due to the drawbacks of the commonly used SPE phases (previously discussed), the main current trends is the study and development of new sorbents materials. These new materials try to fulfill the requirements according to present needs, such higher specific surface area, selectively towards the target analytes, easy manipulation allowing coupling on-line configurations and higher biocompatibility, with the overall objective of the enhancement of the efficiency of the extraction process. Among them, molecularly imprinted polymers (MIPs), restricted access material (RAM), porous graphite carbon (PGC) and mixed-mode polymeric sorbent are attracting much interest [3,27-28].

MIP, which has become more and more popular in recent years, is a technology where recognition sites are created by copolymerization of a target molecule in a macromolecular matrix. This technique will be discussed in detail in section 3.

RAM materials possess a pore size that restricts big molecules from entering the interior extraction phase based on size. They have a bimodal surface topochemistry and enable the simultaneous performance of two different chromatographic processes [29]:

1. size exclusion chromatography (SEC), i.e. macromolecular sample components (>15,000 Dalton) are directly eluted to waste;
2. adsorption chromatography (e.g. reversed-phase chromatography), i.e. low-molecular-weight sample components are bound adsorptively on the internal pore surface.

Because the ability of these phases to exclude proteins, RAMs are the most suitable choice for clean-up biological and food samples. In a recent work, Chico et al. [30], evaluated the SPE-RAM clean- up for tetracyclines analysis in milk and water samples. The RAM clean-up removed large peaks that otherwise appeared in the initial time window of the chromatograms, attributed to proteins in milk samples and humic substances in water samples. Thus, quantification of analytes in real samples, especially of the most polar compounds such as oxytetracycline and tetracycline, was clearly improved.

Porous graphite carbon (PGC) material is manufactured by impregnating a high porosity LC silica gel (to provide the desired pore size) with a phenol-formaldehyde resin.

PGC behaves as a strong reversed-phase stationary phase, even stronger than SPE-18 silica phase which represents the most hydrophobic of the commonly used alkyl substituted silica phases [31]. The retention mechanism of PGC is different from that observed of reversed-phase silicas. The retention mechanism of polar analytes on PGC is a charged-induced interaction of the polar analyte with the polarizable surface of graphite [32]. The strength of interaction between a hydrophobic analyte molecule and the PGC surface largely depends on how well the molecule fit onto the flat graphite surface. PGC has been found to be particularly selective with respect to geometrical isomers and closely related substances. It was found that non-polar analytes were strongly retained on PGC.

Mixed mode polymeric sorbents combine the polymeric skeleton with ion-exchange group. It can be divided into cationic (SPE-MCX) or anionic (SPE-MAX) and as weak or strong ion exchange, depending on the ionic group linked to the resin. The retention mechanism of mixed-mode ion exchange chromatography combines the use of reversed-phase and ion-exchange modes into a single protocol on a single SPE cartridge. The mixed mode sorbents are useful for fractionation of analytes. It can be used to isolate and separate neutral, acidic, and basic compounds from a single complex matrix. Intermediate washes with organic solvent mixtures of appropriate elution strength may be used to isolate neutral compounds, including ionizable analytes in their neutral state. Selective elution of ionically bound analytes may be attained by manipulating the charge of either the analyte (when bound to strong ion exchangers) or the sorbent (for analytes bound to weak ion exchangers).

2.2. Off-line SPE

SPE is widely used in environmental and food analysis in order to clean-up complex matrices and to isolate or/and concentrate target compounds. Two different SPE approaches are currently available: off-line and on-line procedures. In any case, the method development in SPE is accomplished by investigating pH, ionic strength, polarity and flow-rate of the elution solvent and physico-chemicals characteristics of the sorbent bed. Briefly, if the target analytes are polar, normal phase extraction is indicated. When the analytes are less polar, reverse phase separation is advised. While ion exchange SPE extraction is suitable when the analytes are charged.

Some reasons for low sample recovery in SPE are: inappropriate cartridge conditioning, too-strong loading and wash solvent, too large volume (mass) of loaded, and too weak or too small volume of elution mobile phase [16].

The off-line SPE equipment is economical and uncomplicated, thus fully applicable to on-site sampling. This consists in a sorbent material, which come in different packaging (syringe barrels, microtubes-tips and discs), in a solvent system (eluent) and in a vacuum manifold. The most popular packaging format for off-line SPE results a syringe barrels which are easy to handle by using vacuum or positive-pressure manifold. However, it is not easy to control the flow-rate, and care should be taken to prevent the column from drying out prior to sample application. As it could be difficult to elute the analytes of interest from SPE syringe using minimal solvent volume unless organic solvent composition rises up to 100%, special SPE discs are typically used for these purposes. This approach is much quicker as evaporation to dryness and reconstitution are no longer necessary because elution can be performed directly by a mobile phase. A major drawback is the small sample capacity of the discs.

The off-line SPE procedure presents the following weaknesses: it is time consuming, it requires large amount of the organic solvent for the elution, and it could cause a possible loss of analytes during the evaporation steps. In addition off-line SPE provide a large manipulation of the samples thus possibility of contamination, less accuracy and precision can be found.

In spite of all disadvantages, off-line SPE approach remains useful for analyzing complex samples, because of its greater flexibility and whenever elution solvent is not compatible with the subsequent method of analysis [33].

Off-line SPE remain the widely used sample preparation technique for environmental and food analysis. Some of the most recent applications of SPE to environmental and food matrices are summarized in **Table 1**.

Compound	Sample	Detection technique	SPE column	Reference
Food analysis				
Mineral oil saturated hydrocarbons	Vegetable oils	GC-FID	Silver silica gel	[34]
Pesticides	Berry fruits	GC-MS	Envi-Carb + SPE-NH_2	[35]
Anthocyanin	Berry fruits	HPLC-MS	Amberlite XAD7	[36]
Sudan dyes	Hot chili powder	HPLC-DAD	Alumina	[37]
High-intensity sweeteners	Food products (aqueous solution)	HPLC-MS	Chromabond C18ec, Strata-X RP, Bakerbond Octadecyl, Bakerbond SDB-1, Bakerbond SPE Phenyl, Oasis HLB, LiChrolut RP-18, Supelclean LC-18, Discovery DSC-18,Zorbax C18	[38]
Melamine and cyanuric acid	Egg, pork, liver, kidney of pig, shrimp, honey, soybean milk, soybean powder, protein powder, milk and other dairy products	HPLC–MS/MS	Hydrophilic functional gel and cation exchange sorbent	[39]
Environmental analysis				
Nitrobenzene compounds	Lake water, sanitary wastewater, and pond water	GC-MS	Phenothiazine bonded silica (PTZ-Si)	[40]
Polybrominated diphenyl ethers	Food samples (fish, meat and vegetables) environmental samples (soil or sediments)	GC-MS	Florisil; Alumina	[41]
Alkylphenol ethoxylates; steroidal hormones; bisphenol-A;	Wastewater	HPLC-MS	Sep-Pak Vac C_{18}, Oasis HLB, Bond Elut-ENV, Bond Elut Plexa,LiChrolut EN (500 mg)	[42]
Polycyclic aromatic hydrocarbons	Environmental water	GC-MS	Multi-walled carbonphase	[43]

Table 1. Some recent examples about applications on off-line SPE

2.3. On-line SPE

To meet the ever-growing demands for sensitivity, reliability and speed, the continuous development of more efficient methods for both sample pretreatment and analysis is crucial.

SPE technique can be easily coupled on-line to high performance liquid chromatography (HPLC) and gaschromatography (GC) systems. On-line systems are beneficial when the amount of sample is limited, or when very high sensitivity is required. In most cases, even though the use of an automated on-line instrument is quite straightforward, experienced personnel are required for method development and eventual trouble-shooting.

The strong differences among the solid phase extraction principle and the gas chromatography analysis made the on-line combination of SPE and GC more complicated.

Nevertheless, the combination with GC has already been successfully applied in environmental analysis [44-46]. In the on-line SPE-GC the analytes are trapped in a short column (10-20 mm×1-4.6 mm i.d.) packed with a suitable stationary phase (typically C8, C18 or styrene-divinylbenzene copolymer). The SPE procedures are essentially the same as the off-line ones. It involves conditioning of the SPE material before loading of the sample. Since water is not a good solvent for GC, primarily because it hydrolyses the siloxane bonds in GC columns causing deterioration of the column performance, the introduction of water directly to GC should be avoided. After trapping, and before elution of the analytes, the SPE column is often dried with a gas flow, or the extract is dried with a separate drying column packed with copper sulphate or silica to remove water, which is placed after the SPE column. The column can also be heated during the drying process, but this increases the risk of losing volatile analytes. The elution of the analytes is performed with a solvent suitable for the GC injector system.

On the other hand, the methods which combine SPE with HPLC are the most frequently used in environmental and food analysis, mainly to determine polar compounds in water solution. Different systems and configurations are available. The most commonly used approach involves the implementation of a small SPE column within the injection loop of a six-port rotary valve (**Figure 2**).

After conditioning, sample application, and eventual clean-up by means of a high-pressure pump, the SPE column is placed in front of an analytical column by switching the valve into the "inject" position. A sample is thus loaded in this SPE column, whereupon the valve is switched in order to elute the analytes out of the sorbent by the LC mobile phase and transfer them into the analytical column [3]. The SPE column is reusable. However, the reusability can cause a progressive deterioration of the column material and thus, lead to a change in their selectivity and capacity. Moreover, the SPE column must be filled with a sorbent compatible with the sorbent of the analytical column which efficiently traps the analytes. The SPE column should be as small as possible in order to prevent band broadening. Usually, the dimension of stainless steel columns is 30 mm length, 2 mm i.d. and 8 mm length, 3 mm i.d.

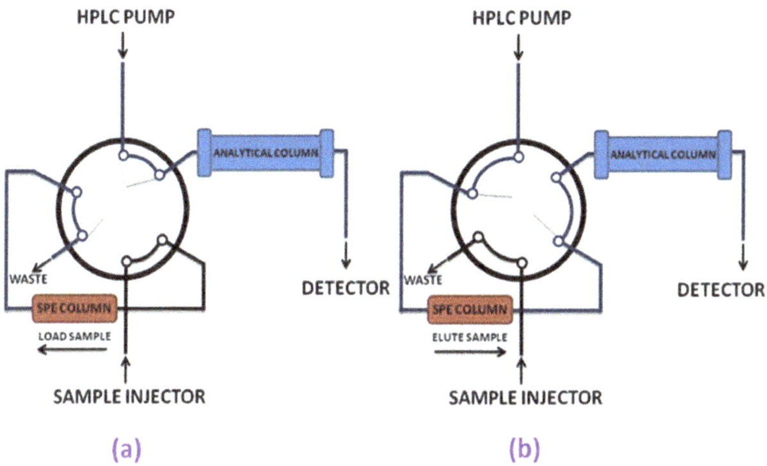

Figure 2. Typical setup of an on-line SPE-LC system with a 2-position 6-port switching valve; (a) Load position and (b) injection position

Recent advances in technology have made the Ultra High Performance Liquid Chromatography (UHPLC)/mass spectrometry (MS) system a perfect candidate for combination with on-line SPE. The on-line SPE-UHPLC/MS allows complete separation of high number of analytes via a single chromatographic run that takes few minutes.

Gosetti et al. [5], applied an automated on-line SPE UHPLC–MS/MS method for the identification and quantification of nine perfluorinated compounds (PFCs) in matrices of environmental, biological and food interest. The SPE protocol was performed by using an anion exchange SPE column (Poros HQ column). The separation of nine PFCs was obtained within 7 min. The limits of detection (LODs) ranged from 3 to 15 ng L^{-1} whereas the limits of quantification (LOQs) from 10 to 50 ng L^{-1}.

The same authors performed a simultaneous determination of thirteen polycyclic aromatic hydrocarbons (PAHs) and twelve aldehydes in cooked food by means of an automated on-line SPE-UHPLC MS. The resolution allowed the separation of four couples of PAH isomers. SPE treatment was made using Strata C18-E column and the extraction procedure was carefully optimized in order to apply the whole methodology to the analysis of different food matrices as salmon, frankfurter, steak, and pork chop, subjected to different cooking modes (smoked, grilled, cooked in oil or in butter). LODs values ranging from 0.028 to 0.768 μg L^{-1} for PAHs and from 0.002 to 0.125 μg L^{-1} for aldehydes were obtained [4].

Previously, PAH, such as naphthalene, biphenyl, acenaphthene, anthracene and pyrene have been determined in natural water by using on-line SPE–HPLC-UV [47]. Fluorocarbon polymer SPE sorbent was used. This application resulted in better extraction selectivity towards PAHs in comparison with several other sorbents and provided no additional peak broadening. Detection limits of method were established as 5 ng L^{-1} (biphenyl), 7 ng L^{-1} (anthracene), 8 ng L^{-1} (acenaphthene), 30 ng L^{-1} (pyrene), 40 ng L^{-1} (naphthalene).

Gallart-Ayala et al. [48], used an automated on-line SPE fast LC–MS/MS method for the simultaneous analysis of bisphenol A (BPA), bisphenol F (BPF), bisphenol E (BPE), bisphenol B (BPB) and bisphenol S (BPS) in canned soft drinks without any previous sample treatment. SPE on-line pre-concentration was performed by using a C18 cartridge. The analysis of all compounds was accomplished in 3 min. Quality parameters of the method were established and the authors obtained a simple, fast, reproducible (RSD values lower than 10%) and accurate (trueness higher than 93%) method for the analysis of bisphenols in canned soft drinks at the ng L^{-1} level using matrix-matched calibration.

Finally, in the current year, Vega-Morales et al. [2] used an on-line SPE-UHPLC-MS/MS to characterize 27 endocrine disrupting compounds (norethindrone, norgestrel, 17-alpha-ethinyloestradiol, etc.) in sewage samples. SPE treatment was performed by using Oasis HLB columns (mixed-mode sorbent). The complete analysis of each sample required less than 4 min and provided satisfactory recoveries (72–110%) and limits of detection in the order of few nanograms per liter (0.3-2.1 ng L^{-1}).

3. Molecularly imprinted polymers (MIPs)

Molecularly imprinted polymers (MIPs) are cross-linked, synthetic polymers with an artificially generated three-dimensional network able to specifically rebind a target analyte, or a class of structural analogues [1]. The principle is that a polymer network is obtained by polymerizing functional and cross-linking monomers around a template molecule. Subsequent removal of the template leaves a cavity with specific recognition sites complementary in shape, size and functional groups to the target analyte (**Figure 3**). These recognition sites can specifically bind target compounds in a similar way that antibodies specifically bind to an antigen, with the advantages of being very selective without suffering from stability problems associated to biological receptors [49]. All these aspects, together with the fact that MIPs synthesis is also relatively easy and cheap when compared with the purification procedure of natural antibodies, have led to a considerable growth of interest in the use of MIPs in several analytical techniques and applications.

Over the last 15 years, MIPs have been successfully applied as stationary phase on liquid chromatography, solid-phase extraction, micro-extraction, capillary electrochromatography, immunoassay determinations, and chemical sensing, with an almost exponential increase in the number of publications [50]. However, it should be pointed out that even if the interest in the area is relatively new, the concept itself has a long history. The earliest documents describing conceptually similar approaches had first been published in the early 1930s [51]. Nonetheless, today's concept of molecular imprinting technology started back in 1972 when the groups of Klotz and Wulff independently presented the first examples of synthetic organic polymers with predetermined ligand selectivities. In both of these studies, MIP synthesis was based on a covalent linkage of the template molecule to the monomers prior to polymerization. Later on, in the early 1980s, the group of Mosbach has reported for the first time a general non-covalent approach for producing organic imprinted polymers [52,53]. This important development has broadened the scope of molecularly imprinting polymers, improving considerably the versatility and the number of possible applications for this type of materials.

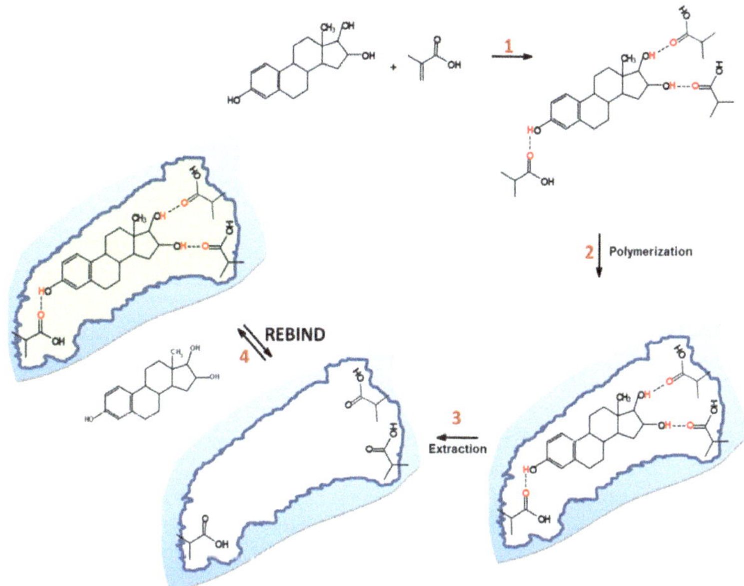

Figure 3. Schematic representation of non-covalent molecular imprinting procedures: (1) complex formation between the template (methacrylic acid) and the functional monomers (estriol), (2) polymerization, (3) template extraction, (4) analyte rebinding.

3.1. Applications of MIPs to SPE

Out of all the MIPs applications, the use of MIPs as selective sorbents for solid-phase extraction (MIP-SPE) represents the most important application area in the field of analytical separation sciences [54]. Solid-phase extraction (SPE) is a well-established method routinely used for clean-up and pre-concentration of analytes in a wide range of environmental, pharmaceutical, agricultural and food analysis [1]. Nevertheless, sorbents used in conventional SPE often lack selectivity resulting in co-extraction of interfering matrix components. Therefore, specificity, selectivity and sensitivity together with high extraction efficiency can be obtained using sorbents based on molecularly imprinted polymers (MIPs) [8].

To assess the potential of MIPs in terms of selectivity, we have compared the ability of MIP-SPE for selective extraction of zearalenone from cereal sample extracts with that of a commercial immunoaffinity column (IAC). **Figure 4** shows the similarity of the behavior of these two types of selective sorbents, resulted in high degrees of clean-up. In both cases, very reliable baselines and similar recoveries were obtained, proving that the high selectivity of immunoaffinity sorbents also can be achieved with molecularly imprinted polymers SPE. Furthermore, previous studies have found MIP-SPE to have a similar selectivity but a higher capacity than commercial IAC columns [49,55].

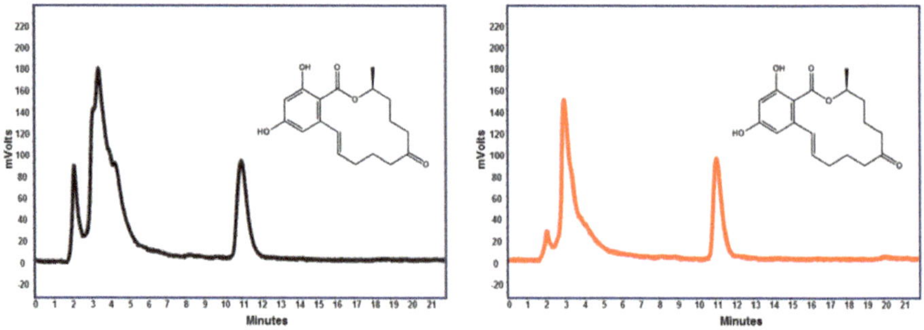

Figure 4. HPLC-FLD chromatograms of wheat sample spiked with ZON at level of 100 μgkg^{-1} after extract clean-up with (—) MIP-SPE (AFFINIMIP® SPE ZEARALENONE; Polyintell) and (—) immunoaffinity column (IAC).

These aspects are highly attractive for matrix clean-up, enrichment and selective extraction of analytes in difficult samples that are very common to food and environmental analyses. Hence, several examples of MIP-SPE applications have been described in the literature, as exemplified in **Table 2**, which presents a selection of the most recently published scientific research.

Compounds	Sample	Analysis	Recovery Rates (%)	Analytical features	Reference
Food Analysis					
Sudan I	Chilli Sauce	HPLC-UV	87.5 to 103.4 %	LOD ≥ 3.3 μg kg^{-1}	[56]
Ochratoxin A	Wheat Samples	MISPE-FLD	92.1 to 104 %	LOD ≥ 1.2 ng mL^{-1}	[57]
Domoic acid	Seafood Samples	HPLC-PDA	93.4 to 96.7%	LOQ ≥ 0.1mg L^{-1}	[58]
Catechins	Tea, Cocoa, Grape	HPLC–PDA–FL	50 to 100%	--	[59]
Zearalenone	Cereal Samples	HPLC-UV	82 to 90%	--	[49]
Mycophenolic acid	Maize	HPLC-MS-MS	49 to 84%	LOD ≥ 0.17 μg kg^{-1} LOQ ≥ 0.57 μg kg^{-1}	[60]
Tetracycline Antibiotics	Egg Samples	HPLC-PDA	91.6 to 107.6%	LOQ ≥ 0.8 ng g^{-1}	[6]
Thiamphenicol	Milk and Honey Samples	HPLC-PDA	92.9 to 99.3%	LOD≥ 0.003 μg mL^{-1} LOD ≥ 0.002 μg g^{-1}	[61]
Environmental Analysis					
Bisphenol A	Ultrapure, Tap, Drinking, River Water Samples	HPLC-FLD	84.7 to 93.8%	LOD ≥ 2.50 pg mL^{-1} LOQ ≥ 8.33 pg mL^{-1}	[62]

Natural and Synthetic Estrogens	River and Tap Water Samples	UHPLC-MS-MS	48 to 106 %	LOD ≥ 4.50 ng L^{-1} LOQ ≥ 14.9 ng L^{-1}	[8]
Pyrethroid Insecticides	Aquaculture Seawater	GC-ECD	86.4 to 96.0%	LOD ≥ 16.6 ng L^{-1} LOQ ≥ 55.3 ng L^{-1}	[63]
Water-Soluble Acid Dyes	Wastewater and Soft Drink Samples	HPLC-PDA	89.1 to 101.3%	LOD ≥ 0.095 μg L^{-1}	[64]
Levonorgestrel	River and WWTP influent and effluent samples	HPLC-UV	79.9 to 132.7 %	--	[65]
Methamidophos	Soil Samples, Tap and River Water Samples	GC-NPD	95.4 to 96.1 %	LOD ≥ 10 ng L^{-1} LOD ≥ 3.8 ng g^{-1}	[66]
Dibutyl Phthalate	Aqueous Environment Samples	GC-MS	94.7 to 101.9%	LOD ≥ 5.49 ng L^{-1}	[67]
Atrazine Herbicide	Aqueous Environment Samples	HPLC-PDA	94 to 99 %	LOD ≥ 80 ng L^{-1}	[68]
Chlorsulfuron	Water, Soil, and Wheat Plant Samples	HPLC-UV	82.3 to 94.7%	--	[69]
Parabens	Soil and Sediment Samples	HPLC-UV	80 to 90 %	LOD < 1 ng g^{-1}	[70]
Fluoroquinolone antimicrobials	Water Samples	HPLC-FLD	62 to 102%	LOD ≥ 1 ng L^{-1}	[7]

Table 2. Some recent applications of MIP-SPE in food and environmental analysis. LOD= Limit of detection; LOQ= Limit of quantification

Regarding the analytical method, MIP-SPE procedure is based on the same main four steps as conventional SPE such as pre-conditioning of the sorbent, sample loading, interferences wash step and elution of the target compounds. Therefore, to obtain optimal recovery rates and selectivity, each step of the extraction procedure must be properly optimized.

MIP-SPE can be basically used in both the reversed phase and normal phase modes. In the normal phase approach, the sample is usually percolated though the MIP-SPE column using the same solvent that was used as porogen for the MIP synthesis. Under this condition, the target analyte develops specific interactions with the monomer residues present in the polymer cavities, resulting in selective adsorption and molecular recognition by MIP due to the well-known solvent "memory" effect [54]

Nonetheless, in some common situations, a loading step based on direct percolation of aqueous sample through the MIP-SPE cartridge is highly desirable since most environmental or biological samples exist in an aqueous matrix [71]. Under this reversed-phase condition, the target analytes as well as non-polar interfering compounds are mainly retained by non-specific hydrophobic interactions. Thus, to generate specific interactions between the target compounds and the MIP and to disrupt the non-specific interactions between the residual monomers located at the surface of the polymer and matrix components, a selective washing step using low-to-medium polarity organic solvents, such as dichloromethane, chloroform, toluene or acetonitrile, is usually required [49]. It should be pointed out, however, that in some cases this selective washing step can be problematic because of the low polarity of common solvents used, which may give rise to miscibility problems and/or losses of the analyte [54]. Consequently, a drying step prior to this organic washing process becomes mandatory [8].

Once matrix interferences are removed, the analytes can be eluted from the column with a pure solvent, solvent that contains a small amount of modifier such as acetic acid or a combination of solvents with different polarities that must possess an elution strength sufficiently high to disrupt the specific interactions of the target analytes with the polymer, in minimal elution volumes.

As example of the successful application of a MIP-SPE compatible with aqueous samples, **Figure 5** shows the HPLC-FLD chromatogram (red line) corresponding to the injection of the elution fraction obtained after the purification of 100mL of Seine river water spiked with 0.5 ng mL^{-1} of 17β-estradiol.

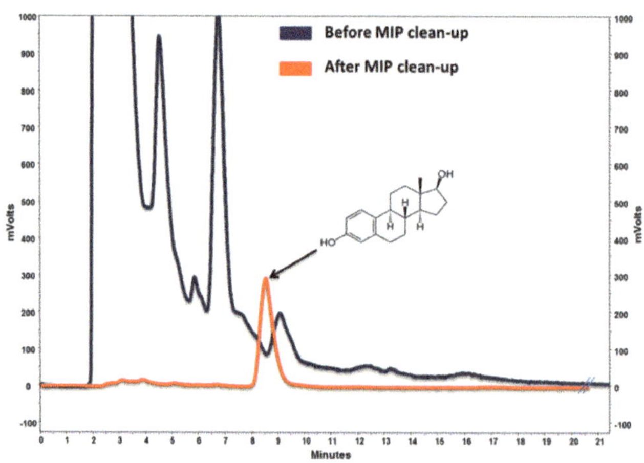

Figure 5. HPLC-FLD chromatograms obtained after extracts clean-up with MIP-SPE (AFFINIMIP® SPE Estrogens; Polyintell) of 100mL of Seine water spiked at 0.5 ng mL^{-1} with 17β-estradiol (—) and before MIP clean-up (—).

The chromatogram obtained (red line) clearly illustrates the efficiency of the MIP-SPE procedure (extraction rate of about 90%) and the advantages of both the concentration and sample clean-up with very low background and no interferences close to the retention time of 17β-estradiol. As a result, the use of such kind of selective sorbent allowed the successful detection of estradiol present at very low concentration without the need for a more selective and sensitive method of data acquisition such as mass spectrometry (MS) detection.

Regarding different operation modes, MIP-SPE has been used both in on-line and off-line modes prior to various detection techniques. Most of the MIP-SPE applications reported so far have been developed in the off-line mode because of its simplicity, ease-of-use and high flexibility. In addition, the drying step required in most of the off-line procedures is not compatible with on-line operations. Nonetheless, in the last years, there has been a considerably increase in the number of applications that use MIPs as sorbent for on-line SPE because provides higher pre-concentration factors with reduced analysis time and sample manipulation.

In the off-line mode, MIP–SPE has been applied for the selective extraction and pre-concentration of a wide range of analytes, such as mycotoxins in cereal samples [55], thiamphenicol in milk and honey samples [61], domoic acid from seafood [58], ofloxacin and lomefloxacin in chicken muscle [72], pyrethroid insecticides in aquaculture seawater [63] and parabens in soil and sediment samples [70]. Furthermore, MIP-SPE has been applied not only for the extraction of single target analyte but also for the simultaneous isolation of a class of structurally related compounds such as catechins or estrogens from real samples [8,59]. Luo et al. [64], developed a sensitive and selective off-line MIP-SPE based method to determine five water-soluble acid dyes in wastewater and soft drink samples. The precision and accuracy of the method were satisfactory and it gave average recoveries between 89.1 and 91.0%. In a similar way, Qi et al. [65], prepared a MIP by conventional bulk polymerization for the extraction of levonorgestrel from water samples. The synthesized MIPs not only displayed high specific recognition for levonorgestrel (recoveries > 79%), but also showed high cross-reactivity values for structurally related contraceptive drugs, suggesting that MIPs could be used as broad specific recognition absorbent.

On-line MIP-SPE protocol has been successfully used to extract benzimidazole fungicides in water samples [73], ochratoxin A in wheat samples [57], and bisphenol A in environmental water samples [74]. An automated on-line SPE using microspherical monodispersed molecularly imprinted particles coupled to HPLC-fluorescence detector was successfully applied to the simultaneous multi-residue analysis of six fluoroquinolone antimicrobials (enrofloxacin, ciprofloxacin, norfloxacin, levofloxacin, danofloxacin, and sarafloxacin) in water samples [7]. In this work, polymer particles prepared via precipitation polymerization were used as SPE sorbent. High recoveries with good precision (RSDs <5%) were obtained for the different fluoroquinolones tested, with values ranging from 91 to 102% in drinking and fish farm water samples. The detection limits were between 1-11 and 1-12 ng L^{-1} for drinking and fish farm water samples, respectively.

On-line MIP-SPE pre-concentration methodology has also recently been used by Jing et al. [6], for the determination of trace tetracycline antibiotics (TCs) in egg samples. This approach affords high-throughput analysis (18 min per sample), and also provides high sensitivity and selectivity with recoveries ranging between 91.6 and 107.6%, showing that efforts should continue to be made in this promising research area.

4. QuEChERS

4.1. QuEChERS procedure

The need for a simple, rapid, cost-effective and multi-residue method able to provide high quality of analytical results led Anastassiades et al. to develop in the years 2001 and 2002 a new sample treatment method called "QuEChERS". Initially, the methodology was developed for the analysis of veterinary drugs (anthelmintics and thyreostats) in animal tissues, but after realizing its great potential in the extraction of polar and particularly basic compounds, it was also tested with great success on pesticide residue analysis in plant material. The detailed method was first published in 2003 [75].

QuEChERS, acronym of "*Quick, Easy, Cheap, Effective, Rugged and Safe*", is a sample preparation technique entailing solvent extraction with acetonitrile and partitioning with magnesium sulfate alone or in combination with other salts followed by a clean-up step using dispersive solid-phase extraction (d-SPE). This last step is performed by adding small amounts of bulk SPE packing sorbents to the extract. This procedure has attracted the attention of pesticides laboratories worldwide and it is the most commonly employed sample treatment methodology for the multi-residue analysis of pesticides in fruits and vegetables [9,10,76-80]. But today, this methodology is not limited to the analysis of pesticides and its use for the extraction of other families of compounds is tremendously increasing.

The different steps on a typical QuEChERS procedure for the multi-residue analysis of pesticides are shown in **Figure 6**.

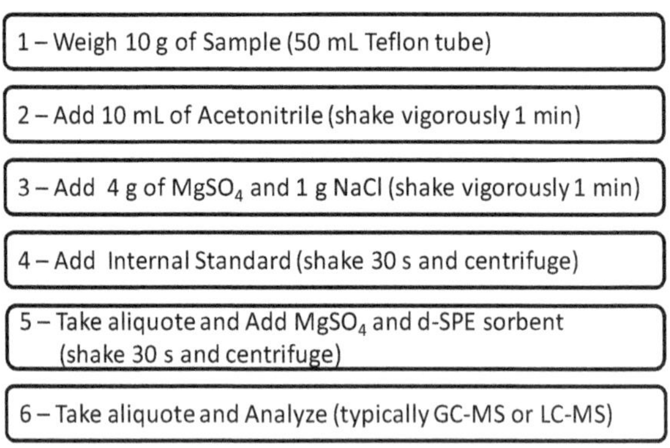

Figure 6. Schematic view of a typical analytical QuEChERS procedure for the analysis of pesticides as described in [75].

The idea of the QuEChERS procedure was to reduce complicated, laborious and time-consuming multi-residue sample treatment methods that required high amount of solvents and were therefore expensive. Moreover, some basic, acidic and very polar compounds cannot be satisfactorily extracted with common multi-residue methods. Thus, in order to cover all these analytes, laboratories have to perform specific analysis, and as a consequence, some of these compounds were not being monitored.

The first two steps of a typical QuEChERS procedure consist in weighing an appropriate amount of sample previously processed and homogenized (for instance 10 g) in a 50 mL Teflon tube (Step 1) and the addition of a solvent for the extraction (Step 2), in general acetonitrile, although the use of other organic solvents such as acetone, THF or ethyl acetate have been described [81].

Then, an extraction-partitioning step takes place by the addition of magnesium sulfate alone or in combination with other salts, generally sodium chloride (Step 3). Acetonitrile is the recommended solvent for QuEChERS because, upon the addition of salts, it is easily separated from water than, for instance, acetone. Ethyl acetate has the advantage of a partial miscibility with water but it can also co-extract lipids and provides lower recoveries during the dispersive SPE. The extraction of lipophilic materials is lower with acetonitrile but this solvent can form two phases with water when samples with high sugar content are manipulated [75]. The addition of salts in Step 3 helps to induce the phase separation. This salting-out effect also influences analyte partition, which of course is also dependent upon the solvent used for extraction. The concentration of salt can also influence the percentage of water in the organic phase and can play an important role in adjusting its polarity. Magnesium sulfate acts as a drying salt to reduce the water phase, thereby helping to improve recoveries by promoting partitioning of the pesticides (or other target compounds) into the organic layer while sodium chloride helps to control the polarity of the extraction solvent. With this, a single extraction-partitioning step is carried out (similarly to an "on-line" approach) which simplifies the necessity of multiple partitioning steps required in other multi-residue methods. Moreover, this extraction-partitioning step is produced by shaking vigorously for a few minutes, thus preventing more time-consuming steps such as sample blending. At this point, internal standards can be added to the system if necessary (Step 4), followed by shaking again the solution and a centrifugation step that help to separate salts. The use of internal standards can minimize the error generated in the multiple steps of the QuEChERS method. Sometimes the use of more than one internal standard is recommended especially with samples with high fat content because the excessive fat can form an additional layer into which the analytes can also partition [82]. Another advantage of QuEChERS procedure is the fact that, once the extraction-partitioning step is carried out, an aliquot of the extract is used for the next steps, minimizing also the separation or the transfer of the entire extracts frequently employed in other multi-residue methods.

Then, a dispersive solid-phase extraction (d-SPE) clean-up procedure is carried out with an aliquot (for instance 1 mL) of the extract which is placed in a vial containing again magnesium sulfate and small amounts of bulk SPE sorbent materials (Step 5). The vial is then shaken vigorously or mixed on a vortex mixer to distribute the SPE material and facilitate the clean-up process. This d-SPE step is quite similar to matrix solid-phase dispersion developed by Barker [83, 84], but in d-SPE the sorbent is added to an aliquot of the extract rather than to the original solid sample. Moreover, small amounts of sorbents are used because only a small portion of the extract is subjected to the clean-up procedure, and compared to conventional SPE clean-up methods, d-SPE is less laborious and time-consuming. Magnesium sulfate is again used in this step as a drying agent to remove water and improve analyte partitioning to provide better clean-up. Primary secondary amine (PSA) is the most common SPE sorbent used in QuEChERS procedure for pesticide analysis. The idea is to use a sorbent able to retain matrix components, but not the analytes of interest. However, depending on the sample matrix, other SPE sorbents can also be used alone or combined with PSA, such as C18, OASIS HLB, and graphitized carbon black sorbents. For instance, for samples with high fat content, PSA mixed with C18 is recommended [85] while for samples with moderate or high levels of chlorophyll and carotenoids (for example carrots), PSA mixed with graphitized carbon black is frequently used [86-88]. After the clean-up, the extract is centrifuged and an aliquot of the supernatant can be concentrated or directly analyzed usually by means of Gas Chromatography-Mass Spectrometry (GC-MS) or Liquid Chromatography-Mass Spectrometry (LC-MS) techniques (Step 6).

Although QuEChERS is quite a simple procedure as can be seen from **Figure 8**, in some cases method development will be necessary depending on the family of compounds to be analyzed, and many modifications over the QuEChERS procedure are being proposed. At the end, compromises will be required to ensure simplicity, speed, broad applicability, high recovery and selectivity. For instance, the control of pH in the extraction step is animportant factor when analyzing pesticides in order to ensure an efficient extraction of pH-dependent compounds such as phenocyalcanoic acids and to minimize degradation of labile pesticides under alkaline or acidic conditions. Buffering with citrate salts has been introduced in the first extraction/partitioning step to adjust the pH at around 5 where most labile pesticides under acidic or alkaline conditions are sufficiently stabilized. The pH control can also be very important in other steps of the QuEChERS procedure to prevent degradation of some compounds. For instance, to improve the stability of alkaline-labile compounds after the PSA clean-up step the final sample extract is slightly acidified by the addition of small amounts of formic acid. Of great importance was the introduction of acetate buffering to achieve a pH value of 6 in order to improve recoveries of pH-dependant analytes [89]. This approach resulted in the official method AOAC 2007.01.

The use of analyte protectants is also often proposed as an optional step previous to GC analysis for those compounds that might tail or breakdown on the capillary GC column interior surfaces, on the inlet liner or on the guard column. A combination of sorbitol, gulonolactone, and ethylglycerol was found to be the most effective analyte protectant to cover the whole range of pesticide compounds [90]. The hydroxyl groups of these protectants can interact with active sites on the chromatographic column and in the flowstream and enhanced the pesticide analyte response. Those protectants are of course not required when LC methods are employed for analysis.

4.2. Applications of QuEChERS

As commented before, QuEChERS procedure is the sample treatment of choice for the multi-residue analysis of pesticides in fruits and vegetables, and many works can be found in the literature dealing with the simultaneous extraction and clean-up of more than 100 pesticides with acceptable recoveries [75,77,79,80,91]. As an example, Woo Lee et al. developed a new QuEChERS method based on dry ice for the determination of 168 pesticides in paprika [91]. For this purpose, extraction was carried out by using 30 mL of acetonitrile and 10 mL of water, and approximately 10 g of dry ice granules were poured and maintained until layer separation. Clean-up was then carried out by using PSA and GCB sorbents. The separation of the sample extract was induced via the sublimation of dry ice, which occurs at -78.5°C at atmospheric pressure (1 atm). After some minutes of dry ice sublimation with the sample extract, the reduced temperatures of acetonitrile and water ranged from -4.0 to -5.0 °C and from -6.0 to -6.5 °C, respectively, and water was then iced and super cooled. The negative temperatures of the two solvents may reduce their entropies and allowed them to separate. As densities of ice and supercooled water were heavier than that of acetonitrile at 0 °C, water changed to ice and the supercooled water was separated with an acetonitrile layer from the mixed solution. This methodology improved the extraction for flonicamid and its metabolites which was not satisfactory enough using citrate-buffering QuEChERS method.

Recently, the use of carbon-based nanoparticles has also been described as clean-up sorbent for the analysis of pesticides [10]. In this work, multi-walled carbon nanotubes (MWCNTs) were proposed as reversed-dispersive solid phase extraction (r-DSPE) material for the analysis of 30 pesticides in fruits and vegetables. The amount of MWCNTs influenced the clean-up performance and the recoveries, but the use of only 10 mg MWCNTs was suitable for cleaning up all analyzed matrices and showed to be a good alternative to PSA sorbent. The method was validated for different matrices such as spinach, orange and cabbage with recoveries in the range of 71-100% for all 30 pesticides. QuEChERS has also been proposed for the extraction and clean-up of pesticide residues in other food matrices such as milk [85]. In this last work, recovery of 14 different pesticides residues in milk was investigated to respect the amounts of sodium acetate, PSA and C18 used on the clean-up step. Recoveries for hydrophilic pesticides such as myclobutanil ranged from 82 to 99% while lower values (<80%) were obtained for lipophilic pesticides because they were partially removed by C_{18} along with other fatty compounds.

But today, one of the most important features of QuEChERS may be its application to other families of compounds in a variety of matrices different than fruits and vegetables. Some examples of the use of QuEChERS for the extraction of compounds other than pesticides are given in **Table 3**. QuEChERS methodology has already been applied to the analysis of polycyclic aromatic hydrocarbons (PAHs) in fish and shrimp samples [92,93]. Forsbeg et al. developed and validated a modified QuEChERS method for the determination of 33 parent and substituted PAHs in high-fat smoked salmon that greatly enhanced analyte recovery compared to traditional QuEChERS procedure [93]. For this purpose, a mixture of acetone, ethyl acetate and isooctane instead of acetonitrile was employed for the extraction, and different kinds of salts were used for partitioning. The proposed modified QuEChERS

Compounds	Sample	QuEChERS procedure		Analysis	Reference
		Extraction-partitioning	Clean-up		
Polycyclic Aromatic Hydrocarbons	High-Fat Salmon	2 mL acetone:ethyl acetate:isooctane (2:2:1 *v/v/v*) + 6 g $MgSO_4$ + 1.5 g $NaC_2H_3O_2$	150 mg $MgSO_4$ + 50 mg PSA	GC-MS	[93]
		2 mL acetone:ethyl acetate:isooctane (2:2:1 *v/v/v*) + 4 g $MgSO_4$ + 1 g NaCl + 1g $NaC_6H_7O_7$ + 0.5 g $Na_2C_6H_8O_8$			
Acrylamide	Foodstuffs	5 ml of hexane (only for high fatty matrices) + 10 mL water + 10 mL acetonitrile + 4 g $MgSO_4$ + 0.5 g NaCl	150 mg $MgSO_4$ + 50 mg PSA	LC-MS or GC-MS	[94]
Veterinary drug residues	Milk	10 ml acetonitrile (1% acetic acid) + 10 ml 0.1M Na2EDTA solution + 4g MgSO4 + 1 g sodium acetate	--	LC-MS/MS	[95]
Persistent organic pollutants: 22 organochlorine pesticides + 7 polychlorinated biphenyls (PCBs)	Fish tissue	10 mL acetonitrile + 10 ml water + 4 g $MgSO_4$ + 1 g NaCl + 0.5 sodium citrate dibasic + 1 g sodium citrate tribasic	pre-frozen step (2 hours) 1 g calcium chloride + 900 mg $MgSO_4$ + 150 mg PSA	GC-MS	[96]
10 mycotoxins	Eggs	10 mL methanol:water (80:20 *v/v*) with 1% acetic acid+ 4 g $MgSO_4$ + 1 g sodium acetate	(C18 or Oasis HLB SPE)	LC-MS/MS	[12]
Phytohormones	Vegetables	10 mL acetonitrile with 1% acetic acid + 4 g $MgSO_4$ + 1 g NaCl + 1g sodium citrate + 0.5 disodium citrate	--	UHPLC-MS/MS	[97]
UV Ink Photoinitiators	Packaged food (baby food, fruit juices, wine)	12 mL acetonitrile + 4 g $MgSO_4$ + 1.5 g NaCl	250 mg $MgSO_4$ + 750 mg PSA	LC-MS/MS	[11]

Table 3. Application of QuEChERS procedure for the extraction of different kind of analytes.

substantially improved average recovery of 15 PAHs by roughly 38% and led to individual gains of 50-125% for some PAHs such as naphthalene and anthracene among others when compared to traditional Soxhlet extraction with hexane. Acrylamide has also been extracted from various food matrices such as chocolate, peanut butter, and coffee [94]. An accurate determination of acrylamide in foodstuffs was possible using QuEChERS since the use of salt and the PSA sorbent increased the selectivity of the method by reducing the content of more polar matrix coextractives. Hexane was required for high fatty samples such as peanut butter.

The extraction of veterinary drugs residues from animal tissues [98] and from milk [95,99,100] has also been described. For instance in [95] the use of QuEChERS was proposed as a fast sample treatment for a rapid screening method in the identification of 21 veterinary drug residues in milk. In this case, 1% acetic acid in acetonitrile together with a 0.1 M Na_2EDTA solution was proposed as extraction solvent, using for partitioning magnesium sulfate and sodium acetate, but no further clean-up step was necessary to attain good results. The analysis of persistent organic pollutants (POPs) such as organochloride pesticides and polychlorinated biphenyls (PCBs) in fish tissues was recently reported using conventional QuEChERS procedure but with the addition of a pre-frozen step for 2 hours at -24°C (by means of a homemade freezing device) before the PSA clean-up step for removal of lipids [96]. After this freezing step, between 60 to 70% of lipids were removed. The reduction of co-extractives increased up to 96% by treatment with calcium chloride and PSA. Extraction of mycotoxins and phytohormones from eggs [12] and vegetables [97], respectively, has also been described using QuEChERS with no further clean-up steps, although in the case of mycotoxins a SPE clean-up step using C18 or Oasis HLB cartridges was sometimes proposed. An interesting application of QuEChERS was recently reported by Gallart-Ayala et al. for the analysis of contaminants migrating into food from food packaging materials [11]. In this case, the extraction of UV Ink photoinitiators such as benzophenone and isopropylthioxanthone (ITX) in several foodstuffs (baby food, fruit juices and wine) packaged in tetra brick containers was performed using a common QuEChERS procedure with PSA sorbent for clean-up. The extraction method proposed showed comparable results in terms of method limits of quantification, run-to-run and day-to-day precisions, and quantification results than a previous SPE method reported for the analysis of ITX, with the advantage of being 12 times faster (per sample).

Summarizing, QuEChERS approach appears to have a bright future not only for the analysis of pesticide residues in foods and other agricultural products but also for the analysis of different families of contaminants either in food or even in other matrices. For instance, QuEChERS has also been proposed in environmental analysis for the extraction of chlorinated compounds from soil samples [101]. The simplicity of its use and the great range of modifications that can be applied, make QuEChERS an ideal extraction procedure to think about when dealing with the extraction of any kind of analytes, and the number of publications using QuEChERS will considerably increase in the future.

5. Turbulent-flow chromatography (TFC)

For a long time the determination of small drug molecules in biological fluids was a very challenging task due to both the complexity of biological samples and the requirement of long chromatographic separations because of the presence of endogenous interferences. The recent implementation of automated on-line extraction procedures has allowed fast sample clean-up in bioanalytical applications, and turbulent-flow (or TurboFlow) chromatography (TFC) appears as one of the most interesting ones in this area [102-105].

The on-line set-up for a typical TFC-LC-MS method is shown in **Figure 7**. TFC methods are based on the direct injection of biological samples without previous extraction or any treatment into a column packed with large particles. These large particles could have some stationary phase bonded to them, adding an additional selectivity to the extraction procedure. Once the sample have been injected (**Figure 7a**) onto a TurboFlow column, an extraction solution is pumped into the column at a high flow rate (between 1.5 to 5.0 mL min^{-1}) generating turbulent flow conditions inside the column (**Figure 7b**). In general, 100% aqueous mobile buffers are used for this purpose. Under these conditions, small analyte molecules are retained via diffusion processes into the particle pores, while big molecules such as proteins are washed out from the column. In this way, the compounds of interest are extracted from the biological matrix and then eluted from the TurboFlow column onto the analytical column with a volume of solvent which was stored in a holding loop or pumped directly from the chromatographyic LC system (**Figure 7c**). In general organic mobile phases or pH buffered solutions are used for the elution of the compounds of interest depending on the chromatographic separation used after extraction, but the elution volume must be at least ten times that of the TurboFlow column in order to guarantee a complete elution. The analytes are released from the TurboFlow column at a considerably lower flow rate than the one used during extraction into the analytical system where they are mixed with the chromatographic mobile phase and introduced into the chromatographic column, being focused into a sharp band at the head of the HPLC column. When the transfer of analytes is complete, the TurboFlow column could be washed for the next extraction while a regular gradient or isocratic elution is taking place in parallel on the analytical column. The optimization of the different on-line extraction steps is crucial and parameters like mobile phase composition, flow rates and extraction time windows will affect recovery or extraction efficiency in general.

The theory of turbulent flow in open tubes has been discovered and studied for decades. Nevertheless, its application to LC packed columns was only patented in 1997 [106]. The challenge at the moment was to design a chromatographic platform using turbulent flow properties to isolate small analytes from macromolecules present in complex matrices such as biological fluids.

TFC has been used mainly in the handling of biological samples containing a large amount of proteins, such as blood plasma [107,108]. For instance, Michopoulos et al. compared the use of TFC for the metabonomic analysis of human plasma with protein precipitation showing that TFC could be effectively used with the benefit that off-line sample handling was significantly reduced [107]. However, the analysis of the data obtained with TFC for human plasma revealed substantial differences in the overall metabolite profiles compared to methanol-precipitated HPLC-MS, probably due to greatly reduced amounts of phospholipids (ca. 10 fold reduction) with TFC methodology compared to protein-precipitated samples. TFC seems to be also more efficient at removing proteins based on their size than restricted access media (RAM) or solid phase extraction (SPE) [1,109].

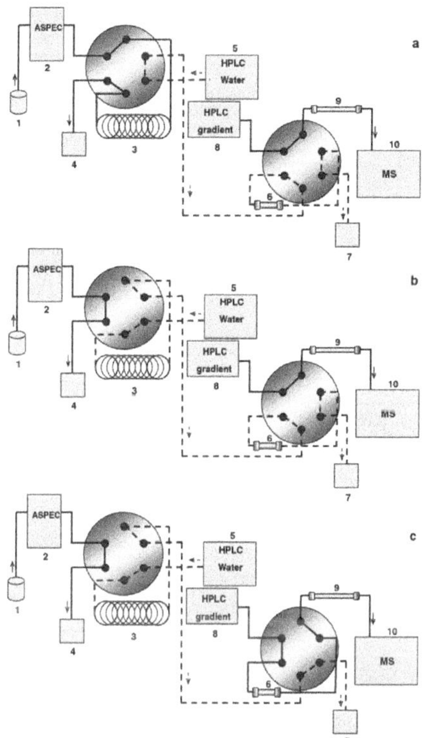

Figure 7. On-line set-up of a typical TFC-LC-MS system. The position of the various valves to perform the sample clean-up and subsequent analysis of the samples is shown. **a)** Filling the sample loop; **b)** transfer from the sample loop to the TFC column followed by washing; **c)** elution from the TFC column to the Chromatographic column using gradient elution. 1) sample; 2) ASPEC system with a syringe pump; 3) sample loop (10 mL); 4) Waste; 5) HPLC Pump (4 mL min^{-1}); 6) TFC column; 7) Waste; 8) HPLC Pump (gradient); 9) HPLC column; 10) MS analyzer. Reproduced from [111].

TFC shows also a big potential in clinical applications, where the increased emphasis on both drug safety and translational biology, e.g. the need to understand how pre-clinical efficacy models are representative of human pharmacology, has considerably modified the expectations for what needs to be measured routinely in biological samples. Moreover, sometimes it is necessary to monitor in the same sample not only the drug levels but also its potential active/reactive metabolites as well as the biomarkers associated with the mechanism of action of the drug. Those biomarkers could span from a very small and polar compounds such as a neurotransmitter to a very large and hydrophobic entity like fatty acids. So, amongst the key analytical challenges with biomarkers is generally the sampling procedure as well as the sample volume available. TFC has some intrinsic capabilities that facilitate the analysis of biomarkers and metabolites [103]. First, it provides high sensitivity assays without the need for high sample volume. In addition, the on-line extraction approach removes the need for lengthy sample preparation procedures, hence reducing sample degradation issues frequently observed in biomarker analysis. As an example, Mueller et al. proposed a fully automated toxicological LC-MSn screening system in urine using on-line extraction with TFC [110].

Very recently the use of TFC couple to tandem mass spectrometry has been reported for the automated analysis of perfluorinated compounds (PFCs) in human hair and urine samples [112]. The method allowed the extraction and analysis of 21 PFCs with recoveries between 60 to 105%.

But today, TFC is being used in other fields of applications, such as food or even environmental analysis. **Table 4** shows some examples of these TFC applications.

Compounds	**Sample**	**TFC column**	**Flow-rate / Injection volume**	**Detection**	**Reference**
Quinolones	Honey	Cyclone HTLC, 50 x 0.5 mm, 60 μm (Thermo Fisher Scientific)	1.5 mL min^{-1} / 160 μL	LC-ESI-MS/MS	[13]
Quinolones (Enrofloxacin and Ciprofloxacin)	Edible tissues (cattle, pig, turkey, rabbit)	Cyclone HTLC, 50 x 1.0 mm, 50 μm (Thermo Fisher Scientific)	5 mL min^{-1} / 20 μL	LC-ESI-MS/MS	[14]
Veterinary drugs	Milk	Cyclone – Cyclone P connected in tandem, 50 x 0.5 mm, 60 μm (Thermo Fisher Scientific)	1.5 mL min^{-1} / 50 μL	LC-ESI-MS/MS	[15]
Flavonoids and resveratrol	Wine	50 x 1.0 mm, 60 μm C18 (Thermo Fisher Scientific)	4 mL min^{-1} / 10 mL	LC-ESI-MS LC-APCI-MS	[111]
PFOS	River water	50 x 1.0 mm, 50 μm C18 (Cohesive Technologies)	1 mL min^{-1} / 1 mL	LC-APPI-MS	[113]
Anti-infectives	Wastewater	50 x 1.0 mm, 50 μm C18 SL (Cohesive Technologies)	3 mL min^{-1} / 1 mL	LC-ESI-MS/MS	[114]
Pesticides	Surface, drinking wáter	50 x 1.0 mm, 35 μm Oasis HLB (Waters)	5 mL min^{-1} / 10 mL	LC-APPI-MS/MS	[115]

Table 4. Relevant examples of the application of turbulent flow chromatography.

In a recently published review dedicated to sample preparation methodologies for the isolation of veterinary drugs and growth promoters from food, Kinsella et al. described turbulent flow chromatography as a technique that eliminates time-consuming sample clean-up, increases productivity and reduces solvent consumption without sacrificing sensitivity [116]. Food matrices have a high content of fat and proteins, which makes TFC an ideal sample treatment technique for the determination of a specific class of contaminants in various matrices such as honey, tissues and milk [117]. Some examples are described in the literature concerning the determination of veterinary drugs such as quinolones in honey and animal tissue [13-14]. For instance, turbulent flow chromatography coupled to LC-MS/MS was proposed for the quantitative high-troughput analysis of 4 quinolones and 12 fluoroquinolones in honey [13]. The manual sample preparation was limited to a simple dilution of the honey test portion with water followed by a filtration. The extract was then on-line purified on a large particle size TFC column where the sample matrix was washed away while the analytes were retained. Recoveries of 85-127% were obtained, while matrix effects were still observed which led to the use of standard addition for calibration. The proposed methodology has also shown good robustness, with over 400 injections of honey extracts without any TFC column deterioration, with the consumption of 44 mL of solvent per sample. The authors described that TFC showed a strong potential as an alternative extraction and clean-up sample method compared to those making use of off-line sample preparation, in terms of both increasing the analysis throughput and obtaining higher reproducibility linked to automation to ensure the absence of contaminants in honey samples. In the case of animal tissues TFC was used for sample preparation in the analysis of two quinolones (enrofloxacin and its metabolite ciprofloxacin) [14]. Sample was extracted with a mixture of acetonitrile/water 1:1 acidified with 0.01% formic acid. Mean recovery rates for the tissues of the different species (cattle, pig, turkey and rabbit) were in the range of 72-105% in a run time of only 4 min.

Presta et al. [111] described the use of TFC coupled to LC-MS for the determination of flavonoids and resveratrol in wines. 10 mL of sample (diluted wine) was passed over the TFC column, after which the retained analytes were separated by reversed-phase LC. The method proved to be fast, non-laborious, robust and sensitive.

Turboflow chromatography has also been described for sample treatment in the screening of eight veterinary drugs in milk [15]. Protein precipitation was induced before analyzing samples of whole, skimmed and semi-skimmed milk samples. While matrix effects – ion suppression and enhancement – were obtained for all analytes, the method has proved to be useful for screening purposes because of its sensitivity (0.1 to 5.2 $\mu g\ L^{-1}$), linearity and repeatability (RSD $\leq$ 12%). As an example, **Figure 8** shows the chromatographic separation of a non-fat milk sample spiked with target veterinary drugs and analyzed by TFC-LC-(ESI)-MS/MS.

This sample treatment technique has also been applied successfully to environmental samples. For instance, anti-infectives analysis in wastewater has been reported with good recovery (86-141%) and limits of quantification (45-122 ng L^{-1}) [114]. Signal distortion, represented as matrix effect, was still observed probably due to the fact that small molecules (below 1000 Da) present in wastewater samples will have affinity for the stationary phase and will not be completely removed in the clean-up step. Takino et al. have minimized the matrix effect observed by using atmospheric pressure photoionization (APPI) instead of electrospray (ESI) as ionization source [113]. In this case, a simple, fast and sensitive LC/APPI-MS method, with automated on-line extraction using TFC was developed for the determination of perfluorooctane sulfonate (PFOS) in river water. TFC columns packed with organic polymers or graphitized carbons were also found to be highly capable for enrichment of trace pesticides from drinking and surface water samples [115].

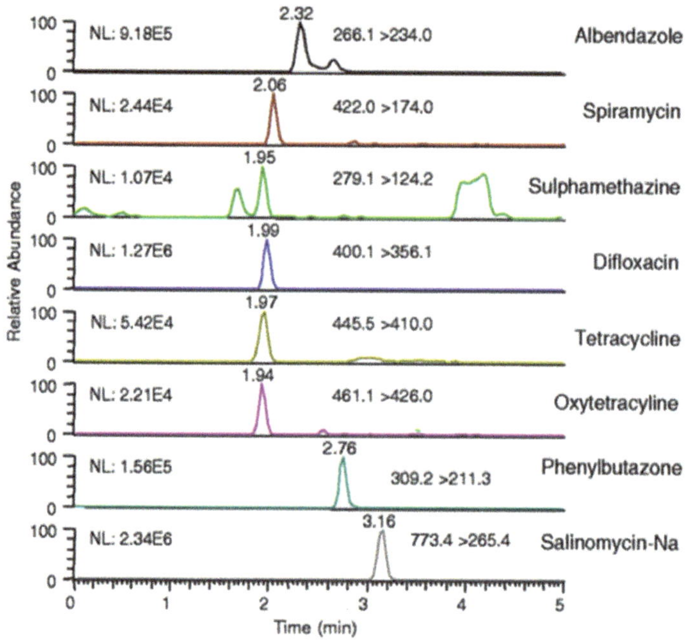

Figure 8. Representative SRM chromatograms of a non-fat milk sample spiked with the mixture of antibiotics standards at 100 μg L^{-1} level and analyzed by TFC-LC-(ESI)-MS/MS. Reproduced from [15] with permission from Springer.

In summary, turbulent flow chromatography appears as a very useful approach for sample treatment because it possesses greater efficiency in removing proteins based on their size than restricted access media or SPE procedures and combines high-throughput and high reproducibility by means of separating analytes from various matrices with reduced sample handling. The advantages of this sample extraction and clean-up procedure is unquestionable in bioanalytical applications, and although not many applications in other fields such as food and environmental analysis are yet available, it will surely become a very useful method in the area of food analysis, especially in matrices with a high content of fat and proteins such as milk.

6. Conclusion

There is an increasing demand for high-throughput chromatographic separations in food and environmental analysis where highly heterogeneous and difficult matrices may be analyzed. Despite the important advances in chromatographic separations, food and environmental matrices are very complex samples, and sample extraction and clean-up treatments are usually required. Therefore, sample treatment is still one of the most important parts of whole analytical method and effective sample preparation is crucial in achieving accurate analytical results. Food and environmental analysis generally requires several steps such as extraction from the sample of interest, removal of co-extracted matrix components, analytes enrichment and their subsequent quantification. Thus, the availability of robust, sensitive, selective and rapid analysis methods is of primary importance. The most recently introduced sample treatment methodologies in food and environmental applications have been discussed in this chapter, such as on-line SPE methods, QuEChERS, MIPs as selective sorbents for SPE, and the use of turbulent-flow chromatography.

On-line SPE is a viable and increasingly popular technique used to improve the sample throughput by reducing sample preparation time and overcome many of the limitations associated with the classical off-line SPE procedure.

For sensitive and selective determination of compounds in very complex matrices, the use of polymers with recognition sites able to specifically bind a particular substance or a group of structural analogues has attracted increase attention due to their outstanding advantages, i.e., high specificity, selectivity and capacity.

QuEChERS appeared as satisfactory, simple, rapid and inexpensive sample extraction and clean-up multi-residue methods especially employed in the analysis of pesticides. However, this methodology is also being successfully employed for the extraction of other families of compounds in food and environmental matrices such as acrylamide, mycotoxins, PAHs and chlorinated compounds.

The use of turbulent-flow chromatography represents a highly attractive and promising approach for removing proteins based on their size better than RAM or SPE procedures. TFC has been satisfactory applied to the direct analysis of complex matrices with reduced or without any sample manipulation and, even not many applications in food and environmental samples are yet available, it will become a very useful method due to its great potential for the analysis of protein- and fat-rich matrices.

Finally, future developments in all areas of analytical sample preparation are expected to continue in order to improve accuracy, sensitivity, specificity, and reproducibility of the sample treatment technique together with reduced analysis time and sample manipulation.

Author details

Paolo Lucci
Department of Nutrition and Biochemistry, Faculty of Sciences, Pontificia Universidad Javeriana, Bogotà D.C., Colombia

Deborah Pacetti and Natale G. Frega
Department of Agricultural, Food and Environmental Sciences, Marche Polytechnic University, Ancona, Italy

Oscar Núñez
Department of Analytical Chemistry, University of Barcelona, Barcelona, Spain

7. References

[1] Núñez, O., Gallart-Ayala, H., Martins, C.P. & Lucci P. (2012) New trends in fast liquid chromatography for food and environmental analysis. J Chromatogr A, 1228:298-323

[2] Vega-Morales, T., Sosa-Ferrera, Z. & Santana-Rodríguez, J.J. (2012). Development and optimisation of an on-line solid phase extraction coupled to ultra-high-performance liquid chromatography - tandem mass spectrometry methodology for the simultaneous determination of endocrine disrupting compounds in wastewater samples. J Chromatogr A, 1230:66 - 76

[3] Rodriguez-Mozaz, S., Lopez de Alda, M.J.& Barceló D. (2007). Advantages and limitations of on-line solid phase extraction coupled to liquid chromatography–mass spectrometry technologies versus biosensors for monitoring of emerging contaminants in water. J. Chromatogr. A, 1152:97-115

[4] Gosetti F., Chiuminatto U., Mazzucco E, Robotti E., Calabrese, G., Gennaro M.C. & Marengo E. (2011). Simultaneous determination of thirteen polycyclic aromatic hydrocarbons and twelve aldehydes in cooked food by an automated on-line solid phase extraction ultra high performance liquid chromatography tandem mass spectrometry. J. Chromatogr. A, 1218:6308-6318

[5] Gosetti F., Chiuminatto U., Zampieri D., Mazzucco E., Robotti, E., Calabrese, G., Gennaro M.C.& Marengo E. (2010). Determination of perfluorochemicals in biological, environmental and food samples by an automated on-line solid phase extraction

[6] Jing, T., Niu, J., Xia, H., Dai, Q., Zheng, H., Hao, Q., Mei, S. & Zhou, Y. (2011) Online coupling of molecularly imprinted solid-phase extraction to HPLC for determination of trace tetracycline antibiotic residues in egg samples. *J Sep Sci.*, 34:1469-76

[7] Rodríguez, E., Navarro-Villoslada, F., Benito-Peña, E., Marazuela, M.D. & Moreno-Bondi M.C. (2011) Multiresidue determination of ultratrace levels of fluoroquinolone antimicrobials in drinking and aquaculture water samples by automated online molecularly imprinted solid phase extraction and liquid chromatography. *Anal Chem.* 83:2046-55.

[8] Lucci, P., Núñez, O. & Galceran, M.T. (2011). Solid-phase extraction using molecularly imprinted polymer for selective extraction of natural and synthetic estrogens from aqueous samples. *J Chromatogr A,* 1218:4828-33.

[9] Wilkowska, A. & Biziuk, M, (2011). Determination of pesticide residues in food matrices using the QuEChERS methodology. *Food Chem.*, 125:803-812.
[10] Zhao, P., Wang, L., Zhou, L., Zhang, F., Kang, S. & Pan, C. (2012). Multi-walled carbon nanotubes as alternative reversed-dispersive solid phase extraction materials in pesticide multi-residue analysis with QuEChERS method.*J. Chromatogr. A,* 1225:17-25.
[11] Gallart-Ayala, H., Nunez, O., Moyano, E. & Galceran, M.T. (2011). Analysis of UV ink photoinitiators in packaged food by fast liquid chromatography at sub-ambient temperature coupled to tandem mass spectrometry. *J. Chromatogr. A,* 1218:459-466.
[12] Frenich, A.G., Romero-Gonzalez, R., Gomez-Perez, M.L. & Vidal, J.L.M. (2011). Multi-mycotoxin analysis in eggs using a QuEChERS-based extraction procedure and ultra-high-pressure liquid chromatography coupled to triple quadrupole mass spectrometry. *J. Chromatogr. A*, 1218:4349-4356.
[13] Mottier, P., Hammel, Y.A., Gremaud, E. & Guy, P.A. (2008). Quantitative High-Throughput Analysis of 16 (Fluoro)quinolones in Honey Using Automated Extraction by Turbulent Flow Chromatography Coupled to Liquid Chromatography-Tandem Mass Spectrometry. *J. Agric. Food Chem.* 56:35-43.
[14] Krebber, R., Hoffend, F.J. & Ruttmann, F. (2009). Simple and rapid determination of enrofloxacin and ciprofloxacin in edible tissues by turbulent flow chromatography/tandem mass spectrometry (TFC-MS/MS). *Anal. Chim. Acta,* 637:208-213.
[15] Stolker, A.A.M., Peters, R.J.B, Zuiderent, R., Di Bussolo, J.M. & Martins, C.P.B. (2010). Fully automated screening of veterinary drugs in milk by turbulent flow chromatography and tandem mass spectrometry. *Anal. Bioanal. Chem.*, 397:2841-2849.
[16] Gilar, M., Bouvier, E.S.P. & Compton, B.J. (2001). Advances in sample preparation in electromigration, chromatographic and mass spectrometric separation methods. *J. Chromatogr. A,* 909:111-135.
[17] Titato, G. & Lanças, F. (2005) Comparison Between Different Extraction (LLE and SPE) and Determination (HPLC and Capillary-LC) Techniques in the Analysis of Selected PAHs in Water Samples. *Journal of Liquid Chromatography & Related Technologies*, 28:3045-3056.
[18] Nováková, L. & Vlcková, H. (2009). A review of current trends and advances in modern bio-analytical methods: chromatography and sample preparation. *Analytica Chimica Acta*, 656:8-35.
[19] Zwir-Ferenc, A. & Biziuk, M. (2006). Solid phase extraction Technique – Trends, opportunity and applicantions. *Polish J. of Environ. Stud.*, 15:677-690.
[20] Masqué, N., Marce, R.M. & Borrull, F. (1998). New polymeric and other types of sorbents for solid-phase extraction of polar organic micropollutants from environmental water. *Trends Anal. Chem.*, 6:384-389.
[21] Masqué, N., Galià, M, Marcé, R.M. & Borrull. F. (1998). New chemically modified polymeric resin for solid-phase extraction of pesticides and phenolic compounds from water. *J. Chromatogr. A*, 803:147-155.
[22] Supelco Bulletin 910 (1998), Guide to Solid Phase Extraction, http://www.sigmaaldrich.com/Graphics/Supelco/objects/4600/4538.pdf

[23] Leon-Gonzalez, M.E. & Perez-Arribas, L.V. (2000). Chemically modified polymeric sorbents for sample Preconcentration. *J. Chromatogr. A* 902:3-16.
[24] Fontanals, N., Marcè R.M. & Borrull F. (2007) New materials in sorptive extraction techniques for polar compounds, *J. Chromatogr. A*, 1152:14–31.
[25] Mazzella, N., Debenest, T., & Delmas, F. (2008). Comparison between the polar organic chemical integrative sampler and the solid-phase extraction for estimating herbicide time-weighted average concentrations during a microcosm experiment. *Chemosphere*, 73:545-550.
[26] Yu, Z.; Liu, B.; Jiang, Z.H. & Zhang, G. (2009). Simultaneous determination of herbicide mefenacet and ist metabolities residues in river water by solid phase extraction and rapid resolution liquid chromatography-mass spectrometry with pre-column derivatization. *J. Chromatogr. A*, 1216:3090-3097.
[27] Pyrzynska Krystyna, (2011). Solid-Phase Extraction for Enrichment and Separation of Herbicides. Herbicides, Theory and Applications, In tech; ISBN: 978-953-307-975-2
[28] Fontanals, N., Cormack, P.A.G., Rosa M., Marcé, R.M. & Borrull, F. (2010). Mixed-mode ionexchange polymeric sorbents: dual-phase materials that improve selectivity and capacity. *Trends in Analytical Chemistry*, 29:765-779.
[29] Mullett, W.M., Walles, M., Levsen, K., Borlak, J. & Pawliszyn J. (2004). *J. Chromatogr. B*, 801:297- 306
[30] Chico, J., Meca, S., Companyó, R., Prat, M.D., &Granados, M. (2008). Restricted access materials for sample clean-up in the analysis of trace levels of tetracyclines by liquid chromatography. Application to food and environmental analysis.*J. Chromatogr. A*, 1181:1-8.
[31] Michel, M. & Buszewski, B. (2009). Porous graphitic carbon sorbents in biomedical and environmental applications. *Adsorption*, 15,193-202.
[32] Pereira L. (2008). Porous Graphitic Carbon as a Stationary Phase in HPLC: Theory and Applications. *Journal of Liquid Chromatography & Related Technologies*,31:1687-1731.
[33] Rodrıguez-Plasencia, F.J., Navarro-Villoslada, F., Perez-Arribas L.V., Leon-Gonzalez, M.E. & Polo-Dıez L.M (1997). Preconcentration of triazine herbicides from water by an ion chromatography column and determination by gas chromatography-mass spectrometry. *J. Chromatogr. A*, 760:314-318.
[34] Moret, S., Barp, L., Grob, K. & Conte L.S. (2011) Optimised off-line SPE–GC–FID method for the determination of mineral oil saturated hydrocarbons (MOSH) in vegetable oils. *Food Chemistry*, 129:1898-1903.
[35] Yang, X., Zhanga, H., Liub, Y., Wangc, J., Zhanga, Y.C. , Donga, A.J., Zhaoa, H.T., Sund, C.H. & Cuia, J. (2011). Multiresidue method for determination of 88 pesticides in berry fruits using solid-phase extraction and gas chromatography–mass spectrometry: Determination of 88 pesticides in berries using SPE and GC–MS. *Food Chem.*, 127:855-865
[36] Denev, D., Ciz, M., Ambrozova, G., Lojek, A., Yanakieva, I. & Kratchanova, M. (2010). Solid-phase extraction of berries' anthocyanins and evaluation of their antioxidative properties. Food Chemistry,123:1055-1061.
[37] Qi, P., Zeng, T., Wen, Z., Liang, X. &Zhang X. (2011) Interference-free simultaneous determination of Sudan dyes in chili foods using solid phase extraction coupled with HPLC–DAD. *Food Chemistry*,125:1462-1467.

[38] Zygler, A., Wasik, A. & Namieśnik, J. (2010). Retention behaviour of some high-intensity sweeteners on different SPE sorbents. *Talanta,* 82:1742-1748.

[39] Deng, X.J., Guo, D., Zhao, S., Han, L., Sheng, J., Yi, X. , Zhou, Y. & Peng, T. (2010).A novel mixed-mode solid phase extraction for simultaneous determination of melamine and cyanuric acid in food by hydrophilic interaction chromatography coupled to tandem mass chromatography. *J. Chromatogr. B,* 878:2839–2844

[40] Peng, X., Zhao, X. & Feng, Y. (2011). Preparation of phenothiazine bonded silica gel as sorbents of solid phase extraction and their application for determination of nitrobenzene compounds in environmental water by gas chromatography–mass spectrometry. *J. Chromatogr. A,* 1218:9314-9320.

[41] Król, S., Zabiegała, B. & Namieśnik, J. (2012) PBDEs in environmental samples: Sampling and analysis. *Talanta* In press. Available online 2 February 2012 In Press, http://dx.doi.org/10.1016/j.talanta.2012.01.048.

[42] Vega-Morales, T., Sosa-Ferrera, Z. &Santana-Rodríguez, J.J. (2010). Determination of alkylphenol polyethoxylates, bisphenol-A, 17α-ethynylestradiol and 17β-estradiol and its metabolites in sewage samples by SPE and LC/MS/MS. *Journal of Hazardous Materials,* 183:701-711.

[43] Ma, J., Xiao, R., Li, J., Yu, J., Zhang, Y. & Chen, L., (2010). Determination of 16 polycyclic aromatic hydrocarbons in environmental water samples by solid-phase extraction using multi-walled carbon nanotubes as adsorbent coupled with gas chromatography–mass spectrometry. *J. Chromatogr. A,* 1217:5462-5469.

[44] Slobodnik, J., Ramalho S., van Baar B.M.L, Louter, A.J.H., Brinkman U.A. (2000). Determination of microcontaminants in sediments by on-line solid-phase extraction/gas chromatography/mass spectrometry. *Chemosphere,* 41:1469-1478.

[45] Hyotylainen, T. & Riekkola, M.L (2004) Approaches for on-line coupling of extraction and chromatography. *Analycal Bioanalytical Chemistry,* 378:1962-1981.

[46] Kerkdijk, H., Mol, H.G.J. & Van der Nagel, B. (2007).Volume Overload Cleanup: An Approach for On-Line SPE-GC, GPC-GC, and GPC-SPE-GC. *Analytical Chemistry,* 79:7975-7983.

[47] Oliferova, L., Statkus, M., Tsysin, G., Shpigun, O. & Zolotov, Y. (2005). On-line solid-phase extraction and HPLC determination of polycyclic aromatic hydrocarbons in water using fluorocarbon polymer sorbents. *Analytica Chimica Acta* Vol. 538, pp 35–40.

[48] Gallart-Ayala, H., Moyano, E. & Galceran, M.T. (2011) Analysis of bisphenols in soft drinks by on-line solid phase extraction fast liquid chromatography tandem mass spectrometry. *Analytica Chimica Acta,*683:2227-233.

[49] Lucci, P., Derrien, D., Alix, F., Pérollier, C. & Bayoudh S. (2010). Molecularly imprinted polymer solid-phase extraction for detection of zearalenone in cereal sample extracts. Anal *Chim Acta,* 672:15-19.

[50] Tse Sum Bui, B. & Haupt, K. (2010) Molecularly imprinted polymers: synthetic receptors in bioanalysis. *Anal Bioanal Chem.,* 398:2481–2492

[51] Andersson, H.S. & Nicholls, I.A. (2000). Chapter 1:A historical perspective of the development of molecular imprinting *Techniques and Instrumentation in Analytical Chemistry,* 23: 1–19

[52] Arshady, R. & Mosbach, M. (1981) Synthesis of substrate-selective polymers by host–guest polymerization. *Macromol. Chem. Phys.-Makromol. Chem.*, 182:687–692.

[53] Andersson, L., Sellergren, B. & Mosbach, K. (1984) Imprinting of amino-acid derivatives in macroporous polymers. *Tetrahedron Letters*, 25:5211–5214

[54] Lasáková, M. & Jandera, P., (2009). Molecularly imprinted polymers and their application in solid phase extraction. *J Sep Sci.*, 32:799-812.

[55] Ali, W.H., Derrien, D., Alix, F., Pérollier, C., Lépine, O., Bayoudh, S., Chapuis-Hugon, F. & Pichon, V. (2010).Solid-phase extraction using molecularly imprinted polymers for selective extraction of a mycotoxin in cereals *J Chromatogr A.*, 1217:6668-73

[56] Zhang, Z., Xu, S., Li, J., Xiong, H., Peng, H., Chen, L. (2012) Selective solid-phase extraction of Sudan I in chilli Sauce by single-hole hollow molecularly imprinted polymers. *J. Agric. Food Chem.*, 60:180-187

[57] Vidal, J.C., Duato, P., Bonel, L. & Castillo, J.R. (2012) Molecularly imprinted on-line solid-phase extraction coupled with fluorescence detection for the determination of ochratoxin A in wheat samples. *Analytical Letters*, 45:51-62

[58] Zhou, W.H., Guo, X.C., Zhao, H.Q., Wu, S.X., Yang, H.H. & Wang, X.R. (2011) Molecularly imprinted polymer for selective extraction of domoic acid from seafood coupled with high-performance liquid chromatographic determination. *Talanta*, 84:777-782

[59] Castro López, Mdel M., Cela Pérez, M.C., Dopico García, M.S., López Vilariño, J.M., González Rodríguez, M.V. & Barral Losada, L.F. (2012). Preparation, evaluation and characterization of quercetin-molecularly imprinted polymer for preconcentration and clean-up of catechins. *Anal Chim Acta.*, 721:68-78

[60] De Smet, D., Kodeck, V., Dubruel, P., Van Peteghem, C., Schacht, E. & De Saeger, S. (2011). Design of an imprinted clean-up method for mycophenolic acid in maize. *J Chromatogr A.*, 1218:1122-30

[61] Li, J., Chen, H., Chen, H. & Ye, Y. (2012). Selective determination of trace thiamphenicol in milk and honey by molecularly imprinted polymer monolith microextraction and high-performance liquid chromatography. *J Sep Sci.*, 35:137-44

[62] Lin, Z., Cheng, W., Li, Y., Liu, Z., Chen, X. & Huang, C. (2012). A novel superparamagnetic surface molecularly imprinted nanoparticle adopting dummy template: An efficient solid-phase extraction adsorbent for bisphenol A. *Anal Chim Acta*, 720:71-6.

[63] Shi, X., Liu, J., Sun, A., Li, D. & Chen, J. (2012) Group-selective enrichment and determination of pyrethroid insecticides in aquaculture seawater via molecularly imprinted solid phase extraction coupled with gas chromatography-electron capture detection. *J Chromatogr A*, 1227:60-6

[64] Luo, X., Zhan, Y., Tu, X., Huang, Y., Luo, S. & Yan, L. (2011) Novel molecularly imprinted polymer using 1-(α-methyl acrylate)-3-methylimidazolium bromide as functional monomer for simultaneous extraction and determination of water-soluble acid dyes in wastewater and soft drink by solid phase extraction and high performance liquid chromatography. *J Chromatogr A*, 1218:1115-21.

[65] Qi, X., Yang, X., Wang, B., Yang, H. & Deng, A. (2012) Development of a broad selective molecularly imprinted polymers-based solid phase extraction of contraceptive drug levonorgestrel from water samples. *Intern. J. Environ. Anal. Chem.*, 92:268–278

[66] Shen, Z.L., Yuan, D., Su, Q.D., Zhang, H., Wang, J., Zhu, J.H. & Liu, Y.M. (2011) Selective solid-phase extraction using molecularly imprinted polymer for analysis of methamidophos in water and soil samples. *Biosci Biotechnol Biochem.*, 75:473-9.

[67] He, J., Lv, R., Cheng, J., Li, Y., Xue, J., Lu, K. & Wang, F. (2010). Preparation and characterization of molecularly imprinted microspheres for dibutyl phthalate recognition in aqueous environment. *J Sep Sci.*, 33:3409-14.

[68] Kueseng, P., Noir, M.L., Mattiasson, B., Thavarungkul, P. & Kanatharana, P. (2009). Molecularly imprinted polymer for analysis of trace atrazine herbicide in water. *J Environ Sci Health B.*, 44:772-80.

[69] Fu, X.W., Wu, Y.J., Qu, J.R. & Yang, H. (2011). Preparation and utilization of molecularly imprinted polymer for chlorsulfuron extraction from water, soil, and wheat plant. *Environ Monit Assess.*, 2011 Jul 30.

[70] Núñez, L., Turiel, E., Martin-Esteban, A. & Tadeo, J.L. (2010) Molecularly imprinted polymer for the extraction of parabens from environmental solid samples prior to their determination by high performance liquid chromatography-ultraviolet detection. *Talanta*. 80:1782-8.

[71] Hu, Y., Liu, R., Zhang, Y. & Li G. (2009) Improvement of extraction capability of magnetic molecularly imprinted polymer beads in aqueous media via dual-phase solvent system. *Talanta*, 79:576-82

[72] Qiao, J., Yan, H., Wang, H. & Lv, Y. (2011) Determination of ofloxacin and lomefloxacin in chicken muscle using molecularly imprinted solid-phase extraction coupled with liquid chromatography. *J Sep Sci.*, 34:2668-73

[73] Zamora, O., Paniagua, E.E., Cacho, C., Vera-Avila, L.E. & Perez-Conde, C. (2009) Determination of benzimidazole fungicides in water samples by on-line MISPE-HPLC. *Anal Bioanal Chem.*, 393:1745-53

[74] Watabe, Y., Kondo, T., Morita, M., Tanaka, N., Haginaka, J. & Hosoya K. (2004) Determination of bisphenol A in environmental water at ultra-low level by high-performance liquid chromatography with an effective on-line pretreatment device. *J Chromatogr A*, 1032:45-9

[75] Anastassiades, M., Lehotay, S.J., Stajnbaher, D. & Schenck, F.J. (2003). Fast and easy multiresidue method employing acetonitrile extraction/partitioning and "dispersive solid-phase extraction" for the determination of pesticide residues in produce. *J. AOAC Int.*, 86:412-431.

[76] Walorczyk, S. & Drozdzynski, D. (2011). Development and validation of a routine multiresidue method for determining 140 pesticides in fruits and vegetables by gas chromatography/tandem quadrupole mass spectrometry. *J. AOAC Int.*, 94:1625-1642.

[77] Wang, J., Chow, W. & Cheung, W. (2011). Application of a Tandem Mass Spectrometer and Core-Shell Particle Column for the Determination of 151 Pesticides in Grains. *J. Agric. Food Chem.*, 59:8589-8608.

[78] Marchis, D., Ferro, G.L., Brizio, P., Squadrone, S. & Abete, M.C. (2012). Detection of pesticides in crops: A modified QuEChERS approach. *Food Control*, 25:270-273.

[79] Kmellar, B., Abranko, L., Fodor, P. & Lehotay, S.J. (2010). Routine approach to qualitatively screening 300 pesticides and quantification of those frequently detected in fruit and vegetables using liquid chromatography tandem mass spectrometry (LC-MS/MS). *Food Addit. Contam.*, 27:1415-1430.

[80] Koesukwiwat, U., Lehotay, S.J., Miao, S. & Leepipatpiboon, N. (2010). High throughput analysis of 150 pesticides in fruits and vegetables using QuEChERS and low-pressure gas chromatography-time-of-flight mass spectrometry. *J. Chromatogr. A*, 1217:6692-6703.

[81] Broszat, M., Welle, C., Wojnowski, M., Ernst, H. & Spangenberg, B. (2010). A versatile method for quantification of aflatoxins and ochratoxin A in dried figs. *J. Planar Chromatogr. --Mod. TLC*, 23:193-197.

[82] Anastassiades, M. (2005) A Mini-Multiresidue Method for the Analysis of Pesticide Residues in Low-Fat Products. Available:
http://www.restek.com/pdfs/quechers_en_oct2005.pdf

[83] Barker, S.A. (2000). Applications of matrix solid-phase dispersion in food analysis. *J. Chromatogr. A*, 880:63-68.

[84] Barker, S.A. (2000). Matrix solid-phase dispersion. *J. Chromatogr. A*, 885:115-127.

[85] Jeong, I.S., Kwak, B.M., Ahn, J.H. & Jeong, S.H. (2012). Determination of pesticide residues in milk using a QuEChERS-based method developed by response surface methodology. *Food Chem.* Vol (in press.):doi:10.1016/j.foodchem.2012.01.004-

[86] Anastassiades, M., Scherbaum, E., Tasdelen, B. & Stajnbaher, D. (2007). Recent developments in QuEChERS methodology for pesticide multiresidue analysis. *Pestic. Chem.* 439-458.

[87] Lee, J.M., Park, J.W., Jang, G.C. & Hwang, K.J. (2008). Comparative study of pesticide multi-residue extraction in tobacco for gas chromatography-triple quadrupole mass spectrometry. *J. Chromatogr. A*, 1187:25-33.

[88] Li, L., Li, W., Qin, D., Jiang, S. & Liu, F. (2009). Application of graphitized carbon black to the QuEChERS method for pesticide multiresidue analysis in spinach. *J. AOAC Int.*, 92:538-547.

[89] Lehotay, S.J., Mastovska, K. & Lightfield, A.R. (2005). Use of buffering and other means to improve results of problematic pesticides in a fast and easy method for residue analysis of fruits and vegetables. *J. AOAC Int.*, 88:615-629.

[90] Anastassiades, M., Mastovska, K. & Lehotay, S.J. (2003). Evaluation of analyte protectants to improve gas chromatographic analysis of pesticides. *J. Chromatogr. A*, 1015:163-184.

[91] Lee, SW., Choi, J.H., Cho, S.K., Yu, H.A., Abd El-Aty, A.M. & Shim, J.H. (2011). Development of a new QuEChERS method based on dry ice for the determination of 168 pesticides in paprika using tandem mass spectrometry. *J. Chromatogr. A*, 1218:4366-4377.

[92] Smoker, M., Tran, K. & Smith, R.E. (2010). Determination of Polycyclic Aromatic Hydrocarbons (PAHs) in Shrimp. *J. Agric. Food Chem.* 58:12101-12104.

[93] Forsberg, N.D., Wilson, G.R. & Anderson, K.A. (2011). Determination of Parent and Substituted Polycyclic Aromatic Hydrocarbons in High-Fat Salmon Using a Modified QuEChERS Extraction, Dispersive SPE and GC-MS. *J. Agric. Food Chem.*, 59:8108-8116.
[94] Mastovska, K. &Lehotay, S.J. (2006).Rapid Sample Preparation Method for LC-MS/MS or GC-MS Analysis of Acrylamide in Various Food Matrices.J. Agric. Food Chem., 54:7001-7008.
[95] Martinez Vidal, J.L., Frenich, A.G., Aguilera-Luiz, M.M. & Romero-Gonzalez, R. (2010). Development of fast screening methods for the analysis of veterinary drug residues in milk by liquid chromatography-triple quadrupole mass spectrometry. *Anal. Bioanal. Chem.* 397:2777-2790.
[96] Norli, H.R., Christiansen, A. & Deribe, E. (2011). Application of QuEChERS method for extraction of selected persistent organic pollutants in fish tissue and analysis by gas chromatography mass spectrometry. *J. Chromatogr. A*, 1218:7234-7241.
[97] Alarcon Flores, M.I., Romero-Gonzalez, R., Frenich, A.G. & Vidal, J.L.M. (2011). QuEChERS-based extraction procedure for multifamily analysis of phytohormones in vegetables by UHPLC-MS/MS. *J. Sep. Sci.*, 34:1517-1524.
[98] Stubbings, G. & Bigwood, T. (2009). The development and validation of a multiclass liquid chromatography tandem mass spectrometry (LC-MS/MS) procedure for the determination of veterinary drug residues in animal tissue using a QuEChERS (QUick, Easy, CHeap, Effective, Rugged and Safe) approach. *Anal. Chim. Acta*, 637:68-78
[99] Whelan, M., Kinsella, B., Furey, A., Moloney, M., Cantwell, H., Lehotay, S.J. & Danaher, M. (2010). Determination of anthelmintic drug residues in milk using ultra high performance liquid chromatography-tandem mass spectrometry with rapid polarity switching. *J. Chromatogr. A*, 1217:4612-4622.
[100] Karageorgou, E.G. & Samanidou, V.F. (2011). Development and validation according to European Union Decision 2002/657/EC of an HPLC-DAD method for milk multi-residue analysis of penicillins and amphenicols based on dispersive extraction by QuEChERS in MSPD format. *J. Sep. Sci.*, 34:1893-1901.
[101] Pinto, C.G., Laespada, M.E.F., Martin, S.H., Ferreira, A.M.C., Pavon, J.L.P. & Cordero, B.M. (2010). Simplified QuEChERS approach for the extraction of chlorinated compounds from soil samples. *Talanta*, 81:385-391.
[102] Rudewicz, P.J. (2011). Turbulent flow bioanalysis in drug metabolism and pharmacokinetics. *Bioanalysis*, 3:1663-1671.
[103] Chassaing, C. & Robinson, S. (2009). Turbulent Flow Chromatography. an Evolving Solution for Bioanalysis. *Chromatography today*, 2:20-24.
[104] Edge, T. (2003). Turbulent flow chromatography in bioanalysis. Handb. *Anal. Sep.*, 4:91-128.
[105] Oberhauser, C.J., Niggebrugge, A.E., Lachance, D., Takarewski, J.J., Pegram, M.M. & Quinn, H.M. (2000). Turbulent-flow LC for LC-MS and LC-MS-MS bioanalysis. *LC-GC*, 18:716-724.
[106] Quinn, H.M. & Takarewski, J.J. (1997). Int Patent Number WO97/16724.
[107] Michopoulos, F., Edge, A.M., Theodoridis, G. & Wilson, I.D. (2010). Application of turbulent flow chromatography to the metabonomic analysis of human plasma: comparison with protein precipitation. *J. Sep. Sci.*, 33:1472-1479.

[108] Xin, G.Z., Zhou, J.L., Qi, L.W., Li, C.Y., Liu, P., Li, H.J., Wen, X.d. & Li, P. (2010). Turbulent-flow chromatography coupled on-line to fast high-performance liquid chromatography and mass spectrometry for simultaneous determination of verticine, verticinone and isoverticine in rat plasma. *J. Chromatogr. B: Anal. Technol. Biomed. Life Sci.*, 878:435-441.

[109] Herman, J.L., Poster n. 72, 27th Montreux Symposium on LC/MS, Montreux, Switzerland, November 2010.

[110] Mueller, D.M., Duretz, B., Espourteille, F.A. & Rentsch, K.M. (2011). Development of a fully automated toxicological LC-MS^n screening system in urine using online extraction with turbulent flow chromatography. *Anal. Bioanal. Chem.*, 400:89-100.

[111] Presta, M.A., Bruyneel, B., Zanella, R., Kool, J., Krabbe, J.G. & Lingeman, H. (2009). Determination of Flavonoids and Resveratrol in Wine by Turbulent-Flow Chromatography-LC-MS. *Chromatographia,* 69:S167-S173

[112] Perez, F., Llorca, M., Farre, M. & Barcelo, D. (2012). Automated analysis of perfluorinated compounds in human hair and urine samples by turbulent flow chromatography coupled to tandem mass spectrometry. *Anal. Bioanal. Chem.*, 402:2369-2378.

[113] Takino, M., Daishima, S. & Nakahara, T. (2003). Determination of perfluorooctane sulfonate in river water by liquid chromatography/atmospheric pressure photoionization mass spectrometry by automated on-line extraction using turbulent flow chromatography. *Rapid Commun. Mass Spectrom.*, 17:383-390.

[114] Segura, P.A., Gagnon, C. & Sauve, S. (2009). Application of turbulent flow chromatography load columns for the on-line analysis of anti-infectives in wastewaters. *Chromatographia*, 70:239-245.

[115] Asperger, A., Efer, J., Koal, T. & Engewald, W. (2002). Trace determination of priority pesticides in water by means of high-speed on-line solid-phase extraction-liquid chromatography-tandem mass spectrometry using turbulent-flow chromatography columns for enrichment and a short monolithic column for fast liquid chromatographic separation. *J. Chromatogr. A,* 960:109-119.

[116] Kinsella, B., O'Mahony, J., Malone, E., Moloney, M., Cantwell, H., Furey, A. & Danaher, M. (2009). Current trends in sample preparation for growth promoter and veterinary drug residue analysis. *J. Chromatogr. A,* 1216:7977-8015.

[117] Marazuela, M.D. & Bogialli, S. (2009). A review of novel strategies of sample preparation for the determination of antibacterial residues in foodstuffs using liquid chromatography-based analytical methods. *Anal. Chim. Acta,* 645:5-17.

Chapter 5

Analytical Methods for Quantification of Drug Metabolites in Biological Samples

Robert Roškar and Tina Trdan Lušin

Additional information is available at the end of the chapter

1. Introduction

The study of the metabolic fate of drugs is an essential and important part of the drug development process, research of drug metabolism pathways, drug drug interactions, drug-herb interactions, influence of genetic polymorphisms and other factors that influence the phase I and/or II metabolism of a drug. Different *in vitro* methods, from subcelullar to organ range, and *in vivo* studies are applied for the clarification of drug metabolism. The analysis of metabolites in complex biological matrices is a challenging task therefore several analytical methods for qualification and quantification of drug metabolites are used. Liquid chromatography coupled with mass spectrometry (LC-MS) has become the most powerful analytical tool for screening and identification of drug metabolites in biological matrices. However, adequate sample preparation is a key prerequisite aspect of successful quantitative and qualitative bioanalysis. Different approaches for metabolite quantification in biological samples from direct quantification, indirect quantification through parent drug after metabolite hydrolysis to quantification supported by using response factors between drug and their metabolites are often used. The most frequently used method for quantification is liquid chromatography coupled to different detectors such as mass spectrometer or UV detector. The LC-MS/MS methods are considered as most appropriate for determination of drugs and their metabolites and are also best suited for high throughput analysis. However, in LC-MS/MS assays, matrix effect and selection of suitable internal standards should be adequately addressed.

2. Background of drug metabolism

The study of the metabolic fate of drugs is an essential and important part of the drug development process. During drug evaluation the research of drug metabolism is of high importance especially when metabolites are pharmacologically active or toxic or when a drug is extensively metabolized [1]. Interindividual differences in drug metabolism also lead to the research of factors that affect drug metabolism [2, 3]. Moreover, a metabolism of toxic substances is also frequently investigated [4].

In early discovery, drug metabolism input provides a basis for choosing chemical structures and lead compounds with desirable drug metabolism and pharmacokinetic (DMPK) or safety profiles [5, 6]. It is the fact that the shift of the rate of drug attrition from 40% in 1990 to 10% in 2000 was due to increased efforts in applying DMPK principles for drug development. Beside traditional drug metabolism research that focuses on absorption, distribution, metabolism and excretion *in vitro* and *in vivo* studies, the knowledge about pharmacogenetics, pharmacogenomics and transporters brought many advances in drug metabolism research [5]. For the feasibility to successfully monitor the drug metabolism, suitable bioanalytical methods have to be developed and validated. Studies of metabolic fate of drugs in living systems may be divided into three areas: 1) elucidation of biotransformation pathways, 2) determination of pharmacokinetics of the parent drug and/or its primary metabolites and 3) identification of chemically-reactive metabolites that are important in drug-induced toxicity [7].

Metabolism is a process of biotransformation when drugs are transformed into a different chemical form by enzymatic reactions. Mainly, metabolism increases drug hydrophilicity and decreases the toxicity and activity of most drugs. On the other hand, the biotransformation reactions could lead to bioactivation of drugs in which case the metabolite is more toxic and/or more active than the parent drug (reactive metabolite formation) [8]. The mechanism of bioactivation of drugs may be classified into following categories: biotransformation to stable but toxic metabolites, biotransformation to electrophiles, biotransformation to free radicals and formation of reactive oxygen metabolites. Additionally, bioactivations are also the transformations of a prodrug, promoiety or bioprecursor prodrug to a more effective metabolite [9]. Prodrug approach is commonly used in order to overcome the poor bioavailability of the active form of the drug. In case when prodrug consists of two pharmacologically active drugs that are coupled together in a single molecule it is called promoiety. Another type of prodrug is a bioprecursor drug which does not contain a carrier or promoiety, but results from a molecular modification of the active agent itself [9].

There are several factors influencing drug metabolism such as genetic, physiologic, pharmacodynamic and environmental factors. CYP2D6, CYP2C19, CYP2C9, CYP3A4, CYP3A5 are enzymes that are responsible for metabolism of many marketed drugs and are also highly polymorphic [10]. Many non-cytochrome P450 drug metabolizing enzymes also play important role in the metabolism of a variety of drugs. Among them polymorphisms of thiopurine methyltransferase (TPMT), butyrylcholinesterase, N-acetyltransferase (NAT) and UDP-glucuronosyltransferase (UGT) influence the metabolism of drugs [11]. Different physiological factors such as age, sex, disease state, pregnancy, exercise, circadian rhythm and starvation lead to the impaired metabolism among subjects and should be taken into consideration when evaluating the drug metabolism. Dose, frequency, route of administration, tissue distribution and protein binding of the drug affect its metabolism. Moreover, environmental factors such as environmental chemicals, co-administered drugs, tobacco, smoking, alcohol drinking and dietary constituents may change not only the kinetics of enzyme reaction but also the whole pattern of metabolism, thereby altering the bioavailability, pharmacokinetics, pharmacologic activity or the toxicity of the drug [10, 11].

3. Drug metabolic pathways

Drugs are metabolized by different reactions that are classified into two groups: phase I and phase II. Phase I reactions include oxidation, reduction and hydrolysis. The function of phase I reactions is to introduce a new functional group within a molecule, to modify an existing functional group or to expose a functional group that is a substrate for phase II reactions. Phase I reactions are responsible for enhancement of drugs' hydrophilicity and consequently facilitate the excretion. Phase II reactions represent conjugating reactions and mainly further increase the hydrophilicity and facilitate the excretion of metabolites from the body [10]. Enzymes that catalyze phase I reactions include microsomal monooxygenases (cytochrome P450, flavin-dependent monooxygenase) and peroxidases, cytosolic and mitochondrial oxidases, reductases and hydrolytic enzymes. Cytochrome P450 enzymes may catalyze aliphatic hydroxylation, N-, O-, S-dealkylation, oxidative dehalogenation, epoxidation [6]. The participation (%) of hepatic CYP450 isoforms in the metabolism of clinically important drugs is as follows: 3A4/5 (36%), 1A1 (3%), 1A2 (8%), 2B6 (3%), 2C8/9 (17%), 2C18/19 (8%), 2D6 (21%), 2E1 (4%) [10]. Flavin-dependent monooxygenase, a flavoprotein, is a microsomal monooxygenase that is not dependent on cytochrome P450. It is capable of oxidizing nucleophilic nitrogen and sulfur atoms [6, 10]. Other typical phase I oxidation enzymes are monoamineoxidase (MAO), diamineoxidase (DAO), cyclooxygenase (COX), alcohol dehydrogenase (ADH), aldehyde dehydrogenase (ALDH), molybdenum hydroxylase (include aldehyde oxidase, xanthine oxidase and xanthine dehydrogenase). In addition to promoting oxidative metabolism, cytochrome P450 enzymes may also catalyze reductive biotransformation reactions for the reduction of azo and nitro compounds to primary amines [10, 12]. Hydrolytic enzymes that consist of non-specific esterases and amidases are also a member of phase I enzymes of metabolism [6, 10].

Phase I reactions may be followed by phase II reactions; however preceding phase I reactions are not a prerequisite. Phase II enzymes are highly capable of polarizing lipophilic drugs through conjugation with a polar substrate that facilitates excretion [13]. Contrary to phase I reactions, phase II reactions demand energy to drive the reaction. Energy is usually consumed to generate a cofactor or an activated intermediate then utilized as co-substrate [6]. Phase II reactions are catalysed by UDP-glucuronosyltransferases (UGT), sulfotransferases (SULT), N-acetyltransferases (NAT), glutathione-S-transferases (GST) and methyltranserases [6, 10, 13]. Of the conjugating reactions glucuronidation, which catalyzes the transfer of glucuronic acid to aliphatic and aromatic compounds, is the most important. UGTs are able to form O-, N- and S-glucuronides and require uridine diphosphate glucuronic acid for glucuronide formation [6, 10]. SULT is the enzyme responsible for the formation of sulfate esters in the presence of co-substrate 3′-phospohoadenosine-5′-phosphosulfate (PAPS). Aromatic amines, hydrazines, sulfonamides and certain aliphatic amines are biotransformed to amides in a reaction catalyzed by N-acetyltransferase and utilize acetyl coencyme A as cofactor [6]. Another important conjugating reaction is a conjugation with glutathione which is present in many cells at high concentrations. Glutathione conjugation captures reactive electrophiles and transforms them to stable, often non-toxic tioethers [6]. Methylation is a process that results in a formation of O-, N- and S-methylated products by the transfer of methyl group from methionine [10].

4. Models for evaluation of drug metabolism

In this chapter different *in vitro* and *in vivo* models for the evaluation of drug metabolism are presented. Advantages and disadvantages of subcellular fractions (microsomes, recombinant enzymes, cytosolic liver fractions, liver S9 fraction), cellular fractions (isolated hepatocytes, immortal cell lines, liver slices, perfused liver), *in vivo* animal and human studies will be presented.

Biotransformation occurs in liver, intestine, kidney, lungs, brain, nasal epithelium and skin. Since liver is the most important organ for drug metabolism [14, 15] the liver-based *in vitro* technologies for evaluation of drug metabolism are presented below. *In vitro* models that range from whole cell system to enzyme preparations are now increasingly applied for quantitative and qualitative assessment in preclinical drug development, post-approval routine checks, identification of metabolic determinant factors and prediction of drug-drug, herb-drug and food-drug interactions [15].

4.1. Recombinant human CYP and UGT enzymes (supersomes, baculosomes)

Recombinant human CYP and UGT enzymes have proven to be a useful tool in *in vitro* biotransformation studies. This *in vitro* model, referred to also as supersomes or baculosomes, is produced by transfection of insect cells with cDNA for human CYP and UGT by baculo virus, namely insect cells lack endogenous CYP and UGT activity. The advantage of this system is that enzyme activity of one single CYP or UGT isoform is expressed and therefore the assessment of individual metabolic enzyme and its contribution to the metabolic pathway could be performed. Additionaly, this *in vitro* system could be used also for the evaluation of drug-drug interactions. Moreover, due to availability of supersomes with different CYP and UGT genotypes, the influence of different polymorphisms on drug biotransformation could be estimated. Currently, all common human CYPs and UGTs co-expressed with NADPH-cytochrome P450 reductase are commercially available. The disadvantage of this *in vitro* model is the latency of glucuronidation because the active site of UGT is shielded behind a hydrophobic barrier. To resolve this problem a pore-forming agents such as alamethicin are used [14-18].

When performing the experiment with supersomes, the experiment with control nontransfected supersomes should be conducted. A NADPH regenerating system (NRS), which consists of β-NADPH, glucose-6-phosphate and glucose-6-phosphate dehydrogenase, or NADPH is required in the incubation for the evaluation of CYP activity and uridine diphospoglucuronic acid (UDPGA) has to be added as a cofactor when evaluating UGT enzyme activity [14-16].

4.2. Human liver microsomes (HLM)

HLM are vesicles of hepatocyte endoplasmic reticulum obtained by differential centrifugation of liver preparations (homogenates) from fresh human liver, liver slices, liver cell lines and primary hepatocytes. This subcellular fraction is a rich source of following enzymes: cytochrome P450s, flavin-monooxigenase (FMO), carboxyl esterases, epoxyde hydrolase and UGTs. Therefore, HLM are most frequently utilized *in vitro* model in drug metabolic profiling and drug interaction studies. Moreover, the influence of specific isoenzymes is studied using liver microsomes in the presence of specific inhibitors. There are interindividual variations in the activity of human liver microsomes; therefore they can be utilized also to study interindividual variability. In case of general estimation of drug metabolism, pooled microsomes from a large bank of individual liver tissues can be used to overcome the influence of interindividual variability. Microsomes from other human organs (intestine, kidney, lung) [19] are also available and are utilized to evaluate extrahepatic metabolism. Additionally, gender-specific microsomes are available for the estimation of gender-based discrepancies in drug biotransformation. In drug discovery process HLM are used for metabolite identification, evaluation of interspecies differences in drug metabolism, prediction of *in vivo* clearance, reaction phenotyping and metabolic pathway identification [14-18, 20].

NADPH or NRS is required in the incubation for the estimation of CYP activity. In order to evaluate the UGT activity UDPGA and alamethicin (pore-forming reagent) are required [14-16].

The advantages of HLM are ease of use, low costs, best-characterized *in vitro* model for estimation of drug biotransformation, easy storage, appropriate for studying of interindividual and population-based variation, long term storage, provide qualitative estimations of *in vitro* drug metabolism, convenient tool for high throughput screening of compounds, appropriate for lead compound optimization studies and drug interaction studies. However, some disadvantages of HLM also exist. HLM are not appropriate for quantitative estimation of drug biotransformation because of absence of enzymes like NAT, GST and SULT and cofactors needed. This limits the expected metabolic competition and formation of some *in vivo* present metabolites. Another drawback is a very difficult assessment of the fraction of drug bound to plasma proteins versus to microsomes which is an important factor in the estimation of *in vivo* biotransformation [14-16, 18].

4.3. Cytosolic fraction

Cytosolic fraction is an *in vitro* model that has not been used very often so far. Like HLM, cytosol is produced by differential centrifugation of liver homogenate. Soluble enzymes of phase II such as NAT, GST, SULT, carboxylesterase, soluble epoxide hydrolase, diamine oxidase, xanthine oxidase and alcohol dehydrogenase are expressed in cytosolic fraction, but only first three are expressed at higher concentration. This *in vitro* model requires cofactors like acetyl coA, dithiothreitol and acetyl coA-regenerating system for NAT, PAPS for SULT, glutathione for GST activity [14-16, 18].

The main advantage is the presence of only aforementioned enzymes at higher concentrations than in liver S9 fraction. The biotransformation by NAT, GST or SULT can be studied separately or in combination depending on the cofactors added. The main disadvantage is the absence of UGT and therefore glucuronidation cannot be studied by this model [14-16, 18].

4.4. S9 fractions

S9 fraction contains both microsomal and cytosolic fractions and consequently expresses a wide range of metabolic enzymes – CYP, FMO, carboxylesterases, epoxide hydrolases, UGT, SULT, methyl transferases, acethyltransferases, GST and others. This *in vitro* model could be employed for metabolic, toxicity and mutagenicity studies. Similar to upper mentioned *in vitro* models the addition of cofactors is needed; NADPH or NRS for CYP, UDPGA for UGT, acetyl coA, dithiothreitol and acetyl coA-regenerating system for NAT, PAPS for SULT and glutathione for GST [14-16, 18, 20].

The main advantage over microsomes and cytosolic fraction is a more complete representation of the metabolic profile due to the presence of phase I and phase II enzymes. However, a disadvantage is the overall lower enzyme activity in the S9 fraction compared to microsomes and cytosol, which may leave some metabolites unnoticed [14-16, 18].

4.5. Cell lines

This *in vitro* model is less popular than other described models due to dedifferentiated cellular characteristics and lack of complete expression of all families of metabolic enzymes. The sources of cell lines are primary tumors of liver parenchyma. Currently available cell lines are Hep G2, Hep 3B, SNU-398, SNU-449, SNU-182, SNU-475, BC2, PLC/PRE/5, C3A, SK-Hep-1 and among them Hep G2 cell line is most frequently used for biotransformation studies. The metabolic activity of liver cell lines is generally low compared to freshly isolated human hepatocytes. Metabolic activity of some metabolic enzymes is even not detected. The problem of low activity could be partly overcome by the pretreatment of cell lines by inducers of various metabolic enzymes. But still the induced activity is below the enzymatic activity in freshly isolated human hepatocytes. Liver cell lines require appropriate culture medium, whose composition significantly influences the metabolic activity. The described *in vitro* model is easy to culture and have stable enzyme concentration. On the other hand, the absence or low expression of most important phase I and phase II drug metabolizing enzymes limits the application of this *in vitro* model. Moreover, metabolites are not easily detected and it is difficult to investigate individual enzymes due to their low expression level [14-16].

4.6. Transgenic cell lines

Transgenic cell line is a cell line that recombinantly expresses human phase I and/or phase II enzymes. All important human CYPs and UGTs have been expressed in this way to overcome the limitations of liver cell lines. Cell lines may be transfected at high efficiency using protoplast fusion. The main advantages are the ease of culturing, high expression of CYP and UGT isoenzymes, possibility to study single enzyme reactions and the influence of one isoenzyme or a combination of a number of isoenzymes. This *in vitro* model can also be used in the study of metabolite structures, pharmacological elucidation and to assess drug-drug interactions. The main drawback is that only one or a few of isoenzymes are expressed, therefore the complete *in vivo* situation cannot be reflected. Moreover, transgenic cell lines are more expensive than other enzyme-based technologies [14-16, 18].

4.7. Hepatocytes

Hepatocytes are well-established, well-characterized and frequently used *in vitro* model in drug biotransformation research. This *in vitro* model could be employed for the evaluation of metabolic stability, metabolite profiling and identification, drug efficacy, hepatic proliferation, hepatotoxicity and drug-drug interactions. Phase I and phase II drug metabolism pathways can be studied by the use of primary hepatocytes and cultured hepatocytes. Like with microsomes interindividual variation can be observed with hepatocytes. This can be overcome by using mixture of hepatocytes from different donors. Cryopreservation of hepatocytes offers many advances in the experimentation, namely activity of most phase I and phase II enzymes is retained.

Primary hepatocytes are obtained by collagenase perfusion of whole liver or a part of liver. This *in vitro* system has strong resemblance of *in vivo* situation due to heterogeneity of enzyme expression in human liver and preservation of drug metabolizing enzymes at *in vivo* levels. Another advantage of primary hepatocytes is the ease of use and high throughput. The important disadvantage is the drop of hepatocytes viability during incubation period (viable 2-4 hrs). Moreover, lack of liver non-hepatocyte cells which may be necessary for cofactor supply, lack of cell polarity, cell-cell and cell-matrix contacts limits the *in vivo* resemblance [14-18, 20].

After isolation, hepatocytes can be cultured in a monolayer in order to prolong the viability to 4 weeks. This characteristic in combination with the prolonged regulatory pathways allows the use of this *in vitro* model in studies of up-regulation or down-regulation of metabolic enzymes. However, cultured hepatocytes gradually lose viability and liver specific function. Many factors influence the morphology and functions of hepatocytes in culture: medium formulation, extracellular matrix, initial cell suspension and density, drug concentrations. Hepatocytes could also be cultured in a sandwich configuration where hepatocytes are placed between two layers of gelled extracellular matrix. This type of culture retains liver hepatocyte specific functions for a longer period [18, 20].

4.8. Liver slices

Liver slices and hepatocytes are the most physiologically relevant *in vitro* techniques used for quantitative and qualitative measurement of hepatic phase I and phase II metabolism of drugs due to full complement of enzymes and cofactors. High-precision tissue slicers (e.g. Krumideck slicer, Brenden-Vitron slicer) are used for the production of liver slices of uniform thickness (less than 250 μm). The advantage of liver slices over hepatocytes lies in the intact structure of liver tissue containing hepatic and non-hepatic cells, normal spatial arrangement and possibility of morphological studies. The described *in vitro* model allows higher throughput compared to isolated perfused liver. Another advantage is the non-requirement for digestive enzymes and consequently the preservation of intact tissue structure. Moreover, no addition of cofactors is needed for enzyme activity. However, some disadvantages of this model are known: decrease of CYP activity in short time due to impaired diffusion of nutrients and oxygen in the liver slice, damaged cells on the outer sides of the slice, inadequate tissue penetration of the test medium, short viability period (5 days), lack of optimal cryopreservation procedures and a need for expensive equipment [14-16, 18, 20].

4.9. Isolated perfused liver

Isolated perfused liver gives an excellent representation of the *in vivo* situation but it is not used frequently due to practical inconveniences. Normally animal liver tissue on a small scale is used, but never human liver tissue. The additional advantages of this *in vitro* model are also three-dimensional architecture, presence of hepatic and non-hepatic cell types, possibility to collect bile. The important disadvantages of this model are: poor reproducibility, functional integrity limited to 3 hours, difficult handling, poor perfusion of cells by nutrients and oxygen, low throughput and no availability of human liver. This model is useful only in case when bile secretion is the subject of research [14-16].

4.10. Animal and human *in vivo* studies

The identity of metabolites present in any matrix of animal or human provides essential information about the biotransformation pathways involved in the clearance of a drug. When the metabolite profiling of a parent drug is similar qualitatively and quantitatively between animal and human, we can assume that potential clinical risks of parent drug and metabolite have been adequately investigated during nonclinical studies. When a difference arises between *in vitro* and *in vivo* findings, the *in vivo* results should always take precedence over *in vitro* studies [21]. The FDA guidance encourages the identification of differences in drug metabolism between animals and humans as early as possible during the drug development process in order to find unique human metabolites and major metabolites [1, 21]. FDA defines that metabolites will need to undergo additional safety evaluation when steady-state systemic exposure to metabolite in humans exceeds 10% of parent drug exposure (disproportionate metabolite) [1].

The results of aforementioned *in vitro* studies can be correlated to *in vivo* situation and vice versa. This multidisciplinary approach of translational medicine yields an insight into complex mechanisms of drug disposition. The principle of translational medicine is presented on raloxifene, a selective estrogen receptor modulator, which exhibits quite large and unexplained interindividual variability in pharmacokinetics and pharmacodynamics [2, 3, 19, 22]. The gained knowledge about drug pharmacokinetics and pharmacodynamics insures a safer and more effective treatment strategy in the clinical setting.

5. Qualitative evaluation of metabolites

The known identity of metabolites is the prerequisite for a suitable metabolic assessment of drugs. Liquid chromatography coupled with mass spectrometry has become the most powerful analytical tool for screening and identification of drug metabolites in biological matrices. A short overview of analytical strategies for identification of metabolites will be provided. More information regarding metabolite identification can be found in following review articles [7, 23-27]. The selection of suitable LC-MS instrumentation is needed for qualitative evaluation of metabolites. Moreover, this issue is also important for quantitative evaluation of metabolites as discussed in section 8. Additionally, some examples for metabolite identification using LC-MS/MS will be provided in this section.

5.1. LC-MS instrumentation

5.1.1. Ionization techniques

A LC-MS ion source has the double role of eliminating the solvent from the LC eluent and producing gas-phase ions from the analyte. The application of atmospheric pressure ionization (API) methods has provided a breakthrough for the LC-MS systems and has brought it to the forefront of analytical techniques. Some ion sources such as API operate at atmospheric pressure where others like electron impact (EI) or chemical ionization (CI) operate in vacuum. While soft API interfaces, in particular electrospray, produce molecular ions with minimal fragmentation, high energy sources like EI mostly generate fragment ions. API techniques are most widely used for metabolite detection, identification and quantification [7, 28] due to the ability to operate at atmospheric pressure, good compatibility with reversed phase chromatography and generation of intact molecule ions at very high sensitivity. All three API techniques: electrospay ionization (ESI), atmospheric pressure chemical ionization (APCI) and atmospheric pressure photoionization (APPI) are complementary.

- **Electrospay ionization** is by far the preferred method for metabolite identification and quantification. It is the most universal technique for introducing the molecules into the gas phase and it is most gentle and therefore likely to yield an intact molecular ions. ESI is ideally suited for polar, ionic and thermally labile compounds such as drug metabolites; in particular glucuronides and others phase II metabolites. This technique requires ionization of analytes within solution prior to introduction into ion source and thus works best for fairy basic or acidic compounds. Depending on the voltage polarity, nebulised droplets trapping the ionized analyte will be positively or negatively charged. The reduction in size caused by solvent evaporation accounts for the increase in charge density in the droplet leading to its explosion when repulsive forces between charges exceed the cohesive forces of the droplet. This process occurs repeatedly until gas phase ions are produced [29]. Ions in solution are emitted into gas phase without application of heat making ESI suitable for analysis of thermo labile compounds. Many parameters, such as analyte and solution characteristics: pKa, analyte concentration, other electrolytes in solution, dielectric constant of the solvent, affect the ion formation process [7]. The effects of several mobile phase additives on the ionization efficiency have been reviewed [30] and will be discussed later (section 8). Depending on the chemical structure of an analyte, multiple-charged molecular ions can be formed, which is optimal for the analysis of biological macromolecules (e.g. proteins). Despite the numerous benefits of ESI, it suffers from a shortcoming in that it is susceptible to ion suppression effects from high concentrations of buffer, salts and other endogenous compounds in matrix solutions [23].
- **Atmospheric pressure chemical ionization** is more suited for less polar compounds. Certain classes of compound such as heavily halogenated analoges and highly aromatic compounds will run readily on APCI while giving no or a weak response on ESI [7]. APCI like ESI produces ions based on the API strategy, but thought a completely different process. Here, the liquid eluent is sprayed into heated chamber [450-550°C) where the high temperature of a nebulizer gas flow causes the immediate evaporation of the solvent and the analyte. In addition to volatility at the applied temperature, thermal stability of the analyte is also a prerequisite for the successful application of APCI (e.g. glucuronides may break down and appear in the form of protonated aglycone [31]. Ionization of analytes takes place in gas phase where due to high flux of electrons from corona discharge needle, solvent molecules initially react with electrons and form ions that produce protonated solvent ions through secondary reactions. These protonated solvent ions then transfer a proton to form protonated analytes. For efficient ionization, the employed mobile phase should be volatile and also amenable to gas phase acid-base reactions. APCI technique is less prone to ion suppression and provides a wider dynamic detection range than ESI due to ionization that occurs mainly in gas phase. Also, typically higher flow rate is used with APCI [1-2 mL/min) then that in conventional ESI (0.1-0.5 mL/min) [23].

- **Atmospheric pressure photoionization** is relatively new ionization method. This technique can be used for ionization of analytes that are not easily ionizable by ESI and APCI. APPI has similar application range as APCI but slightly extended toward nonpolar compounds [32]. The APPI ion source is very similar to APCI source, except the APCI corona discharge needle is replaced by photoionzation lamp. Depending on the analyte proton affinity relative to the composition of the mobile phase, either a radical molecular ion (typically for nonpolar compounds) or a protonated molecular ion (typically for polar compound) is obtained. APPI has a potencial in the analysis of drug metabolites but more research is needed to fully understand the important parameters and factors that affect the ionization efficiency [33].

5.1.2. Mass analyzers

The function of mass analyzer is the separation of ions formed in ionization source according to their different mass-to-charge (m/z) ratios. The quality of mass separation is characterized by the degree to which close m/z values can be separated in the mass analyzer. Mass analyzers are classified regarding resolution into low and high resolution instruments. The later ones are associated with another important parameter, mass accuracy, which allows determination of elemental formula of particular analyte. The selection of suitable analyzer is driven by the purpose of the analysis and the instrument performance but also depends on the instrument availability and cost effectiveness.

- **Triple quadrupole instruments (QQQ)** are the most common mass spectrometers in analytical laboratories, having most often been acquired for their evident strengths in high sensitivity quantitative analysis of known analytes. These instruments have been often applied also for metabolite identification due to wide availability and excellent tandem mass (MS/MS) properties. In QQQ, the first quadrupole filters ions of interest, the second quadrupole also called collision cell fragments these ions and further the fragment ions are filtered by third quadrupole before reaching the mass detector. Such QQQ configuration allows performing different scans such as full scan, product ion scan, precursor ion scan (PI), constant neutral loss scan (CNL), single ion monitoring (SIM) and selected reaction monitoring (SRM) or multiple reaction monitoring (MRM). PI and CNL are particularly useful in metabolite identification since both scanning modes do not require previous knowledge about the molecular weight of metabolites. High sensitivity for quantitative purposes is retained only when working in MRM mode, however, the detection sensitivity decreases dramatically when wide mass range is analyzed in a scanning mode. This is one of the major disadvantages of using QQQ for the screening of drug metabolites.
- **Ion trap instruments (IT)** are like QQQ relatively inexpensive and compatible with wide range of ionization interfaces. These analyzers utilize ion trap chamber where ions are trapped and then selectively ejected from the chamber. Additionally, the resonance excitation applied in the trap provides efficient dissociation of the precursor ions to product ions. IT provides more sensitivity for structural elucidation than QQQ due to

its better sensitivity in full scan mode and efficient dissociation of the precursor ions which allows multiple stages mass spectrometry (MS^n). Recently, to address classical ion traps (called also 3D IT) shortcomings of insufficient ion storage efficiency, capacity and deterioration of the mass spectrum and dynamic response range, linear IT has been developed [25]. The detection sensitivity in linear IT is at least two orders of magnitude higher than that in 3D IT. Because of these advantages, linear IT will probably in near future totally replace old 3D IT [23].

- **Triple quadrupole-linear ion traps (QTrap)** combine sensitive QQQ technology with high capacity of linear IT incorporating high trapping efficiencies. In this instrument, the last quadrupole of QQQ is replaced with a linear ion trap, which operates as a mass resolving quadrupole or a linear ion trap. This provides clearly increased metabolite screening capabilities compared to traditional IT or QQQ. QTrap enables high sensitivity, wide range mass scanning and MS^n together with QQQ capabilities, such as PI, CNL and very high sensitive MRM data acquisition.
- **Time of flight (TOF)** analyzers are the most suitable high resolution mass spectrometers for fast and cost-efficient metabolite identification. TOF are relatively simple and capable of recording all formed ions on a microsecond time scale offering high sensitivity detection. Ions are accelerated from the ion interface to a fixed kinetic energy and then pass through a field-free tube to the detector. The time needed for ion to reach the detector is proportional to its m/z ratio. TOF strength lies in its very high detection sensitivity when acquiring wide range data, enabling the simultaneous detection of data for all metabolites of interest in one run. High mass resolution and mass accuracy (< 3-5ppm) enable reliable and accurate identification of metabolites by determination of elemental formula of a metabolite. Additionally, the very high acquisition speed makes them ideal for fast chromatography [24].
- **Triple quadrupole-time of flight (Q-TOF)** instruments combine first mass filter and collision cell of QQQ with TOF as the second mass analyzer. These instruments can operate as true tandem MS while providing accurate mass of the product ions. Most modern Q-TOFs have good linear response and are therefore also suitable for quantitative purposes. However, TOF instruments have not the ability to perform positive/negative switching in one run [24].
- **Orbitrap** is another high resolution analyzer which is a hybrid composed of a linear IT and Fourier transform mass spectrometer. It is an effective alternative to the TOF instruments used for metabolite profiling. Orbitrap is capable of high sensitivity screening over wide mass range, MS^n and tandem mass spectrometry with accurate mass data for both parent and fragment ion. However, it is not suitable for fast chromatography because it suffers from a slow data acquisition [24].
- **Fourier transform-ion cyclotron resonance (FT-ICR)** is the third high resolution mass analyzer. The high sensitivity, accurate mass measurements, high mass resolution and MS/MS capabilities of FT-ICR make it attractive for structural determination of ions. However, the combined requirement of ultra-high vacuum system, superconducting magnets as well as sophisticated data system place the cost of these instruments beyond the means of most laboratories involved in drug metabolism studies [7].

5.2. Strategies for metabolite identification

MS methodology is the most suitable approach for metabolite identification as commonly low concentrations of drug metabolites are present in complex biological matrices. Appropriate LC-MS instrumentation is clearly critical to both, detection and structural elucidation, although alternative non-MS approaches may also be important in cases when MS data alone are not sufficient. Tandem mass spectrometry instruments are beside their key role for metabolite quantification also well suited for qualitative purposes. Tandem mass spectrometry experiments, which allow different scan mode possibilities, are by far most informational techniques for structural characterization of metabolites [23]. But these experiments require a set of injections to perform full scan and other scan analyses to identify metabolites of interest. The drive to more versatile and powerful instruments which can perform intelligent data dependent experiments has led to newer mass analyzers, such as high resolution Q-TOF instruments, which now dominate the metabolite identification field.

5.2.1. Full scan

The non selective nature of full mass scan acquisition enables detection of practically all ionizable metabolites and giving most complete information in terms of metabolite molecular mass. However, two major disadvantages arise by this approach. Firstly, detection sensitivity using QQQ decreases dramatically when wide mass range is scanned. This obstacle can be overcome by using IT analyzers as its full scan is much more sensitive or even better by using TOF instruments which additionally enable accurate mass determination [23]. In case when only QQQ is available, a practical approach may be applied to improve sensitivity; the whole mass range should be divided to narrow scanning ranges by performing multiple analyses of the same sample. Secondly, other non-metabolite matrix compound may interfere with obtained MS data. A common procedure for metabolite detection involves analysis of test and control samples what then allows subtraction of control sample data. This approach is less successful when complex biological samples, such as plasma and urine, are examined. Expected metabolites in studied samples may be predicted based on biotransformation pathways of parent drugs what enables focused search of these compounds. The most common changes in mass caused by biotransformation are shown in Table 1.

5.2.2. Precursor ion and constant neutral loss scan

PI and CNL are more specific approaches for identification of unknown metabolites. This scan mode is only possible for tandem mass spectrometers and therefore suffers at sensitivity like other QQQ scanning acquisitions. In PI scan mode, the second quadrupole mass filter is set to pass only the selected product ions, while the first quadrupole mass filter scans a range of m/z values. In CNL scan mode, both quadrupoles are scanning m/z values while the m/z difference between the quadrupoles is kept constant. Several phase II metabolites at fragmentation lose a distinct neutral group that can be used for specific identification of these conjugates. Glucuronides, sulfates and glutathione conjugates are often detected by CNL of m/z 176, 80 and 129, respectively. Typical PI for some drug conjugates in negative ionization mode like aliphatic sulfates, sulfonates and phosphates are m/z 97, 81 and 79, respectively [28]. Although PI and CNL provide high selectivity for identification of metabolites, the methods are based on predicted fragmentation behavior of metabolites what depends to some extent also on abilities of the analyst. Therefore, metabolites with unexpected fragmentation can be missed. Nevertheless, in combination with full scan data, PI and CNL is a powerful tool for metabolite identification.

Biotransformation	Change in molecular formula	Change in mass (Da)
Dehydration	- H_2O	-18
Demethylation	- CH_2	-14
Dehydrogenation	- H_2	-2
Hydrogenation	+ H_2	+2
Methylation	+ CH_2	+14
Hydroxylation	+ O	+16
Epoxidation	+ O	+16
S/N-oxidation	+ O	+16
Hydration	+ H_2O	+18
Dihydroxylation	+ O_2	+32
Acetylation	+C_2H_2O	+42
Sulfation	+SO_3	+80
Glucuronidation	+$C_6H_8O_6$	+176
Glutathione conjugation	+$C_{10}H_{15}O_6N_3S$	+305

Table 1. The nominal mass changes in biotransformation of drugs by common metabolic reactions [28, 34]

5.2.3. Product ion scan

Product ion scan is used for structural characterization of the detected metabolites. In product ion mode, a precursor ion (metabolite) is selected in first quadrupole, fragmented in collision cell and the product ions are then scanned in second quadrupole. Structural information is obtained by interpretation of the fragmentation patterns for both metabolite and parent drug. Complete structural characterization of metabolites may be hindered by the absence of useful product ions in tandem mass spectrometry. To obtain more specific structural data, the use of multistage (MS^n) scan by using ion trap instruments can be provided. The selected product ion can be selectively isolated and further fragmented, narrowing the potential sites of modification and providing a more complete assessment of the metabolite structure.

5.2.4. Multiple reaction monitoring

Although the use of PI and CNL data acquisition improves the selectivity of metabolite detection when comparing with full scan acquisition, all three approaches have reduced sensitivity. For this reason, specific MRM screening may serve as alternative approach for metabolite detection. MRM is the most appropriate acquisition method for quantification of analytes. In this mode, the first quadrupole is set to pass only the selected precursor ion that is fragmented in collision cell and usually the most abundant fragment (product ion) is then filtered in a second quadrupole. Monitoring of specific transition for each analyte yields a superior signal-to-noise ratio with significantly higher selectivity. Utilizing metabolism prediction and knowledge of the tandem mass fragmentation of the parent drug, the approach gives a significant increase in sensitivity and enables a wide range of potential MRM transitions to be targeted. Although the possibility to overlook metabolites remains the targeting MRM is a powerful alternative for metabolite detection when sensitivity is an issue. Single ion monitoring is is another option to overcome low sensitivity of QQQ screening techniques. SIM is less specific and sensitive acquisition compared to MRM but may provide advantages when the potential metabolite fragmentation pattern cannot be predicted correctly In this case a multiple SIM transitions of the predicted metabolites are performed, which are set accordingly to the expected nominal mass changes regarding to parent drug (Table 1).

5.2.5. High resolution mass spectrometry

The most widespread analyzer providing high mass accuracy (TOF, Orbitrap, FT-ICR) used in metabolite identification is TOF instrument. The specificity in the detection of metabolites with high resolution is significantly higher than that with unit resolution QQQ or IT instruments where the ion chromatograms can be recorded using a 0.1 mass unit window. The high selectivity provides also better sensitivity for the detection of metabolites. It was reported that detection limits for several drugs were 5-25 times better with accurate mass TOF, than with nominal mass TOF (same unit level than at QQQ) [28]. Accurate mass measurements enable to determine the elemental formula of metabolites. Moreover, exact mass shift enables the establishment of the change in molecular formula of the parent drug. For example, metabolites formed by hydroxylation and dehydrogenation (at same time) are, in this way separated from those formed via methylation, in spite that both reaction increase the molecular weight by 14 (Table 1) [24]. The benefit of reliable accurate mass measurements for structural elucidation of unknown metabolites is therefore extremely high. However, metabolites with the same exact mass cannot be distinguished by analyzers. In this case other approaches are needed. Ion mobility time-of-flight mass spectrometry (IM-MS), which separates ions on the basis of their m/z ratios as well as their interactions with a buffer gas, is very convenient. The main advantage of IM-MS is the potential for separation of metabolite isomers without chromatographic separation which makes it a powerful analytical tool for investigation of complex samples [35].

Q-TOFs are the key high resolution instruments for drug metabolism research. Q-TOF instruments provide sufficient mass resolution (up to 40,000) and accurate mass measurements (below 1 ppm). In addition, they can operate at relatively high scanning rates, which are considered as the main drawback of most of the Orbitrap based instruments. On the other hand, Orbitrap analyzers provide a resolving power of up to 100,000 with mass accuracy below 1 ppm. FT-ICR analyzers provide ultrahigh mass resolving power greater than 200,000 but high purchasing and maintenance cost are beyond financial capabilities of most routine laboratories [27].

5.2.6. Other approaches

Other approaches can be applied to provide specific structural information in cases when MS data are not sufficient to determine metabolite structure. Hydrogen/deuterium exchange LC-MS allows studying mechanisms of MS fragment ion formation and metabolic pathways of drugs, as well as differentiate the structures of isomeric metabolites [7]. Metabolites can be isolated and purified from the incubations, followed by structural analysis by NMR. Alternatively, LC-NMR analysis can be performed on biological samples with minimal sample processing but certain limitations occur with this technique, such as lower sensitivity compared with LC-MS and the requirements of relatively expensive deuterated buffers in mobile phase. More recently, LC-NMR has been coupled with MS which enables simultaneous metabolite structure elucidation [25]. Tentatively identified structure of metabolites may also be synthesized and LC-MS data for these compounds are compared with data from the actual metabolites.

5.3. Examples of metabolite identification

Tandem mass spectrometry is well suited for identification of phase II metabolites [36]. As example for this approach, the elucidation of three raloxifene glucuronides in urine as well as their identity confirmation after bioproduction by using QQQ is provided [37]. Chromatograms of each bioproduced glucuronide standard obtained in ESI positive full scan mode gave only one chromatographic peak where MS spectra of each peak showed strong molecular ions at m/z 650, 650 and 826 for two raloxifene monoglucuronides and diglucuronide, respectively. Nominal mass shift of 176, 176 and 2 x 176 Da compared to parent drug (m/z 474) is characteristic for the structure of monoglucuronide and diglucuronide metabolites (Table 1). Product ion scan showed the same mass spectra for both predicted monoglucuronides: fragmentation of the parent ion m/z 650 to 474 and 112. Product ion spectra confirmed also diglucuronide structure by two subsequent m/z 176 neutral losses from the parent molecular ion (m/z 826), giving fragments of monoglucuronide (m/z 650) and of parent raloxifene (m/z 474) as well as additional m/z 112 fragment of raloxifene (N-ethyl-piperidine). Additionally, constant neutral loss scan (m/z 176) and precursor ion scans (m/z 112 and 474) in urine sample have been performed. The analysis in all three cases gave three distinct peaks in chromatograms at retention times for the diglucuronide, and both monoglucuronides (data not shown, but same retention times as in Figure 1) confirming again the structure of metabolites.

Another important point had to be considered, good chromatography was needed in order to separate both monoglucuronides since they cannot be distinguished based on MS. Representative LC-MS/MS chromatogram, using MRM acquisitions for quantitative purposes of raloxifene and its three metabolites in urine sample is shown in Figure 1.

Identification of bisphenol A glucuronide and deuterated bisphenol A glucuronide in microsomal incubations [4] is another example. Twin peaks of metabolites with known mass difference [14] (Da in this case) are helpful for studying fragmentation paths. Product ion scan in ESI negative ionization mode for bisphenol A glucuronide (m/z 403) showed fragments m/z 227 (bisphenol A), 212 (bisphenol A fragment - loss of CH_3) and 113. The molecular ion of deuterated bisphenol A glucuronide fragments from m/z 417 to m/z 241 (deuterated bisphenol A), 223 (fragment - additional loss of CD_3) and 113. Fragment m/z 113, which is present in both cases represent a glucuronic acid fragment in negative ionization with subsequent loss of H_2O and CO_2.

In case of reactive metabolite studies there are typical approaches to identify glutathione conjugates: increased mass shift 305 Da according to the parent, constant neutral loss of pyroglutamic acid (m/z 129) in the positive ionization mode and/or precursor ion of m/z 272 in the negative ionization mode [34, 38]. Recently, an *in vitro* bioactivation study using these identification approaches has confirmed that bazedoxifene does not show the formation of glutathione conjugates compared to raloxifene what offers an improved safety profile of this third generation drug relative to other available SERMs [39].

The glucuronide metabolites may be also simply verified by using ß-glucuronidase which provides the conversion of the glucuronide to its aglycone (see next section). If the conversion is complete, this approach is valid for determination of the metabolite stock solution concentrations when small amounts of glucuronide standards are obtained or available [37].

However, for more demanding application QQQ is usually not satisfactory. Identification of phase I and phase II metabolites of two antineoplastic agents was demonstrated by use of Q-TOF [40]. In this study, 32 metabolites for dimefluron and 28 metabolites for benfluron were detected in the rat urine within 25 min chromatographic run. The identification of individual biotransformation was performed using high mass accuracy measurements for both full scan and tandem mass spectra by extracted ion chromatograms for expected masses of metabolites together with the information about characteristic neutral loss. Another study compared QQQ, linear IT (QTrap), TOF and Orbitrap instruments for identification of microsomal metabolites of verapamil and amitriptyline [41]. Only TOF found all 28 amitriptyline and 69 verapamil metabolites; both expected and unexpected. The TOF offered sensitivity and high mass resolution and also lowest overall time consumption together with the Orbitrap. Orbitrap also showed good mass resolution but was less sensitive, resulting in some metabolites not being observed. Approaches with QQQ and Q-Trap provided the highest amount of fragment ion data for structural elucidation, but being unable to produce very high important accurate mass data, they suffered from many false negatives and especially with the QQQ from very high overall time consumption.

6. Approaches for metabolite quantification

The demand for analyses of low-level drugs in complex biological samples has increased significantly in last years. New pharmaceuticals have typically high potency, so small doses are given and therefore the detection limits of these drugs and their metabolites are of great importance. Selective and sensitive analytical methods for the quantitative evaluation of these analytes are critical for the successful conduction of pharmacological studies. Metabolite quantification is always required when the metabolite is toxic or pharmacologically active or when the concentration of metabolite reaches or exceeds the parent drug concentration in plasma. Different approaches for metabolite determination in biological samples have been used which can be generally divided to direct quantification, indirect quantification through parent drug after metabolite hydrolysis or quantification supported by using response factors between drug and their metabolites. The key role in the selection of the particular approach is driven by the availability of suitable authentic standards. Hence some examples of metabolites production will be also shown here.

6.1. Direct quantification

Direct quantification is the most appropriate approach for metabolite determination in biological matrices but two major points need to be considered. Firstly, in general metabolites are much more hydrophilic than parent drug, especially glucuronides [34]. That fact has represented a hindrance for direct metabolite determination because chromatographic separation between these polar analytes and interfering matrix components could not be achieved in many cases. However, this problem has been overcome by advent of powerful liquid chromatography-tandem mass spectrometry instruments which allow direct quantification of these metabolites [42]. LC-MS/MS nowadays play predominant role in bioassays for pharmacokinetic and metabolism studies due to its inherent specificity, sensitivity and speed. Secondly, appropriate authentic standards are needed for reliable and accurate quantification in biological samples. Proper validation of analytical methods includes preparation of calibration and control samples in given biological matrices using suitable reference standards. Authentic metabolite standards are often not commercially available, particularly in the case of new drugs or drugs of abuse. Moreover, available metabolites may be very expensive and therefore not accessible for every research group, especially not in academic sphere. Furthermore, stable isotope labeled standards of metabolites, which are most convenient internal standards for LC-MS/MS analyses, are even less available and/or more expensive than unlabeled metabolites.

In such situation question may arise why not quantitate metabolites concentration based on parent drug calibration curve as this standard are freely accessible. Modified structure of metabolites may change the response to quite diverse extent among various liquid chromatography detection systems. Mass spectrometry using atmospheric pressure ionization sources is very prone to this issue as the intensity of the MS signal strongly depends on the analyte even at small structural changes. It has been reported that the response in ESI-MS differed by factor 25 for two oxidative isomeric metabolites with same chemical formula [43] or that no signal in contrary to parent drug has been observed for metabolite in positive ionization ESI. Detection of metabolite was in this case possible only in negative ionization mode [26]. However, with the commonly and easily used UV detection, the metabolites have often the same chromophore as the parent drug (but not always [40], hence giving similar response. But the main limitation of this technique in pharmacokinetic studies lies in not sufficient sensitivity and also in lower selectivity as some compounds does not have UV absorption at a wavelength to distinct it from the background. In contrast to UV, fluorescence and electrochemical detection can be very selective and sensitive. For electrochemical detection the response may also be very dependent on structure, especially for phase I metabolites which usually possess changed oxido-reductive properties compared to parent drug [34]. Beside that both detector systems are very specific what makes them of limited applicability.

Direct quantification can also be performed without suitable standards. For that purpose detectors need to give an equimolar response for all compounds of interest. Additionally, such detectors should be highly sensitive with wide dynamic range, robust and easy to use, compatible with reverse phase gradient elution and not prone to matrix interferences, namely give a response independent of compound [44]. Although there are sophisticated detectors available, few are used routinely for metabolite quantification. Beside radioactivity detector (RAD) which also require suitable standards (radiolabeled compounds) other compound response independent detectors has been recently discussed elsewhere [25, 44-46]. Such approach has become even more important for metabolite evaluation in the light of recently introduced FDA guidelines on metabolites in safety testing, which recommends that all metabolites greater than 10 percent of parent drug should be examined [1]. Some further examples of metabolite quantification using accelerator MS [47], inductively coupled plasma MS [43], chemiluminescene nitrogen detector [48], quantitative NMR [49] and evaporative light-scattering detector [50] are given.

6.2. Quantification using response factor

Prerequisite to make this approach successful is the chromatographic separation of drug and all metabolites. Quantification is based on using LC-MS/MS in combination with detector that gives an equimolar response independent of the compound, usually with RAD. Response ratio of the metabolite to parent drug on RAD is then correlated to response ratio on LC-MS/MS. Low amounts of metabolites and parent drug in samples are measured by sufficiently sensitive LC-MS/MS, where the analysis of higher amounts allows detection on RAD and due to response factor enables calculating of metabolite concentration. The best way to perform analyses is to combine RAD with MS after liquid chromatography with splitting flow in order to obtain peaks of the metabolites and parent at the same retention times on both detectors [51]. RAD is convenient for such analyses because of the large dynamic range but its use is limited by the availability of radiolabeled standards. However, the most straightforward detection technique generally found with LC-MS/MS is UV detection. Metabolites can be (semi)quantified using UV response ratio in cases when the parent drug chromophore offers sufficient selectivity, is not altered by metabolism and the metabolites are well separated from other drug related entities and endogenous compounds [46].

This approach may be also reasonable to quantify metabolites in case of limited amounts of authentic standards. After determination of the response factors, metabolites could be then quantified based on calibration curve of parent drug [19]. A constant response factor is absolutely essential and therefore in such cases response factors should be periodically verified. Using the same instrument and without major instrument breakdowns, the response factor seems to be very stable over long periods [52].

6.3. Indirect quantification

Refer to evaluation of glucuronides and other phase II metabolites. These metabolites are determined by cleavage of conjugates to yield the parent drug, which is subsequently detected. This indirect approach has several limitations, including the risk of incomplete hydrolysis, moderate repeatability and time consuming sample preparation [42]. Another drawback is non-selectivity of this procedure toward study of particular metabolite of interest when distinct drug metabolite conjugates are present in sample, like in case of morphine which is transformed to two isomeric metabolites. Morphine-3-glucuronide is an inactive metabolite but morphine-6-glucuronide possesses even greater pharmacological activity than the parent drug [53]. In such cases this approach is not suitable in pharmacokinetic studies as the overall drug concentration including more metabolites is determined in examined biological fluid. However, in the field of toxicology, doping control or drugs of abuse this information may be even more valuable [54, 55]. Nevertheless, direct quantification of metabolites and their indirect quantification via parent drug after metabolite hydrolysis may give comparable results like in case of buprenorphine metabolites [56].

Cleavage of conjugates can be performed by fast chemical hydrolysis or by gentle but time consuming enzyme hydrolysis. Deconjugation by ß-glucuronidase is the predominantly used approach. Different types of enzymes are commercially available but the most frequently used are ß-glucuronidases from *E. coli* or *Helix pomatia*, sometimes combined with arylsulfatase. ß-glucuronidase from *Helix pomatia* provides the advantage of the cleavage of glucuronide and sulfatate conjugates at same time what is important in the field of toxicology [6]. However, the glucuronidase activity is not as high as at *E. coli*. In order to achieve a successful enzyme hydrolysis it is crucial to pay attention on several factors, such as temperature, pH, enzyme origin, enzyme concentration and incubation time [57]. However, cleavage with ß-glucuronidase is not always preferential as for acyl glucuronides (ester conjugates) where alkaline hydrolysis is more suitable [55]. Acid hydrolysis may also be sometimes the better possibility for other glucuronide types [58]. N-glucuronides (primary, secondary and N-hydroxylated amines) are hydrolyzed under mild acidic conditions but quartenary ammonium glucuronides under basic conditions [25]. Additionally, enzymatic hydrolysis of acyl glucuronides may be hindered due to acyl migration what leads to ß-glucuronidase resistant derivates [59]. Nevertheless, if the ß-glucuronidase treatment is successful for the metabolite of interest, this procedure should be the method of choice.

Another aspect for quantification using this approach has been shown recently [60]. Different benzodiazepines were determined via their metabolites by using acid hydrolysis of urine samples. The parent drug and all metabolites, conjugated as well as non-conjugated (I phase metabolites), were converted to corresponding benzophenone under studied conditions. Such approach reduces the specificity but at same time the overall sensitivity of the method increases, which makes such method suitable for drug abuse monitoring.

6.4. Metabolite production

Alternative approach for direct quantification is to obtain authentic metabolite standards. The chemical synthesis is mainly suitable for achieving phase I metabolites, like O-demetylation, N-demetylation, N-oxidation, carbonyl reduction and other. However, synthesis of the phase II metabolites can be cumbersome and stereochemically demanding and hence go beyond possibilities of most laboratories [31]. Versatile alternative to chemical synthesis is enzyme-assisted *in vitro* production of these metabolites using liver homogenates, liver microscale cultures, cell culture lines or microbial systems where each of these methods has its specific drawbacks [61, 62]. Raloxifene, which is metabolized to two distinct monoglucuronides and one diglucuronide, is an illustrative example for in-house production of authentic standards. Glucuronide yield by chemical synthesis was very low and not sufficient enough to characterize those metabolites. On contrary, the biosynthesis with recombinant human UGT enzymes turned out to be successful in converting parent drug to both monoglucuronides [63]. In last attempt the bioproduction of all three metabolites could be accomplished by using the microorganism *Streptomyces sp* [37]. For more detail about raloxifene *in vitro* metabolism refer to [19]. Availability of both metabolite standards – unlabeled and stable isotope-labeled internal standards is even more important for reliable quantification using LC-MS/MS. Stable isotope labeled metabolites can be obtained by microsomal incubation of labeled drug, of course if it is available and not too expensive [4]. The alternative approach is to use a labeled UDP-glucuronic acid as cofactor in bioproduction of metabolites [62].

Moreover, metabolites can be isolated from urine after oral administration and after purification and characterization they can be used as standards. Bisphenol A glucuronide and its deuterated glucuronide were isolated from rat urine [64]. Recently published work dealing with microsomal bioproduction of the same metabolites [4] revealed some drawbacks of the isolation approach. Beside ethical considerations, the yield of both standards was much lower from animal samples (microgram scale) than microsomal incubates (milligram scale). Additionally, urine as matrix requires also more extensive purification procedure in order to obtain highly pure standards. However, in cases where metabolites cannot be produced by proposed *in vitro* models the *in vivo* biological samples are then the only media for isolation of those metabolites.

7. Sample preparation

Adequate sample preparation is a key aspect of quantitative bioanalysis and it is usually the most time consuming part of analyses. Interfering matrix compounds, such as proteins, lipids, salts, other endogenous and background compounds, should be removed in sample pretreatment, not only to avoid column clogging and instrument soiling, but also to improve the sensitivity, selectivity and reliability of analyses. Selection of an appropriate preparation procedure depends upon metabolite characteristics, their expected concentrations, the sample size and matrix, and the availability of analytical techniques for analyte quantification. Insufficiently treated samples may cause interfering peaks when using spectroscopic detection techniques such as UV-absorbance or fluorescence. However, analyses by LC-MS/MS are less prone to sample matrix and therefore usually require less pretentious sample clean up. Commonly and widely applied sample preparation techniques include protein precipitation (PP), liquid-liquid extraction (LLE) and solid-phase extraction (SPE). Manual operations associated with sample treatment may be very labor intensive and time consuming and that could be avoided with automation in 96-well plate format or direct sample injection followed by on-line extraction methods.

7.1. Protein precipitation

Protein precipitation (PP) is simple and straightforward method widely used in bioanalysis of plasma samples. It is accomplished by using organic solvent (typically acetonitrile or methanol) or an acid (typically perchloric or trichloroacetic acid). It is followed by centrifugation to separate proteins from liquid supernatant and additionally, supernatant is sometimes diluted with chromatographically compatible solvent (e.g. mobile phase eluent). Supernatant can be directly injected or pre-concentrated after evaporation and reconstitution. Although only proteins are removed, other endogenous compounds remain which can still cause interferences such as matrix effect in mass spectrometry analyses. PP offers a generic and fast sample preparation technique that can be easily automated. The method has been also extended to quantification of drugs and metabolites from whole blood [65]. The same sample preparation technique in not suitable only for plasma but can be transferred to other biological samples such as urine. Moreover, the absence of proteins in these matrices allows direct injection without sample pretreatment. Nevertheless, it is advisable to dilute and filter or centrifuge the samples to reduce matrix effect and to remove eventually present particles [42]. Many examples for metabolite determination using PP, mainly in serum and urine, can be found in recently published review articles [66-68]. PP is also the most convenient method for less complex biological matrices in pharmacokinetic studies, such as hepatocytes [69] or microsomes [4]. In this case protein precipitation by ice-cold methanol (triple volume) at the same time terminates the incubation reaction and introduces internal standard to the final sample.

7.2. Liquid-liquid extraction

To obtain a sensitive analysis for a complex biological media (plasma, urine) liquid-liquid extraction (LLE) or solid phase extraction (SPE) are often required instead of PP. LLE sometimes gives better sample clean up showing less matrix effect in comparison with SPE [70]. Additionally, LLE is in general simpler and may be applicable to almost all laboratories using large variety of available solvents. LLE is also less expensive and flexible as several samples may be prepared in parallels. On the other hand emulsion formation, mutual solubility of analytes in both phases or large volumes of flammable and/or toxic solvents should be considered. In recently published comprehensive overview of methods for measurement of antidepressants and their metabolites in biofluids, many examples of extraction including LLE conditions can be found [67]. Offline methodologies are often very tedious and time consuming, and the risk of sample loss and/or contamination is high. Lack of automation possibilities is therefore another important LLE drawback. However, several research groups have developed different approaches to solve mixing and phase separation problems typically seen in a 96-well plate LLE method [71]. A semi automated LLE procedure using 96-well plates was reported [72].

7.3. Solid-phase extraction

SPE has become very popular and is nowadays considered as a basic technique in many laboratories for sample preparation of drugs and their metabolites from biological matrices. SPE offers several advantages over LLE, including higher recoveries, no problems with emulsions, less solvent consumption and a smaller sample volume requirement. Moreover, automation of sample treatment with high speed and feasibility for treatment of numerous samples at one time is possible. However, a drawback often associated with SPE is their high dead volume, which can lead to loss of sample and may cause dilution of applied samples. SPE column lot production variability or column blockage due to sample viscosity or precipitation may also occur. Columns can be supplied as individual units for manual use and also in 96-well plate format for use with robotic sample processors. The column dead volume has been overcome with a novel 96-well SPE plate that was designed to minimize elution volume (< 25μL). The evaporation and reconstitution step that is usually required in SPE is avoided due the concentration ability of the sorbent [71].

SPE is based on chromatographic separation such as liquid chromatography. Wide variety of cartridge types and solvents make SPE procedure suitable for many polar or nonpolar analytes. The extraction procedure can be a generic protocol or can be optimized if better sample clean up is desired. Beside classical reverse phase (e.g. C8 or C18) also polymer reverse phase (e.g. divinylbenzene, N-vinylbenzene), polymer ion-exchange (e.g. weak or strong anion/cation-exchange) or mixed mode ion-exchange sorbents are available. Polymeric reverse phase materials possess both hydrophilic and lipophilic properties and are capable of capturing polar analytes such as drug metabolites [66]. Another advantage regarding silica based phase is ease of use, since there is no need to keep those phases moistened to maintain interaction. Mix mode ion-exchange and ion-exchange sorbents are even more convenient since strongly retained ionic metabolites allow rigorous washing of cartridge (e.g. 100% methanol) achieving cleaner sample with less matrix interferences [73].

There are now commercially available protein precipitation devices in plate format that allow PP within the plate whilst also removing phospholipids (HybridSPE™ and Waters Ostro™). This novel semi automated sample clean up procedure includes combination of PP and SPE. Proteins in sample are firstly precipitated with organic solvent, then transferred to SPE and directly injected into the analytical instrument. Method is simple, fast and almost free from phospholipids [74]. This sample preparation approach has been successfully applied in metabolism studies of various drugs [75].

7.4. On-line SPE

The on-line SPE offers speed, high sensitivity by the pre-concentrating factor, and low extraction cost per sample, but typically require the use of program switch valves and column re-configurations [71]. Biological samples can be directly injected into liquid chromatographic system without any sample preparations except for aliquoting samples, adding the internal standard and sometimes sample diluting and/or centrifugation. On-line SPE is considered as another dilute and injection approach like protein precipitation, however, it provides cleaner extract with reduced chance for matrix effect. Commonly used columns for on-line SPE are packed with large particles (typically > 20μm) of stationary material, such as polymeric and silica based, which work based on reversed phase, ion-exchange or mixed mode of separation. The combination of large particle size in these narrow bore columns (typically 50x1 mm) and fast flow (typically 3-5 mL/min), called also as turbulent flow chromatography, promotes the rapid removal of proteins with simultaneous retention of the small-molecular analytes of interest. After flushing all the proteins to waste, the direction of the flow is switched; the analytes are back-flushed onto the analytical column for chromatographic separation and detection. Fully integrated home-built or commercial systems enable eluting analytes from the extraction column onto analytical column in narrow bands. That allows multiple injections onto analytical column prior to elution into the instrument detector resulting in better sensitivity [76]. Most on-line SPE approaches use column-switching to couple with the analytical column as well as additional HPLC pump. Various instrument setups and column dimensions can be configured for the fast analysis of drugs and their metabolites in biological matrix at the ng/mL levels or lower [71].

Typically, on-line SPE columns can withstand few hundred injections of diluted plasma or urine samples what depends on the injection volume and sample matrix [77]. Beside mentioned SPE sorbents for turbulent flow chromatography, restricted access materials (RAM), monolithic materials and disposable SPE cartridges are available. The working principle of RAM phases is to isolate macromolecules from the target small molecules in biological samples based upon their particle sizes and also due the chromatographic interaction. The proteins, that are unable to penetrate into the hydrophobic pores and the hydrophilic outer layer of particles, are first eluted to waste, the smaller molecules penetrate into pores and are additionally retained through the hydrophobic forces [78]. RAM columns may be used either in single column mode, being extraction (SPE) and analytical column at same time, or extraction column in combination with second analytical column. Single column mode approach shows simplicity but is limited due to chromatographic separation power [79]. Monolith phases as extraction sorbents for sample treatment looks promising and has been reviewed recently [77]. Monolith columns may be very convenient as single column mode for high throughput method in LC-MS/MS analysis [80].

8. Analytical methods for metabolite quantification

LC-MS/MS has become the predominant bioanalysis method for pharmacokinetic and metabolism studies due to its inherent specificity, sensitivity and speed. A literature survey of analytical methods for metabolite determination in biological samples undoubtedly confirms that fact. However, HPLC coupled with other detector systems or other separation techniques is often used. As an example, analytical methods for determination of antidepressants and their metabolites [67] are shown. HPLC coupled to different detectors (73%), among them the most popular being mass spectrometry (35%) and UV detection (24%), is the most frequently used analytical method. Applications of electrophoretic and gas chromatography methods for analysis of antidepressants and their metabolites in biofluids have seldom been published in literature (13 and 9%, respectively). Since the data were collected in time frame 2000-2010 [67], the frequency of LC-MS/MS methods is believed to be growing and is nowadays significantly higher because mass spectrometers are lately more accessible. In this section the most frequently used separation techniques as well as detectors will be overviewed with emphasis on LC-MS/MS.

8.1. Liquid chromatography

Good chromatographic separation is prerequisite for reliable and accurate quantification of metabolites in the biological samples. Baseline resolution must be achieved when liquid chromatography is coupled to non-MS detector. Although extensive chromatographic separation using LC-MS/MS is often not necessary, for certain cases, adequate resolution between drugs and various metabolites is required to avoid mass spectrometric interferences. Different metabolites may share the same MRM transition, such as hydroxylate metabolites [81] or glucuronides [37]. An example is shown in Figure 1. Additionally, unstable metabolites, such as N-oxides or glucuronides may be converted to parent drug by in-source dissociation or thermal degradation [79] or in collision cell (ion channel cross-talk). Interferences with endogenous compounds should also be avoided as matrix effect may appear (see 9.1.).

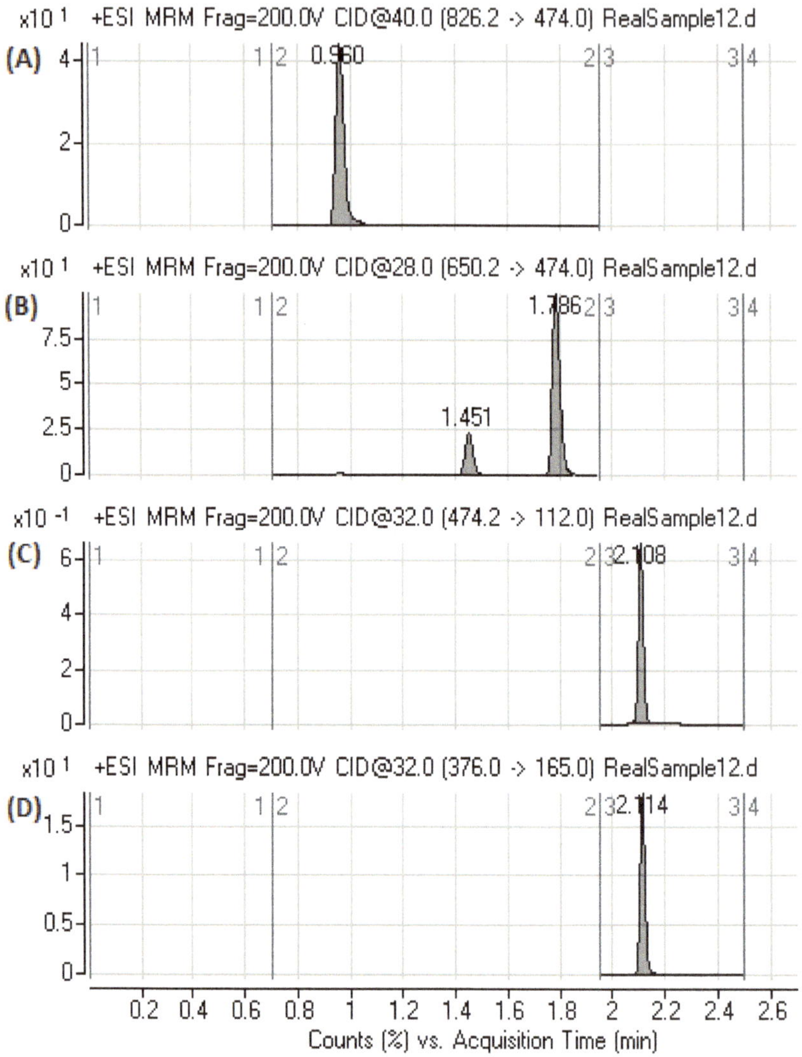

Figure 1. LC-MS/MS chromatogram of urine sample from a patient receiving raloxifene. MRM transitions represent (A) raloxifene diglucuronide, (B) two raloxifene monoglucuronides (C) parent raloxifene, (D) haloperidol as internal standard. For analysis conditions refer to [37].

8.1.1. Reversed phase chromatography

Reversed phase chromatography is most widely used technique in analysis of drugs and their metabolites due to its extensive application to most small molecules which are separated by their degree of hydrophobic interaction with the stationary phase. In most cases, metabolic changes lead to an increased polarity of the metabolite (strong shift for glucuronides and other phase II metabolites as also demonstrated in Figure 1) and therefore decreased retention on this stationary phase in relation to the parent drug [34]. The use of gradient elution is usually required to perform analysis of parent drug and polar metabolites. Common HPLC methods typically use a combination of water and either methanol or acetonitrile containing nonvolatile buffers, such as phosphate buffer and other inorganic additives as mobile phase. However, these nonvolatile additives cannot be recommended for LC-MS/MS because of possible MS contamination and also strong ion suppression effect. Volatile additives are used instead, such as formic or acetic acid (0.1% or lower) or ammonium acetate/formiate (2-10 mM) as salts. In order to maintain consistent chromatographic conditions, the pH of the mobile phase should be two units above or below pKa. C18 column is most commonly used. In some cases for polar metabolites short-chain bonded phases, such as C8, phenyl or cyano are more appropriate. Another effective way to resolve the retention issue is to add ion-paring reagent into mobile phase. The formed neutral ion pars increase retention and also improve peak shape. Among different ion-paring reagents trifluoroacetic acid and other perfluorated acids for basic analytes and for instance nucleoside phosphates for acidic analytes are appropriate for LC-MS/MS analyses [79]. These additives, especially trifluoroacetic acid, should be used at low concentrations because they cause ion suppression.

8.1.2. Hydrophilic interaction chromatography (HILIC)

HILIC using low aqueous/high organic mobile phase is emerging as a valuable supplement to the reversed phase chromatography for the retention of polar analytes [82]. An appropriate amount of water (usually 5-15%) in the mobile phase is suggested for maintaining a stagnant enriched water layer on the surface of the polar stationary phase where the analytes partite. HILIC separates compounds by eluting with strong organic mobile phase against a hydrophilic stationary phase where elution is driven by increasing the water content in the mobile phase [83]. Although some column companies are marketing columns specific for HILIC, most columns used with normal phases, such as pure silica or cyano columns, can operate in HILIC conditions. The highly volatile organic mobile phases, such as methanol and acetonitrile provide low column backpressure and also increased ionization efficiency for MS detection. It has been reported that the ionization responses for basic and acidic polar compounds were enhanced by 5-8 fold in the positive ionization mode and up to 20-fold in the negative ionization mode by the HILIC LC-MS/MS methods as compared to the reversed phase LC-MS/MS method [84]. Low back-pressure allows higher flow rates and may be used for shortening run times, up to several times [85]. Another advantage of HILIC is the possibility to inject higher volumes of organic solvent onto the column without impairing peak shapes. Therefore, evaporation and reconstitution step of organic extracts after extraction procedure could be omitted making improvement in sample preparation automation and throughput [86].

8.1.3. Chromatographic approaches for polar metabolites

Metabolites, in particular glucuronides, have typically higher polarity than their parent drugs (Figures 1 and 2). This is the reason that classical reversed phase chromatography (e.g. C18) is sometimes not sufficient enough to maintain appropriate chromatographic retention of these analytes. In such cases already mentioned approaches like short-chain bonded reversed phases and ion-paring reagents (8.1.1.) or HILIC (8.1.2.) may be used. Additionally, mixed-mode columns with an embedded ion-paring group in the reversed phase stationary phase provide the capability for both ion-exchange and hydrophobic interactions in the mobile phase to retain ionizable polar analytes. The mixed-mode column allows retaining hydrophobic analytes by the reversed phase mechanism and hydrophilic analytes by the ion exchange mechanism at higher organic content in the mobile phase [87]. Normal phase chromatography may also be used for retention of polar analytes but due to limited amount of water allowed in the mobile phase, normal phase chromatography interfaced with MS requires complex pretreatment steps for biological samples and therefore has much fewer applications than reversed phase LC-MS/MS [88].

The use of special packing material known as porous graphic carbon (PGC) is another alternative to achieve retention and separation of polar analytes. PGC chromatography commonly employs water, acetonitrile and methanol as the mobile phase but provides markedly greater retention and selectivity for polar analytes than reversed phase columns. For analyte elution PGC normally requires larger organic content in the mobile phase than reversed phase chromatography what consequently results in favorable sensitivity with MS detection [79, 83, 89].

Derivatization of polar analytes results in the reduction of polarity and is therefore another possibility to enhance the chromatographic retention. But this approach is disadvantaged as it is not going toward high throughput, especially in case when the primary purpose of the derivatization is not the detection or stability improvement of the analyte.

8.2. Strategies for high-throughput improvement in liquid chromatography

Current trend in pharmaceutical analysis is the reduction of the analysis time and the increase in sample throughput without sacrificing the separation selectivity. High-throughput bioanalytical assays are typically based on LC-MS/MS but may also be successfully extended to classical HPLC analyses. Approaches to achieve faster analyses include sample preparation (on-line automation or offline semi automation, section 7) and fast liquid chromatography. The later may be in general improved by three approaches: smaller particle size, shorter columns and higher mobile phase flow rates.

8.2.1. *Ultra-high performance liquid chromatography (UHPLC)*

Reducing the particle diameter from 5.0 µm to 1.7 µm will, in principle, result in a 3-fold increase of efficiency, 1.7-fold increase in resolution, a 1.7-fold in sensitivity, and 3-fold increase in speed [79]. For fast analyses using sub-2µm particle column dimensions are typically 50x2 mm. An additional benefit of UHPLC is the low consumption of mobile phase, where it saves at least 80% compared to HPLC [90]. The high back-pressure resulting in decreased particle size need appropriately designed chromatographic system that would withstand such high pressure (instruments nowadays up to 1200 bars) and also provide at least possible extra column effects. To prevent clogging, manufacturers of UHPLC recommend filtration of both samples and solvents through 0.2 µm filter. Advantages as enhanced separation efficiency, short analysis time and high detection sensitivity make UHPLC coupled with MS/MS an even more powerful analytical support in pharmacokinetic studies [4].

8.2.2. *Core-shell column*

An emerging alternative to porous particles are porous layer beads, known as core-shell or fused-core particles. The high separation efficiency of core-shell particles is a result of a faster analyte mass transfer from the mobile phase to outer porous layer of the particle. The improved dynamics of analyte movement through these columns result in higher effective peak capacities and separation efficiencies comparable to those fully porous sub-2µm but with advantage of lower back-pressure [91]. This technology is comparable to UHPLC in terms of chromatographic performance but demands neither expensive UHPLC instrumentation nor new laboratory protocols [88]. Commonly available columns, such as Ascentis, Poroshell and Kinetex, use different stationary phases and particle sizes (e.g. Kinetex 1.7 and 2.6 µm) and are widely used with classical HPLC instruments, also in our laboratories. Core-shell columns in combination with UHPLC-MS/MS exhibit excellent performance, as demonstrated in quantification of raloxifene and its three glucuronides [37].

8.2.3. *Monolithic chromatography*

The use of single rod monolith column is an alternative approach to the chromatographic columns packed with fine particles. The high permeability allows the use of higher flow rates and therefore shorter chromatographic runs, as demonstrated for the separation of bupropion metabolites in 23 seconds or for methylphenidate and its metabolite in 15 seconds [71].

High flow rates may require flow splitting before entering MS. An attractive approach using monolith separation is to combine it with high flow on-line extraction, which allows fast extraction and separation of samples [77]. Current limitations in the application of these columns are the small pH range [2-8], poor temperature resistance, limited column dimensions and stationary phases (C8 and C18) as well as higher costs due to higher mobile phase consumption.

8.3. Other separation techniques

Gas chromatography with mass spectrometry (GC-MS) is most useful for the analysis of trace amounts of organically extractable, non-polar, volatile compounds and highly volatile compounds that may undergo headspace analysis. The GC-MS analysis of polar compounds, such as metabolites, from biological matrices requires analyte extraction into a volatile organic solvent either directly or after chemical derivatization, which typically enhances the volatility of previously non-volatile organic compounds [25]. Most analytes need extensive time-consuming sample preparation including derivatization to become stable, volatile and amenable to the ionization technique. This drawback in throughput necessitated the direction of GC-MS to LC-MS. LC-MS has an advantage over GC-MS method in drug metabolism studies, particularly for low dosed and large drugs, and of course for the analysis of phase II metabolites. However, GC-MS may also have advantages, especially in clinical and forensic toxicology or doping control [54]. GC-MS has been frequently applied for quantification of glucuronides in biological samples but only after treatment with ß-glucuronidase in order to obtain parent drug before analysis [92]. The GC-MS technique is receiving wider acceptance in various classes of antidepressant agents, representing 6% of overall analytical methods for determination of antidepressants and their metabolites [67].

Capillary electrophoresis (CE) is another separation method for quantification of metabolites. This method offers very high resolution capability, high efficiency and short time of analysis. Moreover, CE in many instances can have distinct advantages over HPLC in terms of simplicity, rapid method development, solvent saving and minimal sample requirement [10-30 nL injected) making this technique very interesting for rapid and practical analyses in the biomedical field. However, the main disadvantage is low sensitivity. For this reason, application of CE for analysis of antidepressants and their metabolites is not so widely reported [67]. Applicability of CE using UV-absorbance or mass spectrometry detection was reported for determination of tamoxifene and its phase I metabolites [68].

8.4. Mass spectrometry

Currently, the QQQ using single or multiple reaction monitoring is most often used for quantitative analysis of small molecules in the pharmaceutical industry. QQQ or single stage MS, operating in SIM, is not anymore recommended for reliable bioanalytical quantification, because it suffers from insufficient selectivity in comparison with MRM. SIM can provide the selected ion at certain m/z value, but the matrix or impurity interferences may occur at the same m/z value. Beside lower selectivity, SIM shows also much lower sensitivity in comparison to MRM due to much higher background noise. However, in some specific cases of good chromatographic resolution and the absence of matrix interferences, SIM may be considered as an alternative quantification method. Occasionally, due to the nature of dissociation pathways, resulting in low molecular weight product ions, radial ejection preceding dissociation and/or charge stripping, reliable precursor → product ion transitions cannot be established. Alternatively, a precursor → precursor scan for reducing noise can be employed [70].

Also other analyzers, such as ion traps and TOF, have been widely and increasly used for metabolite quantification. Especially hybrid instruments which combine a QQQ (Q1 and colission cell) and ion trap or TOF (Qtrap, Q-TOF). These instruments can operate as true tandem mass spectrometry and are usually applied for this purposes. Q-TOF can also operate as TOF and thus provide accurate mass measurments. Additonaly, high resolution of TOF instruments allows the resolution of chromatographic peak from background interferences achieveing better sensitivity. However, it does apear that QQQ using MRM remain about three to five times more sensitive than TOF [93].

The selection of an appropriate ionization technique depends on the analyte characteristics, such as the structure, polarity or molecular weight. In LC-MS/MS analysis three atmospheric pressure ionization techniques cover the whole range of compound polarities and molecular weight: ESI, APCI and APPI. Moreover, the polarity mode can be chosen according to the acidic, neutral or basic properties of the analytes. If the right choice of ionization technique and the polarity mode is not so obvious, all available possibilities should be considered in order to obtain the best response for tested analytes. The softest ionization technique, ESI, is the method of choice for polar and ionic compounds. The advantage of soft ionization is in providing reliable information about molecular weight of the phase II metabolites in comparison to other ionization techniques [42]. For parent drug and phase I metabolites with a lower polarity, APCI and APPI may provide better ionization efficiency and sensitivity [34]. APPI has a similar application range as APCI, but slightly extended toward nonpolar compounds [32]. ESI is generally more susceptible to matrix ionization suppression than APCI [94]. In case of neutral steroids or other poorly ionizable analytes, derivatization can be employed in order to increase detection sensitivity, but additionally the chromatographic retention enhancement of such derivatizated analytes may therefore provide less matrix effect. On the other hand, the disadvantage of derivatization lies in an additional time consuming step for sample preparation [30]. The adjustment of the mobile phase for improving analyte response is much easier compared to derivatization. The effect of mobile phase on ESI efficiency is not well understood and hence the behavior of an analyte in different mobile phase conditions cannot be routinely predicted. Various factors can affect the ionization of analytes in ESI, such as pH, mobile phase additives, flow rate, solvent composition and concentration of electrolytes. It is recommended to evaluate the use of additives (e.g. formic acid, ammonium acetate) and organic modifiers in mobile phase to maximize the ionization efficiency of the analyte, which is highly dependent on its chemical structure. Acidic conditions often promote positive mode ionization of basic compounds and conditions, which are slightly below neutral, neutral or basic, promote negative ionization of acidic compounds.

A dramatic difference in the ESI response can be found even when acetonitrile is replaced by methanol in mobile phase. It was reported that an analyte gave only weak ESI response in the positive ionization mode in mobile phase containing acetonitrile with formic acid and/or ammonium acetate. But replacement of acetonitrile with methanol in mobile phase gave approximately 25-fold higher response. On the other hand, for the same analyte, mobile phases containing acetonitrile or methanol gave about the same response in negative ionization [79]. Another interesting example is analysis of bisphenol A and its metabolite in biological samples. In order to gain the highest possible sensitivity for bisphenol A and bisphenol A glucuronide, LC-MS/MS conditions were optimized. ESI ionization source operating in negative ionization mode was selected for further optimization of mobile phase. It was found that substitution of acetonitrile/water with methanol/water as mobile phase increased response of parent by approximately two-fold but at the same time decreased response of its metabolite by approximately three-fold. Acetonitrile was selected as organic modifier because metabolite quantification is the main concern of metabolism studies. Additionally, sufficiently high sensitivity is needed for metabolite determination as low concentrations are expected in such studies [4].

8.5. Other detection techniques

UV, fluorescence or electrochemical detectors are usually coupled with liquid chromatograph for determination of drugs and their metabolites. Total analysis time of these methods is often long because baseline chromatographic separation is required for quantification purposes. In terms of reproducibility and robustness, UV and fluorescence detection have an advantage over mass spectrometry. However, methods are less sensitive and specific what requires extensive and time-consuming sample preparation compared to mass spectrometry.

Before the advent of mass spectrometry, UV was the primary detection technique used in pharmacokinetics for quantification of drugs and their metabolites in biological matrices. Although robust, reliable, simple and easy to use, UV detection provides relatively poor sensitivity, especially when the compound of interest has no significant chromophore [44]. However, HPLC coupled with UV detection is still widely applicable for determination of drugs and their metabolites in biological samples [95-97].

In contrast to UV, fluorescence or electrochemical detection can be a very selective and sensitive detection technique. These detectors can extend the sensitivity by 1-3 orders of magnitude if the analyte exhibits, or can be readily derivatized to exhibit, fluorescence or electroactivity [7]. Some drugs such as morphine have good fluorophores which allows its detection without derivatization. For direct determination of morphine and its two glucuronides assays based on liquid chromatography with different detector systems (UV, fluorescence, electrochemical, MS) has been reported. Limits of quantifications for both metabolites were comparable for MS and fluorescence detection but were as expected higher for UV detection [98]. Oxidation vulnerability and native fluorescence properties of most biogenic amines may explain the long history of their quantification by these conventional HPLC detection methods. However, LC-MS/MS methods are rapidly emerging due to its specificity, sensitivity and high throughput [99]. Electrochemical detection is also very suitable for determination of antioxidants, such as ascorbic acid or glutathione, in biological samples [100].

When analytes do not exhibit fluorescence, electroactivity or have poor UV detection, derivatization can be performed to enhance their detection. In addition, chromatographic retention is enhanced by derivatization what is a very convenient in analysis of polar drug metabolites. Derivatization is an additional step in sample preparation where consideration regarding the stability of derivatizated analyte to solvolysis and thermal degradation need to be addressed. Nevertheless, fluorescence detection is still widely used [67, 68, 101].

9. Aspects of analytical quality

LC-MS/MS is currently considered as the method of choice for quantitative analysis of drugs and their metabolites. The advantages of using this technique in MRM mode due to high sensitivity, selectivity and speed allow developing high throughput methods with little or no sample preparation and minimal chromatographic retention. However, matrix effect may have a significant impact on such LC-MS/MS analyses [94, 102, 103]. Therefore, the evaluation of matrix effect as well as strategies for its elimination or minimization needs to be adequately addressed. Another important parameter for analytical quality is the selection of an appropriate internal standard for the compensation of possible loss of analytes during sample clean up and variations in instrument performance. Other LC-MS/MS issues, such as ion channel cross-talk and carry-over should also not be overlooked. Moreover, metabolite instability may have an influence on the analytical performance and will be additionally addressed here.

9.1. Matrix effect

Matrix effect (ME) is a term that describes any changes in the MS response of analyte that can lead to either a reduced response (ion suppression) or an increased response (ion enhancement) of the LC-MS system. ME is caused by molecules originating from the sample matrix or mobile phase that co-elute with the analyte of interest and therefore interfere with the ionization process in the MS ion source. Several approaches have been proposed to evaluate ME. Among them the post column infusion technique is widely used. Use of this qualitative evaluation technique allows the determination of the matrix effect of endogenous components in blank matrix. During analysis of blank matrix, analyte response is monitored to provide information where in the chromatographic run interferences between the analyte and matrix compounds occur. ME is illustrated as response deviation in the otherwise flat response time trace of the continuously post-column infused analyte [104]. This approach is very useful during method development because it provides information on the retention times where ME has to be expected, which can later be avoided for analyte of interest by optimizing chromatographic conditions

For quantitative estimation of ME another well recognised approach is more suitable. Matuszewski et al. reported practical approach for the assessment of the absolute and relative ME as a part of validation of bioanalytical LC-MS/MS methods [94]. The difference in response between neat solution sample and post-extraction spiked sample is called the absolute ME, while difference between various lots of post-spiked samples is called the relative ME. As such will an absolute ME primary affect the accuracy and relative ME will affect the precision of the method. The determination of a relative ME is much more important than the determination of absolute ME in the evaluation and validation of bioanalytical method in biofluids [94]. The relative ME caused by interindividual variability in the sample matrix is assessed based on at least 5 lots of different matrices. Relative ME can be expressed as a coefficient of variation at particular concentration level or calculated based on slope lines. For the method to be considered reliable and free from the relative ME, the calculated coefficient of variation of determined slopes in different sources of matrices should not exceed 3-4% [105].

ME is known to be both component and matrix dependent. It was demonstrated that matrix induced ion suppression is especially important for early eluting compounds, such as polar metabolites. Typically, ME more strongly influences lower than higher analyte concentrations [71]. The main source of the commonly observed ME of plasma samples is believed to be endogenous phospholipids and proteins. The lysophospholipids which normally elute earlier in reversed phase chromatography are more likely to cause matrix effects compared to the later eluting phospholipids in spite of the larger concentrations of the latter in plasma [106]. Phospholipids cause ion suppression in both, positive and negative ESI modes and must be removed or resolved chromatographically.

To remove or reduce ME, modification to the sample extraction methodology (SPE or LLE instead of PP) and improved chromatographic separation must be performed. The majority of ME occur in the solvent front of a chromatographic run and if the analytes can be retained to some degree, matrix effects can be minimized. Suitable sample preparation and chromatographic conditions are linked together and form the basis of developing a successful and robust quantitative LC-MS/MS method [102, 103]. Another consideration when dealing with ME is selection of ionization interface. APCI is generally considered to be less prone to ion suppression compared with ESI [94, 105]. However, assay sensitivity and thermal stability of the analyte should be evaluated for eventual APCI application. Reducing the flow rate [20 μl/min or below) directed to ESI source by post column splitting may also reduce or completely remove the ion suppression [107]. Additionally to other approaches UHPLC technology in combination with polymeric mixed-mode SPE and appropriate mobile phase pH may provide significant advantages for reducing ME [73]. Mobile phase additives such as triethylamine and trifluoroacetic acid can also lead to ion suppression. The use of other reagents such as formic or acetic acid, trifluoroacetic acid in conjunction with 10 mM ammonium acetate or addition of 1% propionic acid to the mobile phase may overcome the ME of trifluoroacetic acid. Triethylamine may be replaced with other ion paring reagent such as hexylamine [70].

However, the most efficient way to eliminate the influence of ME on the accuracy and precision of a quantitative analytical method is the use of stable isotope labeled analogs as internal standards [105].

9.2. Selection of internal standard

The selection of a suitable internal standard (IS) is one of the key parameters for establishing a successful LC-MS/MS method. Usually, stable isotope, such as ^{2}H (D, deuterium), ^{13}C, ^{15}N or ^{17}O, labeled standards are most appropriate for compensation of possible loss of analytes during sample clean up and variations in instrument performance (typically caused by matrix effect) since their physicochemical characteristics are practically identical to that of unlabeled analyte. In general, a stable isotope labeled IS is considered to be ideal, since it shows almost same behavior to the analyte of interest in sample preparation, chromatography as well as in ionization process [102]. However, issues like isotopic purity of IS, cross-contamination and cross-talk between MS ion channels, as well as IS stability and isotopic integrity of the label in biological fluid and during sample preparation should be carefully addressed [94].

Mass difference between analyte and IS should be at least 3 Da to avoid signal contribution of the natural isotopes to the signal of IS. Although deuterated IS are most frequently used, several disadvantages need to be considered in some cases, such as different extraction recoveries and retention times between such IS and analyte or exchange of deuterium atoms by hydrogen atoms [108]. For example, differences in retention times for deuterium labeled (d_{16}) and unlabeled bisphenol A compounds is shown in Figure 2, where both, labeled metabolite and parent compound eluted slightly before their unlabeled analogs. Interestingly to mention, deuterium labeled bisphenol A (d_{16}) has all 16 hydrogen atoms substituted by deuterium atoms but is actually d_{14} labeled (mass shift 14 Da in MRM transition, Figure 2) because two deuterium atoms from the functional group (–OD) are easily exchangeable by hydrogen atoms. The observable chromatographic retention time shifts for deuterated analogs depend on the number of deuterated atoms in structure. Deuterated analogs have no retention time shifts up to approximately six deuterium atoms [34] but when more deuterium atoms are included in structure, such as ten, the retention time shift may be significant (up to 1.2 min) [108]. Therefore, other stable isotope labeled IS, such as ^{13}C, are considered as more appropriate [108].

The concentration of IS used for sample preparation is also important and should be approximately at the middle of calibration curve. Optimization of IS concentration is critical to avoid ion suppression by co-eluting analyte leading to standard curve non-linearity [70]. Even if stable labeled isotope IS is used, ME should still be investigated. Namely if ion suppression significantly reduces the signal of both, analyte and IS, the signal to noise ratio may decrease to a point where accuracy and precision may be negatively affected [102].

Problems may occur when more than one compound is determined in the same analytical method. A number of labeled IS identical to number of analyzed compounds would in this case be required [102]. This is not always practically feasible, especially not for stable isotope labeled drug metabolites as their availability is very limited. In such cases, to assure the suitability of the quantification method for determination of drugs and their metabolites, ME evaluation should be carefully addressed [37]. On the other hand, example for simultaneous determination of parent compound and its metabolite using both labeled IS [4] is shown in Figure 2. It is preferred that the labeled IS product ion used for the MRM transition retains the stable isotope moiety as for m/z 403 → 227 (bisphenol A-glucuronide) versus 417 → 241 (deuterated bisphenol A-glucuronide) in Figure 2.

However, labeled ISs are not always available or can be very expensive. As an alternative, structural analogues can be used, with consideration of the structural similarities between the IS and the analyte. To be suitable, the ionization of the analogue must be compared with analyte and should preferably co-elute with the analyte (Figure 1). The selected analog IS should not correspond to any metabolic product of analyte, such as hydroxylated or N-dealkylated metabolites [108]. Nevertheless, in many cases analog ISs or structurally unrelated ISs are not able to compensate the ME [94, 105]. In such situation, other more rigorous approaches to reduce or eliminate ME should be applied (see 9.1.).

9.3. LC-MS pitfalls

Metabolite may produce a molecular ion that is identical to parent ion through in source conversion. Typical metabolites for such conversion are glucuronides, sulfates or N-oxides [79]. N-oxide conversion may also serve for identification purposes as this transformation takes place in APCI source and represent another potential way to differentiate N-oxide from hydroxylated metabolites (both with exact mass) since the later usually do not undergo thermal deoxygenation [23]. Therefore, an inadequate chromatographic separation between parent drug and such metabolites will result in over estimation of parent drug concentration in the presence of these isomeric compounds. The same situation may happen also by ion channel cross-talk in MRM mode, which means that fragment issued from the other scanned transition is still present in collision cell. Metabolites, such as diverse glucuronides, that give the same product ion as parent drug, have to be therefore chromatographically separated. However, for co-eluting compounds, such as analytes and their IS, the absence of cross-talk has to be demonstrated [37].

Carry-over, which is the appearance of an analyte signal in a blank injection subsequently to analysis of high concentration samples, is also a common problem in LC-MS/MS methods. This problem occurs due to retention of analytes by adsorption on active surfaces of the autosamples, solvent lines, extraction columns (e.g. online SPE) or the analytical column. Most carry-over problems can be minimized by an appropriate choice of injector wash solutions and methods, by proper choice of mobile phase and tubings, and by choice of suitable stationary phase and proper variation in the solvent strength [70]. Carry-over becomes prominent after injection of analyte at high concentration and should be assessed during the validation of the method. Carry-over is usually expressed as percentage of detected analyte in blank sample regarding to its limit of quantification [4] and may significantly lower the sensitivity of the method [109].

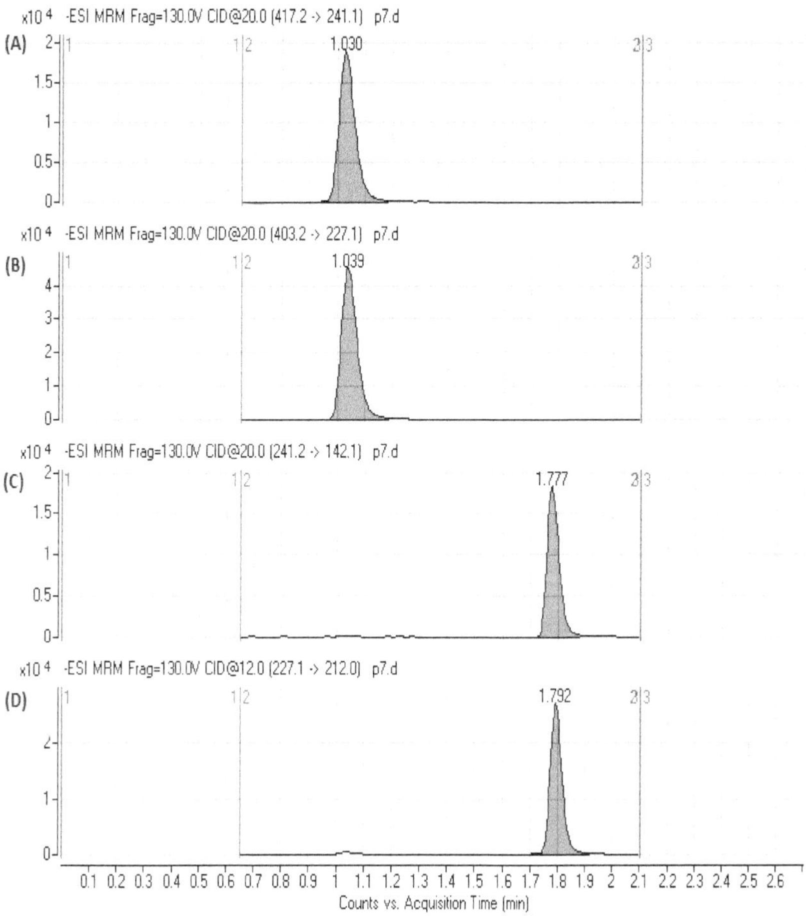

Figure 2. LC-MS/MS chromatogram in MRM mode for metabolite (BPAG) and parent bisphenol A (BPA) of a typical microsomal incubation. (A) mass transition for $BPAG_{d16}$ (internal standard for BPAG), (B) mass transition for BPAG, (C) mass transition for BPA_{d16} (internal standard for BPA), (D) mass transition for BPA. For analysis conditions refer to [4]

9.4. Metabolite stability

For drugs and metabolites that are unstable, with one converting to other, conditions used during sample preparation and analysis must be optimized in order to minimize such conversion and to achieve accurate quantification of the drug and metabolite. Common factors that affect drug and drug metabolite stability in biological matrices include temperature, light, pH, oxidation and enzymatic degradation [110]. Acyl glucuronides (O-ester conjugates) are probably the most commonly encountered problematic metabolites in bioanalysis. Acyl glucuronides are unstable and hydrolyze to release aglycone under neutral and alkaline conditions. Different acyl glucuronides have been shown to have different rates of hydrolysis. However, mildly acidic conditions (pH 3-5) should be the most desirable pH for minimizing the reaction in biological samples [59, 111]. Acyl glucuronides (1-O-acyl glucuronides) are also susceptible to internal migration under both, physiological and alkaline conditions. The rate of migration resulting in 2-, 3- and 4-O-acyl glucuronides increases with increasing pH and temperature. Such isomeric glucuronides are not susceptible to hydrolysis by ß-glucuronidases and may compromise the quantification of metabolites and total parent drug when indirect quantification approach via conjugate cleavage is applied [111]. Similar to O-glucuronides, N-glucuronides can be converted back to parent drug under acidic/basic/neutral pH conditions or at elevated sample processing temperature, but this instability is largely compound dependent. For example olanzapine glucuronide can be readily cleaved under acidic conditions, clozapine and cyclizine glucuronides are unstable at range pH 1-3 and doxepin glucuronide at pH 11 [112]. Lactone is another commonly unstable metabolite function group which may be converted to its open ring hydroxy acid drug. Lactone metabolites, such as atorvastatin metabolite, require optimization of the sample pH (typically 3-5) in order to minimize the hydrolysis of lactone metabolite back to parent drug [113].

N-oxides are also unstable in solutions and biological samples during sample preparation, especially under strong acidic or basic conditions. Light exposure may further accelerate the decomposition of these metabolites. Other compounds are also susceptible to photodegradation such as, catechols, nisoldipine, rifampin and their metabolites and should be protected from light during sample preparation and analysis [112].

10. Conclusion

Current drug discovery efforts have been focused on identification of drug metabolism and pharmacokinetic issues at the earliest possible stage in order to reduce the attrition rate of drug candidates during the developmental phase. Metabolic fate of drugs can be responsible for problems associated with their bioavailability, interindividual variability, drug-drug interactions, pharmacologic activity or the toxicity. Different *in vitro* methods, from subcelullar to organ range, and *in vivo* studies are applied for the clarification of drug metabolism. Among them microsomes and hepatocytes are the most frequently utilized *in vitro* models in drug metabolic profiling and drug interaction studies. For the successful monitoring of the drug metabolism, suitable bioanalytical methods have to be developed and validated. Liquid chromatography coupled with mass spectrometry has become the most powerful analytical tool for identification and quantification of drug metabolites.

The known identity of metabolites is the prerequisite for a suitable metabolic assessment of drugs. Appropriate LC-MS instrumentation is clearly critical to both, detection and structural elucidation. Tandem mass spectrometry instruments are beside their key role for metabolite quantification also well suited for qualitative purposes as they enable different scan possibilities (constant neutral loss, precursor ion, product ion) for structural characterization of metabolites. However, the drive to more versatile and powerful instruments which can perform intelligent data dependent experiments and accurate mass measurements has led to newer high resolution mass analyzers, such as Q-TOF instruments, which now dominate the metabolite identification field.

Direct quantification of metabolites in biological samples is the most appropriate approach, but also others approaches such as indirect quantification through parent drug after metabolite hydrolysis or quantification supported by using response factors may be used which primary depends on the availability of suitable authentic standards. Analytical methods for metabolite quantification are based on liquid chromatography or other separation techniques coupled with various detector systems where LC-MS/MS plays predominant role in bioassays for pharmacokinetic and metabolism studies due to its inherent specificity, sensitivity and speed.

In order to support metabolism experiments in a timely manner, the use of high throughput methods for the analysis of drugs and their metabolites in biological samples has become an essential part, especially in the most time consuming sample preparation. Trend is going toward semi automated off-line sample treatment in 96-well plate format or on-line SPE after direct injection of samples. Additionally, other high throughput approaches can be introduced. Ultra-high performance liquid chromatography with small particles and monolithic chromatography offer improvements in speed, resolution and sensitivity compared to conventional chromatographic techniques. Hydrophilic interaction chromatography (HILIC) or other specialized columns suitable for polar metabolites are emerging as a valuable supplement to classical reversed phase chromatography.

The main advantage of LC-MS/MS allows development of high throughput methods with little or no sample preparation and minimal chromatographic retention. However, matrix effect may have a significant impact on such analyses. Matrix effect issue is frequently underestimated and should be adequately addressed. Not without reason, matrix effect have been called the Achilles heel of quantitative LC-ESI-MS/MS [103]. The use of stable isotope labeled analog as internal standard is the most efficient way to reduce matrix effect. But normally, additional approaches to reduce or eliminate matrix effect are needed.

Abbreviations

DMPK, drug metabolism and pharmacokinetic; CYP, cytochrome P450; UGT, UDP-glucuronosyltransferase; NAT, N-acetyltransferase; GST, glutathione-S-transferase; SULT, sulfotransferase; UDPGA, uridine diphospoglucuronic acid; PAPS, phospohoadenosine phosphosulfate; HLM, human liver microsomes; LC-MS, liquid chromatography-mass spectrometry; LC-MS/MS, liquid chromatography-tandem mass spectrometry; GC-MS, gas chromatography-mass spectrometry; CE, capillary electrophoresis; LC-NMR, liquid chromatography-nuclear magnetic resonance; HPLC, high performance liquid chromatography; UHPLC, ultra-high performance liquid chromatography; HILIC, hydrophilic interaction chromatography; PGC, porous graphic carbon; RAD, radioactivity detector; MS, mass spectrometry; MS/MS, tandem mass spectrometry; API, atmospheric pressure ionization; ESI, electrospay ionization; APCI, atmospheric pressure chemical ionization; APPI, atmospheric pressure photoionization; QQQ, triple quadrupole; IT, ion trap; QTrap, triple quadrupole-linear ion trap; TOF, time of flight; Q-TOF, triple quadrupole-time of flight; FT-ICR, fourier transform-ion cyclotron resonance; IM-MS, ion mobility mass spectrometry; MRM, multiple reaction monitoring; SRM, selected reaction monitoring; SIM, single ion monitoring; CNL, constant neutral loss scan; PI, precursor ion scan; PP, protein precipitation; LLE, liquid-liquid extraction; SPE, solid-phase extraction; RAM, restricted access materials; ME, matrix effect; IS, internal standard.

Author details

Robert Roškar and Tina Trdan Lušin
University of Ljubljana, Faculty of Pharmacy, Slovenia

11. References

[1] FDA. Guidance for Industry, Safety Testing of Drug Metabolites. In: Services USDoHaH, editor.2008.

[2] Trdan Lušin T, Stieger B, Marc J, Mrhar A, Trontelj J, Zavratnik A, et al. Organic anion transporting polypeptides OATP1B1 and OATP1B3 and their genetic variants influence the pharmacokinetics and pharmacodynamics of raloxifene. Journal of Translational Medicine 2012;10[76].

[3] Trdan Lušin T, Mrhar A, Stieger B, Kullak-Ublick GA, Marc J, Ostanek B, et al. Influence of hepatic and intestinal efflux transporters and their genetic variants on the pharmacokinetics and pharmacodynamics of raloxifene in osteoporosis treatment. Translational research. 2012.

[4] Trdan Lušin T, Roskar R, Mrhar A. Evaluation of bisphenol A glucuronidation according to UGT1A1*28 polymorphism by a new LC-MS/MS assay. Toxicology. 2012 Feb 6;292[1]:33-41.

[5] Yengi LG, Leung L, Kao J. The evolving role of drug metabolism in drug discovery and development. Pharm Res. 2007 May;24[5]:842-58.

[6] Dekant W. The role of biotransformation and bioactivation in toxicity. EXS. 2009;99:57-86.
[7] Kamel A, Prakash C. High performance liquid chromatography/atmospheric pressure ionization/tandem mass spectrometry (HPLC/API/MS/MS) in drug metabolism and toxicology. Curr Drug Metab. 2006 Dec;7[8]:837-52.
[8] Walsh JS, Miwa GT. Bioactivation of drugs: risk and drug design. Annu Rev Pharmacol Toxicol. 2011 Feb 10;51:145-67.
[9] Rautio J, Kumpulainen H, Heimbach T, Oliyai R, Oh D, Jarvinen T, et al. Prodrugs: design and clinical applications. Nat Rev Drug Discov. 2008 Mar;7[3]:255-70.
[10] Williams DA. Drug Metabolism. In: Troy D, editor. Foye's Principles of Medicinal Chemistry. fifth edition ed. Baltimore: Lippincott Williams & Wilkins; 2002. p. 174-233.
[11] Ma Q, Lu AY. Pharmacogenetics, pharmacogenomics, and individualized medicine. Pharmacol Rev. 2011 Jun;63[2]:437-59.
[12] Deenen MJ, Cats A, Beijnen JH, Schellens JH. Part 2: pharmacogenetic variability in drug transport and phase I anticancer drug metabolism. Oncologist. 2011;16[6]:820-34.
[13] Deenen MJ, Cats A, Beijnen JH, Schellens JH. Part 3: Pharmacogenetic variability in phase II anticancer drug metabolism. Oncologist. 2011;16[7]:992-1005.
[14] Brandon EF, Raap CD, Meijerman I, Beijnen JH, Schellens JH. An update on in vitro test methods in human hepatic drug biotransformation research: pros and cons. Toxicol Appl Pharmacol. 2003 Jun 15;189[3]:233-46.
[15] Fasinu P, Bouic PJ, Rosenkranz B. Liver-based in vitro technologies for drug biotransformation studies - a review. Curr Drug Metab. 2012 Feb 1;13[2]:215-24.
[16] Asha S, Vidyavathi M. Role of human liver microsomes in in vitro metabolism of drugs-a review. Appl Biochem Biotechnol. 2010 Mar;160[6]:1699-722.
[17] Ekins S, Ring BJ, Grace J, McRobie-Belle DJ, Wrighton SA. Present and future in vitro approaches for drug metabolism. J Pharmacol Toxicol Methods. 2000 Jul-Aug;44[1]:313-24.
[18] Hariparsad N, Sane RS, Strom SC, Desai PB. In vitro methods in human drug biotransformation research: implications for cancer chemotherapy. Toxicol In Vitro. 2006 Mar;20[2]:135-53.
[19] Trdan Lušin T, Trontelj J, Mrhar A. Raloxifene glucuronidation in human intestine, kidney, and liver microsomes and in human liver microsomes genotyped for the UGT1A1*28 polymorphism. Drug Metab Dispos. 2011 Dec;39[12]:2347-54.
[20] Jia L, Liu X. The conduct of drug metabolism studies considered good practice (II): in vitro experiments. Curr Drug Metab. 2007 Dec;8[8]:822-9.
[21] Liu X, Jia L. The conduct of drug metabolism studies considered good practice (I): analytical systems and in vivo studies. Curr Drug Metab. 2007 Dec;8[8]:815-21.
[22] Trontelj J, Marc J, Zavratnik A, Bogataj M, Mrhar A. Effects of UGT1A1*28 polymorphism on raloxifene pharmacokinetics and pharmacodynamics. Br J Clin Pharmacol. 2009 Apr;67[4]:437-44.
[23] Ma S, Chowdhury SK, Alton KB. Application of mass spectrometry for metabolite identification. Curr Drug Metab. 2006 Jun;7[5]:503-23.

[24] Tolonen A, Turpeinen M, Pelkonen O. Liquid chromatography-mass spectrometry in in vitro drug metabolite screening. Drug Discov Today. 2009 Feb;14[3-4]:120-33.
[25] Prakash C, Shaffer CL, Nedderman A. Analytical strategies for identifying drug metabolites. Mass Spectrom Rev. [Review]. 2007 May-Jun;26[3]:340-69.
[26] Wright P. Metabolite identification by mass spectrometry: forty years of evolution. Xenobiotica. 2011 Aug;41[8]:670-86.
[27] Meyer MR, Maurer HH. Current applications of high-resolution mass spectrometry in drug metabolism studies. Anal Bioanal Chem. 2012 Feb 16.
[28] Kostiainen R, Kotiaho T, Kuuranne T, Auriola S. Liquid chromatography/atmospheric pressure ionization-mass spectrometry in drug metabolism studies. J Mass Spectrom. 2003 Apr;38[4]:357-72.
[29] Kebarle P. A brief overview of the present status of the mechanisms involved in electrospray mass spectrometry. J Mass Spectrom. 2000 Jul;35[7]:804-17.
[30] Gao SM, Zhang ZP, Karnes HT. Sensitivity enhancement in liquid chromatography/atmospheric pressure ionization mass spectrometry using derivatization and mobile phase additives. J Chromatogr B. [Review]. 2005 Oct;825[2]:98-110.
[31] Keski-Hynnila H, Kurkela M, Elovaara E, Antonio L, Magdalou J, Luukkanen L, et al. Comparison of electrospray, atmospheric pressure chemical ionization, and atmospheric pressure photoionization in the identification of apomorphine, dobutamine, and entacapone phase II metabolites in biological samples. Anal Chem. 2002 Jul 15;74[14]:3449-57.
[32] Cai Y, Kingery D, McConnell O, Bach AC, 2nd. Advantages of atmospheric pressure photoionization mass spectrometry in support of drug discovery. Rapid Commun Mass Spectrom. 2005;19[12]:1717-24.
[33] Marchi I, Rudaz S, Veuthey JL. Atmospheric pressure photoionization for coupling liquid-chromatography to mass spectrometry: A review. Talanta. [Review]. 2009 Apr;78[1]:1-18.
[34] Holcapek M, Kolarova L, Nobilis M. High-performance liquid chromatography-tandem mass spectrometry in the identification and determination of phase I and phase II drug metabolites. Anal Bioanal Chem. 2008 May;391[1]:59-78.
[35] Kanu AB, Dwivedi P, Tam M, Matz L, Hill HH, Jr. Ion mobility-mass spectrometry. J Mass Spectrom. 2008 Jan;43[1]:1-22.
[36] Levsen K, Schiebel HM, Behnke B, Dotzer R, Dreher W, Elend M, et al. Structure elucidation of phase II metabolites by tandem mass spectrometry: an overview. J Chromatogr A. [Review]. 2005 Mar;1067[1-2]:55-72.
[37] Trdan T, Roskar R, Trontelj J, Ravnikar M, Mrhar A. Determination of raloxifene and its glucuronides in human urine by liquid chromatography-tandem mass spectrometry assay. J Chromatogr B Analyt Technol Biomed Life Sci. 2011 Aug 1;879[23]:2323-31.
[38] Ma S, Zhu M. Recent advances in applications of liquid chromatography-tandem mass spectrometry to the analysis of reactive drug metabolites. Chem Biol Interact. 2009 Apr 15;179[1]:25-37.

[39] Trdan Lušin T, Tomašić T, Trontelj J, Mrhar A, Peterlin-Mašič L. In vitro bioactivation of bazedoxifene and 2-[4-hydroxyphenyl)-3-methyl-1H-indol-5-ol in human liver microsomes. Chem Biol Interact. 2012 Mar 10;197[1]:8-15.
[40] Jirasko R, Holcapek M, Nobilis M. Identification of phase I and phase II metabolites of benfluron and dimefluron in rat urine using high-performance liquid chromatography/tandem mass spectrometry. Rapid Commun Mass Spectrom. 2011 Aug 15;25[15]:2153-62.
[41] Rousu T, Herttuainen J, Tolonen A. Comparison of triple quadrupole, hybrid linear ion trap triple quadrupole, time-of-flight and LTQ-Orbitrap mass spectrometers in drug discovery phase metabolite screening and identification in vitro - amitriptyline and verapamil as model compounds. Rapid Commun Mass Spectrom. [Article]. 2010 Apr;24[7]:939-57.
[42] Ketola RA, Hakala KS. Direct analysis of glucuronides with liquid chromatography-mass spectrometric techniques and methods. Curr Drug Metab. 2010 Sep;11[7]:561-82.
[43] de Vlieger JSB, Giezen MJN, Falck D, Tump C, van Heuveln F, Giera M, et al. High temperature liquid chromatography hyphenated with ESI-MS and ICP-MS detection for the structural characterization and quantification of halogen containing drug metabolites. Anal Chim Acta. [Article]. 2011 Jul;698[1-2]:69-76.
[44] Wright P, Miao Z, Shilliday B. Metabolite quantitation: detector technology and MIST implications. Bioanalysis. 2009 Jul;1[4]:831-45.
[45] Nedderman AN. Metabolites in safety testing: metabolite identification strategies in discovery and development. Biopharm Drug Dispos. 2009 May;30[4]:153-62.
[46] Leclercq L, Cuyckens F, Mannens GS, de Vries R, Timmerman P, Evans DC. Which human metabolites have we MIST? Retrospective analysis, practical aspects, and perspectives for metabolite identification and quantification in pharmaceutical development. Chem Res Toxicol. 2009 Feb;22[2]:280-93.
[47] Lappin G, Seymour M. Addressing metabolite safety during first-in-man studies using ^{14}C-labeled drug and accelerator mass spectrometry. Bioanalysis. [Review]. 2010 Jul;2[7]:1315-24.
[48] Deng Y, Wu JT, Zhang H, Olah TV. Quantitation of drug metabolites in the absence of pure metabolite standards by high-performance liquid chromatography coupled with a chemiluminescence nitrogen detector and mass spectrometer. Rapid Commun Mass Spectrom. 2004;18[15]:1681-5.
[49] Espina R, Yu LN, Wang JY, Tong Z, Vashishtha S, Talaat R, et al. Nuclear Magnetic Resonance Spectroscopy as a Quantitative Tool To Determine the Concentrations of Biologically Produced Metabolites: Implications in Metabolites in Safety Testing. Chem Res Toxicol. [Article]. 2009 Feb;22[2]:299-310.
[50] Fries HE, Evans CA, Ward KW. Evaluation of evaporative light-scattering detection for metabolite quantification without authentic analytical standards or radiolabel. J Chromatogr B Analyt Technol Biomed Life Sci. 2005 May 25;819[2]:339-44.
[51] Tozuka Z, Aoyama S, Nozawa K, Akita S, Oh-Hara T, Adachi Y, et al. Comprehensive Quantitative and Qualitative Liquid Chromatography-Radioisotope-Mass Spectrometry

Analysis for Safety Testing of Tolbutamide Metabolites Without Standard Samples. J Pharm Sci. [Article]. 2011 Sep;100[9]:4024-36.

[52] Nilsson LB, Eklund G. Direct quantification in bioanalytical LC-MS/MS using internal calibration via analyte/stable isotope ratio. J Pharm Biomed Anal. 2007 Feb 19;43[3]:1094-9.

[53] Osborne R, Thompson P, Joel S, Trew D, Patel N, Slevin M. The analgesic activity of morphine-6-glucuronide. Br J Clin Pharmacol. 1992 Aug;34[2]:130-8.

[54] Meyer MR, Maurer HH. Current status of hyphenated mass spectrometry in studies of the metabolism of drugs of abuse, including doping agents. Anal Bioanal Chem. 2012 Jan;402[1]:195-208.

[55] Maurer HH. Analytical toxicology. EXS. 2010;100:317-37.

[56] Al-Asmari AI, Anderson RA. Comparison of Nonhydrolysis and Hydrolysis Methods for the Determination of Buprenorphine Metabolites in Urine by Liquid Chromatography-Tandem Mass Spectrometry. J Anal Toxicol. [Article]. 2008 Nov-Dec;32[9]:744-53.

[57] Tsujikawa K, Kuwayama K, Kanamori T, Iwata Y, Ohmae Y, Inoue H, et al. Optimized conditions for the enzymatic hydrolysis of alpha-hydroxytriazolam-glucuronide in human urine. J Health Sci. [Article]. 2004 Jun;50[3]:286-9.

[58] Costa S, Barroso M, Castanera A, Dias M. Design of experiments, a powerful tool for method development in forensic toxicology: application to the optimization of urinary morphine 3-glucuronide acid hydrolysis. Anal Bioanal Chem. [Article; Proceedings Paper]. 2010 Apr;396[7]:2533-42.

[59] Regan SL, Maggs JL, Hammond TG, Lambert C, Williams DP, Park BK. Acyl glucuronides: the good, the bad and the ugly. Biopharm Drug Dispos. 2010 Oct;31[7]:367-95.

[60] Roskar R, Sollner Dolenc M. Determination of benzodiazepines in urine via benzophenone derivatives using liquid chromatography-tandem mass spectrometry. Arh Hig Rada Toksikol. 2010 Dec 1;61[4]:381-8.

[61] Soars MG, Mattiuz EL, Jackson DA, Kulanthaivel P, Ehlhardt WJ, Wrighton SA. Biosynthesis of drug glucuronides for use as authentic standards. J Pharmacol Toxicol Methods. 2002 May-Jun;47[3]:161-8.

[62] Dragan CA, Buchheit D, Bischoff D, Ebner T, Bureik M. Glucuronide production by whole-cell biotransformation using genetically engineered fission yeast Schizosaccharomyces pombe. Drug Metab Dispos. 2010 Mar;38[3]:509-15.

[63] Trontelj J, Bogataj M, Marc J, Mrhar A. Development and validation of a liquid chromatography-tandem mass spectrometry assay for determination of raloxifene and its metabolites in human plasma. J Chromatogr B Analyt Technol Biomed Life Sci. 2007 Aug 15;855[2]:220-7.

[64] Volkel W, Colnot T, Csanady GA, Filser JG, Dekant W. Metabolism and kinetics of bisphenol a in humans at low doses following oral administration. Chem Res Toxicol. 2002 Oct;15[10]:1281-7.

[65] Koseki N, Nakashima A, Nagae Y, Masuda N. Simultaneous quantitative determination of cyclosporine A and its three main metabolites (AM1, AM4N and AM9] in human

blood by liquid chromatography/mass spectrometry using a rapid sample processing method. Rapid Commun Mass Spectrom. 2006;20[5]:733-40.
[66] Nakamura M. Analyses of benzodiazepines and their metabolites in various biological matrices by LC-MS(/MS). Biomed Chromatogr. [Review]. 2011 Dec;25[12]:1283-307.
[67] Plenis A, Baczek T. Modern chromatographic and electrophoretic measurements of antidepressants and their metabolites in biofluids. Biomed Chromatogr. 2011 Jan;25[1-2]:164-98.
[68] Teunissen SF, Rosing H, Schinkel AH, Schellens JH, Beijnen JH. Bioanalytical methods for determination of tamoxifen and its phase I metabolites: a review. Anal Chim Acta. 2010 Dec 17;683[1]:21-37.
[69] Hakala KS, Link M, Szotakova B, Skalova L, Kostiainen R, Ketola RA. Characterization of metabolites of sibutramine in primary cultures of rat hepatocytes by liquid chromatography-ion trap mass spectrometry. Anal Bioanal Chem. 2009 Feb;393[4]:1327-36.
[70] Bakhtiar R, Majumdar TK. Tracking problems and possible solutions in the quantitative determination of small molecule drugs and metabolites in biological fluids using liquid chromatography-mass spectrometry. J Pharmacol Toxicol Methods. 2007 May-Jun;55[3]:262-78.
[71] Xu RN, Fan L, Rieser MJ, El-Shourbagy TA. Recent advances in high-throughput quantitative bioanalysis by LC-MS/MS. J Pharm Biomed Anal. 2007 Jun 28;44[2]:342-55.
[72] Li W, Luo S, Smith HT, Tse FL. Simultaneous determination of midazolam and 1'-hydroxymidazolam in human plasma by liquid chromatography with tandem mass spectrometry. Biomed Chromatogr. 2007 Aug;21[8]:841-51.
[73] Chambers E, Wagrowski-Diehl DM, Lu Z, Mazzeo JR. Systematic and comprehensive strategy for reducing matrix effects in LC/MS/MS analyses. J Chromatogr B Analyt Technol Biomed Life Sci. 2007 Jun 1;852[1-2]:22-34.
[74] Pucci V, Di Palma S, Alfieri A, Bonelli F, Monteagudo E. A novel strategy for reducing phospholipids-based matrix effect in LC-ESI-MS bioanalysis by means of HybridSPE. J Pharm Biomed Anal. 2009 Dec 5;50[5]:867-71.
[75] Ardjomand-Woelkart K, Kollroser M, Li L, Derendorf H, Butterweck V, Bauer R. Development and validation of a LC-MS/MS method based on a new 96-well Hybrid-SPE((TM))-precipitation technique for quantification of CYP450 substrates/metabolites in rat plasma. Anal Bioanal Chem. [Article]. 2011 Jun;400[8]:2371-81.
[76] Herman JL. The use of turbulent flow chromatography and the isocratic focusing effect to achieve on-line cleanup and concentration of neat biological samples for low-level metabolite analysis. Rapid Commun Mass Spectrom. [Article]. 2005;19[5]:696-700.
[77] Saunders KC, Ghanem A, Boon Hon W, Hilder EF, Haddad PR. Separation and sample pre-treatment in bioanalysis using monolithic phases: A review. Anal Chim Acta. 2009 Oct 12;652[1-2]:22-31.
[78] Hsieh Y, Korfmacher WA. Increasing speed and throughput when using HPLC-MS/MS systems for drug metabolism and pharmacokinetic screening. Curr Drug Metab. 2006 Jun;7[5]:479-89.

[79] Jemal M, Xia YQ. LC-MS Development strategies for quantitative bioanalysis. Curr Drug Metab. 2006 Jun;7[5]:491-502.

[80] Zang X, Luo R, Song N, Chen TK, Bozigian H. A novel on-line solid-phase extraction approach integrated with a monolithic column and tandem mass spectrometry for direct plasma analysis of multiple drugs and metabolites. Rapid Commun Mass Spectrom. 2005;19[22]:3259-68.

[81] Berginc K, Trdan T, Trontelj J, Kristl A. HIV Protease Inhibitors: Garlic Supplements and First-pass Intestinal Metabolism Impact on the Therapeutic Efficacy. Biopharm Drug Dispos. [Article]. 2010 Nov-Dec;31[8-9]:495-505.

[82] Kolmonen M, Leinonen A, Kuuranne T, Pelander A, Ojanpera I. Hydrophilic interaction liquid chromatography and accurate mass measurement for quantification and confirmation of morphine, codeine and their glucuronide conjugates in human urine. J Chromatogr B. [Article]. 2010 Nov;878[29]:2959-66.

[83] Hsieh Y. Potential of HILIC-MS in quantitative bioanalysis of drugs and drug metabolites. J Sep Sci. 2008 May;31[9]:1481-91.

[84] Naidong W, Shou W, Chen YL, Jiang X. Novel liquid chromatographic-tandem mass spectrometric methods using silica columns and aqueous-organic mobile phases for quantitative analysis of polar ionic analytes in biological fluids. J Chromatogr B Biomed Sci Appl. 2001 Apr 25;754[2]:387-99.

[85] Shou WZ, Chen YL, Eerkes A, Tang YQ, Magis L, Jiang X, et al. Ultrafast liquid chromatography/tandem mass spectrometry bioanalysis of polar analytes using packed silica columns. Rapid Commun Mass Spectrom. 2002;16[17]:1613-21.

[86] Song Q, Naidong W. Analysis of omeprazole and 5-OH omeprazole in human plasma using hydrophilic interaction chromatography with tandem mass spectrometry (HILIC-MS/MS)--eliminating evaporation and reconstitution steps in 96-well liquid/liquid extraction. J Chromatogr B Analyt Technol Biomed Life Sci. 2006 Jan 2;830[1]:135-42.

[87] Hsieh Y, Duncan CJ, Liu M. A mixed-mode liquid chromatography-tandem mass spectrometric method for the determination of cytarabine in mouse plasma. J Chromatogr B Analyt Technol Biomed Life Sci. 2007 Jul 1;854[1-2]:8-12.

[88] Hsieh Y, Korfmacher W. The role of hyphenated chromatography-mass spectrometry techniques in exploratory drug metabolism and pharmacokinetics. Curr Pharm Des. 2009;15[19]:2251-61.

[89] Hsieh Y, Duncan CJ, Brisson JM. Porous graphitic carbon chromatography/tandem mass spectrometric determination of cytarabine in mouse plasma. Rapid Commun Mass Spectrom. 2007;21[5]:629-34.

[90] Chen S, Kord A. Theoretical and experimental comparison of mobile phase consumption between ultra-high-performance liquid chromatography and high performance liquid chromatography. J Chromatogr A. 2009 Aug 21;1216[34]:6204-9.

[91] Kortz L, Helmschrodt C, Ceglarek U. Fast liquid chromatography combined with mass spectrometry for the analysis of metabolites and proteins in human body fluids. Anal Bioanal Chem. 2011 Mar;399[8]:2635-44.

[92] Dekant W, Volkel W. Human exposure to bisphenol A by biomonitoring: methods, results and assessment of environmental exposures. Toxicol Appl Pharmacol. 2008 Apr 1;228[1]:114-34.
[93] Williamson LN, Bartlett MG. Quantitative liquid chromatography/time-of-flight mass spectrometry. Biomed Chromatogr. 2007 Jun;21[6]:567-76.
[94] Matuszewski BK, Constanzer ML, Chavez-Eng CM. Strategies for the assessment of matrix effect in quantitative bioanalytical methods based on HPLC-MS/MS. Anal Chem. 2003 Jul 1;75[13]:3019-30.
[95] Yamada T, Mino Y, Yagi T, Naito T, Kawakami J. Rapid simultaneous determination of voriconazole and its N-oxide in human plasma using an isocratic high-performance liquid chromatography method and its clinical application. Clin Biochem. [Article]. 2012 Jan;45[1-2]:134-8.
[96] Fox D, O'Connor R, Mallon P, McMahon G. Simultaneous determination of efavirenz, rifampicin and its metabolite desacetyl rifampicin levels in human plasma. J Pharm Biomed Anal. [Article]. 2011 Dec;56[4]:785-91.
[97] Ahmad L, Iqbal Z, Nazir S, Shah Y, Khan A, Khan MI, et al. Optimization and validation of HPLC-UV method for simultaneous determination of omeprazole and its metabolites in human plasma: effects of various experimental conditions and parameters. J Liq Chromatogr Relat Technol [Article]. 2011;34[15]:1488-501.
[98] Kaushik R, Levine B, LaCourse WR. A brief review: HPLC methods to directly detect drug glucuronides in biological matrices (Part I). Anal Chim Acta. [Review]. 2006 Jan;556[2]:255-66.
[99] de Jong WH, de Vries EG, Kema IP. Current status and future developments of LC-MS/MS in clinical chemistry for quantification of biogenic amines. Clin Biochem. 2011 Jan;44[1]:95-103.
[100] Vovk T, Bogataj M, Roskar R, Kmetec V, Mrhar A. Determination of main low molecular weight antioxidants in urinary bladder wall using HPLC with electrochemical detector. Int J Pharm. [Article; Proceedings Paper]. 2005 Mar;291[1-2]:161-9.
[101] Nobilis M, Vybiralova Z, Szotakova B, Sladkova K, Kunes M, Svoboda Z. High-performance liquid chromatographic determination of tiapride and its phase I metabolite in blood plasma using tandem UV photodiode-array and fluorescence detection. J Chromatogr B. [Article]. 2011 Dec;879[32]:3845-52.
[102] Van Eeckhaut A, Lanckmans K, Sarre S, Smolders I, Michotte Y. Validation of bioanalytical LC-MS/MS assays: evaluation of matrix effects. J Chromatogr B Analyt Technol Biomed Life Sci. 2009 Aug 1;877[23]:2198-207.
[103] Taylor PJ. Matrix effects: the Achilles heel of quantitative high-performance liquid chromatography-electrospray-tandem mass spectrometry. Clin Biochem. 2005 Apr;38[4]:328-34.
[104] Bonfiglio R, King RC, Olah TV, Merkle K. The effects of sample preparation methods on the variability of the electrospray ionization response for model drug compounds. Rapid Commun Mass Spectrom. [Article]. 1999;13[12]:1175-85.

[105] Matuszewski BK. Standard line slopes as a measure of a relative matrix effect in quantitative HPLC-MS bioanalysis. J Chromatogr B Analyt Technol Biomed Life Sci. 2006 Jan 18;830[2]:293-300.

[106] Xia YQ, Jemal M. Phospholipids in liquid chromatography/mass spectrometry bioanalysis: comparison of three tandem mass spectrometric techniques for monitoring plasma phospholipids, the effect of mobile phase composition on phospholipids elution and the association of phospholipids with matrix effects. Rapid Commun Mass Spectrom. 2009 Jul;23[14]:2125-38.

[107] Kloepfer A, Quintana JB, Reemtsma T. Operational options to reduce matrix effects in liquid chromatography-electrospray ionization-mass spectrometry analysis of aqueous environmental samples. J Chromatogr A. 2005 Mar 4;1067[1-2]:153-60.

[108] Stokvis E, Rosing H, Beijnen JH. Stable isotopically labeled internal standards in quantitative bioanalysis using liquid chromatography/mass spectrometry: necessity or not? Rapid Commun Mass Spectrom. 2005;19[3]:401-7.

[109] Volkel W, Bittner N, Dekant W. Quantitation of bisphenol A and bisphenol A glucuronide in biological samples by high performance liquid chromatography-tandem mass spectrometry. Drug Metab Dispos. 2005 Nov;33[11]:1748-57.

[110] Briscoe CJ, Hage DS. Factors affecting the stability of drugs and drug metabolites in biological matrices. Bioanalysis. 2009 Apr;1[1]:205-20.

[111] Shipkova M, Armstrong VW, Oellerich M, Wieland E. Acyl glucuronide drug metabolites: toxicological and analytical implications. Ther Drug Monit. 2003 Feb;25[1]:1-16.

[112] Li WK, Zhang J, Tse FLS. Strategies in quantitative LC-MS/MS analysis of unstable small molecules in biological matrices. Biomed Chromatogr. [Review]. 2011 Jan-Feb;25[1-2]:258-77.

[113] Jemal M, Ouyang Z, Chen BC, Teitz D. Quantitation of the acid and lactone forms of atorvastatin and its biotransformation products in human serum by high-performance liquid chromatography with electrospray tandem mass spectrometry. Rapid Commun Mass Spectrom. 1999;13[11]:1003-15.

Chapter 6

Using High Performance Liquid Chromatography (HPLC) for Analyzing Feed Additives

Jolanta Rubaj, Waldemar Korol and Grażyna Bielecka

Additional information is available at the end of the chapter

1. Introduction

Feed additives have been widely used in animal nutrition. Recommendations concerning using feed additives, their categories and the description of requirements related to such additives can be found in the regulation (EU) No 1831/2003 [1]. Detailed regulations oblige the entities launching feed additives on the market to specify the methods used to analyse active substances of additives for the needs of official feed control. Official feed control is implemented in order to monitor adequate and safe use of feed additives in animal nutrition. Moreover, controlling feed production in this respect results in improving the quality and safety of animal products for consumers.

HPLC methods have been widely used in the analyses of feed additives, such as vitamins, feed colorants, antioxidants, amino acids and coccidiostats in preparations, premixes and feed mixtures. It is relatively simple to analyse preparations as they are usually composed of a particular active substance and a carrier. A premix is more complex feed consisting of a combination of a dozen different feed additives on a mineral (calcium carbonate) or organic (wheat bran) carrier. Complete feed mixtures used in animal nutrition, produced by combining premixes with feed materials are often greased and subjected to further hydro- or barothermal processing, e.g. pelleting, extrusion or expanding. In order to counteract decomposition of the active substance, feed additives are secured by protective coating, e.g. vitamin A, canthaxanthin, which enhances their durability in the feed matrix. A specific protection of feed additives by protective coating, thermal processing, greasing the feed, varied composition of feed materials in mixtures may hinder the transfer of the active substance into a solution during extraction and purification of the extract.

The key issue becomes selection of a chromatographic column (in a normal phase or reversed phase), mobile phase, detector, as well as optimization the conditions of chromatographic separation. In examining feed additives with the use of HPLC methods the most frequently used types are spectrophotometric detection (UV-VIS), detection with the help of diode array and fluorescent detection. The choice of optimal parameters for chromatographic separation is done during validation of the method. The analyses in this respect should be accompanied by an assessment of the method's robustness [2]. A practical way to verify the precision of a method in a laboratory (repeatability) and in interlaboratory studies (reproducibility) is taking advantage of Horwitz equation [3].

Using HPLC methods for examining feed additives was the subject matter of numerous studies on the basis of which official methods of analysing certain feed additives were developed. The studies presented the basic validation parameters for the methods of examining the content of fat-soluble vitamins [4-7], water-soluble vitamins [8,9], coccidiostats [4,10-12], and other feed additives, amino acids, methionine hydroxy analog and antioxidants [13,4,8]. However, in case of carotenoids such as canthaxanthin or apocarotenoic acid ester official methods of examining these additives are still based on spectrophotometric measurement rather than on HPLC methods [13,8].

New requirements have been introduced regarding the validation parameters for the methods of analysing feed additives, e.g. those listed in the regulation No 882/2004 [14], taking into consideration, among others, the uncertainty of measurement. It is necessary to determine the uncertainty of measurement with a particular method in order to interpret adequately the result of examining feed additives in feedingstuffs and to assess acceptable tolerance in compliance with the requirements of the regulation No 939/2010 [15]. The new requirements in this area should be taken into account while validating the methods of testing feed additives in order to solve the problem of interpreting the results.

The aim of the present work was to offer a review of HPLC methods used for analysing active substances in certain feed additives, with regard to current requirements defined in the regulations. In some justified cases the results of the authors' own studies were presented, as well as the procedures for determining vitamins B_1 and B_2, canthaxanthin and methionine hydroxy analog (MHA). Some validation parameters were presented, such as the limit of the method's quantification (LOQ), linearity of the calibration curve, repeatability, within-laboratory reproducibility (intermediate precision), recovery and the uncertainty of measurement. Also, the results of verifying the developed methods and laboratories participating in proficiency testing (PT) were demonstrated. The ways of quality assurance of the tests in reference to HPLC methods were discussed. The work presents the method of assessing combined standard uncertainty of measurement with the use of experimental approaches based on within-laboratory reproducibility and calculations for the bias of the method on the basis of CRM studies or PT results.

2. A review of HPLC methods used to determine feed additives

Feed additives are commonly used in animal nutrition, e.g. in order to supplement the animals' requirement for nutrients (amino acids), useful micro components (vitamins), to prevent invasive diseases (e.g. coccidiosis – coccidiostats), to reduce oxidation processes of feed's components (antioxidants), to enhance the dietary value and quality in food products of animal origin (amino acids, carotenoids– egg yolk coloration).

In case of some vitamins and antioxidants their maximum content in feed mixtures was determined, e.g. for vitamins A and D_3, antioxidants (ethoxyquin, BHA, BHT), carotenoids (canthaxanthin, apocarotenoic acid ester and others). Maximum contents are subject to official control in reference to their conformity with the requirements related to the safety of feedingstuffs. Additionally, a feed manufacturer is obliged to declare the content of feed additives on the label of a premix or feed mixture. Thus, it is necessary to have access to analytical methods for testing the content of feed additives in a wide range of concentrations in the preparations containing additives, premixes and feed mixtures.

Table 1 presents examples of well-known HPLC methods for examining fat-soluble vitamins in feedingstuffs, including the official methods accepted by the European Commission. A commonly used method of preparing a sample for analysing the content of vitamin A is alkaline hydrolysis during which gelatin/sugar cross-linked beadlets which protect vitamin A in the form of retinol acetate are solved and then purified by liquid-liquid extraction. An interesting option for purifying vitamin A extracts from feed mixture with the use of the SPE technique was presented by Fedder & Ploger [7]. The step of alkaline hydrolysis is also used while determining vitamins E and D_3. Chromatographic separation of vitamins, except for vitamin D_3 where preliminary separation and fraction collection are necessary [5], does not present any serious problems. However, a problem may be posed by the quality and durability of a standard, as well as poor precision resulting from a too low of analytical weight [16]. It is necessary to verify vitamin A standards with the use of a spectrophotometric method [4].

The official methods of determining the content of water-soluble vitamins, such as B_1, B_2 and B_6 are based on spectrophotometric or fluorometric methods [8,13]. The results of analyses using these methods may be biased with errors due to some interferences from other substances in the variable feed matrix. Recently HPLC methods to determine vitamins B_1, B_2, B_6, nicotinic acid and nicotinamide in mineral preparations and mixtures [8], as well as vitamin B_1 in feed mixtures and premixes [9] were published (Table 2). Due to the high limit of quantification for vitamin B_1 amounting to 5 mg/kg according to Italian Official Method [9], it cannot be used for analysing vitamin B_1 in typical feed mixtures to which it is normally added at the amount of 2-4 mg/kg. Moreover, the method quoted above makes it possible to examine vitamin B_1 added to feedingstuffs but not the total content of this vitamin, regarding its presence in feed materials. It is thus necessary to have access to chromatographic methods enabling the examination of water-soluble vitamins present in feed materials and added in the form of feed additives. The procedures of HPLC methods of vitamins B_1 and B_2 developed during the authors' own studies are presented later in the chapter [17,18].

Analyte, matrix reference	Extraction, extract clean up and chromatography parameters	Performance parameters
Vitamin A in feedingstuffs and premixes, Commission Regulation, 2009 [4]	Sample hydrolyze with ethanolic KOH, extraction into light petroleum, evaporation and dissolution in methanol, reversed phase HPLC, C_{18} column (250 x 4 mm) 5 μm or 10 μm packing , mobile phase: methanol and water 98+2 (v/v), fluorescence (or UV) detector: excitation 325 nm; emission 475 nm or UV detector (325 nm)	LOQ=2000 IU/kg; SD_r (%): 3.0-8.1; SD_R (%): 6.2-20.0;
Vitamin E in feedingstuffs and premixes, Commission Regulation, 2009 [4]	Sample hydrolyze with ethanolic KOH solution, extraction into light petroleum, evaporation and dissolution in methanol, reversed phase HPLC, C_{18} column (250 x 4 mm) 5 μm or 10 μm packing , mobile phase: methanol and water 98+2 (v/v), fluorescence detector: excitation 295 nm; emission 330 nm or UV detector (292 nm)	LOD=2 mg/kg; LOQ=10 mg/kg; SD_r (%): 2.2-4.1; SD_R (%): 4.8-12.7
Vitamin D_3 in feedingstuffs and premixes [5]	Feed saponification and extraction with diethyl ether; evaporation and solvation in methanol; reverse phase preparative chromatography; eluat collection with vitamin D_3 , evaporation and next solvation in n-hexan or isooctane; normal phase chromatography, column 250 mm x 4 mm, Si-60, 5 μm packing; UV detection at 264 nm; mobile phase: preparative column: methanol -water (92+8)), analytical column – n-hexan – dioxan - isopropanol (94.5+5+0.5)	LOQ=1000 IU/kg; RSD_r to 5000 IU/kg: 1000 IU/kg; 5000-20000 IU/kg:20% 20000-100000 IU/kg: 15%; >100000 IU/kg: 10%
Vitamin K_3 in feedingstuffs, premixes and feed additives [6]	Sample extraction with chloroform, transfer vitamin K substances to free menadion; clean-up with Celite and sodium sulphate anhydrous; normal phase chromatography, Si-60 column 250 mm x 4 mm, 10 μm packing; UV detection at 251 nm	LOQ=0.5 mg/kg RSD_r at 1 mg/kg: 10%; SD_r at 8 mg/kg: 4%; SD_r at the level >2500 mg/kg: 3%;
Vitamin A and E, feedingstuffs [7]	Sample hydrolyze with ethanolic KOH solution; clean-up on SPE column; elution in ethyl acetate, evaporation and dissolution in methanol; reverse phase chromatography, ODS2 column 250 mm x 4.6 mm, 5 μm packing; UV detection at 325 nm (vitamin A) and 292 nm (vitamin E).	Range: Vitamin A=1250-20000 U/kg; Vitamin E=3-300 mg/kg RSD_{ip}= 21% (vit. A), 11% vit. E; Rec: vit. A - 80%, vit. E - 110%

Table 1. HPLC methods for the analysis of fat-soluble vitamins in feeds

Analyte, matrix reference	Extraction, extract clean up and chromatography parameters	Performance Parameters
Vitamin B_1, B_2, B_6, NA, NSA in premixes and mineral feeds [8]	Extraction with methanol -Titriplex solution, clean-up on membrane filter 0.45 μm, reverse phase HPLC coupled to UV or diode array detector, column Nucleosil 250 x 3.0 mm, 5 μm packing; mobile phase: mixture of water solution of acetonitrile and acetic acid	Range, mg/kg: B1= 320-7940; B2= 868-15990; B6=627-11530; NA=4520-77850; NSA= 3665-61230; RSD_r(%) = 2.1-5.1; RSD_R(%) = 4.2-30.2
Vitamin B_1 in feedingstuffs and premixes [9]	Extraction with methanol ; clean-up on SPE, reverse phase HPLC, coupled to a fluorescence detector, excitation at 360 nm, emission at 430 nm	LOQ=5 mg/kg; Range, mg/kg: 7 – 484; RSD_r(%)=4.2-4.7 RSD_R(%)=5-13 Rec.(%)=88-97

B_1 – thiamine; B_2 – riboflavin, B_6 – pirydoxin; NA- nicotin acid; NSA – nicotinamid;

Table 2. HPLC methods for the analysis of water-soluble vitamins in feeds

Numerous methods have been developed to examine coccidiostats in feedingstuffs with the use of high performance liquid chromatography. Examples of such methods are presented in Table 3. Satisfactory precision of such methods has been obtained, in conformity with that calculated from the Horwitz equation [3] and with the requirements of the Commission's Decision [19], which enables analyzing coccidiostats at the levels declared by manufacturers. Due to the hazard of cross-contamination with the remains of coccidiostats found in non-target mixed feeds on the production line and the risk of carry-over the remains of contamination onto the products of animal origin, it is necessary to continue lowering the limit of methods' quantification in order to control safe use of coccidiostats.

Table 4 presents HPLC methods for testing other feed additives, such as antioxidants, amino acids, methionine hydroxy analog. The official AOAC method for determining ethoxyquin was verified in testing pet food and meat meal [13]. Due to the determination of maximum content of antioxidants in feed mixtures for animals used for food production there is a necessity to check this method in testing typical feed mixtures and premixes. Amino acids are present in typical feed materials as components of proteins. In order to determine amino acids in feed materials it is necessary to subject proteins to hydrolysis and next to separate amino acids using ion-exchange chromatography and apply derivatization. With intensive animal production it is necessary to supplement the deficiency of amino acids, such as lysine, methionine, threonine and tryptophan. New feed additives have been registered recently, such as arginine, valine and cysteine. The official AOAC methods make it possible to determine mainly the composition and content of amino acids in feedingstuffs after hydrolysis [13], yet validation parameters of the determination methods have not been defined for all synthetic amino acids. Moreover, the precision parameters of the method used to determine amino acids with sodium metabisulphite or the hydrobromic acid method were in many cases unsatisfactory, which was confirmed by the values of the Horwitz ratio higher than 2, e.g. in a feed mixture for broiler chickens the HorRat (H) values amounted to 1.7-3.6, with mean 2.5, while satisfactory H values are within the range of 0.5 > H > 2. This requires further studies with the use of high performance liquid chromatography in order to determine the total content of amino acids after hydrolysis and added amino acids.

Analyte, matrix reference	Extraction, extract clean up and chromatography parameters	Performance Parameters
Halofuginone, medicated feeds [10,4]	Ethyl acetate extraction, purification by ion-exchange chromatography, reversed phase HPLC with UV detection at 243 nm, C_{18} column (300 x 10 mm) 10 µm packing, mobile phase: mixture of acetonitrile and ammonium acetate buffer solution	LOQ = 1 mg/kg ; RSD_r (%): 2.0-4.7; Rec.(%) 75.3-98.0; at the level of 3 mg/kg
Lasalocid, monensin, salinomycin and narasin, poultry feed [12]	Methanol extraction without clean-up, derivatization with 2,4-dinitrophenylhydrazide (DNP) in acid medium at 55 °C, ODS column (150 x 4.6 mm, 5 µm); eluent: methanol – 1.5% aqueous acetic acid (90:10, v/v), UV detection at 305/392 nm	LOQ = 40 mg/kg conc. range 50-150 mg/kg; RSD_r (%): 4-10; Rec. 85-100%
Lasalocid, poultry feeds, premixes [11,4]	Extraction into acidified (HCl) methanol, agitation in ultrasonic bath at 40 ºC, filtration through a 0.45 µm filter, reversed phase HPLC, C_{18} column (125 x 4 mm) 5 µm packing , mobile phase: mixture of phosphorus buffer solution and methanol 5+95 (v/v), fluorescence detector: excitation 310 nm; emission 419 nm	LOD=5 mg/kg; LOQ=10 mg/kg; RSD_r (%): 2.1-5.4; RSD_R (%): 5.0-10.7; Rec : feed ≥ 80%; premixes ≥ 90%
Robenidine, feedingstuffs, premixes [4]	Extraction into acidified (HCl) methanol, clean-up on an aluminum oxide column; reversed phase HPLC, UV detection at 317 nm; C_{18} column (300 x 4 mm) 10 µm packing; mobile phase: mixture of acetonitrile and sodium and potassium phosphate solution	LOQ=5 mg/kg SD_r (%): 3.3-5.4; SD_R (%): 9.7-10.1; Rec. for blanc sample ≥ 85%
Diclazuril, feedingstuffs, premixes [4]	Extraction with acidified methanol with internal standard; purification on C_{18} solid phase extraction cartridge (SPE), evaporation and dissolution in DMF; reversed phase gradient HPLC, Hypersil ODS column, 100 mm x 4.6 mm, 3 µm packing; mobile phase: (1) aqueous solution of ammonium acetate and tetrabutyl-ammonium hydrogen sulphate, (2) acetonitrile, (3) methanol	LOD=0.1 mg/kg; LOQ=0.5mg/kg; SD_r (%): 1.9-17.3; SD_R (%): 7.4-18.6; Rec. for blanc sample ≥ 80%

SD_r - standard deviation of repeatability; SD_R - standard deviation of reproducibility; rec. – recovery; LOD – limit of determination; LOQ – limit of quantification; DMF – N,N-dimethylformamide;

Table 3. HPLC methods for the analysis of coccidiostats in feeds

Analyte, matrix reference	Extraction, extract clean up and chromatography parameters	Performance Parameters
Ethoxyquin in pet-food and meat meal [13]	Extraction with acetonitrile without clean-up, reversed phase HPLC, C_{18} column (250 x 4.6 mm) 5 µm packing , mobile phase: acetonitrile and 0.01 M ammonium acetate (70 + 30, v/v); fluorescence detector: excitation 360 nm; emission 432	Method range: 0.5-300 mg/kg; SD_r (%): 4.5-32; SD_R (%): 4.5-55; Rec. 60-83%
Phenolic anti-oxidants* in fats [13]	Extraction with acetonitrile, extract is concentrated and diluted with 2-propanol; reversed phase gradient HPLC, C_{18} column with guard column; mobile phase: (1) 5% acetic acid in water, (2) acetonitrile-methanol (1 +1, v/v)	Method range: 10-200 mg/kg; SD_r (%): 2.1-11.5; SD_R (%): 2.7-21.5; Rec. 83-103%
Tryptophan in feedingstuffs and premixes[4]	For total tryptophan alkaline hydrolise with saturated barium hydroxide solution; for free tryptophan extraction under mild acid conditions; reversed phase HPLC with fluorescence detector, excitation 280 nm, emission 356 nm; C_{18} column (125 x 4 mm) 3 µm packing; mobile phase: acetic acid and 1,1,1 trichloro-2-methyl-2-propanol solution, pH 5.00	Feedingstuffs: SD_r (%): 1.6-1.9; SD_R (%): 2.2-6.3; Feed materials: SD_r (%): 0.8-1.3; SD_R (%): 4.1-5.1;
Amino acids in feeds** [13]	Performic acid oxidation of the sample to oxidize cystine and methionine; amino acids liberation from protein by hydrolysis with 6 M HCl; dilution with sodium citrate buffer; amino acid separation on ion-exchange chromatograph with ninhydrin post-column derivatisation	Broiler feed: SD_r (%): 1.1-4.7; SD_R (%): 6.0-19.8; HorRat: 1.7-3.6 ~ 2.5
MHA in feedingstaffs and premixes [8]	Extraction with water solution of acetonitrile; reversed phase HPLC with UV detection at 210 nm; RoSil-NH_2 column (250 mm x 4.6 mm, 5 µm packing) with guard column; mobile phase: acetonitrile with phosphoric acid solution (23+77)	LOD=0.2 g/kg LOQ=0.5 g/kg

*PG-propyl gallate; THBT – 2,4,5-trihydroksybutyrophenone; TBHQ – *tetr*-butylhydroquinone; NDGA – nordihydroguaiaretic acid; Ionox 100 – 2,6 di-*tert*-butyl-4-hydroxymethylphenol BHA- 3-*tert*- butyl-4-hydroxyanisole; BHT – 3,5-di-*tert*-butyl—4-hydroxytoluene; OG, DG – octyl and dodecyl gallate

**Sodium metabisulphite method and hydrobromic acid method not applicable to determination of tyrosine and tryptophan; acid hydrolysis method not applicable for methionine, cysteine and tryptophan; MHA - methionine hydroxy analog

Table 4. HPLC methods for the analysis of other feed additives in feeds

3. A description of feed matrix and active substances in feed additives

The difficulty in determining certain feed additives is related with their low stability. In order to obtain a more durable form, resistant to the manufacturing conditions of feed mixtures, the additives are secured by protective coating. This concerns primarily vitamins A and D_3 , as well as feed colorants, such as canthaxanthin.

Vitamin A is produced in the form of gelatin-and-sugar beadlets or fat beadlets. Each beadlet contains ca. 0.5-0.6 μg of vitamin A, as calculated for retinal (ca. 2 IU). The distribution of beadlets in the feed is not equal and the feed enriched in vitamin A tends to segregate vitamin beadlets during the process of manufacturing and transporting the feedingstuff, especially in case of loose products. On the other hand, pelleting feed mixtures or subjecting them to other barothermal processes, such as extrusion or expanding reduces vitamin segregation, yet it lowers their durability at the same time. Ultimately, the unequal distribution of vitamins in feed may affect the precision and accuracy of results of analyses. Grinding the samples may improve the distribution of vitamin A, yet it will also increase the risk of its oxidation. Vitamin A is chemically unstable and its content and biological activity are reduced along with the presence of oxygen from the air, light, humidity, inorganic acids, choline hydrochloride, microelements and peroxides created in the processes of fat oxidation. It is recommended that samples should be ground immediately prior to an analysis into 1 mm particles. Further grinding of the sample before determining the content of vitamins may lead to their decomposition. A useful guideline regarding the preparation of samples for analyses, including the analyses of unstable feed additives such as vitamins, is provided by the currently issued ISO/FDIS International Standard 6498 [20].

Vitamins are protected against oxidation by antioxidants (e.g. ethoxyquin) which are added to the materials creating beadlets. Some forms which are physically and chemically stable are created in this way, e.g. the oleic form of vitamin D_3 and more stable compounds (menadione bisulfite, thiamine mononitrate and riboflavin phosphate). Feed additives used in feed manufacturing are introduced on mineral carriers or on wheat bran. The most frequently used mineral carrier is fodder chalk. In case of extracting additives from premixes with the use of diluted acids the influence of the carrier on the conditions of extraction should be considered. In such a situation it is recommended that the robustness of a method to slight changes in the analytical procedure or a change of matrix should be verified [2]. In analyzing feedingstuffs the interfering agent is fat which often occurs in significant amounts on feed mixtures (up to 10%). In high-fat samples containing more than 0.25g of fat in an analytical weight while determining fat-soluble vitamins, additional soaps are formed in the saponification process, which hinder the separation of the examined analyte.

It was possible to resolve the problem of interfering substances after using HPLC methods. However, the diversity of matrices and inhomogeneity of feedingstuffs pose numerous analytical problems while determining feed additives, such as vitamins or carotenoid colorants. The biggest difficulty is related to proper clean-up of the extract and selection of adequate conditions for chromatographic separation.

In case of vitamins, while examining the relevance of the producer's declaration and interpreting the result of the examination, one should be aware of the effect brought about by numerous factors on the content of vitamins in feedingstuffs. There were detailed studies in this respect conducted by Coelho [21]. Table 5, based on Coelho's article, presents the method of estimating the summary influence of different factors, such as the type of a premix and the time passed since the moment of its production (column 2), the type of conditions of hydro- and barothermal processing (column 3), feed storage time (column 4) on vitamin retention. The product of the factors in columns 2, 3 and 4 (expressed as a fraction) will let us estimate the retention of a particular vitamin in a particular feed (column 5).

Vitamin	Vitamin, premix (Coelho [21], Table 8)	Pelleting temp., (Coelho[21], Table 11)	Feed storage time (Coelho [21], Table 10)	Total vitamin reten. %
1	2	3	4	5
	2 months	96 °C	2 weeks	2 x 3 x 4
A beadlet	90	88	98	78
D_3 beadlet	91	91	99	82
E acetate 50%	92	91	99	83
Thiamine	77	82	99	63
Riboflavin	91	84	99	76
B_{12}	96	95	100	90
Ca pantothenate	87	84	99	72
Biotin	89	84	99	74
Niacin	90	86	99	77

Table 5. Vitamin stability in premixes and feeds (%), (Coelho, 2002)

Similarly, the authors observed in their own studies that the conditions and premix storage time in the laboratory affected the content of vitamins A and E. Fractioned samples of premix, each weighing 100 g, were stored for 8 months at room temperature (22 °C), in a fridge (5 °C) and in a freezer (-18 °C) (Table 6). In the samples stored at room temperature the content of vitamins after 8 months of storage was reduced to as little as 3% of the initial value. The analyses of vitamin content should be performed immediately after receiving the samples by the laboratory, otherwise the samples should be stored in a fridge until the analyses can be done.

Item	Vitamin retention, %*								
	Month								
	0	1	2	3	4	5	6	7	8
Vitamin A, initial value 2794000 IU/kg:									
- room 22 °C	100	94	82	64	32	13	6	4	3
- refrigerator 5 °C	100	102	99	94	94	93	93	92	85
- freezer -18 °C	100	100	100	100	99	100	98	99	96
Vitamin E, initial value 15.56 g/kg:									
- room 22 °C	100	98	96	94	94	97	95	94	90
- refrigerator 5 °C	100	99	98	100	98	100	100	98	90
- freezer -18 °C	100	99	97	100	99	100	100	100	94

* expanded uncertainty (k=2) of the result of examining is 16% for vitamin A and 12% for vitamin E

Table 6. The results of laboratory retention of vitamins A and E, %

4. Extraction and extract purification

A sample for a quantitative analysis should be prepared in such a way that the isolation of a selected analyte and removal of interfering substances is possible. The condition necessary for adequate quantitative determination in liquid chromatography is eliminating any possibility of coelution and minimization the drift of a basic line. Extractions of the analyzed substances were done classically by shaking the sample with a solvent. In case of vitamins B_1 and B_2 and carotenoids, such as canthaxanthin and apocarotenic acid ester, extraction was performed in an ultrasonic bath. In order to purify the extract, aluminum oxide (e.g. canthaxanthin), celite and anhydrous sodium sulphate (e.g. vitamin K_3) were used and PTFE and Nylon (PA) (0.45 μm or 0.20 μm) syringe filters were applied before injecting on the chromatographic column. Syringe (hydrophobic) PTFE filters are used in case of solutions with high acid and base content, whereas nylon (hydrophilic) filters are used with aqueous and organic solutions. Filtration of extracts is necessary as it prolongs durability of a column due to eliminating permanent contamination which blocks the column's intake and increases back pressure. If needed, the analyte may be concentrated by evaporating the solvent. When there is a risk that the studied analyte becomes oxidized, evaporation is done in neutral gas, e.g. nitrogen or argon.

The removal of gelatin-and-sugar beadlets protecting vitamins A and D_3 is done at the stage of saponification and transforming the vitamins into alcoholic forms. Douša & Břicháč [16] demonstrated that saponification in standard conditions did not affect the results of analyses. In case of canthaxanthin enzymatic hydrolysis through adding trypsin and pepsin is used. While determining the total content of thiamine and/or riboflavin in feedingstuffs (endogenic and added) at the stage of preliminary preparation an ultrasonic bath and also enzymatic hydrolysis (taka-diastase) are applied.

5. Optimization the conditions of chromatographic separation

The method of high performance liquid chromatography (HPLC) with flouorometric detection or a diode array is characterized by sufficient selectivity and sensitivity required to determine feed additives. Each of the developed and verified procedures includes precisely defined stages of hydrolysis and extraction which make it possible to determine the total or added analyte in a sample. Chromatographic separation of the analyzed mixture is affected by the properties of chromatographic arrangement. The developed analytical methods took advantage of chromatography in a regular and reversed arrangement of phases. In the reversed phase methanol and water or acetonitrile and water were used. In the standard arrangement the mobile phase was hexane or chloroform. In the majority of the developed methods presented in this chapter isocratic elution was applied, except for methionine hydroxy analog where gradient elution was used. Adequate separation was achieved through the use of ODS column with 18 atoms of carbon in the chain (C_{18}) and columns with 8 atoms of carbon (C_8) in the alkyl chain in chromatography with reversed phase (RP). In chromatography with the normal phase arrangement columns filled with silica gel were used. The identification and content of analyte was examined with the method of absolute calibration (with external model), analyzing separately the sample and the model and identifying the peaks with the help of retention values, comparing retention time of the identified substance with the retention time of a standard, chromatographed in identical conditions. Examples of chromatograms for the standard extract and the sample of the examined analyte (riboflavin) with the use of fluorescent detector are presented below (Fig.1, Fig.2).

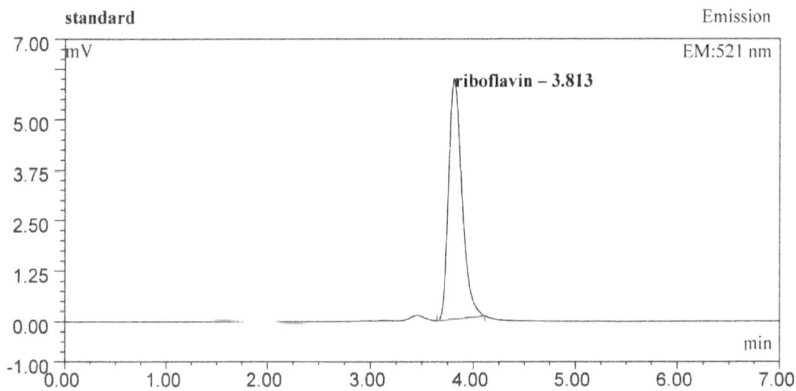

Figure 1. Characteristic chromatograms of riboflavin: chromatogram of standard solution

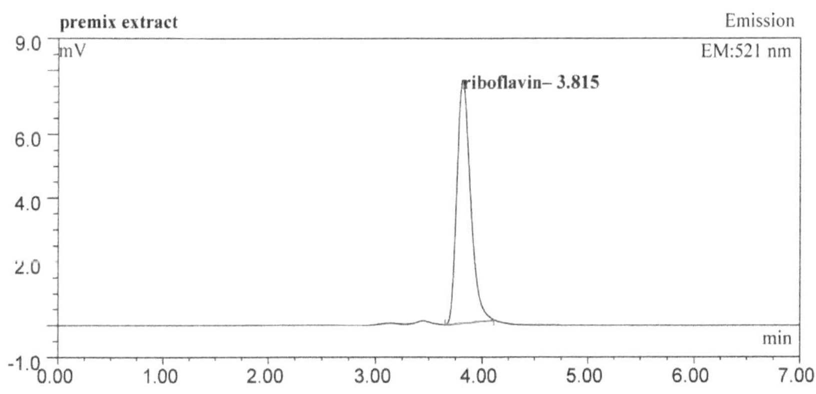

Figure 2. Characteristic chromatograms of riboflavin: chromatogram of premix extract

6. Selected methods of testing feed additives

Selected methods of testing feed additives presented below were validated with the help of a high pressure liquid chromatograph (Dionex P-680) with fluorescence detector (Dionex RF 2000) or with a diode array.

Vitamin B_1 (Thiamine)

The following procedure used for determination of thiamine in premixes and compound feeds was elaborated on the basis of the article published by Rubaj *et al.* [18].

Principle

Vitamin B_1 is extracted with hydrochloric acid of 0.1 mol/l and next oxidized to thiochrome and marked with the use of high performance liquid chromatography (HPLC) with a fluorescence detection.

Reagents and Solvents. All reagents and solvents should be of analytical grade: chloroform; methanol; hydrochloric acid, c=0.1 mol/l; trichloroacetic acid, 50%; sodium hydroxide, 15%; water saturated isobutanol; potassium hexacyanoferrate (III) ; vitamin B_1 standard, taka-diastase, sodium acetate.

Apparatus: laboratory shaker, centrifuge, water bath with Allin condenser, HPLC set with fluorescence detector.

Procedure

Thiamine was extracted from the examined feed sample with 0.1 M hydrochloric acid at 100°C for 30 minutes. In case of compound feedingstuffs 10% taka-diastase solution was added to the samples, and then samples were incubated at 37°C for 17 hours. Afterwards thiamine was oxidized to thiochrom by 1% alkaline $K_3Fe(CN)_6$.

Chromatography

Column	25 cm x 4.6 mm
Stationary phase	LichroCart 250-4, Lichrospher100 NH_2 (5µm)
Mobile phase	Chloroform and methanol, 90+10 (v/v)
Column temperature	25 ºC
Flow rate	2.0 ml/min
Injection	20 µl
Detector	Fluorescence, Ex λ= 365, Em λ=435
Calculation	External standard, peak area, linear regression

Special Comment

This method was applied for the quantification of total content of thiamine in compound feedingstuffs as well as added thiamine in the form of hydrochloric or nitrate salt.

Vitamin B_2 (Riboflavin)

The following procedure for determination of riboflavin in premixes and compound feeds was elaborated on the basis of the article published by Rubaj et al. [17].

Principle

Riboflavin was extracted from a feed sample in autoclave with 0.1M sulphuric acid. The ester bonds with phosphoric acid were hydrolyzed by the Taka-diastase enzyme. Riboflavin content was determined by high performance liquid chromatography (HPLC) with reversed-phase and usage of fluorescence detection.

Reagents and Solvents. All reagents and solvents should be of analytical grade: methanol for HPLC; sulphuric acid, 0.1 mol/l; sulphuric acid, 30%; sodium hydroxide, 0.5 mol/l; sodium acetate, 2mol/l; acetic acid, 99.5%; citric acid; taka-diastase, 10% suspension; vitamin B_2 standard.

Apparatus: autoclave, ultrasound bath, HPLC set with fluorescence detector.

Procedure

Riboflavin was extracted from the examined feed sample with 0.1M sulphuric acid, and that solution was boiled for 15 min. at temperature from 110°C to 120°C. After cooling to the room temperature, the whole volume of hydrolysed sample was transferred to a 100 ml measuring flask. Next taka-diastase suspension was added to the flask, which was then placed into a water bath at 45°C for 20 min. The enzymatic reaction was stopped by adding sulphuric acid. The sample solution was next chilled to room temperature, and the volume was corrected to 100 ml by adding 0.1 mol/l sulphuric acid. Afterwards, samples were mixed and filtrated. Extract clean-up was done by adding methanol to the sample and filtration through syringe filter before injection on the column.

Chromatography

Column	25 cm x 4.6 mm
Stationary phase	C_{18}
Mobile phase	Methanol and citric acid 0.2 mg/l (30:70 v/v). That solution was mixed with methanol with ratio 1:1
Column temperature	25 ºC
Flow rate	0.8 ml/min
Injection	20 µl
Detector	Fluorescence, Ex λ= 453, Em λ=521
Calculation	External standard, peak area, linear regression

Special Comment

Vitamin B_2 is sensitive to light, hence all the activities were conducted without any access of day light (by using amber glass flask or flask covered by aluminum foil).

Canthaxanthin

The following procedure for determination of canthaxanthin in premixes and compound feeds was elaborated on the basis of the article published by Rubaj et al. [22].

Principle

The principle of this method is based on the hydrolysis of a powdered formulation of canthaxanthine with trypsin and pepsin in water solution of ammonia and its purification on the aluminium oxide filled column. The canthaxantine content is determined by high performance liquid chromatography (HPLC) in normal phase with usage of DAD detector.

Reagents and Solvents: trypsin 200 FIP – U/g; pepsin 700 FIP – U/g; ammonia; n – hexane; diethyl ether; 99.8% ethyl alcohol; acetone; aluminum oxide, neutral, activity 1; canthaxanthin standard; chloroform. All reagents and solvents should be of analytical grade.

Apparatus: ultrasonic bath; vacuum rotary evaporator, HPLC set with Array's diode detector.

Procedure

The sample is hydrolyzed with an aqueous solution of ammonia at the presence of trypsin and pepsin following extraction with ethyl alcohol and diethyl ether. Purification occurs on the aluminum oxide filled column. The extract prepared in this way should be evaporated, dissolved in the mobile phase, filtered through syringe filters and dosed on the column.

Chromatography

Column	4.6 x 250 mm
Stationary phase	LiChrospher Si 60
Mobile phase	n-hexane: acetone 86:14 (v/v)
Column temperature	25 ºC
Flow rate	1.3 ml/min
Injection	20 μl
Detector	DAD λ=446 nm
Calculation	External standard, peak area, linear regression

Methionine hydroxy analog (MHA)

The procedure of analyzing methionine hydroxy analog was developed on the basis of the work by Matyka *et al.* [25] and the official VDLUFA method [8].

Principle

Methionine hydroxy analog is extracted from the sample by 10% acetonitrile, and next hydrolyzed with potassium hydroxide and determined by high performance liquid chromatography (HPLC) with reversed phase and UV detection.

Reagents and Solvents: acetonitrile; orthophosphoric acid; solution for extraction: acetonitrile - water 10+90 (V/V); solution for hydrolysis: 50% potassium hydroxide (w/v); phosphoric acid, 0.01mol/l.

Apparatus: centrifuge, HPLC set, diode array detector.

Procedure

Extract methionine hydroxy analog from the feed, with the use of 10% acetonitrile. After centrifuging, perform hydrolysis with potassium hydroxide and next with a solution of orthophosphoric acid. Filter the supernatant through syringe filters and inject on the column.

Chromatography

Column	25 cm x 4.6 mm
Stationary phase	C_{18}, LiChrospher
Mobile phase	Eluent 1 : acetonitrile - phosphoric acid 10+90 (v/v)
	Eluent 2 : acetonitrile - phosphoric acid 23+77 (v/v)
Column temperature	25 °C
Flow rate	0.8 ml/min
Injection	20 µl
Detector	UV, λ=214 nm
Calculation	External standard, peak area, linear regression

Special Comment

- If the degree of MHA polymerization in the feed mixture is high and depolymerization in the conditions presented in the analytical procedure is incomplete it is necessary to increase the amount of hydroxide taken for hydrolysis and the amount of phosphoric acid for neutralization, keeping constant proportions, and next to take into account the change in the amount of the solution after hydrolysis, while calculating MHA content.
- The quality of separation on the chromatographic column depends on pH of the mobile phase. When acidity increases retention time for MHA is reduced. The excess of phosphoric acid in the injected solution after depolymerization reduces the time of MHA retention.

7. Validation parameters of the methods used to analyze feed additives

A significant element in verifying a chemical method, including chromatography, is its validation. Validation is a confirmation through examining and presenting some objective evidence that some particular requirements regarding the intended application have been fulfilled. The basic validation parameters include: calibration linearity, the limit of detection and quantitative determination, precision (repeatability, indirect precision, reproducibility), recovery and uncertainty.

Calibration linearity is defined as a relationship between the readings of the measuring device and the concentration of a particular component, in conformity with the regression equation: bx + a = y. The measure of linearity is Pearson's linear correlation coefficient (r) for parameters with regular distribution. The scale presented below is adopted to estimate the correlation coefficient: 0.0-0.2: very weak relationship; 0.2-0.4: weak relationship; 0.4-0.6: moderate relationship; 0.6-0.8: strong relationship; 0.8-1.0: very strong relationship.

In case of feed additives discussed in the present chapter, determined with HPLC methods, external calibration was used.

In case of chromatographic methods the value of limit of detection (LOD) may be determined on the basis of the obtained chromatogram of blanc sample, as the threefold value of a noise signal. To do this, it is necessary to determine the level of noise, by measuring on the chromatogram the range of signal change near retention time of examined analyte. With chromatographic methods, the bottom limit of the method's application may be also regarded as the content of the analyzed component, which is equal to the lowest concentration of the standard used for calibration.

Analyte	Matrix	LOQ	CV_r %	CV_{ip} %	Rec. %	Linear range
Vitamin A	Feedingstuff	1000 IU/kg	1.6	4.0	96.0	7.0-70 IU/ml; r=0.999
	Premixture		1.4	2.0	95.2	
Vitamin E	Feedingstuff	6.0 mg/kg	2.0	2.0	96.7	0.05-0.3 mg/ml; r=0.999
	Premixture		1.0	2.0	96.4	
Vitamin K_3	Feedingstuff	1.0 mg/kg	6.4	-	100.9	0.046-4.62 μg/ml; r=0.999
	Premixture		5.7	-	99.4	
	Preparation		1.9	-	101.2	
Vitamin D_3	Premixture	200 IU/g	1.4	1.7	99.3	1.06-10.68 μg/ml; r=0.999
	Preparation		1.3	1.3	98.4	
Vitamin B_1	Feedingstuff	1.0 mg/kg	5.6		98.9	0.2-1.0 μg/ml; r=0.999
	Premixture		3.7		102.3	
Vitamin B_2	Feedingstuff	1.0 mg/kg	3.4	5.1	98.0	0.17-0.67 μg/ml; r=0.999
	Premixture		2.3	6.2	98.3	

CV_r – coefficient of variation; CV_{ip} – intermediate precision; rec. - recovery

Table 7. Validation parameters obtained for selected feed additives – vitamins in feeds

Analyte	Matrix	LOQ	CV_r %	CV_{ip} %	Rec. %	Linear range
Cantha-xanthin	Feedingstuff	1.0 mg/kg	4.7	7.9	97.3	0.7-8.5 μg/ml; r=0.999
	Premixture		3.3	6.0	98.2	
Tryptophan	Feedingstuff	10.0 mg/kg	4.1	4.0	94.9	12.5-100 nmol/l; r=0.999
	Preparation		1.0	1.4	99.7	
Ethoxyquin	Feedingstuff	0.5 mg/kg	2.0	6.0	99.0	0.01-0.07 μg/ml; r=0.999
MHA	Feedingstuff	0.05%	2.8	-	96.7	0.05-0.45 mg/ml; r=0.997

Table 8. Validation parameters obtained for other feeds

During the validation process in a laboratory the precision of a method is determined through examining such parameters as repeatability and within-laboratory reproducibility (intermediate precision). Within-laboratory reproducibility may be calculated on the basis of control charts or from the range between parallel results of an analysis (replications) of a feed additive, in compliance with the Nordtest [23] handbook. For two or more replications for the analyses of an analyte in each sample it is necessary to calculate the mean value, the difference between measurements (range), relative difference in % and next mean relative difference (%) for all samples of a particular type of feed. The mean range divided by the coefficient (for two replications d = 1.128) makes the standard deviation of within - laboratory reproducibility. In order to verify the method's precision the Horwitz ratio named HorRat (H), may be used which is the ratio of the relative standard deviation of reproducibility SDR_t calculated from the Horwitz formula $SDR_t = 2\ C^{-0.15}$, where C stands for concentration expressed as a dimentionless mass fraction (e.g. 1 mg/kg = 10^{-6}). In order to adjust it to the conditions of repeatability, target standard deviation SDR_t is multiplied by 0.50 ($RSD_r = 0.5\ RSD_R$), [3]. Satisfactory values of the HorRat making the measurement of precision are included in the range of $0.5 < H < 2$ [3]. In case of participating in interlaboratory tests and obtaining satisfactory results, it is possible to include the precision parameters obtained in these analyses. The accuracy of a method may be determined by calculating the recovery degree or examining certified reference material, CRM.

8. Quality assurance and an uncertainty of result

Each laboratory should possess a program of quality assurance of its analyses within good laboratory practice. In case of chromatographic methods steering the quality may be implemented through performing one or more of the activities listed below:

- regular examinations of control samples;
- regular check-ups of the standard for each examination or series of analyses of labile feed components,
- checking b curve slope coefficient from the equation bx + a=y,
- analyzing overlapping samples (e.g. a solution of the sample for measurement, prepared and analyzed on the previous day),
- analyzing the blanc sample and fortified sample,
- examining certified reference materials, reference materials,
- participation in native and international proficiency testing.

Control material may be provided by certified reference material, CRM (matrix + analyzed substance), material from proficiency testing with a value assigned, enriched material prepared in the laboratory (fortified sample) and control material with recognized content of the tested and stable in time component, previously determined in the laboratory.

In compliance with the recommendations of the EN ISO/IEC 17025:2005 [24] standard and requirements defined in some regulations, in order to assess and interpret the result of a test, it is necessary to use the uncertainty of measurement. We hardly ever know the real content of the analyte and the result of the test is biased with an error. Hence, the idea of "uncertainty of measurement" has been introduced which is defined as "a parameter associated with the result of a measurement, that characterises the dispersion of the values that could reasonably be attributed to the measurand" [26]. The EN ISO/IEC 17025:2005 standard recommends at point 5.4.6.2 that testing laboratories should possess and make use of procedures for assessing the uncertainty of measurement. To assess the uncertainty of methods used to analyze feeds the most frequently used approach is the model, consistent with the GUM [26] guidebook, which consists of finding the components of uncertainty and uses the law of error propagation. Using this particular approach to assess uncertainty, it is possible to obtain an underestimated value in case when we do not consider all the components. Other reasons for underestimating uncertainty during validation include a situation when while assessing uncertainty repeatability instead of within-laboratory reproducibility is taken into account or we often forget to consider the bias. Moreover, uncertainty assessment is done during validation when well-ground typical samples are analyzed in a short time, new standards are prepared, the apparatus is controlled (standard conditions). During routine activities we analyze various matrices and obtaining homogeneity is not easy in case of some samples. The conditions mentioned above may affect underestimation of uncertainty. That is why it is important to have a possibility to verify uncertainty with the help of other approaches.

New opportunities concerning verification and assessment of uncertainty can be found in the Eurolab [27] technical report, the Nordtest [23] handbook and in the paper [28] which recommend the use of experimental approaches in assessing uncertainty of laboratory methods, in particular:

- within-laboratory experimental approach based on within -laboratory reproducibility and the assessment of the method's bias, following CRM,
- within -laboratory experimental approach based on within -laboratory reproducibility and the assessment of the method's bias, following participation in PT/ILC.

Using the within-laboratory experimental approach in assessing uncertainty of measurement (u) within -laboratory reproducibility (s) is considered as a measure of precision, as well as the bias (b), in accordance with the equation below following the Eurolab [27] technical report.

$$\sqrt{{s_{Rw}}^2 + b^2} = u \tag{1}$$

Precision of a research procedure in a laboratory is determined during validating the method or on the basis of measurements noted in control charts. Validation usually includes determining standard deviation of within-laboratory reproducibility s_{rw} or standard deviation of within -laboratory reproducibility s_{Rw} which is sometimes called intermediate precision. Bias is determined by means of standard deviation of measurement in relation to the reference value, e.g. while examining certified reference materials, using reference methods. The share of bias (b) in uncertainty of the measurement is determined with the help of mean deviation of measurements (Δ), uncertainty of reference material (u_{ref}) and precision of measurements during examining that bias (s). The value of the expression s^2 / n with a bigger number of measurements is low and can be omitted:

$$\sqrt{\Delta^2 + u^2_{ref} + \frac{s^2}{n}} = b \tag{2}$$

In practice, different measurements result in different values of bias. In such a case the data may be combined and common assessment of a value of bias (u_w) may be determined as a function of the measured value or, for typical data of matrices and levels, according to the formula below:

$$\sqrt{\frac{\sum (bias_i)^2}{n}} = \Delta \qquad (3)$$

When certified reference materials are lacking (a frequent situation) and when no other analyses of bias have been performed in the laboratory (e.g. prior to applying the reference method) bias can be estimated on the basis of proficiency testing, PT.

A laboratory participating in PT may use the results of such tests in order to assess uncertainty of measurement for the testing method/procedure used. Similarly to determining uncertainty in within-laboratory experimental approach, the uncertainty of measurement (u) is equal to the root of the sum of squared values of standard deviation for within-laboratory reproducibility s_{Rw} and the bias (b), which can be calculated from the formulas 1,2 and 3.

With this approach two components of uncertainty are obtained from different sources. Precision is determined on the basis of the authors' own validation data (within-laboratory reproducibility), from the range or on the basis of measuring control charts (in-house). The bias is determined on the basis of PT results. Estimating the bias on the basis of a single participation in PT may have a limited range and should be treated as preliminary. If the data from several PTs are available (a wider range of matrices and concentrations) the assessment of the bias may be referred to the complete measurement range.

The results of analysing uncertainty on the basis of experimental approaches using the results of the authors' own results are presented below along with, for comparison, expanded uncertainties estimated with the help of Horwitz formula RSD_R (%) = $2C^{-0.15}$.

Additive	Feed	s_w (%)	bias (%)	u (%)	U = 2 ·u (%)	U (%) *
Vitamin A	Feedingstuffs	4.0	12.4	13.1	26.2	23.8
	Premixes	2.0	7.2	7.5	15.0	11.8
Vitamin E	Feedingstuffs	1.0	9.0	9.1	18.2	16.1
	Premixes	2.0	6.1	6.4	12.8	8.2
Vitamin B_1	Feedingstuffs	5.6	6.7	8.7	17.4	26.8
	Premixes	3.7	6.7	7.6	15.2	18.6
Vitamin B_2	Feedingstuffs	6.52	3.16	7.2	14.4	24.0
	Premixes	5.09	3.16	6.0	12.0	10.7

*Expanded uncertainty for the HorRat value H=1 calculated from the Horwitz' formula RSD_R=2 $C^{-0,15}$; U (%) = 2 RSD_R

Table 9. Results of uncertainty evaluation for some feed additives in compound feeds and premixes

9. Conclusion

The chapter presents a brief review of the methods used for determining feed additives by means of high proficiency liquid chromatography, HPLC. The authors presented their own research procedures and special attention was given to the preparation of samples for testing, extraction, extract purification, chromatographic separation and the basic elements of method validation and quality control

Using HPLC for testing fat-soluble vitamins in feed materials, mixtures and premixes enabled us to replace colorimetric methods and to eliminate bias, such as the positive error of vitamin A determination related to the presence of carotenoids in the analyzed feed. The problem of low precision of examining certain vitamins, e.g. vitamin A, in feed mixtures is often unrelated to the method of determination, but rather to non-homogenous distribution of vitamin A in the feed related to its being secured against losing activity, due to protective coating. This problem may be solved by preparing the analytical weighed amount of sufficiently high mass and grinding the sample immediately prior to determination procedure to particles sized 1 mm.

Progress in the area of examining the content of water-soluble vitamins is also related to introducing the methods of liquid chromatography. The authors included their own procedures of analyzing vitamins B_1 and B_2, thiamine and riboflavin, with the use of HPLC methods and gave their characteristic parameters which meet the current requirements regarding the assessment of content and interpretation of results. These methods may be used especially to examine low content of thiamine and riboflavin, endogenic and added, in feed materials and mixtures.

HPLC methods have been widely used for testing cocciodiostats in feed preparations, premixes and mixtures. They contributed to improving the safety of using these additives, controlling concordance with manufacturer's declaration and not exceeding the maximum content in feed mixtures, as well as controlling the withdrawal period. Without liquid chromatography with mass spectrometry (LCMS) it wouldn't be possible to analyze effectively the remains of coccidiostats in the tissues and food products of animal origin. To reduce the risk of cross contamination in non-target feeds maximum content values for coccidiostats were determined recently at the level from 0.01 mg/kg (diclazuril) to 1.25 mg/kg (narasin, monensin), [29]. This created a need to develop some test methods adequate for the level of acceptable cross contamination and verifying them in interlaboratory tests. Future research will focus on checking the LCMS method for this particular purpose.

The official methods of separating and determining amino acids in feedingstuffs [13,8] are based mainly on ion exchange chromatography. However, in examining free amino acids (amino acids used as additives: lysine, methionine, threonine, tryptophan, valine, arginine and cysteine) HPLC methods are becoming increasingly more popular as they make the analyses shorter in time. In some cases a HCLP method is the only solution, e.g. while determining methionine hydroxy analog, verified in the authors' own studies. The need to perform a large number of analyses in a shorter time determines the direction of future studies of amino acids in feedingstuffs and using ultra-performance liquid chromatography, UPLC, for this purpose.

In the testings of feed colorants the most frequently used means were spectrophotometric methods [13,8]. The diversity of feed products and the resulting changeability of matrix, as well as determining the maximum content of colorants in feed mixtures, were the reasons for searching for new methods of examining colorants, including HPLC. An example of such a method in reference to canthaxanthin and a procedure based on the authors' own research is quoted in the present work. Future research in this respect will use the LCMS method to a higher degree as it enables detecting and determining several feed colorants in a single sample in view of cis-trans steroisomers.

Author details

JolantaRubaj*, Waldemar Korol and GrażynaBielecka
National Research Institute of Animal Production, National Feed Laboratory, Poland

10. References

[1] Regulation (EC) No 1831/2003 of the European Parliament and of the Council of 22 September 2003 on additives for use in animal nutrition. OJ L 268 from 18.10.2003, 29-43. http://eur-lex.europa.eu
[2] Dejaegher B, Heyden Y (2007) Ruggedness and robustness testing. Journal of Chromatography A, 1158:138-157.
[3] Horwitz, W. & Albert, R. (2006). The Horwitz Ratio (HorRat): A useful index of method performance with respect to precision. Journal of AOAC International, Vol. 89(No. 4): 1095-1109.
[4] Commission Regulation (EC) No 152/2009 of 27 January 2009 laying down the methods of sampling and analysis for the official control of feed. OJ L 54 from 26.02.2009, 1-130. http://eur-lex.europa.eu
[5] Manz, U. & Philipp, K. (1988). Determination of Vitamin D_3 in Complete Feeds and Premixes with HPLC, in Analitycal Methods for Vitamins and Carotenoids in Feed, Roche Publication Index, Basle, No. 1841: 8-11.
[6] Manz, U. & Maurer, R. (1982). Method for the determination of vitamin K_3 in premixes and animal feedstuffs with the aid of high performance liquid chromatography. International Journal for Vitamin and Nutrition Research, Vol.52: 248-252.
[7] Fedder, R. & Plöger, A.(2005). Solid-phase extraction of vitamins A and E from animal feeds: A substitute for liquid-liquid extraction. Journal of AOAC International, Vol. 88 (No.6):1579-1582.
[8] VDLUFA-Methodenbuch. Band III - 5 Erg. (2004) - 6 Erg. (2006). Die chemische Untersuchun von Futtermitteln. VDLUFA – Verlag, Darmstadt.
[9] Decreto 20 febbraio 2006. Approvazione del metodo ufficiale di analisi per la deteminazione della vitamina B1 negli alimenti per animali – Supplemento n.19 http://gazette.comune.jesi.an.it/2006/50/11.htm
[10] Analytical Methods Committee. (1983). Determination of halofuginone hydrobromide in medicated animal feed. Analyst, Vol. 108: 1252-1256.
[11] Analytical Methods Committee. (1995). Determination of lasalocid sodium in poultry feeds and premixes. Analyst, Vol. 120: 2175-2180.
[12] Dusi, G. & Gamba, V. (1999). Liquid chromatography with ultraviolet detection of lasalocid, monensin, salinomycin and narasin in poultry feeds using pre-column derivatisation. Journal of Chromatography A, Vol. 835: 243-246.
[13] Horwitz, W. & Latimer, Jr., G.W. (2011). Official Methods of Analysis of AOAC International, 18th Edition, Revision 4 , AOAC International, Gaithersburg, Maryland, USA.
[14] Regulation (EC) No 882/2004 of the European Parliament and of the Council of 29 April 2004 on official controls performed to ensure the verification of compliance with feed

* Corresponding Author

and food law, animal health and animal welfare rules. OJ L 165 from 30.04.2004, 1-141. http://eur-lex.europa.eu

[15] Commission Regulation (EU) No 939/2010 of 20 October 2010 amending Annex IV to Regulation (EC) No 767/2009 on permitted tolerances for the compositional labeling of feed materials or compound feed as referred to in Article 11(5). OJ L 277 from 21.10.2010, 4-7. http://eur-lex.europa.eu

[16] Douša, M. & Břicháč, J. (2010). HPLC determination of vitamin in the animal feeds – the sources of errors and their reduction. Acta Universitatis Palackianae Olomucensis, Vol.47: 1-12.

[17] Rubaj, J., Bielecka, G., Korol, W. & Kwiatek, K. (2008). Determination of Riboflavin in Premixture and Compound Feed by Liquid Chromatography Method. Bulletin of The Veterinary Institute in Pulawy. Vol. 52(No.4): 619-624.

[18] Rubaj, J., Korol, W., Bielecka, G. & Kwiatek, K. (2008). Determination of Thiamine in Premixture and Compound Feed by Liquid Chromatography Method. Bulletin of The Veterinary Institute in Pulawy Vol. 52(No.3): 435-440.

[19] Commission Decision of 12 August 2002 implementing Council Directive 96/23/EC concerning the performance of analytical methods and the interpretation of results. OJ L 221 from 17.8.2002, 8-36. http://eur-lex.europa.eu

[20] International Standard FDIS 6498 (2012). Animal Feeding Stuffs – Guidelines for sample preparation.

[21] Coelho, M. (2002). Vitamin stability in premixes and feeds – A practical approach in ruminants diets. Proceedings 13th Annual Florida Ruminant Nutrition Symposium, 127- 145.

[22] Rubaj, J., Bielecka, G., Korol, W. & Kwiatek, K. (2009). Determination of Canthaxanthin in Premixture and Compound Feed by Liquid Chromatography Method. Bulletin of The Veterinary Institute in Pulawy Vol. 53(No.4): 707-712.

[23] Handbook for Calculation of Measurement Uncertainty in Environmental Laboratories. Nordtest TR 537, Version 3, 2008. URL:www.nordicinnovation.net

[24] Standard EN ISO/IEC 17025:2005 General requirements for the competence of testing and calibration laboratories.

[25] Matyka, S., Rubaj, J., Korol, W., Bielecka, G. (2006). Industrial feed mixtures with an addition of methionine hydroksy analog. Electronic Journal of Polish Agricultural Universities, Agricultural Engineering, Vol. 9, Issue 1, URL: www.ejpau.media.pl/series/whole/engineering/index.html

[26] Guide to the Expression of Uncertainty in Measurement (GUM). BIPM, IEC, IFCC, ISO, IUPAC, OIML. International Organization of Standardization, Geneva Switzerland, 1st Edition 1993.

[27] Eurolab Technical Report 1/2007. Measurement uncertainty revisited: Alternative approaches to uncertainty evaluation, Eurolab, Paris 2007.

[28] Korol, W., Rubaj, J., Bielecka, G. (2008) An interpretation system of feed analyses based on inter-laboratory comparisons. Krmiva. Vol. 50 (No.6), 319-327.

[29] Commission Regulation (EU) No 574/2011 of 16 June 2011 amending Annex I to the Directive 2002/32/EC of the European Parliament and of the Council as regards maximum levels for nitrite, melamine, Ambrosia spp. And carry-over of certain coccidiostats and histomonostats and consolidating Annexes I and II thereto. OJ L 159 from 17.06.2011, 7-24.. http://eur-lex.europa.eu

Chapter 7

Analysis of Surfactants in Environmental Samples by Chromatographic Techniques

Juan M. Traverso-Soto, Eduardo González-Mazo and Pablo A. Lara-Martín

Additional information is available at the end of the chapter

1. Introduction

Synthetic surfactants are among the most produced and used organic compounds worldwide. They are a wide range of chemicals characterized by their amphiphilic nature. Thus, their molecules consist of an hydrophilic / polar head group (either charged or uncharged) and an hydrophobic / nonpolar hydrocarbon tail. As a consequence, surfactants show solubility in polar and nonpolar liquids, ability to form micelles, adsorption to phase boundaries and reduction of the surface tension of water. They are economically important due to their specific properties that allow using them as washing, wetting, emulsifying and dispersing agents. Therefore, surfactants are mainly used in the formulation of detergents, personal care products, paints, textiles, pesticide formulations, pharmaceutical, and many other products [1, 2]. Many different types of these compounds have been synthesized, although they can be classified into three main groups according to their charge: (1) anionics, (2) non-ionics, and (3) cationics (Figure 1); the first and second groups accounting for the highest production volumes. Thus, the European Committee of Organic Surfactants and their Intermediates (CESIO) reported a production of 1200 ktons of anionic and 1400 ktons of non-ionic surfactants in Europe in 2010, which represents 90% of the total European production of surfactants.

Linear alkylbenzene sulfonates (LAS), alkyl ethoxysulfates (AES) and alkyl sulfates (AS) are the most widely used anionic surfactants. LAS are commercially available as a mixture containing homologues with alkyl chains ranging from 10 to 14 carbon units, and isomers resulting from the different attachment positions of the phenyl group along that chain (Figure 1a). The chemical structure of AS comprises a $C_{12\text{-}16}$ alkyl chain with a terminal sulfate group. AES share the same structure than AS but they also have a variable number of ethylene oxide (EO) units (Figure 1b). All these compounds are commonly employed in household and laundry detergents, hand dishwashing liquids, shampoos, and other

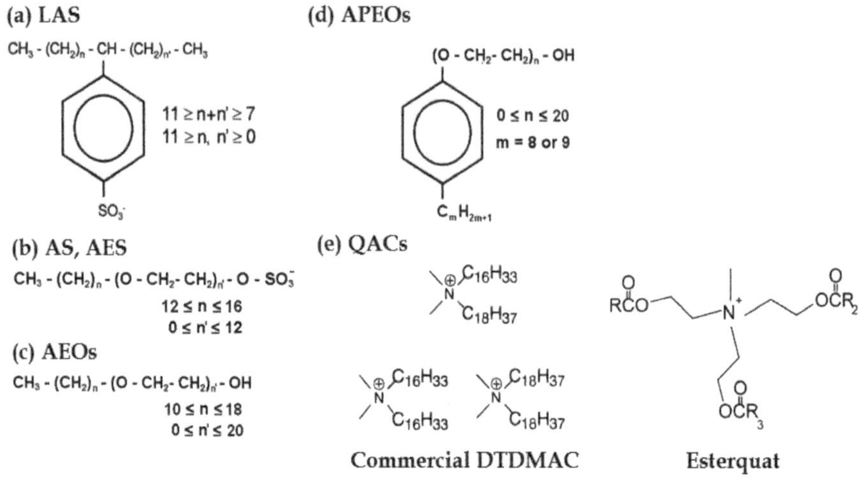

Figure 1. Chemical structures of (a) linear alkylbenzene sulfonates (LAS), (b) alkyl ethoxysulfates (AES), (c) alcohol polyethoxylates (AEOs), (d) alkylphenol polyethoxylates (APEOs), and (e) quaternary ammonium-based compounds (QACs).

personal care products [3-5]. Among the nonionic surfactants, alcohol polyethoxylates (AEOs) are currently produced in the greatest volume (e.g., 747000 tons in Europe in 2000), and alkylphenol ethoxylates (APEOs) are in second place by volume as a consequence of the restrictions on their use in recent years, due to the estrogenicity showed by some of their degradation intermediates [6-8]. AEOs are a mixture of homologues having from 12 to 18 carbon atoms in their alkyl chain, which is connected via an ether bond to an ethylene oxide (Figure 1c). APEOs are mixtures of a wide range of ethoxymers (from 1 to 20 EO units), and isomers, depending on the degree of branching of the alkyl chain (Figure 1d). Both, AEOs and APEOs, are widely employed in domestic and industrial applications [9] (e.g., detergents, emulsifiers, wetting and dispersing agents, industrial cleaners, textile, pulp and paper processing). Finally, quaternary ammonium-based compounds (QACs) are the main class of cationic surfactants, being constituted of at least one hydrophobic hydrocarbon chain linked to a positively charged nitrogen atom, and other alkyl groups which are mostly short-chain substituents such as methyl or benzyl groups (Figure 1e). Major uses of QACs are as fabric softeners and antiseptic agents in laundry detergents as well as other industrial uses [2]. Since the 1960's, the most commonly used active ingredient in fabric softeners has been dehydrogenated tallow dimethyl ammonium chloride (DTDMAC), with industry-wide European annual volumes exceeding 32000 tons through 1990. However, esterquat surfactants were introduced into the European market in the early 1990's because, due to their structure, they were more accessible to hydrolysis and biodegradation than DTDMAC. Hence, most fabric conditioners marketed now are comprised of esterquat types, with a volume of 130000 tons/year used in detergent products in the European Union [10-13].

Once used, the major fraction of synthetic surfactants are disposed down the drain to sewers, where it has been estimated that 50% by volume is degraded, 25% sorpted to suspended solids and 25% dissolved [14, 15]. Later, these chemicals are commonly removed between 81 and 99.9% in wastewater treatment plants (WWTPs) [16-18], although they are frequently detected in sewage effluents showing concentrations up to 872 μg/L for LAS [16], between 0.24 and 3 μg/L for AES [19], up to 4 μg/L for QACs [20], and from 0.2 to 23 μg/L for AEOs and APEOs [21, 22]. Secondary treatment in active sludge units is considered the most important process to eliminate surfactants through aerobic biodegradation, but a considerable fraction is also removed by sorption / precipitation in sludges originated from several decantations (Figure 2) (e.g., 15-37% of total LAS [14, 15, 23] and more than 90% of nonylphenol [24]). These sludges are also a potential source of contamination for soils, groundwater and adjacent rivers as they tend to contain high concentrations of organic contaminants and are often used in agriculture after anaerobic digestion. High levels of surfactants have been measured in treated sludges: up to 5400 mg/kg dry weight for LAS [15, 25], from 119.3 to 380.5 mg/kg for APEOs, AEOs and AES [25], and up to 5870 mg/kg for QACs [26, 27]. Any environmental compartment (surface waters, sediment, biota...) is susceptible of being contaminated by these compounds and/or their degradation metabolites [2, 28]. As example, a considerable number of studies have reported the presence of LAS in surface waters [29-31] at levels typically ranging from less than 1 ng/L to several hundreds of μg/L respectively, depending on the distance from urban wastewater discharge sources and the type of wastewater treatment. Available studies about the presence, environmental behavior and distribution of non-ionic surfactants are mainly focused on NPEOs (nonylphenol polyethoxylates, which are the major fraction of APEOs). Concentrations of these compounds have been reported in surface waters all around the world: <0.1 to 100 μg/L in rivers in Mexico [32], Holland [33], Japan [34] and Taiwan [29], and from <1 to 38.5 μg/L in coastal waters of United States [35], Italy [30], Spain [36] and Israel [37]. Levels of surfactants in surface sediments are usually higher by several orders of magnitude than those measured in water due to their moderate to high sorption capacity. Thus, the presence of LAS [31, 38-40] and NPEOs [22, 41-44] has been widely detected in sediments, with levels ranging from less than 1 to more than 200 mg/kg and from less than 0.1 to 28.5 mg/kg respectively. Available data concerning to the presence of aliphatic anionic (AES) and nonionic (AEOs) surfactants, as well as cationic surfactants of any class, are rather limited. There are only a few papers about the occurrence of AEOs [36, 40, 43, 44] and AES [19, 31] in sediments, showing levels ranging from <0.1 to 23 mg/kg. Some authors have also measured concentrations between <0.1 and 72 μg/L for AES [19, 31] and AEOs [36, 40, 43] in surface waters. QACs have been measured at levels ranging from less than 2 μg/L in surface waters [20] to more than 100 mg/kg in sediments [45, 46].

Summarizing, huge volumes of surfactants are used every day, entering the environment, where these compounds and/or their degradation products may cause damage depending on their concentrations. Therefore, it becomes necessary developing reliable analytical methodologies that allow determining the levels of surfactants in environmental matrices, which may be complicated due several reasons. First, surfactants are often sold as commercial mixtures which can comprise hundreds of different homologues, isomers

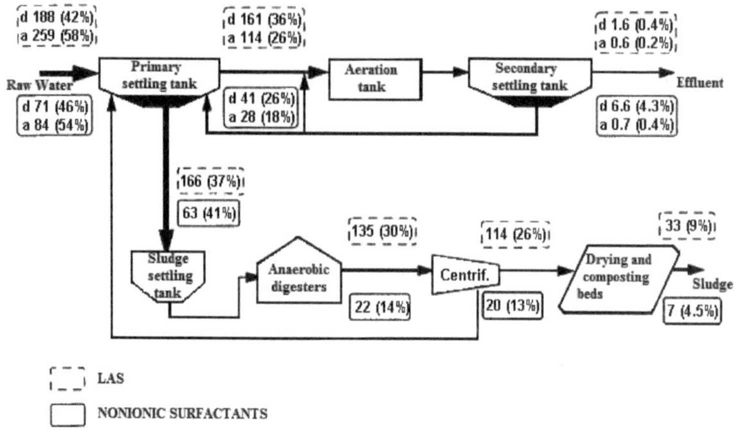

Figure 2. Flowsheet and sampling points of a wastewater treatment plant (WWTP). Figure shows mass balance of dissolved (d) and adsorbed/precipitated (a) of LAS and non-ionic surfactants. Absolute amount (average value in kg/day) and percentage with respect to raw water (adapted from reference [15]).

and/or ethoxymers with different physico-chemical properties. Separation and quantification of these components require the use of chromatographic techniques, mainly gas chromatography (GC) and liquid chromatography (LC). Achieving a successfully identification of every component in the mixture is also desirable for a better understanding of their environmental behavior as they may suffer differential degradation or sorption. There is also an additional challenge when dealing with target compounds that tend to be present at trace levels. In these cases it is necessary to develop reliable extraction, purification, and preconcentration protocols in order to remove as many interferences as possible before analysis without sacrificing high recovery values. Some of the techniques used to this end are also based on chromatographic techniques, such as solid phase extraction (SPE), directly derived from column chromatography. Thus, in this chapter we present the main problems posed by analysis of surfactants in environmental samples from two points of view:

- Isolation and/or preconcentration of surfactants from different types of samples;
- Separation, identification and quantification of analytes in properly prepared extracts.

2. Sample pre-treatment

Correct sampling of environmental samples is indispensable to provide representative information of the environmental compartments from which they are taken and, on the other hand, it is important to preserve the target compounds during storage [1]. Generally, water samples are often immediately preserved upon collection by the addition of biocides such as formaldehyde up to a concentration of 4% [17, 47], chloroform or sodium azide [20], or by filtering through a 0.45 μm membrane filter [16]. Then, aliquots are kept at low temperatures and are often analyzed within a short period of time (48 h) in order to minimize the biodegradation of surfactants. Solid samples (sewage sludges, soils or sediments) are also kept at low temperature once they are collected to avoid any degradation of the analytes during the transport to the laboratory. Later, they are usually dried in a heater [47], at room temperature [25], or frozen at -20 ºC and later freeze-dried [48]. Once dried, samples are milled and strained through a sieve to a particle size of less than 2 mm, and then stored at 4 to -20 ºC for further extraction and analysis.

Surfactants are often found at trace levels (ppb or less) in environmental matrices, frequently below the limits of detection of most analytical techniques. Therefore, it is necessary to carry out not only their extraction but also their isolation and preconcentration to achieve proper identification and quantification. Those methodologies more commonly used at this preliminary stage are commented next for both solid and aqueous samples.

2.1. Extraction from solid samples

For several decades now, Soxhlet extraction and solid-liquid extraction (SLE) have been the most commonly used techniques for the extraction of surfactants and many other organic compounds from solid matrices. These methodologies are cheap and allow simple extraction, although they have also several disadvantages, including the large volume of solvent needed (from 150 to 500 mL [27, 49, 50]), the long time required, which can take 4-18 hours per sample [20, 51, 52], and the production of toxic liquid wastes. Soxtec is an alternative extraction method based on Soxhlet, but the addition of several boiling and rinsing steps reduces the extraction time to 45 min and solvent consumption to 50-100 mL [53, 54]. Application of ultrasounds followed by centrifugation or filtration to separate extracts from solid matrices is another cheap option for extracting surfactants due to the high extraction efficiency in a short time [40, 55, 56]. On the other hand, it also shows the same problems than SLE and Soxhlet extraction (high volume of organic solvents and toxic wastes). Table 1 summarizes the conditions used for the extraction of surfactants employing these conventional extraction techniques. As example, LAS and their degradation intermediates, sulfophenyl carboxylic acids (SPCs), have been extracted from sediments using methanol (MeOH) as solvent [50, 52, 57, 58] by means of Soxhlet extraction and SLE. For APEOs and their metabolites, methodologies have been similar to those used for LAS [30, 59, 60], although methanol tends to be substituted by other less polar solvents (e.g., hexane [51] or dichloromethane (DCM) [61]), in order to enhance the extractability of the more hydrophobic compounds such as nonylphenol (NP). With respect to the extraction of aliphatic surfactants (AEOs and AES) and their main degradation products (polyethylenglycols, PEGs) from solid matrices, most authors have employed methanol [19, 40, 62, 63] and dichloromethane [55, 64] during Soxhlet or Soxtec extraction, SLE and sonication. There are still a few studies dealing with the application of all these techniques for extraction of QACs [26, 27, 65], but acidified methanol is used as solvent in most cases.

Analytes	Matrix	Method	Solvent	Other conditions	Clean-up	Ref.
LAS, AES, AS	Sediment	Soxhlet, PLE	MeOH	Time: 5 h, Temperature: 125 ºC Pressure: 1500 psi	SPE (C18)	[62]
NPEO, NP	Sediment	Soxhlet, Sonication	Hexane/ isopropanol, Hexane/ acetone	Time: 18 h, Not spec.	SPE (cyanopropil)	[51]
QAC	Sediment, sludge	Soxhlet	MeOH/ HCl	Time: 18 h	LLE ($CHCl_3$, water)	[20]
NPEO, OPEO, AEO	Sludge	SLE	DCM	Time: 2 h	-	[64]
LAS	Soil	Soxtec	MeOH	Time: 45 min	-	[54]
NPEO, OPEO, NP, OP, AEO	Sludge	Soxtec	MeOH	Time: 45 min	SPE (C18)	[53]
LAS, SPC, NPEO, NPEC, AEO, PEG	Sediment	Sonication	MeOH	Time: 30 min x 3	SPE (HLB)	[40]
QAC	Sediment	Sonication	MeOH/ HCl	Time: 1 h x 3	LLE ($CHCl_3$, water) + SPE (anion exchange)	[65]
$NP_{1-3}EO$, $OP_{1-3}EO$	Sediment	Sonication	MeOH	Time: 7 min	SPE (aminopropyl silica) + LC column (C18)	[60]

Table 1. Overview of conventional extraction techniques applied to the extraction of surfactants and their metabolites from solid samples.

New extraction methods have been developed within the last decade not only to save time, but also to reduce solvent consumption without losing extraction efficiency. Table 2 shows some examples of the application of new techniques for the extraction of surfactants from solid environmental matrices. Microwave-assisted extraction (MAE) is suitable for the extraction of different anionic [66] and non-ionic [67] surfactants from sediments and sludges. Extractions are often achieved quickly at 120 ºC using low solvent volumes (mainly MeOH [68] or DCM/MeOH [69]). Another advantage of MAE is that it can also be combined with Soxhlet extraction [70] in order to increase its efficiency. Less solvent demand and higher extraction rates compensate the high initial cost of acquiring a MAE unit. Pressurized fluid extraction (PFE), also known as accelerated solvent extraction (ASE) or pressurized liquid extraction (PLE), is a technique based on the use of high temperatures (100-200 ºC) and pressures (1500-3000 psi) to prevent solvents from boiling and to increase the kinetics of extraction. Therefore, PLE allows a faster extraction of organic compounds from solid samples (15-20 min per sample) with a lower uptake of organic solvent than more conventional techniques and without sacrificing high recovery values. Recently anionic [62, 71], cationic [72] and non-ionic [47, 73, 68] surfactants have been extracted using PLE and methanol or mixtures containing hexane, acetone, acetonitrile (ACN) or even water as solvents. However, Petrovic et al. [74] observed the volatilization of some APEOs and their metabolites under these conditions, so they suggested keeping the extraction temperature under 60 ºC in this case. Supercritical fluid extraction (SFE) is another extraction technique that has been recently applied to the extraction of ionic and non-ionic surfactants, using CO_2 [75] or water [25] instead of toxic organic solvents to carry out the extraction in a short time and without requiring further clean-up steps [26]. Sometimes, the mobile phase (CO_2, H_2O) is modified with the addition of low molecular weight alcohols (e.g., MeOH) to improve the efficiency in the extraction of polar or ionic compounds [76, 77]. However, better extraction recoveries for nonpolar compounds, possibility of using water as extraction solvent and automation of PLE has result in a lower interest of using SFE instead PLE for extraction of surfactants.

Analytes	Matrix	Method	Solvent	Other conditions	Clean-up	Ref.
LAS, SPC, AES, AS, NPEO, APEC, AEO	Sediment	PLE	MeOH	Time: 15 min Temperature: 120 ºC Pressure: 1500 psi	SPE (C18)	[47]
LAS, SPC	Soil	PLE	MeOH/H_2O	Time: 15 min Temperature: 120 ºC Pressure: 1500 psi	SPE (C18)	[71]
$NP_{1-5}EO$, $OP_{1-5}EO$, NP, OP	Sediment	PLE	Acetone/hexane	Time: 15 min Temperature: 100 ºC Pressure: 1500 psi	SPE (aminopropyl silica)	[73]
QAC	Sediment	PLE	ACN/H_2O	Temperature: 120 ºC Pressure: 1500 psi	SPE (polymeric)	[72]
NPEO, NP	Sediment	Soxhlet, PLE, MAE	MeOH	Time: 10 h, Time: 10 min Temperature: 100 ºC Pressure: 1500 psi, Time: 20 min	LC column (alumina)	[68]
LAS	Sludge	SFE	CO_2	Time: 15 min	Not required	[75]
QAC	Sludge	SFE	CO_2/MeOH	Time: 45 min	LLE ($CHCl_3$, water) + LC column (anion exchange)	[77]
LAS, AS, AES, AEO, NPEO, NPEC, AP	Sludge	SFE	Water	Time: 27 min	SPE (carbograph 4)	[25]
NP, OP	Sediment	MAE	DCM/MeOH	Time: 25 min	SPE (polyestyrene-divinylbenzene)	[69]
LAS	Sludge	MAE	MeOH	Time: 10 min	Not required	[66]

Table 2. Overview of modern extraction techniques applied to the extraction of surfactants and their metabolites from solid samples.

2.2. Purification and preconcentration

There is a wide variety of techniques to carry out purification and preconcentration of extracts from solid samples, as well as aqueous samples, before proceeding with analysis of surfactants and their degradation metabolites. Liquid-liquid extraction (LLE) is among the first techniques that have been widely applied for the extraction of ionic and non-ionic surfactants. Target compounds are isolated from the sample according to their relative solubilities in two different immiscible o partially miscible liquid phases, usually water and an organic solvent. Several cationic [20, 65] and anionic compounds [78, 79] have been extracted from aqueous samples using chloroform, whereas dichloromethane [80] and ethyl acetate [81] have been used to isolate non-ionic surfactants from water. The main advantage of LLE is that it can be used to determine total concentration of these compounds in water in spite of their solid particle matter level. However, the tendency of surfactants to concentrate at phase boundaries leads to the formation of emulsions, and phase separation during LLE becomes very difficult. This can be avoided by the formation of liphopilic ion pairs between surfactants and ion-pair reagents [1] (e.g., disulphine blue dyes or LAS for cationic surfactants [65, 77, 82], methylene blue [78, 79, 83] or methylene green for anionic surfactants [84], modified Dragendorff reagent for non-ionic [81]).

Nowadays, solid-phase extraction (SPE) is the most extended purification and preconcentration technique for surfactants. LLE requires large amounts of sample (100-500 mL) and high consumption of toxic organic solvents, while SPE is generally faster and needs lower sample and solvent volumes (7-100 mL and 5-20 mL, respectively). Briefly, SPE consists on passing the aqueous sample or extract (mobile phase) through a specific material (solid phase) that retains analytes whereas water, salts and other interferences are discarded. Later, target compounds can be eluted from the solid phase using a minimal amount of solvent (few milliliters) so a clean and low volume extract is obtained. Table 3 shows general information about protocols developed for the isolation of surfactants using both SPE and LLE. SPE has been widely applied to isolate anionic surfactants from aqueous samples. More specifically, octadecylsilica (C18) has been used as the main solid phase to extract LAS and their degradation products (SPCs) from water samples [62, 71], while methanol is commonly employed as elution solvent. Due to the negative charge of these analytes, strong anionic-exchange (SAX) resins have been also employed, alone or combined with C18, for a better purification [52, 85], using a mixture of methanol and hydrochloric acid as elution solvent [57, 58]. Lowering the pH of the sample and/or adding significant amounts of salts such as sodium chloride [52, 71] (salting-out effect) is also convenient to improve the retention of most polar components (e.g., SPCs). Other authors have preferred using graphitized black carbon (GBC) [29, 86] or polystyrene-divinylbezene SDB-1 cartridges [87] instead, also showing good extraction recoveries. Other anionic surfactants (AES and AS) have been successfully isolated by octadecylsilica [62, 85] and GBC [25, 88] SPE cartridges from river, marine and wastewater samples, as well as sludge and sediment extracts. Regarding non-ionic surfactants, a wide variety of different protocols has been developed to extract AEOs and APEOs and their degradation products from liquid samples. Thus, GBC [89, 90] and silica (C2 to C18) cartridges [53, 91, 92], along with methanol, dichloromethane, ethyl acetate and/or acetonitrile as elution solvents, have been employed, sometimes combined with strong cationic-exchange (SCX) and SAX cartridges [90, 92] for the removal of potential anionic and cationic interferences as non-ionic compounds are not retained due to their neutral charge. Additionally, octadecylsilica has been applied to extract the most hydrophobic group of NPEOs metabolites, constituted by NP and short chain oligomers ($NP_{1-3}EOs$), using mainly methanol, acetone or dichloromethane as elution solvents [60, 73]. SAX disks have been also used instead of conventional SPE cartridges to isolate nonylphenol polyethoxycarboxylates (NPECs), NPEO polar degradation products, from sludge extracts. Cassani and co-workers [93] also employed disks (C18) for determination of AEOs in sludge samples and wastewaters. Overall, most authors employ C18 [47, 55] and GBC cartridges [25, 88] because they are suitable for simultaneous isolation of a wide range of anionic (LAS and AES) and non-ionic (AEOs and NPEOs) surfactants, as well as their polar metabolites (PEGs, NPECs and SPCs), in a single stage by fractional elution using mixtures of hexane, dichloromethane, methanol, acetone and ethyl acetate. New polymeric materials are also currently being tested for the extraction of these compounds [51, 73, 69]. Thus, the hydrophilic-lipophilic copolymer Oasis HLB has been presented by Lara-Martín and co-workers [40] as an alternative for the simultaneous isolation of LAS, NPEOs, their carboxylated metabolites (SPCs and NPECs), and AEOs and their polar degradation intermediates (PEGs) from liquid samples in one single purification step. On the other hand, research on the isolation of cationic surfactants using SPE from water samples [94-97] and sediment extracts [72] is more limited. Nonpolar silica sorbents (e.g., C18) are not suitable for QACs because the strong interaction of these compounds with the silanol groups results in very broad elution bands [98]. This issue has been partially solved employing neutral polymeric sorbents instead [72, 96], although better results are obtained using sodium dodecyl sulphate (SDS) hemimicelles attached to alumina or anion exchange resins [94, 95, 97]. Despite this, LLE is still considered to be more effective than SPE for extraction of cationic surfactants from liquid samples [27].

In the past few years, advances in SPE have led to new related techniques such as matrix solid-phase dispersion (MSPD), which is used to extract and purify target compounds simultaneously from solid matrices. In the case of surfactants, this extraction protocol has been mainly applied to fish samples [99], where aliquots are taken and mixed with octadecyl silica in a column, in order to isolate LAS and SPCs, as well as non-ionic surfactants. Afterwards, strong non-polar solvents (e.g., hexane) and methanol are used to remove fats in a first clean-up stage and to extract surfactants after another elution, respectively. There are other simple and low cost extraction techniques which reduce the time needed for sample preparation, and decrease or eliminate solvent consumption [100-111]. As example, dispersive liquid-liquid microextraction (DLLME) is a novel method based on the migration of analytes to a cloudy solution, caused by the dispersion of the extraction solvent (low soluble in water, e.g., chloroform) as very fine droplets due to the appropriate mixture with a dispersant (soluble in water, e.g., acetone) in the aqueous sample. Then, these dispersed fine particles of the extraction phase containing analytes are sedimented in the bottom of a test tube by centrifugation [100]. The main difficulties associated with DLLME are the vulnerability of solvent drop to physic forces and automation issues. This problem of physical instability could be solved by the application of hollow fiber membranes which are impregnated by an organic solvent (e.g., 1-octanol) and placed into the water sample for

Analytes	Sample volume	Method	Solid phase	Isolation conditions	Ref.
QAC	100 mL	LLE	-	Solvent: chloroform (15 mL) Washing: water Ion-par reagent: Patent Blue V	[82]
LAS, SPC	500-1000 mL	LLE	-	Solvent: chloroform (3 x 4 mL) Ion-par reagent: methylene green	[84]
$NP_{1\text{-}2}EO$, NP, OP	300 mL	LLE	-	Solvent: DCM (300 mL)	[80]
QAC	20 mL	LLE	-	Solvent: chloroform (5 mL) Ion-par reagent: disulphine blue	[79]
LAS, AES, AS	10-200 mL	SPE	C18+SAX (LAS) C2 (AES, AS)	1. Conditioning: MeOH, water 2. Washing: MeOH/water 3. Elution: MeOH + HCl/MeOH 1. Conditioning: MeOH/isopropanol, water 2. Washing: water 3. Elution: MeOH/isopropanol	[85]
AEOs	50-2000 mL	SPE	C2+SCX+ SAX	1. Conditioning: Not spec. 2. Fractionation: ACN 3. Fractionation: MeOH/ethyl acetate/water	[92]
QAC	50 mL	SPE	Alumina	1. Passing solution with SDS 2. Elution: MeOH	[95]
NPEO, NPEC	100 mL	SPE	GBC	1. Conditioning: DCM, DCM/formic acid, MeOH, acidified water 2. Washing: MeOH/water, MeOH 3. Elution: DCM/formic acid	[89]
LAS, SPC	25-250 mL	SPE	C18+SAX	1. Conditioning: MeOH, water 2. Washing: water, acidified water 3. Elution: MeOH + acidified MeOH	[52]
QAC	100 mL	SPE	Strata-X	1. Conditioning: ACN, water 2. Washing: water 3. Elution: ACN/acetic acid/water	[96]
LAS, NPEO, NPEC, AEO, PEG, NP, OP	200 mL	SPE	C18	1. Conditioning: MeOH, water 2. Fractionation: hexane/DCM 3. Fractionation: MeOH/DCM	[55]

Table 3. Overview of LLE and SPE techniques used for clean-up and preconcentration of surfactants from environmental samples.

equilibrium extraction of the target compounds. Finally, the fiber is removed from the sample and extracted analytes are desorbed by diffusion into a different solvent (e.g., MeOH) [101]. This technique has been recently applied to isolation of cationic [101, 102], non-ionic [103] and anionic surfactants [104] from aqueous samples. Solid-phase microextraction (SPME) and stir-bar sorptive extraction (SBSE) can be also considered for rapid isolation of surfactants. Both techniques are based on the diffusion of analytes from the sample directly, without requiring any organic solvent, into a fiber or bar made of a specific polymer. The amount of polymer changes from 0.5 μL in SPME fibers up to 300 μL in SBSE bars, therefore improving the sensitivity of target compounds. Different types of fibers have been tested during application of SPME for isolation of anionic (polydimethylsiloxane (PDMS) [105], polyacrylate (PA) [106]) and non-ionic surfactants (carbowax/template resin (CWAX/TR) [107], PDMS/divinylbenzene (DVB), PA [108]). So far, the use of SBSE is limited to the extraction of NP and octylphenol (OP) from liquid samples [109]. Once they are captured by the polymer in the fiber or the bar, analytes are released by heat in the injection port of GC systems (thermal desorption), or by reduced amount of solvents before injection on LC systems (liquid desorption).

3. Separation, identification and quantification of synthetic surfactants

Over the last decades, analysis of surfactants in environmental samples has been carried out using several instrumental techniques. So far, spectrophotometric, potentiometric titrametration (PT) and tensammetric tecniques have been optimized to measure the total content of ionic [112-114] and non-ionic surfactants [115, 116], although their sensitivity and/or specifity tend to be low compared to chromatographic techniques coupled to several types of detectors. Generally, one of the main applications of spectrophotometric techniques has been routine environmental analysis due to their quickness and simplicity. They involve the formation of ion associates of analytes with specific ion-pair reagents and their extraction into appropriated organic solvents. After phase separation, the absorbance of the organic phase is measured. However, despite the advantages described above, the use of spectrophotometry generates very toxic wastes (e.g., chloroform) and is only limited to the analysis of total amount of surfactants [117-119]. PT and tensammetric techniques [120, 121] are based on the changes in electric properties caused by the presence of analytes in environmental samples. They can be only applied to the determination of total ionic and nonionic compounds, being impossible for both techniques to discriminate among individual components from surfactant mixtures. Besides, there are also issues associated with reproducibility and signal stability [113]. Nowadays, it is necessary to go beyond quantification of the total concentration of target analytes and, in most cases, chromatographic techniques (gas chromatography, GC, or high-performance liquid chromatography, HPLC) coupled to various types of detectors are preferred to separate and identify each individual compound from surfactant mixtures.

3.1. Gas chromatography

Less frequently used than HPLC for analysis of surfactants, the main drawback of GC is that all anionic and non-ionic compounds and their metabolites need to be derivatized with specific agents to solve sensitivity, separation or volatilization issues before injecting them into the system. Most commonly used derivatizing agents are trifluoroethanol [29, 58], diazomethane [84], N,O-bis(trimethylsilyl)trifluoro acetamide (BSTFA) [53, 59, 122], acetic anhydride [61, 109] and hydrogen bromide [90], among other reactants. In any case, some low molecular mass metabolites of non-ionic surfactants (NP and short-chain NPEOs) have been analyzed directly by GC [67, 80] as they are volatile enough, although better results can be obtained if derivatization is performed. There are also some advantages in using GC over HPLC. Thus, GC columns have a better capability for achieving complete separation of homologues and isomers of many surfactants after derivatization. This may be a key aspect for those studies on the biodegradability or toxicity of surfactants such as LAS or NPEO, which can change depending on the length of the alkyl chain and/or the position of the phenyl ring [123] (Figures 3a and b). In most cases, anionic and non-ionic surfactants have been separated by nonpolar capillary columns containing 5%-phenyl-95%-dimethylpolysiloxane (e.g., HP-5 [58, 75, 67], SE-54 [76], DB-5 [84, 103, 109]), and a mobile phase comprised of high purity helium as carrier gas with a flow rate from 0.58 to 3.4 mL/min. Regarding cationic surfactants, the application of GC to their separation and analysis has not been mentioned in any paper so far [1]. Table 4 describes general information about some analytical protocols for determination of anionic and nonionic surfactants by means of GC in environmental samples.

Several types of detectors can be used after gas chromatography for the analysis of target compounds, such as flame-ionization detectors (FID), which were used for the analysis of

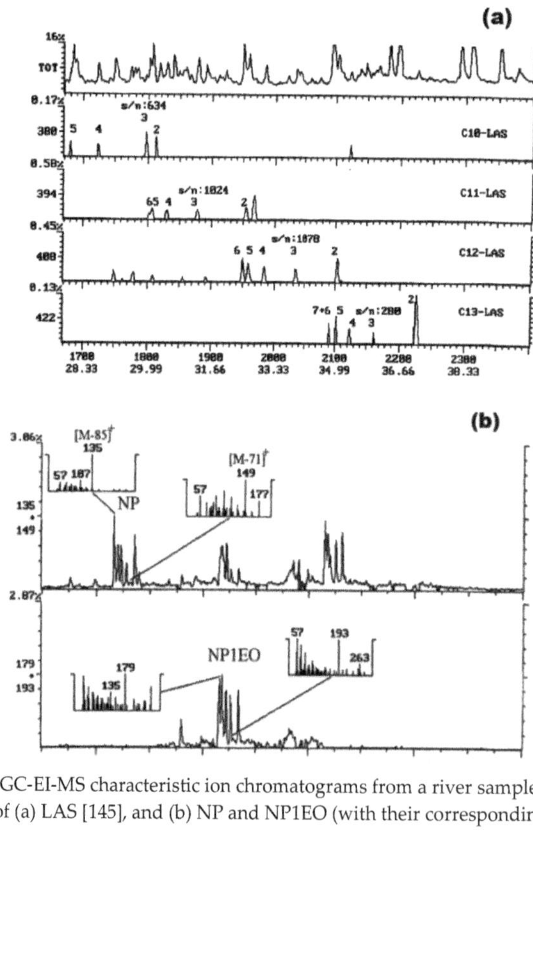

Figure 3. Selected GC-EI-MS characteristic ion chromatograms from a river sample, showing resolution of the derivatives of (a) LAS [145], and (b) NP and NP1EO (with their corresponding mass spectra) [144].

anionic surfactants in water samples [124]. Nowadays, single quadrupole (MS) or tandem mass spectrometers (MS-MS) are commonly preferred because they allow unequivocal identification of analytes by measuring their parent masses and displaying specific fragmentation patterns after their ionization and rupture, respectively. Hence, there are several papers dealing with the analysis of anionic and non-ionic surfactants using GC coupled to MS [69, 105, 108] or MS-MS [125]. Target compounds can be detected by electron impact (EI) or chemical ionization (CI), being more widely used the first mode, although higher sensitivity may be reached using CI for analysis of some anionic compounds.

Target compounds	Matrix	Sample preparation	Recovery (%)	Mobile phase	Column	Detection	LOD/ MLD	Ref
LAS	Wastewater, seawater	Ion pair SPME Derivatization in GC injection port (tetrabutyl ammonium)	-	-	BPX5 (capillary column, 30 m, 0.25 mm ID, 0.25 μm)	EI(+)-MS	0.16-0.8 ng/mL	[105]
$NP_{1\text{-}3}EO$, NP $NP_{1\text{-}3}EC$ LAS, SPC	River water, wastewater	SPE (without derivatization) Derivatization (C_3H_7OH/ CH_3COCl) Derivatization ($SOCl_2$/ CF_3CH_2OH)	81-90 (NP) 75-112 (LAS, SPC)	-	DB-5 (capillary column, 30 m, 0.25 mm ID, 0.25 μm)	EI/CI(+)-MS	≤0.01 μg/L (LOQ)	[29, 143, 144]
AEOs (C_{12}-C_{15})	Wastewater, river water	SPE Derivatization (HBr)	65-102	Helium	Rtx-1 (capillary column, 60 m, 0.25 mm ID, 0.25 μm)	EI(-)-MS	0.001-0.01 mg/L	[90]
$NP_{1\text{-}2}EO$, NP	Marine sediment	MAE, SPE (without derivatization)	100	Helium (2 mL/min)	HP-5 (capillary column, 30 m, 0.25 mm ID, 0.25 μm)	EI(+)-MS	100 ng	[67]
NP, OP	River water	DLLME Derivatization in situ (methyl chloroformate)	88.3-106.7	Helium (1 mL/min)	DB-5 (fused silica capillary column, 30 m, 0.25 mm ID, 0.25 μm)	EI(+)-MS	0.002-0.03 μg/L	[103]
NP,OP	River water, sediment	MAE, SPE Derivatization (BSTFA)	77-109	Helium (1 mL/min)	HP-5 (capillary column, 30 m, 0.25 mm ID, 0.25 μm)	EI(+)-MS-MS	0.01-0.1 ng/L 0.08-0.14 ng/g	[125]

Table 4. Key aspects of GC analysis of surfactants in different environmental matrices.

3.2. Liquid chromatography

High-performance liquid chromatography (HPLC) is currently the most commonly used technique for separation and analysis of commercial mixtures of surfactants in the environment, mainly due to its advantages over GC because HPLC is suitable for determining non volatile analytes from low to high molecular weight and derivatization is unnecessary in most cases. Reverse-phase columns, mainly RP-18 [47, 96, 106] and RP-8 [52, 95], are often employed for chromatographic separations of anionic, non-ionic, cationic surfactants and their degradation products. Mobile phases are solvent mixtures containing deionized water, acetonitrile and/or methanol. Separation can be improved by adding some additives (e.g., ammonium acetate (AMAC), triethylamine) to the mobile phase, as well as acetic (AA) or formic acid (FA) as modifiers [72, 71, 89]. There are also a few works showing efficient separation of NPEOs ethoxymers, some of their metabolites [126] and QACs [101] by amino-silica or cyanopropyl normal phase columns, although the elution order is reversed (more hydrophobic compounds, such as NP, elute first and NPEOs last). In these cases, stronger non-polar solvents (e.g., hexane, chloroform and isopropanol) are preferred. Additionally, some researchers have used new stationary phases that are specific for the separation of ethoxylated surfactants. As example, Lee Ferguson and co-workers [60, 127] tested a mixed-mode HPLC system using a column packed with a polymeric phase capable of separating NPEO and NP components by both size-exclusion and reversed-phase adsorption mechanisms (Figure 4a). Other authors have also applied this technique with some modifications to quantify OP and octylphenol ethoxylates (OPEOs) in environmental samples [73]. Alternative packing materials containing hydrophobic (alkyl chains) and hydrophilic (amide) functional groups to improve the simultaneous separation of cationic, anionic and non-ionic surfactants have also been occasionally employed [128].

Some surfactant classes (e.g., LAS and NPEOs) and their metabolites are still good candidates, due to the presence of an aromatic ring in their molecular structure, to be analyzed by the first quantitative methods based on the use of HPLC coupled to ultraviolet (UV) or fluorescence detectors (FL) [68, 66, 126, 129, 130]. The presence of a benzene group also facilitates the use of UV for identifying some specific cationic surfactants such as benzalkonium chlorides (BACs) [97]. Moreover, HPLC coupled to FL detector was employed by Natkae and co-workers [131] to achieve partial separation of positional isomers and obtain information on the alkyl chain distributions of LAS in river water samples. However, aliphatic surfactants (e.g., AEOs and AES) have not been monitored so much due to their lack of UV absorbance or fluorescence. Prior derivatization using phenyl-isocyanate [132], naphthyl isocyanate and naphthyl chloride (NC) [88, 133], among others, must be carried out. Nowadays, however, this kind of surfactants, along with LAS, NPEOs and many other organic microcontaminants, are preferably determined by HPLC-MS, which offers several advantages over other detectors such as sensitivity, selectivity, and simultaneous identification and confirmation of multiple analyte classes by means of their molecular weight, retention time and mass spectra. In this sense, considerable progress has been achieved in the environmental analysis of surfactants over the last decade due to the development of atmospheric pressure ionization (APCI) or electrospray ionization (ESI) interfaces that allow coupling HPLC to MS. Before this, mass spectrometry was used only for identification of a wide range of surfactants from their mass spectra by flow-injection analysis (FIA) [134].

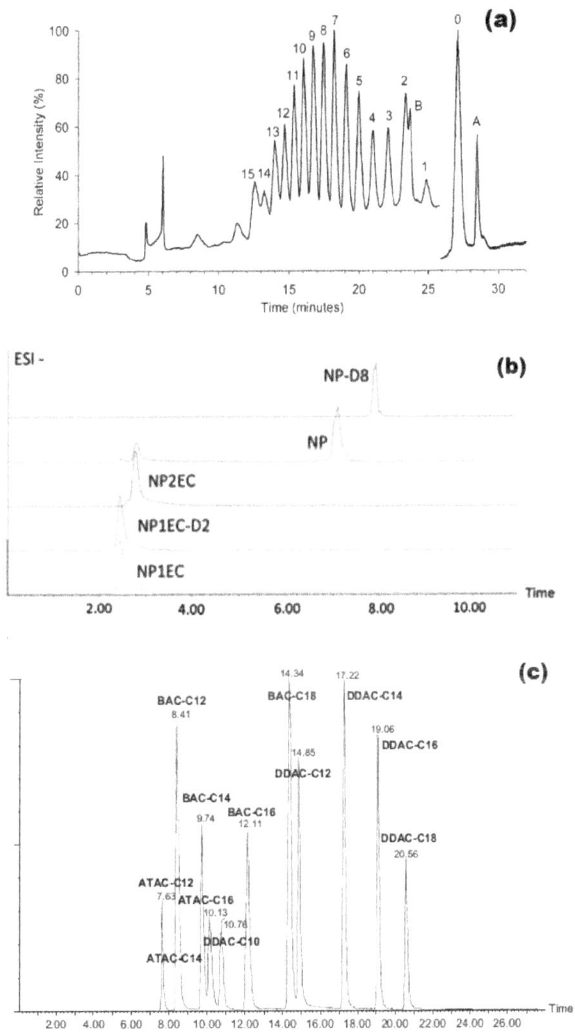

Figure 4. (a) Mixed-mode HPLC-ESI-MS total current ion chromatogram of NPEOs (A= NP, B=n-NP_3EO, 0=NP, 1=NP_1EO, etc.) from a sediment sample, switching MS polarity from positive to negative ion mode at retention time 25.8 min [60]; (b) UPLC-ESI-MS-MS extracted ion chromatograms showing the occurrence of NPEO metabolites in a sediment sample [44]; and (c) HPLC-MS-MS chromatogram of a standard solution of QACs [141].

Among different types of mass analyzers used for the identification and quantification of surfactants, there are several authors that have employed single quadrupole HPLC-MS systems operating in selected ion monitoring (SIM) mode [25, 55]. However, isobaric interferences may lead to sensitivity and resolution issues, which have been commonly solved by means of triple quadrupole [27, 106, 135] or ion trap MS detectors [47, 94]. In recent years, both techniques, especially the first one, have been the main tool for trace analysis of surfactants and many other organic contaminants because their respective MS-MS (triple quadrupole) and MS^n (ion trap) capabilities allow scanning for daughter ions, increasing sensitivity and selectivity (especially for analysis of environmental samples which contain compounds showing the same molecular ions and retention times than those for selected analytes) [72] (Figures 4b and c). As example, discrimination and quantification of the 20 positional isomers of LAS was achieved recently by Lunar and co-workers [136] by monitoring specific fragment ions resulting from the benzylic cleavage of the carbon alkyl chain on both sides of the LAS phenyl group. As a drawback of this type of MS detectors, there is a limited number of predetermined ions that can be monitored during a single experiment and, although less frequent than in single quadrupole MS, interferences may lead to overestimation in the concentration of target compounds. Time-of-flight (ToF) LC-MS systems are less commonly used than other MS analyzers for environmental analysis of surfactants, but their full scan spectral sensitivity in a wide mass range and accurate mass measurement allow the identification and quantification of a large number of target, non-target surfactants and their metabolites in all kinds of matrices [40, 65, 137], constituting a recent alternative to address the issues mentioned above. Occasionally, hybrid systems like quadrupole time-of-flight (Q-ToF) detectors have been applied to determine a wide range of surfactants and some of their degradation products, such as alkylphenols and their carboxylates, in textile wastewaters [138], or to identify for the first time the molecular structure of LAS anaerobic degradation metabolites [139], although due to their high cost and relatively lower sensitivity compared with HPLC-MS-MS they are not often used for routine analysis of these compounds in environmental samples.

Table 5 provides general information about some analytical procedures aimed to the determination of surfactants in different environmental matrices by HPLC-MS (and some other detectors). So far, LAS and SPCs have been determined in both freshwater [71] and marine environments [40] using several kinds of MS detectors coupled to HPLC under negative ion (NI) mode due to the presence of a sulfonate group. Quasi-molecular ions $[M-H]^-$ and a characteristic fragment m/z = 183 were used for their identification and quantification. AES have also been monitored in aquatic systems [62, 140] in a similar way, but m/z = 97, corresponding to $HO-SO_3^-$, was selected as the main fragment ion. On the other hand, identification of QACs relies upon measurement of their molecular ions (M^+) in positive ionization mode (PI), and further confirmation can be achieved by mass measurement of main characteristic ions such as m/z = 60 for alkyltrimethylammonium chlorides (ATACs) [141] or m/z = 91 for BACs [94]. Non-ionic surfactants lack charge or acid/base functional groups, so the most widely used option for ionization of ethoxylated compounds, such as NPEOs and AEOs, is to form adducts as the oxygen atoms in the polyethoxylate chain can donate their free electrons to a selected cation agent and the flexible structure of the chain

Target analytes	Matrix	Sample treatment	Recovery (%)	Mobile phase	Column	Detection	LOD/ MLD	Ref
BAC, DADMA C	Marine sediment	Sonication, LLE, SPE	98-118 (DADMA C)	ACN/H_2O, isopropanol, FA, AMAC	Luna C18 (150 mm, 2mm, 5 μm)	ESI(+)-ToF-MS	0.1-2.6 ng/g (LOQ)	[65]
LAS, SPC	Seawater, marine sediment	Soxhlet, SPE	75-105	MeOH/H_2O, H_2O, tetraethyl ammonium hydrogensulfa te	LiChrosorb RP-8 (250 mm, 4.6 mm, 10.6 μm)	FL	0.2-0.4 μg/L 5-10 μg/Kg	[52]
NPEO, NP	Marine sediment, sewage	Sonication, SPE LLE	64-127 (sediment)	H_2O, MeOH, sodium acetate	MSpak GF-310 4D filtration column (150 mm, 4.6 mm)	ESI-MS NPEO (ESI+) NP (ESI-)	0.78-37.3 ng/g	[127]
LAS, SPC, NPEO, $NP_{1-2}EC$, AEO, PEG	Sewage, marine sediment, seawater, s. solids	Sonication, SPE	26-117 (AEOs, PEGs) 60-108 (NPEOs, $NP_{1-2}EC$) 37-101 (LAS, SPC)	ACN, H_2O, FA, ammonium formate	Luna C18 (150 mm, 2mm, 5 μm)	ESI-ToF MS LAS, SPC, $NP_{1-2}EC$ (ESI-) NPEO, AEO (ESI+)	0.1-11.8 ng/L 0.1-23.7 μg/Kg	[40]
AEO, $NP_{1-3}EO$, NP, $NP_{1-2}EC$	Marine sediment	Sonication, SPE	34-88	ACN, H_2O, FA, ammonium formate	Purospher STAR RP-18 UHPLC column (50 mm, 2 mm, 1.8 μm)	ESI(+)-MS-MS	<0.1-27.3 ng/g	[44]
BAC, ATAC, DADMA C	River sediment, sludge	Soxhlet, LLE	67-95	ACN/H_2O, isopropanol, FA, AMAC	Luna C18 (150 mm, 2mm, 5 μm)	ESI(+)-MS MS	0.6-5 μg/Kg (LOQ)	[27]
QAC	River water, sewage	Microporou s membrane liquid-liquid extraction	-	Chloroform, ethanol, ammonia, heptanoic acid	Cyanopropyl column (250 mm, 2mm)	UV	0.7-5 μg/L	[101]

Table 5. Key aspects of HPLC analysis of surfactants in different environmental matrices.

allows the molecule to "wrap" itself around that cation [64]. Thus, sodium acetate [60, 74], ammonium acetate [89, 142] or different acids [53] are commonly added to the samples or to the mobile phase to increase the MS response of NPEOs and AEOs and to stabilize the generation of $[M+Na]^+$, $[M+NH_4]^+$ or $[M+H]^+$ ions, among others. Additionally, this ability to form different adducts can be used to obtain multiple confirmation points in full-scan mode [40]. Another advantage of MS compared to other detectors is that several types of surfactants can be analyzed within a single run (e.g., NPEOs and AEOs can be separated, using an adequate gradient, and later analyzed under PI [53, 64]). Most recent methodologies allow simultaneous determination of anionic and non-ionic surfactants and their metabolites in environmental samples [47, 55].

Today, mass spectrometry is often combined with ultra-performance liquid chromatography (UPLC), which uses sub-2-μm column particles that provide enhanced separation, faster analysis, and improved sensitivity over HPLC, boosting laboratory efficiency by saving time and decreasing solvent consumption. Most researchers have started to benefit from this combination, although there are still a few examples on its use for analysis of surfactants. So far, UPLC-Q-ToF-MS has been used for structural elucidation of SPC isomers [139] and for environmental screening of several anionic and non-ionic surfactants in wastewater [138]. UPLC-MS-MS [44] has allowed achieving fast analysis (less than 10 min per sample) of NPEO metabolites and AEOs at trace levels in aquatic environments.

4. Conclusion

The assessment of the behavior and final fate of synthetic surfactants in the environment is a crucial matter due to the huge volumes of these chemicals that are discharged into aquatic and terrestrial ecosystems. A significant number of analytical protocols have been developed over the last decades aimed to the individual or simultaneous extraction, isolation and determination of different types of surfactants in environmental samples. Nowadays, the most widely used sample preparation protocols are based on SPE, directly derived from column chromatography. However, the trend is to research on new techniques, such as SPME or SBSE, aimed to reduce, or even eliminate, solvent consumption, as well as saving money by using reusable fibers and bars rather than disposable cartridges. Regarding the separation, identification and quantification of surfactants, HPLC-MS and, to a lesser extent due to the non volatility of most analytes, GC-MS, are the main tools currently employed as they allow for determination of every single homologue, ethoxymer and/or isomer from surfactant mixtures in different environmental matrices (solids, water and biota). Most recently, different classes of time-of-flight and triple quadrupole mass spectrometers have started to be combined with UPLC, which provides enhanced separation, faster analysis, higher confidence, and lower detection limits than more conventional HPLC-MS or HPLC-MS-MS approaches, as well as improves identification of unknown surfactant metabolites and other non target compounds within the same run.

5. List of abbreviations

AA, Acetic acid; ASE, Accelerated solvent extraction; ACN, Acetonitrile; AEOs, Alcohol polyethoxylates; AES, Alkyl ethoxysulfates; AS, Alkyl sulfates; ATACs, Alkyltrimethylammonium chlorides; AP, Alkylphenol; APEOs, Alkylphenol polyethoxylates; APEC, Alkylphenol polyethoxycarboxylate; AMAC, Ammonium acetate; APCI, Atmospheric pressure ionization; BACs, Benzalkonium chlorides; BSTFA, N,O-bis(trimethylsilyl)trifluoro acetamide; CWAX/TR, Carbowax/template resin; CI, Chemical ionization; DTDMAC, Dehydrogenated tallow dimethyl ammonium chloride; DADMAC, Dialkyldimethylammonium chlorides; DCM, Dichloromethane; DLLME, Dispersive liquid-liquid microextraction; EI, Electron impact ionization; ESI, Electrospray ionization; EO, Ethylene oxide; CESIO, European Committee of Organic Surfactants and their Intermediates; FID, Flame-ionization detectors; FIA, Flow-injection analysis; FL, Fluorescence detectors; FA, Formic acid; GC, Gas chromatography; GBC, Graphitized black carbon; HPLC, High-performance liquid chromatography; HLB, Hydrophilic-lipophilic balance; LOD, Limit of detection; LOQ, Limit of quantification; LAS, Linear alkylbenzene sulfonates; LC, Liquid chromatography; LLE, Liquid-liquid extraction; MS, Mass spectrometry; MSPD, Matrix solid-phase dispersion; MeOH, Methanol; MLD, Method limit detection; MAE, Microwave-assisted extraction; NC, Naphthyl chloride; NI, Negative ionization; NP, Nonylphenol; NPEOs, Nonylphenol polyethoxylates; NPECs, Nonylphenol polyethoxycarboxylates; OP, Octylphenol; OPEOs, Octylphenol polyethoxylates; PA, Polyacrylate; PEGs, Polyethylenglycols; SDB, Polystyrene-divinylbezene; PDMS/DVB, Polydimethylsiloxane/divinylbenzene; PI, Positive ionization; PT, Potentiometric titrametration; PFE, Pressurized fluid extraction; PLE, Pressurized liquid extraction; Q-ToF, Quadrupole time-of-flight; QACs, Quaternary ammonium-based compounds; SIM, Selected ion monitoring; SDS, Sodium dodecyl sulphate; SLE, Solid-liquid extraction; SPE, Solid phase extraction; SPME, Solid phase microextraction; SBSE, Stir-bar sorptive extraction; SAX, Strong anionic-exchange; SCX, Strong cationic-exchange; SPCs, Sulfophenyl carboxylic acids; SFE, Supercritical fluid extraction; MS-MS, Tandem mass spectrometry; ToF, Time-of-flight; UPLC, Ultra performance liquid chromatography; UV, Ultraviolet detectors; WWTPs, Wastewater treatment plants.

Author details

Juan M. Traverso-Soto, Eduardo González-Mazo and Pablo A. Lara-Martín
Departamento de Química Física, Facultad de Ciencias del Mar y Ambientales, Campus de Excelencia Internacional del Mar (CEI·MAR), Universidad de Cádiz, Puerto Real, Spain

6. References

[1] Thiele B (2005) Analysis of Surfactants in Samples of the Aquatic Environment. In: L.M.L. Nollet, editor. Chromatographic Analysis of the Environment. Boca Raton: CRC Press. pp 1173–1198.

[2] Ying G.G (2006) Fate, behavior and effects of surfactants and their degradation products in the environment. Environ. Int. 32: 417-431.
[3] Environmental Risk Assesment (2009) Linear Alkylbenzene Sulfonate (LAS). In: Human and Environmental Risk Assesment on Ingredients of European Household Cleaning Products. Brussels.
[4] Environmental Risk Assesment (2004) Alcohol Ethoxysulfates (AES). In: Human and Environmental Risk Assesment on Ingredients of European Household Cleaning Products. Brussels.
[5] Environmental Risk Assesment (2002) Alkyl Sulfates. In: Human and Environmental Risk Assesment on Ingredients of European Household Cleaning Products. Brussels.
[6] Jobling S.J, Sumpter J.P (1993) Detergent components in sewage effluent are weakly oestrogenic to fish: An in vitro study using rainbow trout (*Oncorhynchus mykiss*) hepatocytes. Aquat. Toxicol. 27: 361–72.
[7] Purdom C.E, Hardiman P.A, Bye V.V.J, Eno N.C, Tyler C.R, Sumpter J.P (1994) Estrogenic Effects of Effluents from Sewage Treatment Works. Chem. Ecol. 8: 275–85.
[8] Sonnenschein C., Soto A.M (1998) An Updated Review of Environmental Estrogen and Androgen Mimics and Antagonists. J. Steroid Biochem. Molec. Biol. 65: 143–150.
[9] Human and Environmental Risk Assesment (2009) Alcohol Ethoxylates (AEOs). In: Human and Environmental Risk Assesment on Ingredients of European Household Cleaning Products. Brussels.
[10] Madsen T, Buchardt Boyd H, Nylén D, Rathmann Pedersen A, Petersen G.I, Simonsen F (2001) Environmental and Health Assessment of Substances in Household Detergents and Cosmetic Detergent Products. In: Environmental Project No. 615, Miljøprojekt. Danish Environmental Protection Agency.
[11] Berenbold H (1994) Rinse Softeners-current situation in Europe. International News on Fats, Oils and Related Materials 5(1): 82-84.
[12] Giolando S.T, Rapaport R.A, Larson R.J, Federle T.W, Stalmans M, Masscheleyn P (1995) Environmental fate and effects of DEEDMAC: a new rapidly biodegradable cationic surfactant for use in fabric softeners. Chemosphere 30(6): 1067– 83.
[13] Environmental Risk Assesment (2008) Esterquats. In: Human and Environmental Risk Assesment on Ingredients of European Household Cleaning Products. Brussels.
[14] Berna J.L, Ferrer J, Moreno A, Prats D, Ruiz Bevia F (1989) The fate of LAS in the environment. Tenside Surf. Det. 26: 101-107.
[15] Prats D, Ruiz F, Vazquez B, Rodríguez-Pastor M (1997) Removal of Anionic and Nonionic Surfactants in a Wastewater Treatment Plant with Anaerobic Digestion. A Comparative Study. Wat. Res. 31: 1925-1930.
[16] González S, Petrovic M, Barceló D (2004) Simultaneous extraction and fate of linear alkylbenzene sulfonates, coconut diethanol amides, nonylphenol ethoxylates and their degradation products in wastewater treatment plants, receiving coastal waters and sediments in the Catalonian area (NE Spain). J. Chromatogr. A 1052: 111-120.
[17] Morrall S.W, Dunphy J.C, Cano M.L, Evans A, McAvoy D.C, Price B.P, Eckhoff W.S (2006) Removal and environmental exposure of alcohol ethoxylates in US sewage treatment. Ecotoxicol. Environ. Saf. 64: 3-13.

[18] Clara M, Scharf S, Scheffknecht C, Gans O (2007) Ocurrence of selected surfactants in untreated and treated sewage. Water Res. 41: 4339-4348.
[19] Sanderson H, Dyer S.D, Price B.B, Nielsen A.M, van Compernolle R, Selby M, Stanton K, Evans A, Ciarlo M, Sedlak R (2006) Ocurrence and weight-of-evidencce risk assessment of alkyl sulfates, alkyl ethoxysulfates, and linear alkylbenzene sulfonates (LAS) in river water and sediments. Sci. Total Environ. 368: 695-712.
[20] Kreuzinger N, Fuerhacker M, Scharf S, Uhl M, Gans O, Grillitsch B (2007) Methodological approach towards the environmental significance of uncharacterized substances - quaternary ammonium compounds as an example. Desalination 215: 209-222.
[21] Eadsforth C.V, Sherren A.J, Selby M.A, Toy R, Eckhoff W.S, McAvoy D.C, Matthijs E (2006) Monitoring of environmental fingerprints of alcohol ethoxylates in Europe and Canada. Ecotoxicol. Environ. Saf. 64: 14-29.
[22] Loyo-Rosales J.E, Rice C.P, Torrents A (2007) Octyl and nonylphenol ethoxylates and carboxylates in wastewater and sediments by liquid chromatography/tandem mass spectrometry. Chemosphere 68: 2118-2127.
[23] Cavalli L, Gellera A, Landone A (1993) LAS removal and biodegradation in a wastewater treatment plant. Environ. Toxicol. Chem. 12: 1777-1788.
[24] Ahel M, Giger W, Koch M (1994) Behaviour of Alkylphenol Polyethoxylate Surfactants in the Aquatic Environment-I. Ocurrence and Transformation in Sewage Treatment. Wat. Res. 28: 1131-1142.
[25] Bruno F, Curini R, Di Corcia A, Fochi I, Nazzari M, Samperi R (2002) Determination of surfactants and some of their metabolites in untreated and anaerobically digested sewage sludge by subcritical water extraction followed by liquid chromatography-mass spectrometry. Environ. Sci. Technol. 36:4156– 61.
[26] Fernandez P, Alder A.C, Suter M.J.F, Giger W (1996) Determination of the Quaternary Ammonium Surfactant Ditallowdimethylammonium in Digested Sludges and Marine Sediments by Supercritical Fluid Extraction and Liquid Chromatography with Postcolumn Ion-Pair Formation. Anal. Chem. 68: 921– 9.
[27] Martínez-Carballo E, González-Barreiro C, Sitka A, Kreuzinger N, Scharf S, Gans O (2007) Determination of selected quaternary ammonium compounds by liquid chromatography with mass spectrometry. Part II. Application to sediment and sludge samples in Austria. Environ. Pollut. 146: 543-547.
[28] Holt M.S (2000) Sources of Chemical Contaminants and Routes into the Freshwater Environment. Food Chem. Toxicol. 38: S21–S27.
[29] Ding W.H, Tzing Shin-Haw, Lo Jun-Hui (1999) Ocurrence and Concentrations of Aromatic Surfactants and their Degradation Products in River Waters of Taiwan. Chemosphere 38: 2597-2606.
[30] Marcomini A, Pojana G, Sfriso A, Quiroga Alonso José-Maria (2000) Behavior of Anionic and Nonionic Surfactants and their Persistent Metabolites in the Venice Lagoon, Italy. Environ. Toxicol. Chem. 19: 2000–2007.

[31] Corada-Fernández C, Lara-Martín P.A, Candela L, González-Mazo E (2011) Tracking sewage derived contamination in riverine settings by analysis of synthetic surfactants. J. Environ. Monit. 13: 2010.
[32] Belmont M.A, Ikonomou M, Metcalfe C.D (2006) Presence of nonylphenol ethoxylate surfactants in a watershed in central Mexico and removal from domestic sewage in a treatment wetland. Environ. Toxicol. Chem. 25: 29–35.
[33] Jonkers N, Laane R.W.P.M, de Voogt P (2003) Fate of nonylphenol ethoxylates and their metabolites in two Dutch estuaries: evidence of biodegradation in the field. Environ. Sci. Technol. 37: 321-327.
[34] Maruyama K, Yuan M, Otsuki A (2000) Seasonal changes in ethylene oxide chain length of poly(oxyethylene)alkylphenyl ether nonionic surfactants in three main rivers in Tokyo. Environ. Sci. Technol. 34: 343-348.
[35] Lee Ferguson P, Iden C.R, Brownawell B.J (2001) Distribution and fate of neutral alkylphenol ethoxylate metabolites in a sewage-impacted urban estuary. Environ. Sci. Technol. 35: 2428-2435.
[36] Petrovic M, Rodríguez Fernández-Alba A, Borrull F, Marce R.M, González-Mazo E., Barceló D (2002) Ocurrence and Distribution of Nonionic Surfactants, their Degradation Products, and Linear Alkylbenzene Sulfonates in Coastal Waters and Sediments in Spain,. Environ. Toxicol. Chem. 21: 37–46.
[37] Zoller U (2006) Estuarine and coastal zone marine pollution by the nonionic alkylphenol ethoxylates endocrine disrupters: Is there a potential ecotoxicological problem?. Environ. Int. 32: 269–272.
[38] Lara-Martín P.A, Petrovic M, Gómez-Parra A, Barceló D, González-Mazo E (2006) Presence of surfactants and their degradation intermediates in sediment cores and grabs from the Cadiz Bay area. Environ. Pollut. 144: 483–491.
[39] León V.M, Sáez M, González-Mazo E., Gómez-Parra A (2002) Ocurrence and distribution of linear alkylbenzene sulfonates and sulfophenylcarboxylic acids in several Iberian littoral ecosystems. Sci. Total Environ. 288: 215-226.
[40] Lara-Martín P.A, González-Mazo E, Brownawell B.J (2011) Multi-residue method for the analysis of synthetic surfactants and their degradation metabolites in aquatic systems by liquid chromatography-time-of-flight-mass spectrometry. J. Chromatogr. A 1218: 4799-4807.
[41] Bester K, Theobald N, Schröder H.Fr (2001) Nonylphenols, nonylphenol-ethoxylates, linear alkylbenzenesulfonates (LAS) and bis (4-chlorophenyl)-sulfone in the German Bight of the North Sea. Chemosphere 45: 817-826.
[42] Isobe T, Takada H (2004) Determination of degradation products of alkylphenol polyethoxylates in municipal wastewaters and rivers in Tokyo. Japan. Environ. Toxicol. Chem. 23: 599-605.
[43] Lara-Martín P.A, Gómez-Parra A, González-Mazo E (2008) Sources, transport and reactivity of anionic and non-ionic surfactants in several aquatic ecosystems in SW Spain: A comparative study. Environ. Pollut. 156: 36-45.
[44] Lara-Martín P.A, González-Mazo E, Brownawell B.J (2011) Enviromental analysis of alcohol ethoxylates and nonylphenol ethoxylate metabolites by ultra-performance

liquid chromatography-tandem mass spectrometry. Anal. Bioanal. Chem. DOI: 10.1007/s00216-011-5449-6.

[45] Lara-Martín P.A, Li X, Bopp R.F, Brownawell B.J (2010) Ocurrence of Alkyltrimethylammonium Compounds in Urban Estuarine Sediments: Behentrimonium As a New Emerging Contaminant. Environ. Sci. Technol. 44: 7569-7575.

[46] Li X, Brownawell B.J (2010) Quaternary Ammonium Compounds in Urban Estuarine Sediment Environments – A Class of Contaminants in Need of Increased Attention?. Environ. Sci. Technol. 44: 7561-7568.

[47] Lara-Martín P.A, Gómez-Parra A, González-Mazo E (2006) Development of a method for the simultaneous analysis of anionic and non-ionic surfactants and their carboxylated metabolites in environmental samples by mixed-mode liquid chromatography-mass spectrometry. J. Chromatogr. A 1137: 188-197.

[48] Krogh K.A, Mogensen B.B, Halling-Sorensen B, Cortés A, Vejrup K.V, Barceló D (2003) Analysis of alcohol ethoxylates and alkylamine ethoxylates in agricultural soils using pressurized liquid extraction and liquid chromatography-mass spectrometry. Anal. Bioanal. Chem. 376: 1089-1097.

[49] Jonkers N, Laane R.W.P.M, de Voogt P (2005) Sources and fate of nonylphenol ethoxylates and their metabolites in the Dutch coastal zone of the North Sea. Mar. Chem. 96: 115-135.

[50] Selberg A, Budashova J, Tenno T (2007) Column study of the leaching and degradation of anionic surfactants in oil-polluted soil. Proc. Estonian Acad. Sci. Chem. 56: 87-97.

[51] Shang D.Y, Ikonomou M.G, Macdonald R.W (1999) Quantitative determination of nonylphenol polyethoxylate surfactants in marine sediment using normal-phase liquid chromatography-electrospray mass spectrometry. J. Chromatogr. A 849: 467-482.

[52] León V.M, González-Mazo E, Gómez-Parra A (2000) Handling of marine and estuarine samples for the determination of linear alkylbenzene sulfonates and sulfophenylcarboxylic acids. J. Chromatogr. A 889: 211-219.

[53] Jeannot R, Sabik H, Sauvard E, Dagnac T, Dohrendorf K (2002) Determination of endocrine-disrupting compounds in environmental samples using gas and liquid chromatography with mass spectrometry. J Chromatogr. A 974: 143-159.

[54] Sablayrolles C, Montréjaud-Vignoles M, Silvestre J, Treilhou M (2009) Trace Determination of Linear Alkylbenzene Sulfonates: Application in Artificially Polluted Soil-Carrots System. Int. J. Anal. Chem. 2009: 1-6.

[55] Petrovic M, Barceló D (2000) Determination of Anionic and Nonionic Surfactants, Their Degradation Products, and Endocrine-Disrupting Compounds in Sewage Sludge by Liquid Chromatography/Mass Spectrometry. Anal. Chem. 72: 4560-4567.

[56] Sun H.F , Takata A, Hata N, Kasahara I, Taguchi S (2003) Transportation and fate of cationic surfactant in river water. J. Environ. Monit. 5: 891-895.

[57] Holt M.S, Bernstein S.L (1992) Linear Alkylbenzenes in Sewage Sludges and Sludge Amended Soils. Wat. Res. 26: 613-624.

[58] Reiser R, Toljander H.O, Giger W (1997) Determination of Alkylbenzenesulfonates in Recent Sediments by Gas Chromatography/Mass Spectrometry. Anal. Chem. 69: 4923-4930.

[59] Li D, Oh Jae-Ryoung, Park J (2003) Direct extraction of alkylphenols, chlorophenols and bisphenol A from acid-digested sediment suspension for simultaneous gas chromatographic-mass spectrometric analysis. J. Chromatogr. A 1012: 207-214.

[60] Lee Ferguson P, Iden C.R, Brownawell B.J (2000) Analysis of Alkylphenol Ethoxylate Metabolites in the Aquatic Environment Using Liquid Chromatography-Electrospray Mass Spectrometry. Anal. Chem. 72: 4322-4330.

[61] Bennie D.T, Sullivan C.A, Lee H.-B, Peart T.E, Maguire R.J (1997) Ocurrence of alkylphenols and alkylphenol mono- and diethoxylates in natural waters of the Laurentian Great Lakes basin and the upper St. Lawrence River. Sci. Tot. Environ. 193: 263-275.

[62] Lara-Martín P.A, Gómez-Parra A, González-Mazo E (2006) Simultaneous extraction and determination of anionic surfactants in waters and sediments. J. Chromatogr. A 1114: 205-210.

[63] Chiron S, Sauvard E, Jeannot R (2000) Determination of nonionic polyethoxylate surfactants in wastewater and sludge samples of sewage treatment plants by liquid chromatography-mass spectrometry. Analusis 28: 535-542.

[64] Cohen A, Klint K, Bøwadt S, Persson P, Jönsson J.A (2001) Routine analysis of alcohol and nonylphenol polyethoxylates in wastewater and sludge using liquid chromatography-electrospray mass spectrometry. J. Chromatogr. A 927: 103-110.

[65] Li X, Brownawell B.J (2009) Analysis of Quaternary Ammonium Compounds in Estuarine Sediments by LC-ToF-MS: Very High Positive Mass Defects of Alkylamine Ions as Powerful Diagnostic Tools for Identification and Structural Elucidation. Anal. Chem. 81: 7926-7935.

[66] Villar M, Callejon M, Jimenez J.C, Alonso E, Guiraum A (2007) Optimization and validation of a new method for analysis of linear alkylbenzene sulfonates in sewage sludge by liquid chromatography after microwave-assisted extraction. Anal. Chim. Acta 599: 92-97.

[67] Bartolomé L, Cortazar E, Raposo J.C, Usobiaga A, Zuloaga O, Etxebarria N, Fernandez L.A (2005) Simultaneous microwave-assisted extraction of polycyclic aromatic hydrocarbons, polychlorinated biphenyls, phthalate esters and nonylphenols in sediments. J. Chromatogr. A 1068: 229-236.

[68] Croce V, Patrolecco L, Polesello S, Valsecchi S (2003) Extraction of Nonylphenol and Nonylphenol Ethoxylates from River Sediments: Comparison of Different Extraction Techniques. Chromatographia 58: 145-149.

[69] Stuart J.D, Capulong C.P, Launer K.D, Pan X (2005) Analysis of phenolic endocrine disrupting chemicals in marine samples by both gas and liquid chromatography-mass spectrometry. J. Chromatogr. A 1079: 136-145.

[70] Morales-Muñoz S, Luque-García J.L, Luque de Castro M.D (2004) Screening method for linear alkylbenzene sulfonates in sediments based on water Soxhlet extraction assisted

by focused microwaves with on-line preconcentration/derivatization/detection. J. Chromatogr. A 1026: 41-46.

[71] Eichhorn P, López O, Barceló D (2005) Application of liquid chromatography-electrospray-tandem mass spectrometry for the identification and characterization of linear alkylbenzene sulfonates and sulfophenyl carboxylates in sludge-amended soils. J. Chromatogr. A 1067: 171-179.

[72] Ferrer I, Furlong E.T (2002) Accelerated Solvent Extraction Followed by On-Line Solid-Phase Extraction Coupled to Ion Trap LC/MS/MS for Analysis of Benzalkonium Chlorides in Sediment Samples. Anal. Chem. 74: 1275-1280.

[73] Loyo-Rosales J.E, Schmitz-Afonso I, Rice C.P, Torrents A (2003) Analysis of Octyl- and Nonylphenol and Their Ethoxylates in Water and Sediments by Liquid Chromatography/Tandem Mass Spectrometry. Anal. Chem. 75: 4811-4817.

[74] Petrovic M, Lacorte S, Viana P, Barceló D (2002) Pressurized liquid extraction followed by liquid chromatography-mass spectrometry for the determination of alkylphenolic compounds in river sediment. J. Chromatogr. A 959: 15-23.

[75] Field J.A, Miller D.J, Field T.M, Hawthorne S.B, Giger W (1992) Quantitative Determination of Sulfonated Aliphatic and Aromatic Surfactants in Sewage Sludge by Ion-Pair/Supercritical Fluid Extraction and Derivatization Gas Chromatography/Mass Spectrometry. Anal. Chem. 64: 3161-3167.

[76] Field J.A, Reed R.L (1999) Subcritical (Hot) Water/Ethanol Extraction of Nonylphenol Polyethoxy Carboxylates from Industrial and Municipal Sludges. Environ. Sci. Technol. 33: 2782-2787.

[77] Breen D.G.P.A, Horner J.M, Bartle K.D, Clifford A.A, Waters J, Lawrence J.G (1996) Supercritical Fluid Extraction and Off-Line HPLC Analysis of Cationic Surfactants from Dried Sewage Sludge. Wat. Res. 30: 476-480.

[78] Jurado E, Fernández-Serrano M, Núñez-Olea J, Luzón G, Lechuga M (2006) Simplified spectrophotometric method using methylene blue for determining anionic surfactants: Applications to the study of primary biodegradation in aerobic screening tests. Chemosphere 65: 278-285.

[79] Roslan R.N, Hanif N.M, Othman M.R, Azmi W.N.F.W, Yan X.X, Ali M.M, Mohamed C.A.R, Latif M.T (2010) Surfactants in the sea-surface microlayer and their contribution to atmospheric aerosols around coastal areas of the Malaysian peninsula. Mar. Poll. Bull. 60: 1584-1590.

[80] Nagarnaik P.M, Mills M.A, Boulanger B (2010) Concentrations and mass loadings of hormones, alkylphenols, and alkylphenol ethoxylates in healthcare facility wastewaters. Chemosphere 78: 1056-1062.

[81] Wyrwas B, Szymanski A, Lukaszewski Z (1998) Tensammetric determination of non-ionic surfactants combined with the BiAS separation procedure Part 3. Determination in the presence of hydrocarbons. Talanta 47: 325-333.

[82] Idouhar M, Tazerouti A (2008) Spectrophotometric Determination of Cationic Surfactants Using Patent Blue V: Application to the Wastewater Industry in Algiers. J. Surfact. Deterg. 11: 263-267.

[83] Koga M, Yamamichi Y, Nomoto Y, Irie M, Tanimura T, Yoshinaga T (1999) Rapid Determination of Anionic Surfactants by Improved Spectrophotometric Method Using Methylene Blue. Anal. Sci. 15: 563-568.

[84] Akyüz M (2007) Ion-pair extraction and GC-MS determination of linear alkylbenzene sulphonates in aqueous environmental samples. Talanta 71: 471-478.

[85] Matthijs E, Holt M.S, Kiewiet A, Rijs G.B.J (1999) Environmental Monitoring for Linear Alkylbenzene Sulfonate, Alcohol Ethoxylate, Alcohol Ethoxy Sulfate, Alcohol Sulfate, and Soap. Environ. Toxicol. Chem. 18: 2634-2644.

[86] Crescenzi C, Di Corcia A, Marchiori E, Samperi R, Marcomini A (1996) Simultaneous Determination of Alkylbenzenesulfonates and Dialkyltetralinsulfonates in Water by Liquid Chromatography. Wat. Res. 30: 722-730.

[87] Bruzzoniti M.C, De Carlo R.M, Sarzanini C (2008) Determination of sulfonic acids and alkylsulfates by ion chromatography in water. Talanta 75: 734-739.

[88] Pojana G, Cassani G, Marcomini A (2004) Determination, aerobic biodegradation and environmental occurrence of aliphatic alcohol polyethoxylate sulfates (AES). Int. J. Environ. Anal. Chem. 84: 729-738.

[89] Houde F, DeBlois C, Berryman D (2002) Liquid chromatographic-tandem mass spectrometric determination of nonylphenol polyethoxylates and nonylphenol carboxylic acids in surface water. J. Chromatogr. A 961: 245-256.

[90] Fendinger N.J, Begley W.M, McAvoy D.C, Eckhoff W.S (1995) Measurement of Alkyl Ethoxylate Surfactants in Natural Waters. Environ. Sci. Technol. 29: 856-863.

[91] Petrovic M, Díaz A, Ventura F, Barceló D (2001) Simultaneous Determination of Halogenated Derivatives of Alkylphenol Ethoxylates and Their Metabolites in Sludges, River Sediments, and Surface, Drinking, and Wastewaters by Liquid Chromatography-Mass Spectrometry. Anal. Chem. 73: 5886-5895.

[92] Dunphy J.C, Pessler D.G, Morrall S.W (2001) Derivatization LC/MS for the Simultaneous Determination of Fatty Alcohol and Alcohol Ethoxylate Surfactants in Water and Wastewater Samples. Environ. Sci. Technol. 35: 1223-1230.

[93] Cassani G, Tibaldi F, Donato G, Andriollo N (2011) Determination of Alcohol Ethoxylates derivatized with Napthoyl Chloride, in Waste Water Treatment Plant Influent, Effluent and Sludge Samples by Liquid Chromatography Mass Spectrometry. J. Surfact. Deterg. 14: 139-150.

[94] Merino F, Rubio S, Pérez-Bendito D (2003) Solid-Phase Extraction of Amphiphiles Based on Mixed Hemimicelle/Admicelle Formation: Application to the Concentration of Benzalkonium Surfactants in Sewage and River Water. Anal. Chem. 75: 6799-6806.

[95] Merino F, Rubio S, Pérez-Bendito D (2004) Evaluation and Optimization of an On-Line Admicelle-Based Extraction-Liquid Chromatography Approach for the Analysis of Ionic Organic Compounds. Anal. Chem. 76: 3878-3886.

[96] Bassarab P, Williams D, Dean J.R, Ludkin E, Perry J.J (2011) Determination of quaternary ammonium compounds in seawater samples by solid-phase extraction and liquid chromatography-mass spectrometry. J. Chromatogr. A 1218: 673-677.

[97] Autry J.K, Vaught E.G, Conte E.D (2005) Preconcentration of bezalkonium chloride using sodium dodecyl sulfate attached to a strong anion exchange resin. Microchem. J. 80: 25-29.

[98] Luque N, Merino F, Rubio S, Pérez-Bendito D (2005) Stability of benzalkonium surfactants on hemimicelle-based solid-phase extraction cartridges. J. Chromatogr. A 1094: 17-23.

[99] Tolls J, Haller M, Sijm D.T.H.M (1999) Extraction and Isolation of Linear Alkylbenzenesulfonate and Its Sulfophenylcarboxylic Acid Metabolites from Fish Samples. Anal. Chem. 71: 5242-5247.

[100] Zgoła-Grześkowiak A, Kaczorek E (2011) Isolation, preconcentration and determination of rhamnolipids in aqueous samples by dispersive liquid-liquid microextraction and liquid chromatography with tandem mass spectrometry. Talanta 83: 744-750.

[101] Norberg J, Thordarson E, Mathiasson L, Jönsson J.A (2000) Microporous membrane liquid-liquid extraction coupled on-line with normal-phase liquid chromatography for the determination of cationic surfactants in river and waste water. J. Chromatogr. A 869: 523-529.

[102] Hultgren S, Larsson N, Nilsson B.F, Jönsson J.A (2009) Ion-pair hollow-fiber liquid-phase microextraction of the quaternary ammonium surfactant dicocodimethylammonium chloride. Anal. Bioanal. Chem. 393: 929-937.

[103] Luo S, Fang L, Wang X, Liu H, Ouyang G, Lan C, Luan T (2010) Determination of octylphenol and nonylphenol in aqueous sample using simultaneous derivatization and dispersive liquid-liquid microextraction followed by gas chromatography-mass spectrometry. J. Chromatogr. A 1217: 6762-6768.

[104] Bidari A, Ganjali M.R, Norouzi P (2010) Development and evaluation of a dispersive liquid-liquid microextraction based test method for quantitation of total anionic surfactants: advantages against reference methods. Cent. Eur. J. Chem. 8: 702-708.

[105] Alzaga R, Peña A, Ortiz L, Bayona J.M (2003) Determination of linear alkylbenzensulfonates in aqueous matrices by ion-pair solid-phase microextraction-in-port derivatization-gas chromatography-mass spectrometry. J. Chromatogr. A 999: 51-60.

[106] Rico-Rico A, Droge S.T.J, Widmer D, Hermens J.L.M (2009) Freely dissolved concentrations of anionic surfactants in seawater solutions: Optimization of the non-depletive solid-phase microextraction method and application to linear alkylbenzene sulfonates. J. Chromatogr. A 1216: 2996-3002.

[107] Boyd-Boland A.A, Pawliszyn J.B (1996) Solid-Phase Microextraction Coupled with High-Performance Liquid Chromatography for the Determination of Alkylphenol Ethoxylate Surfactants in Water. Anal. Chem. 68: 1521-1529.

[108] Pan Y.-P, Tsai S.-W (2008) Solid phase microextraction procedure for the determination of alkylphenols in water by on-fiber derivatization with N-tert-butyl-dimethylsilyl-N-methyltrifluoroacetamide. Anal. Chim. Acta 624: 247-252.

[109] Kawaguchi M, Sakui N, Okanouchi N, Ito R, Saito K, Nakazawa H (2005) Stir bar sorptive extraction and trace analysis of alkylphenols in water samples by thermal

desorption with in tube silylation and gas chromatography-mass spectrometry. J. Chromatogr. A 1062: 23-29.

[110] Coll C, Martínez-Máñez R, Marcos M.D, Sancenón F, Soto J (2007) A Simple Approach for the Selective and Sensitive Colorimetric Detection of Anionic Surfactants in Water. Angew. Chem. Int. Ed. 46: 1675-1678.

[111] Coll C, Ros-Lis J.V, Martínez-Máñez R, Marcos M.D, Sancenón F, Soto J (2010) A new approach for the selective and sensitive colorimetric detection of ionic surfactants in water. J. Mater. Chem. 20: 1442-1451.

[112] Ródenas-Torralba E, Reis B.F, Morales-Rubio A, de la Guardia M (2005) An environmentally friendly multicommutated alternative to the reference method for anionic surfactant determination in water. Talanta 66: 591-599.

[113] Khaled E, Mohamed G.G, Awad T (2008) Disposal screen-printed carbon paste electrodes for the potentiometric titration of surfactants. Sens. Actuators B 135: 74–80.

[114] Szymanski A, Wyrwas B, Jesiołowska A, Kazmierczak S, Przybysz T, Grodecka J, Łukaszewski Z (2001) Surfactants in the River Warta: 1990-2000. Pol. J. Environ. Stud. 10: 371–377.

[115] Yokoyama Y, Okabe T, Kubo H, Sato H (2005) Spectrophotometric Determination of Polyoxyethylene Nonionic Surfactants in Ground Waters Using Triple-Stage Solid-Phase Extraction Followed by an Improved Ferric Thiocyanate Complexation Method. Microchim. Acta 149: 287–293.

[116] Bas B, Jakubowska M (2007) Estimation of non-ionic, surface-active substances in aqueous solutions by means of the Controlled Growth Mercury Electrode. Anal. Chim. Acta 592: 218–225.

[117] Li S, Zhao S (2004) Spectrophotometric determination of cationic surfactants with benzothiaxolyldiazoaminoazobenzene. Anal. Chim. Acta 501: 99–102.

[118] Yamamoto K, Oka M, Murakami H (2002) Spectrophotometric determination of trace ionic and non-ionic surfactants based on a collection on a membrane filter as the ion associate of the surfactant with Erythrosine B. Anal. Chim. Acta 455: 83–92.

[119] Zhu Z, Li Z, Hao Z, Chen J (2003) Direct spectrophotometric determination of alkylphenol polyethoxylate nonionic surfactants in wastewater. Wat. Res. 37: 4506–4512.

[120] Szymanski A, Wyrwas B, Szymanowska M, Lukaszewski Z (2001) Determination of Short-Chained Poly(ethyleneglycols) and Ethylene Glycol in Environmental Samples. Wat. Res. 35: 3599-3604.

[121] Lizondo-Sabater J, Martínez-Máñez R, Sancenón F, Seguí M.J, Soto J (2008) Ion-selective electrodes for anionic surfactants using a cyclam derivative as ionophore. Talanta 75: 317-325.

[122] Jiménez-Díaz I, Ballesteros O, Zafra-Gómez A, Crovetto G, Vílchez J.L, Navalón A, Verge C, de Ferrer J.A (2010) New sample treatment for the determination of alkylphenols and alkylphenol ethoxylates in agricultural soils. Chemosphere 80: 248-255.

[123] Moldovan Z, Avram V, Marincas O, Petrov P, Ternes T (2011) The determination of the linear alkylbenzene sulfonate isomers in water samples by gas-chromatography/mass spectrometry. J. Chromatogr. A 1218: 343-349.

[124] McEvoy J, Giger W (1986) Determination of Linear Alkylbenzenesulfonates in Sewage Sludge by High-Resolution Gas Chromatography/Mass Spectrometry. Environ. Sci. Technol. 20: 376-383.
[125] Hibberd A, Maskaoui K, Zhang Z, Zhou J.L (2009) An improved method for the simultaneous analysis of phenolic and steroidal estrogens in water and sediment. Talanta 77: 1315-1321.
[126] Ahel M, Giger W (1985) Determination of Alkylphenols and Alkylphenol Mono- and Diethoxylates in Environmental Samples by High-Performance Liquid Chromatography. Anal. Chem. 57: 1577-1583.
[127] Lee Ferguson P, Iden C.R, Brownawell B.J (2001) Analysis of nonylphenol and nonylphenol ethoxylates in environmental samples by mixed-mode high-performance liquid chromatography-electrospray mass spectrometry. J. Chromatogr. A 938: 79-91.
[128] Olkowska E, Polkowska Z, Namiesnik J (2012) Analytical procedures for the determination of surfactants in environmental samples. Talanta 88: 1-13.
[129] Cantarero S, Zafra-Gómez A, Ballesteros O, Navalón A, Vílchez J.L, Verge C, De Ferrer J.A (2011) Matrix Effect Study in the Determination of Linear Alkylbenzene Sulfonates in Sewage Sludge Samples. Environ. Toxicol. Chem. 30: 813-818.
[130] Cortazar E, Bartolomé L, Delgado A, Etxebarria N, Fernández L.A, Usobiaga A, Zuloaga O (2005) Optimisation of microwave-assisted extraction for the determination of nonylphenols and phthalate esters in sediment samples and comparison with pressurized solvent extraction. Anal. Chim. Acta 534: 247-254.
[131] Nakae A, Tsuji K, Yamanaka M (1980) Determination of Trace Amounts of Alkylbenzenesulfonates by High-Performance Liquid Chromatography with Fluorimetric Detection. Anal. Chem. 52: 2275-2277.
[132] Kiewiet A.T, van der Steen J.M.D, Parsons J.R (1995) Trace Analysis of Ethoxylated Nonionic Surfactants in Samples of Influent and Effluent of Sewage Treatment Plants by High-Performance Liquid Chromatography. Anal. Chem. 67: 4409-4415.
[133] Zanette M, Marcomini A, Marchiori E, Samperi R (1996) High-performance liquid chromatographic-fluorescence determination of aliphatic alcohol polyethoxylates and polye(thylene glycol)s in aqueous samples. J. Chromatogr. A 756: 159-174.
[134] Schröder H.F (2001) Tracing of surfactants in the biological wastewater treatment process and the identification of their metabolites by flow injection-mass spectrometry and liquid chromatography-mass spectrometry and –tandem mass spectrometry. J. Chromatogr. A 926: 127-150.
[135] Jhanke A, Gandrass J, Ruck W (2004) Simultaneous determination of alkylphenol ethoxylates and their biotransformation products by liquid chromatography/electrospray ionization tandem mass spectrometry. J. Chromatogr. A 1035: 115-122.
[136] Lunar L, Rubio S, Pérez-Bendito D (2004) Differentation and quantification of linear alkyl benzenesulfonate isomers by liquid chromatography-ion-trap mass spectrometry. J. Chromatogr. A 1031: 17-25.

[137] Bernhard M, Eubeler J.P, Zok S, Knepper T.P (2008) Aerobic biodegradation of polyethylene glycols of different molecular weights in wastewater and seawater. Wat. Res. 42: 4791-4801.

[138] González S, Petrovic M, Radetic M, Jovancic P, Ilic V, Barceló D (2008) Characterization and quantitative analysis of surfactants in textile wastewater by liquid chromatography/quadrupole-time-of-flight mass spectrometry. Rapid Commun. Mass Spectrom. 22: 1445-1454.

[139] Lara-Martín P.A, Gómez-Parra A, Sanz J.L, González-Mazo E (2010) Anaerobic Degradation Pathway of Linear Alkylbenzene Sulfonates (LAS) in Sulfate-Reducing Marine Sediments. Environ. Sci. Technol. 44: 1670-1676.

[140] Krogh K.A, Vejrup K.V, Mogensen B.B, Halling-Sørensen B (2002) Liquid-chromatography-mass spectrometry method to determine alcohol ethoxylates and alkylamine ethoxylates in soil interstitial water, ground water and surface water samples. J. Chromatogr. A 957: 45-57.

[141] Martínez-Carballo E, Sitka A, González-Barreiro C, Kreuzinger N, Fürhacker M, Scharf S, Gans O (2007) Determination of selected quaternary ammonium compounds by liquid chromatography with mass spectrometry. Part I. Application to surface, waste and indirect discharge water samples in Austria. Environ. Pollut. 145: 489-496.

[142] Droge S.T.J, Sinnige T.L, Hermens J.L.M (2007) Analysis of Freely Dissolved Alcohol Ethoxylate Homologues in Various Seawater Matrixes Using Solid-Phase Microextraction. Anal. Chem. 79: 2885-2891.

[143] Ding W.-H, Tzing S.-H (1998) Analysis of nonylphenol polyethoxylates and their degradation products in river water and sewage effluent by gas chromatography-ion trap (tandem) mass spectrometry with electron impact and chemical ionization. J. Chromatogr. 824: 79-90.

[144] Ding W.-H, Lo J.-H, Tzing S.-H (1998) Determination of linear alkylbenzenesulfonates and their degradation products in water samples by gas chromatography with ion-trap mass spectrometry. J. Chromatogr. A 818: 270-279.

Chapter 8

Application of HPLC Analysis of Medroxyprogesterone Acetate in Human Plasma

Saksit Chanthai and Thanee Tessiri

Additional information is available at the end of the chapter

1. Introduction

1.1. Steroid hormones and hormonal contraceptive drugs

Generally, progestogens, i.e. progesterone in combination with oestradiol esters, are used primarily in the treatment of menstrual irregularities to maintain endangered pregnancies. Their derivatives are orally active medroxyprogesterone acetate including hydroxyprogesterone caproate, chlormadinone acetate, megestrol acetate. Naturally occurring glucocorticoids are often used in medicine such as cortisone, hydrocortisone, prednisone and prednisolone (Fig. 1). Depot-medroxyprogesterone acetate (Depot-MPA) is commercially available and commonly known as a hormonal contraception used by women. It is a synthetic steroid hormone, which mimics natural progesterone, but its contraceptive activity has been shown to be about 30 times higher. Depot-MPA is widely used by intramuscular administration for long-term contraception [1-3].

Research concerning its pharmacokinetics has been performed in order to improve the hormonal activity of the drug formulation [2,4]. This will help to standardize the dosage of the drug.

1.2. Determination of MPA in hormonal contraceptive drugs

Analysis of MPA in plasma or serum has been performed to study its pharmacokinetics and to monitor its residual levels in cancer patients after oral or intramuscular administration [4,5]. Numerous recent researches for this analysis have been reported concerning the method sensitivity and selectivity in association with various instruments including GC, HPLC and others [6-15]. However, each of which had their advantages and disadvantages. Therefore, development and validation of the method for MPA analysis in spiked blood plasma using internal standard by HPLC method with specific derivatizing agents and clean-up by solid-phase extraction have been carried out. The method sensitivity obtained is suitable for pharmacokinetics study of MPA.

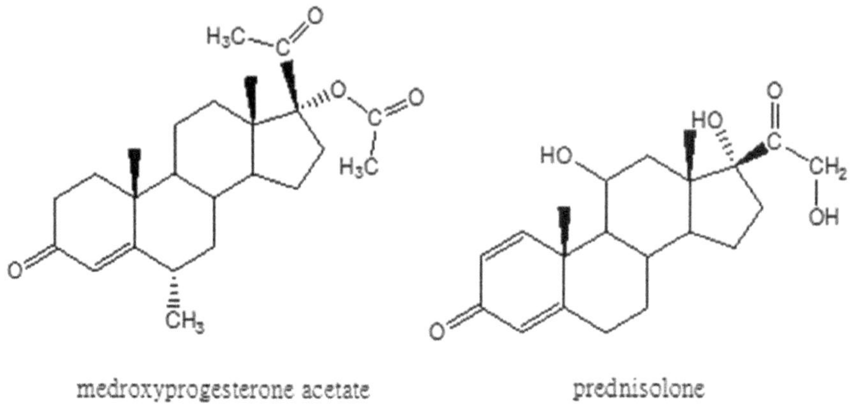

Figure 1. Chemical structure of medroxyprogesterone acetate and prednisolone

Normally, the levels of MPA found in plasma range from 1.75 to 9.0 ng/mL [5]. Formerly, the measurement of MPA in a plasma sample was mostly carried out by radioimmunoassay (RIA) [6]. Trace analysis of MPA in dog plasma was achieved using cyclohexane extraction, followed by heptafluorobutylic anhydride derivatization and gas chromatography with electron capture detector (GC-ECD). However, this procedure could not be applied for MPA analysis in human plasma due to its high matrix interference. Attempts have been made to optimize the methods for MPA analysis using small amounts of plasma samples [7,8]. Quantitative analysis of MPA in plasma by HPLC with ultraviolet (UV) detection was also carried out, but did not give satisfactory detection limit [9,10]. The high sensitivity of the GC-MS has also been used to monitor the MPA level in human serum [11]. This procedure used trifluoroacetic anhydride derivatization of the extracted portion after SPE and gave the detection limit down to 0.5 ng/mL.

Although, HPLC-UV analysis of MPA in plasma has been performed, but its sensitivity was about 5 ng/mL [12]. In order to enhance both selectivity and sensitivity, recent method development and validation of MPA analysis has been focused on HPLC separation with various detection systems. HPLC with chemiluminescence detection was used to trace MPA in serum via 4-(N,N-dimethylaminosulphonyl) -7-hydrazino-2,1,3-benzoxadiazole as a fluorogenic agent [13]. Recently, MPA analysis in plasma sample has been conducted by liquid chromatography-electrospray ion trap mass spectrometry (LC-MS/MS) after liquid phase extraction and this gave 10 times higher sensitivity than GC-MS [11,14].

From these reviews, both GC-ECD and HPLC-UV did not give sensitivity high enough for MPA analysis in pharmacokinetic studies. In clinical aspects, the RIA method has a rather high sensitivity but sometimes gives positive result due to metabolite interference. Both GC-MS and LC-MS/MS are methods of choice with high selectivity and sensitivity, but may be considerably inconvenient for routine analysis due to the high instrument cost. The HPLC with amperometric detector (AD) would be a highly sensitive tool. From the molecular structure, MPA is a ketone steroid in which the ketone group can undergo reduction. Since the reduction is normally easily interfered by oxygen molecule dissolved in mobile phase, this may be a problem [15]. Hydrazines including 2,4-dinitrophenyl hydrazine (DNPH) have been introduced as an electroactive labeling reagent for carbonyl compounds [16]. Until now there is no report of the derivatization of MPA with DNPH (Fig. 2). From the chemical structures condensation reaction can occur between the keto group of MPA and DNPH to give the MPA-hydrazone which can be detected by HPLC-AD. The optimum conditions of DNPH derivatization with MPA were studied. Therefore, development and validation of the method for MPA analysis in spiked blood plasma using prednisolone (P) as internal standard (I.S.) by HPLC-AD after derivatization with DNPH and clean-up by solid-phase extraction have been carried out. The method sensitivity obtained is suitable for pharmacokinetic study of MPA.

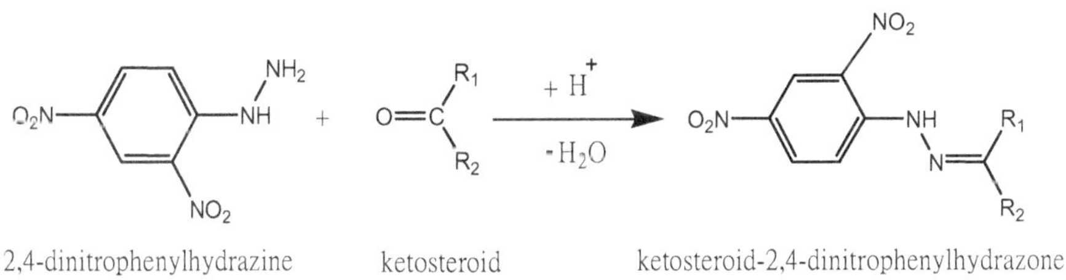

Figure 2. Derivatizing reaction between 2,4-dinitrophenylhydrazine and ketosteroids giving their ketosteroid-2,4-dinitrophenylhydrazones as the derivatized products.

2. HPLC-AD analysis of MPA in human plasma: A case study

2.1. Reagents, standard solution and plasma sample

Medroxyprogesterone acetate (MPA), 99.27%, used as working standard was obtained from A.N.B. Laboratory Co. Ltd. (Thailand). Predinisolone was an analytical reagent (A.R.) grade from Sigma (Germany). 2,4-Dinitrophenylhydrazine (Fluka, Switzerland) was also of A.R. grade. Methanol (MeOH, Merck, Germany), acetonitrile (ACN, Merck, Germany), di-potassium hydrogen phosphate-3-hydrate (BDH, England) and potassium dihydrogen phosphate (BDH, England) were of HPLC grade. Other chemicals used were also A.R. grade including hydrochloric acid (Merck, Germany).

Standard stock solution (500 mg/mL) of MPA was prepared in methanol. Dilute standard solutions of MPA were also prepared in methanol. The 2,4-dinitrophenylhydrazine stock solution used as a labeling reagent (1,000 mg/mL) was prepared by dissolving the hydrazine in 1.2 mL concentrated HCl prior to adjusting the final volume to 25 mL with methanol.

Stock phosphate solution (2 M) was prepared from sodium dihydrogen phosphate monohydrate ($NaH_2PO_4.H_2O$) in deionized water. A buffer solution of 30 mM potassium dihydrogen phosphate (KH_2PO_4), pH 3.0 was made by dissolving KH_2PO_4 in deionized water and adjusting to pH 3.0 with 6 M orthophosphoric acid (H_3PO_4). Mobile phase was prepared and used to optimize the separation of the mixture of MPA-DNPH and P-DNPH derivatives by amperometric detection (HPLC-AD). The suitable solvent system was composed of ACN : MeOH : 30 mM KH_2PO_4 (pH 3.0) in the ratio of 39 : 39 : 22 by volume according to Snyder et al. [17]. The mobile phase was filtered through nylon membrane (0.45 µm, 47 mm) and degassed for 15 min in ultrasonic bath before use. Sep-pak C_{18} cartridge, 100 mg (Waters Associates, U.S.A.) was used.

Human blood plasma samples were kindly obtained from the Blood Bank, Srinagarin Hospital, Khon Kaen University, Khon Kaen, Thailand.

2.2. Instruments and apparatus

HPLC system (Perkin Elmer, U.S.A.) used in this study included LC200 HPLC pump with electrochemical detector (Coulochem II, ESA, U.S.A.). The standard analytical cell (ESA 5011, U.S.A.) consisted of porous graphite working electrode, palladium reference electrode and platinum counter electrode. The analytical column used was stainless steel tube packed with Hypersil ODS, 5 µm particle size (125 x 4.0 mm i.d., Agilent, U.S.A.). Integrator model1022 (Perkin Elmer, U.S.A.) was used. UV-Visible spectrophotometer (Cecil 3000, England) was also used. Analytical balance (AE 200) and pH meter (Delta 350) were from Mettler (U.S.A.). SPE manifold-12 place vacuum manifold (Lida, Germany) was used. Autopipette (Eppendorf, Germany) with volume adjustments was used throughout the experiment.

2.3. Method development and validation

i. Derivatization of MPA with DNPH

Study on the hydrazine derivatization of MPA was carried out using 2,4-dinitrophenylhydrazine (DNPH). The standard solution of MPA (10 mg/mL) was reacted with an excess amount of DNPH for an hour. The MPA-DNPH derivative was then injected into HPLC with detection at 375 nm [18]. The MPA-DNPH derivatives obtained from reacting 0.1 mg MPA with various concentrations (12.5, 25.0, 75.0 and 100.0 mg/mL) of DNPH solution in a 2.5 mL capped vial were used to study the effect of DNPH concentration and reaction time.

Prednisolone (P) was also tested for derivatization with DNPH and was used as I.S. There should be a single peak and no overlap with any interfering peaks in the sample. The P-DNPH derivative was obtained from the reaction of 0.10 g DNPH in 10 mL MeOH and 0.48 mL concentrate HCl with 0.05 g prednisolone [19]. The reaction mixture was placed on a heated water bath at 60°C for 30 min, allowed to cool down to room temperature and deionized water (60 mL) was added until the derivative product had completely precipitated. The derivative product was filtered through a glass filter and washed with 5 mL 2 M HCl twice, followed by 50 mL deionized water. It was then recrystallized in a mixture of chloroform and methanol prior to use.

ii. Solid-phase extraction of MPA-DNPH derivative

The clean up method using solid-phase extraction (SPE) was found to be a critical step for plasma sample both before and after derivatization with DNPH. Spiked standard plasma (MPA 6 ng/mL) was extracted by Sep-pak C_{18} cartridge using a mixture of MeOH : H_2O as extraction solvent. The C_{18} cartridge was first washed with 2 mL MeOH and followed by 2 mL water. The spiked plasma (2 mL) was added into the cartridge and then washed with 2 mL water, followed by 250 mL 50%(v/v) MeOH (3x), and the final elution was made with 1 mL MeOH. The MPA extract portion was kept for derivatizing with DNPH. The solution of DNPH containing 0.1 mg was added into the MPA extract portion and allowed to stand for 30 min, followed by 0.9 mL water and 0.2 mL solution of P-DNPH (10 ng/mL). The sample of the derivative mixture was then introduced into the cartridge followed by 3 times rinsing with 1 mL 50%(v/v) MeOH. The cartridge containing MPA- and P-DNPH products was washed by 250 mL 50%(v/v) MeOH (3x) and again allowed to stand for 15 min before elution by 1 mL MeOH. The elute was then dried over a stream of N_2, re-dissolved in 60 mL MeOH, then 40 mL MeOH added and 60 mL was injected into the HPLC system.

iii. Analysis of MPA-DNPH by RP-HPLC-AD

The separation of both MPA- and P-DNPH derivatives was carried out on Hypersil ODS column by HPLC-AD (0.85 V) using an electrolyte solvent system as mobile phase with a flow-rate of 1 mL/min. The mobile phase of ACN : MeOH : 30 mM KH_2PO_4 (pH 3.0) (30 : 39 : 22, v/v/v) was used in this experiment after optimization of buffer concentration, pH and organic modifier. Both MPA standard solution and MPA spiked plasma sample were prepared at the concentrations of 0.5, 1, 2, 4, 6 and 8 ng/mL.

iv. Method sensitivity of MPA analysis in spiked plasma sample

Various concentrations of MPA spiked in a plasma sample with 10 ng/mL P-DNPH as I.S. were extracted by Sep-pak C_{18} cartridge, and run by HPLC-AD (0.85 V) [20,21]. Limit of detection (LOD) is the concentration of MPA giving a peak height 3 times the baseline noise (3SD) and the limit of quantitation (LOQ) is defined as 10SD. The standard curve was obtained from the concentrations of 0.5, 1, 2, 4, 6, 8 and 10 ng/mL. The accuracy and precision including recovery were determined at three concentration levels (1, 4 and 8 ng/mL) of the spiked plasma standard.

3. Results and discussion

For method development, studies on derivatization of some hydrazines with MPA had been carried out in acidic solution [15]. It was found that DNPH is a suitable reagent. Many internal standards having similar core structure which is derivatized with DNPH were also investigated by the same manner [19]. Prednisolone was found to be the most suitable ketosteroid and its derivatized solid product was prepared. It was eluted along with the MPA-DNPH and was used as I.S. throughout the experiment.

Effects of derivatization time and concentration of DNPH for MPA analysis were studied. The DNPH concentrations used were 12.5, 25, 75 and 100 mg/mL. MPA (100 ng/mL) was used to react with various concentrations of DNPH in the ratio of 1 : 1 at room temperature and their reaction times were also done up to 180 min (Fig. 3). It was found that 75 mg/mL of DNPH gave the highest peak area within 120 min derivatization time. In this study, the derivatization time of 30 min wasenough, since the signal output was not significantly different from that of 120 min derivatization.

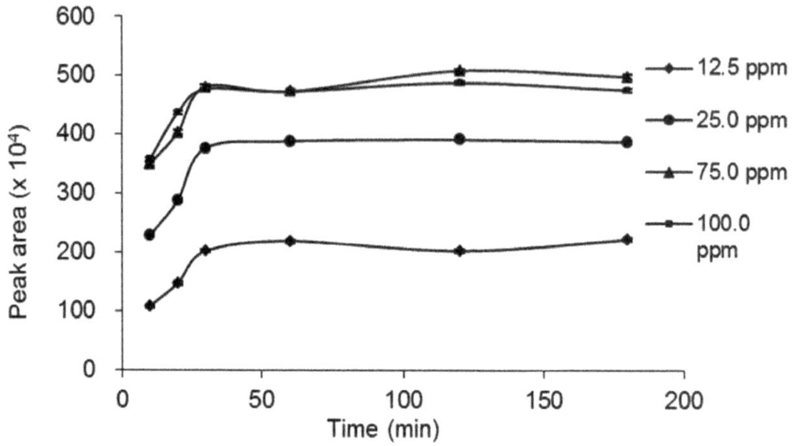

Figure 3. Effects of reaction times and concentrations on the derivatization of DNPH with 100 ng/mL MPA at room temperature.

The effect of temperature for 30 min derivatization was also investigated using 100 ng/mL MPA and 75 mg/mL DNPH in acidic solution. There seemed to be no effect of temperature, except a linear increase of peak area when the reaction was performed at room temperature (Fig. 4). Thus, the reaction was rapidly completed at room temperature.

Concerning on RP-HPLC-AD analysis, the optimization conditions of the HPLC-AD for MPA-DNPH analysis were then investigated. Effects of pH (3-6) and buffer concentration (10-40 mM) on mobile phase (ACN : phosphate buffer) were carried out. It was found that the optimum buffering system was at pH 3.0 and 30 mM phosphate buffer (data not shown). In addition, the ratios of organic solvents were investigated and a good separation was obtained with the mobile phase of ACN : MeOH : 30 mM phosphate buffer (pH 3.0) at the ratio of 39 : 39 : 22 by volume.

Figure 5 shows the hydrodynamic voltammogram (HDV) of MPA- and P-DNPH derivatives. The optimum potential that gave the highest peak area was around 0.8-0.9 V for both derivatized products, and 0.85 V was used in this study.

For method validation, the method limits for detection and quantitation were found to be 0.2 and 1.0 ng/mL with lower 15% precision and 80 - 120% accuracy, respectively. In Table 1, the results for intra-day and inter-day precision were 5.05 and 11.5 %RSD, respectively. The intra-day accuracy ranged from 102.2 - 109.5%. The average recovery was found to be 102.9 $\pm$ 4.4%. It was found that this electrochemical detection was considerably selective and gave a working range of 1.0 - 10.0 ng/mL with linear regression: $y = (3.17 \pm 0.12)x - (0.67 \pm 0.03)$, $r^2 = 0.9985$ for $n = 5$ with %RSD = 2.34 – 7.96.

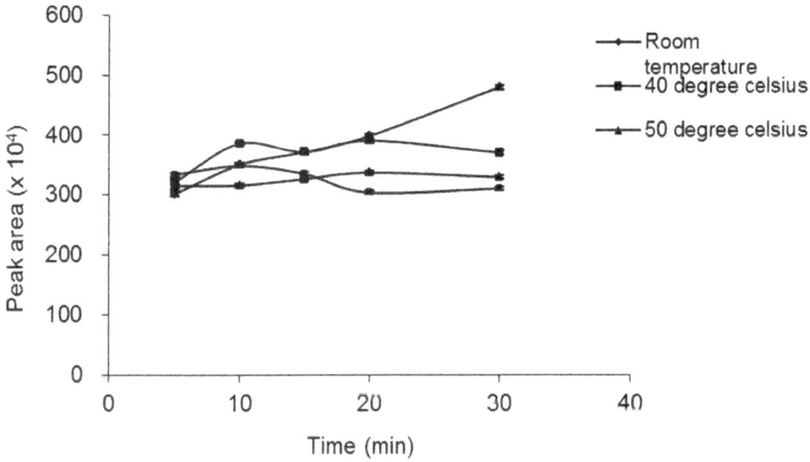

Figure 4. Effect of temperatures on the derivatization reaction of DNPH (75 mg/mL) and MPA (100 ng/mL) between 5 min and 30 min.

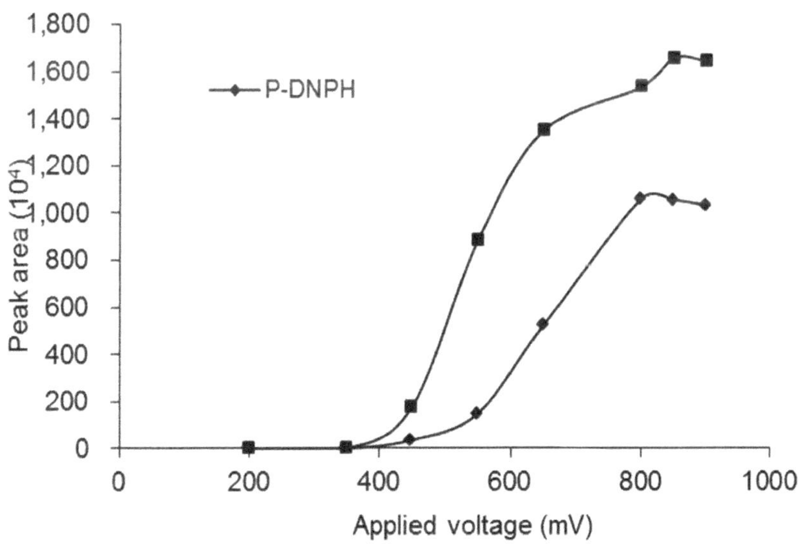

Figure 5. Effect of applied voltammetric potentials on the peak areas of MPA- and P-DNPH derivatives.

Concentration[a] (ng/mL)	Intra-day RSD (%)	Inter-day[b] RSD (%)	Recovery[c] (%)
1.0	7.76	15.4	107.3
4.0	5.06	9.62	98.55
8.0	2.35	9.55	102.9
Mean ± SD	5.05 ± 2.7	11.5 ± 3.4	102.9 ± 4.4

Table 1. Intra-day and inter-day precision and method recovery of MPA-DNPH derivative.
[a]$n = 5$
[b]*within three consecutive days*
[c]*Mean of three determinations*
RSD, relative standard deviation

Since the method sensitivity of MPA analysis was the main objective of this study, it has been shown that the clean-up step, using the SPE method for MPA spiked in human plasma sample, was necessary probably due to both the reagent used and the sample interference matrices (Fig. 6, a-d). The solution of MPA-DNPH derivative products with excess DNPH was not clean enough, so it must be added through the Sep-pak cartridge before being separated on the HPLC-AD system. Without the clean-up step, there might be trouble for the HPLC column used, which may affect the analytical sensitivity as well. This is the first report of the use of DNPH as an electroactive labeling reagent for MPA analysis. Development and validation of the HPLC-AD of the derivatized product were then carried out. The MPA-DNPH was completely separated from the I.S. within a suitable analysis time (Fig. 6, a-d).

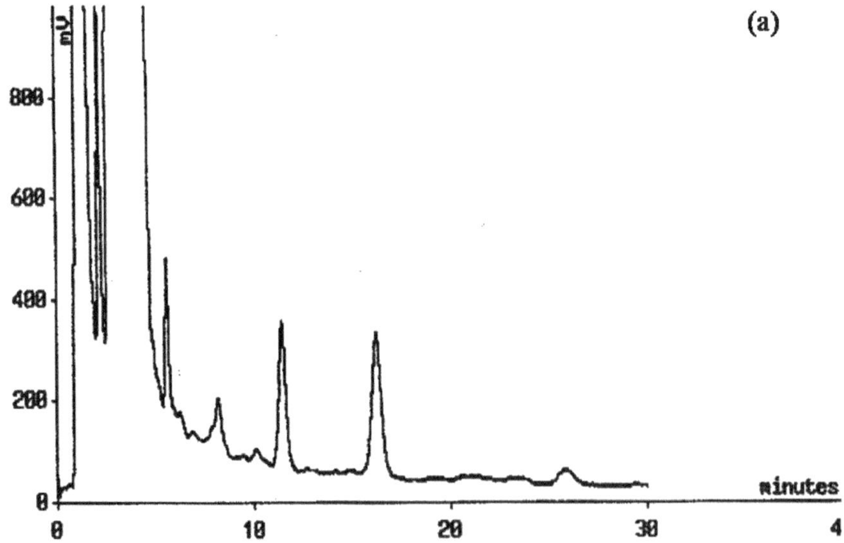

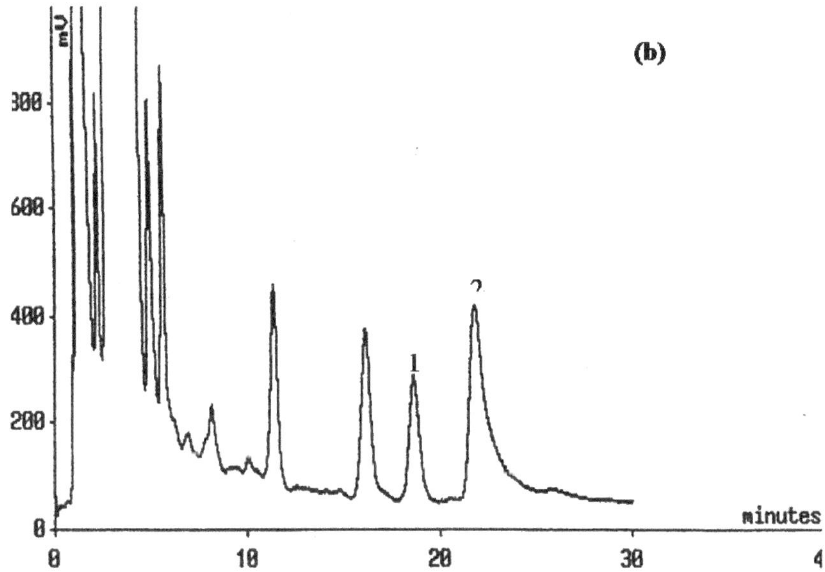
mV
(b)
800
600
400
200
0
2
1
minutes
0
10
20
30
4

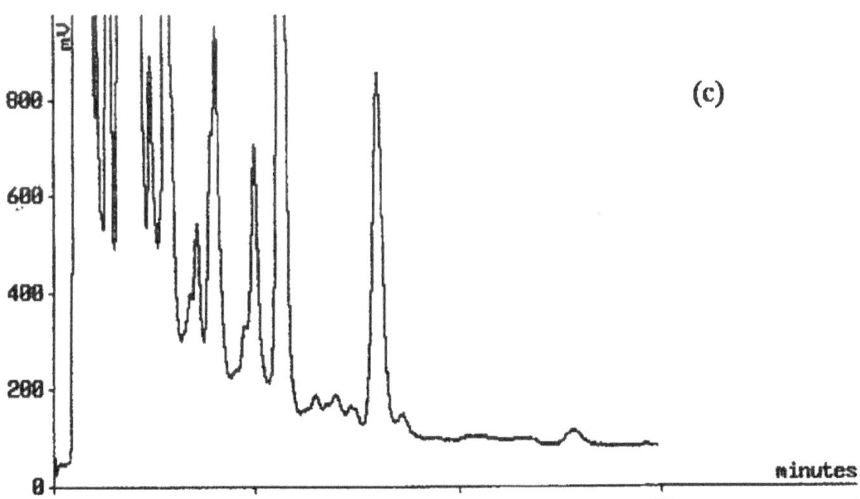
mV
(c)
800
600
400
200
0
minutes
0
10
20
30
4

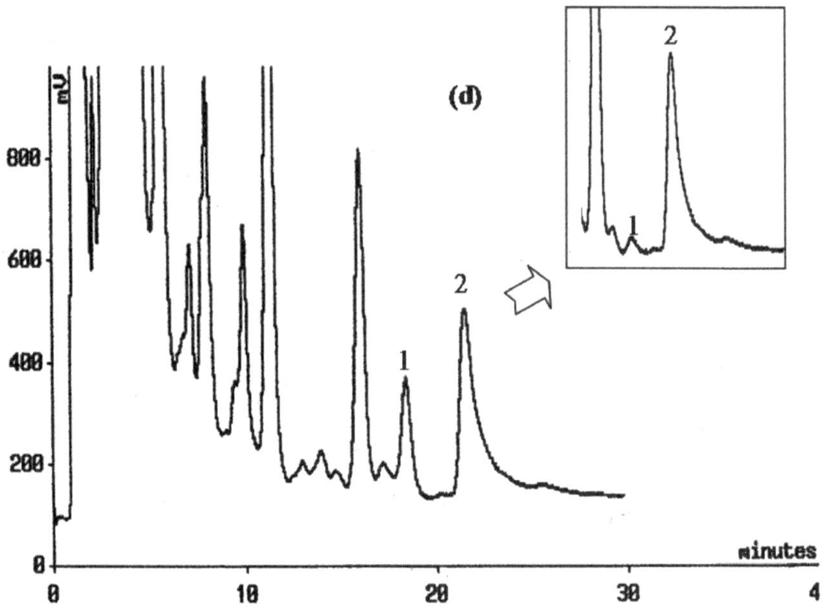

Figure 6. Chromatogram obtained from (a) blank reagent. (b) standard solution of (1) MPA-DNPH (10 ng/mL) and (2) P-DNPH as I.S. (c) blank of blood plasma sample, (d) blood plasma spiked with (1) MPA-DNPH (10 ng/mL) and (2) P-DNPH as I.S. (inset, (1) MPA-DNPH, 1 ng/mL).

Instrument	Extraction method	Sample size (mL)	LOD (ng/mL)	Reference
GC-ECD[a]	liquid phase	4	1-5	[7,8]
GC-MS[b]	solid phase	1	0.5	[11]
LC-MS/MS[c]	liquid phase	1	0.05	[14]
HPLC-UV[d]	solid phase	2	4	[9]
HPLC-PO-Cl[e]	liquid phase	0.1	9	[13]
HPLC-AD[f]	solid phase	2	0.2	This work

Table 2. Some analytical methods and limits of detection of MPA analysis in plasma samples.

[a]*electron capture detection of MPA derivative*

[b]*mass spectrometric detection of MPA derivative*

[c]*tandem mass spectrometry*

[d]*ultraviolet (254 nm) detection of MPA*

[e]*peroxyoxalate chemiluminescence ($\lambda_{ex}/\lambda_{em}$: 480/570 nm) detection of MPA derivative*

[f]*voltammetric (0.85 V) detection of MPA-DNPH derivative*

LOD, limit of detection

The plasma sample size used in this study was still rather high (2 mL) for clinical aspects when compared with other reports [11,13,14], giving LOD and LOQ of 0.2 and 1.0 ng/mL, respectively. Thus, the developed method with electrochemical detection has sufficiently high sensitivity when compared with other methods, except for the tandem MS, as shown in

Table 2. It is, however, satisfactory for pharmacokinetic study of trace MPA in plasma sample, especially in the case of contraceptive administration with very low dose. Since it is rapid procedure and inexpensive, this technique is suitable for routine analysis of the MPA content in most pharmaceutical products and blood samples.

4. Conclusion

A specific and sensitive method is presented for the analysis of medroxyprogesterone acetate (MPA) in human plasma using reversed phase high-performance liquid chromatography (HPLC) with amperometric detection. The blood sample spiked with trace amount of MPA was cleaned up to remove natural interfering matrices by solid-phase extraction (SPE). The MPA extract was then derivatized with 2,4-dinitrophenylhydrazine (DNPH) as an electroactive agent. The MPA-DNPH derivative was re-extracted using SPE prior to analysis by reversed phase HPLC. Quantitative analysis of the MPA-DNPH using prednisolone-DNPH as an internal standard were optimized on a Hypersil ODS column using acetonitrile : methanol : 30 mM phosphate buffer, pH 3.0 (39 : 39 : 22, v/v/v) as mobile phase at a flow-rate of 1.0 mL/min. It was found that the method was selective and gave linear calibration curve for a concentration range of 1.0 - 10.0 ng/mL for 2 mL spiked plasma samples. The relative standard deviation (RSD) of inter-day precision for a period of three validation days was 11.5 $\pm$ 3.4% for all concentration used. The RSD of intra-day precision (n = 5) was 5.05 $\pm$ 2.7% with accuracy (n = 5) of 102.3 $\pm$ 7.4%. The average recovery was found to be 102.9 $\pm$ 4.4%. The correlation coefficient of the calibration curve was 0.9985. The limits of detection and quantitation were found to be 0.2 and 1.0 ng/mL, respectively. Using DNPH as a derivatizing agent can enhance both selectivity and sensitivity of MPA in plasma and is suitable for routine analysis.

5. Abbreviations

ACN:	Acetonitrile
Depot-MPA:	Depot-medroxyprogesterone acetate
DNPH:	2,4-Dinitrophenylhydrazine
GC-ECD:	Gas chromatography-electron capture detection
GC-MS:	Gas chromatography-mass spectrometry
HCl:	Hydrochloric acid
HDV:	Hydrodynamic voltammogram
HPLC-UV:	High-performance liquid chromatography with ultraviolet detection
HPLC-PO-Cl:	High-performance liquid chromatography with peroxyoxalate chemiluminescence
I.S.:	Internal standard
LC-MS/MS:	Liquid chromatography-mass spectrometry/mass spectrometry
LOD:	Limit of detection
LOQ:	Limit of quantitation
MeOH:	methanol

MPA:	Medroxyprogesterone acetate
MPA-DNPH:	Medroxyprogesterone acetate-2,4-Dinitrophenylhydrazone
mM:	Millimolar
mg/mL:	Milligram per milliliter
M:	Molar
ng/mL:	Nanogram per milliliter
ODS:	Octadecylsilane
P-DNPH:	Prednisolone-2,4-Dinitrophenylhydrazone
RP-HPLC-AD:	Reversed phase high-performance liquid chromatography with amperometric detection
RSD:	Relative standard deviation
SD:	Standard deviation
SPE:	Solid-phase extraction

Author details

Saksit Chanthai
Corresponding Author
Department of Chemistry, Faculty of Science, Khon Kaen University, Khon Kaen, Thailand

Thanee Tessiri
Academic and Research Services Unit, Faculty of Pharmaceutical Sciences, Khon Kaen University, Khon Kaen, Thailand

Acknowledgement

The authors gratefully acknowledge the financial support from the National Research University (NRU) project, Khon Kaen University and the Center for Innovation in Chemistry (PERCH-CIC), Commission on Higher Education, Ministry of Education, Thailand, and the Hitachi Scholarship Foundation, Tokyo, Japan.

6. References

[1] Liskin L & Quillin W (1983) Long-acting progestins-promise and prospects. *Population Reports* 2, K-18.

[2] Fotherby K, Koetsawang S & Mathrutham M (1980) Pharmacokinetic study of different doses of Depoprovera. *Contraception* 22(5), 527-35.

[3] Cromer B A, Stager M, Bonny A, Lazebnik R, Rome E, Ziegler J & Debanne S M (2004) Depot medroxyprogesterone acetate, oral contraceptives and bone mineral density in a cohort of adolescent girls. *J Adolescent Health* 35, 434-41.

[4] Ohtsu T, Fujii H, Wakita H, Igarashi T, Itoh K & Imoto S (1998) Pharmacokinetic study of low-versus high-dose medroxyprogesterone acetate (MPA) in women. *Cancer Chemother Pharmacol* 42, 1-8.

[5] Shimanker K, Saxena B N & Fotherby K A (1978) Radioimmunoassay for serum medroxyprogesterone acetate. *J Steroid Biochem* 9(4), 359-63.

[6] Hesselius I & Johansson E D (1981) Medroxyprogesterone acetate (MPA) plasma levels after oral and intramuscular administration in a long-term study. *Acta Obstet Gynecal Scand Suppl* 101, 65-70.

[7] Kaiser D G, Carison R G & Kirton K T (1974) GLC [Gas-liquid chromatography] determination of medroxyprogesterone acetate in plasma. *J Pharm Sci* 83(3), 420-24.

[8] Rossi E, Pascale A D, Negrini P, Frigerio H & Castegnaro E (1979) Quantitative gas-liquid chromatographic determination of medroxyprogesterone acetate in human plasma. *J Chromatogr* 169, 416-21.

[9] Milano G, Carle G, Rence N, Boublil J L & Namer M (1982) Determination of medroxyprogesterone acetate in plasma by high-performance liquid chromatography. *J Chromatogr* 232, 413-17.

[10] Read J, Mould G & Stevenson D (1985) Simple high-performance liquid chromatographic method for the determination of medroxyprogesterone acetate in human plasma. *J Chromatogr* 341, 437-44.

[11] Dikkeschei L D, Veclen H V, Nagel G T, Willemse P H B & Walthers B G (1985) Specific and sensitive determination of medroxyprogesterone acetate in human serum by gas chromatography-mass spectrometry. *J Chromatogr* 345, 1-10.

[12] Gamberini G & Ferioli V (1988) Determination of medroxyprogesterone acetate in human plasma by high-performance liquid chromatography. *Att Dellta Societa dei Naturalistie Matematici di Modena* 118, 119-24.

[13] Uzu S, Imai K, Nakashima K & Akiyama S (1992) Determination of medroxyprogesterone acetate in serum by HPLC with peroxyoxalate chemiluminescence detection using a fluorogenic reagent, 4-(*N,N*-dimethylaminosulphonyl)-7-hydrazino-2,1,3-benzoxadiazole. *J Pharm and Biomed Anal* 10, 979-84.

[14] Kim S M & Kim D H (2001) Quantitative determination of medroxyprogesterone acetate in plasma by liquid chromatography / electrospray ion trap mass spectrometry. *Rapid Communications in Mass Spectro* 15(21), 2041-2.

[15] Bond A M, Hollenkamp A F, Thompson S B, Bourne A R, Huf P A & Watson T G (1988) Analytical and mechanistic aspects of the electrochemical oxidation of keto steroids derivatized with phenylhydrazine, (4-nitrophenyl)hydrazine, and (2,4-dinitrophenyl)hydrazine. *Anal Chem* 60, 1023-27.

[16] Shimada K J, Matsue T & Shimada K Z (1999) Reagents for electrochemical detection. In: Modern Derivatization Methods for Separation Sciences (edited by T. Toyo Oka), pp 191-211. John Wiley & Sons: New York, U.S.A.

[17] Snyder L R, Glajch J L & Kirkland J J (1988) Practical HPLC method development. Wiley-Interscience, New York, U.S.A.

[18] Puputti E & Lehtonen P (1986) High-performance liquid chromatographic separation and diode-array spectroscopic identification of dinitrophenylhydrazone derivatives of carbonyl compounds from whiskies. *J Chromatogr* 353, 163-68.

[19] Reich H, Grane K F & Sanfilippo S J (1953) The reaction of steroid ketones with 2,4-dinitrophenylhydrazine. *J Org Chem* 18, 818-32.

[20] Yang Q, Hartmann C, Verbeke J J & Massart D L (1995) Method development and validation for the determination of mineral elements in food and botanical materials by capillary electrophoresis. *J Chromatogr* 717, 415-25.

[21] Hartmann C, Verbeke J J, Massart D L & McDowall R D (1998) Validation of pharmaceutical and biomedical chromatographic methods. *J Pharm and Biomed Anal* 17, 193-218.

Chapter 9

Instrumental Analysis of Tetrodotoxin

Manabu Asakawa, Yasuo Shida, Keisuke Miyazawa and Tamao Noguchi

Additional information is available at the end of the chapter

1. Introduction

Tetrodotoxin (TTX), a pufferfish ("fugu" in Japanese) toxin named after its order name Tetraodontiformes, is the toxic principle of puffer fish poisoning. This toxin ($C_{11}H_{17}N_3O_8$; a molecular weight of 319) is one of the most potent nonproteinaceous toxins as well as the best-known marine natural toxins (Figure 1). In Japan, pufferfish have been a traditional food for many years, and since people have become accustomed to eating them, cases of TTX poisoning are frequent. It poses a serious hazard to public health. These cases have occurred on a regular basis not only in Japan but also in Asia for a number of years, sporadically resulting in severe poisoning or even death. On the other hand, the Japanese are aware of the its toxicity and have devised methods to reduce TTX levels especially in the liver. However, TTX poisoning incidents continue to occur in Japan. Since there is no antidote for the toxin, patient mortality is very high. Judging from statistics provided by the Japanese Ministry of Health, Labour and Welfare, the number of deaths due to puffer poisoning has steadily declined, from more than 10 cases every year between 1960 and 1981 to less than 10 cases with low mortality every year since 1982, generally with low mortality. This decline is probably due to not only strict adherence to government regulations but also an increase in cultured puffer rather than a decrease in wild puffer. The toxicosis is characterized by the onset of symptoms in the victim. Treatment of the illness is mainly based on the symptoms of the patient. More fruitful treatment can be provided if the causative toxin is identified. In 1950, TTX was isolated for the first time as a crystalline prism from toxic pufferfish ovaries by Yokoo [1]. Its structure was elucidated by three groups in 1964 [2-4]. TTX is a powerful and specific sodium channel blocker [5]. When ingested by humans, it acts to block the sodium channels in the nerve cells and skeletal muscles, and to thereby block excitatory conduction, resulting in the occurrence of typical symptoms and signs such as respiratory paralysis and even death in severe cases. The lethal potency is 5000 to 6000 MU/mg. One MU (mouse unit) is defined as the amount of toxin required to kill a 20g male mouse within 30 min after intraperitoneal administration, and the minimum lethal dose (MLD) for humans is estimated to be approximately 10,000 MU equivalent to 2 mg of pure TTX crystals [6]. Many derivatives of TTX have been found, although their toxicities vary widely. As seen in Figures 1, 2 TTX is a heterocyclic guanide compound whose chemical structure has been characterized. Various TTX derivatives from pufferfish and other TTX-bearing organisms have been identified to date as a result of recent progress in the instrumental analysis of TTX (Figure 3). In marine pufferfish species, toxicity is generally highest in the liver and ovary, whereas in brackish and freshwater species, toxicity is higher in the skin [7-13]. TTX was long believed to be present only in the pufferfish. In 1964, Mosher *et al.* detected TTX in California newt *Taricha torosa,* which was the first TTX-containing organism other than pufferfish [14]. Since then, the distribution of TTX has spread to animals other than pufferfish. The toxin has been detected in a tropical goby *Yongeichthys criniger* [15], atelopid frogs of Costa Rica *Atelopus* spp.[16], blue-ringed octopus *Hapalochlaena maculosus* [17], and several species of carnivorous gastropods such as trumpet shell *Charonia sauliae*[18], ivory shell *Babylonia japonica* [19], frog shell *Tutufa lissostoma* [20] as shown in Table 1. In addition, some species of starfish on which these gastropods prefer to feed also contain TTX [21]. The trumpet shell *Charonia sauliae* accumulates TTX by ingesting toxic starfish, supporting the hypothesis that the TTX of pufferfish is not endogenous, but is introduced via the food chain. The exact origin of TTX in the food chain, however, remains unknown. Because the ecologic environments of TTX-bearing animals apparently have no common factor other than being closely related to an aquatic system, bacteria (omnipresent organisms that commonly inhabit the aquatic system), are implicated as the primary source of TTX. Toxic crabs, flatworms, horseshoe crabs, ribbon worms and arrow worms were also added to the list of TTX-bearing animals. In Japan, TTX is assayed by the official method using mice [22]. It requires ddY strain male mice, but no special instrumentation. This method is simple and convenient but not so sufficiently accurate, and provides no information on the composition of the toxin, nor is it able to distinguish TTX from other neurotoxins such as paralytic shellfish poison (PSP). In addition, animal rights activists across the world are strongly opposed to bioassays using live animals, including mice. Thin layer chromatography (TLC) and electrophoresis are useful means for TTX detection, but they are not suitable for TTX determination. With this background, attempts have been made to develop analytical methods using high performance liquid chromatography (HPLC) in Japan. Detection and determination of TTX are essential not only for diagnosis and treatment purposes, but also for issuing quarantines and public awareness. Quantitative and/or qualitative detection of TTX in a sample is/are performed by a few instrumental analysis methods. The toxin has long been believed to occur exclusively in pufferfish. However, owing to the recent outstanding progress in instrumental methods for the analysis of TTX, its distribution and accumulation in various aquatic organisms have been established. In addition, this toxin has been detected in many other vertebrates and invertebrates. A few intestinal bacteria of TTX-bearing animals were found to produce TTX. This suggested that the accumulated TTX in these animals was being passed along the food web having been acquired from such TTX-producing bacteria. In this section, although there are many aspects of TTX research with respect to treatment and prevention, biologic distribution, sources, infestation mechanism, detection methods, chemistry and pharmacology, the focus is to provide an overview of the instrumental analysis of TTX and present chromatographic methods for the isolation of TTX. Rapid and accurate analysis of a mixture and its analogs occurring in a variety of marine organisms is becoming increasingly important from the standpoint of public health, since food poisoning from the ingestion of these toxins is often fatal to a human. There are also increasing demands for chemical assays of TTX for the study of its biosynthetic and metabolic pathways, which remain unknown. In an attempt to protect consumers from TTX-intoxication, the mouse bioassay has historically been the universally applied tool to determine the toxicity level in monitoring programs. This bioassay, however, shows low precision and requires a continuous supply of mice of a specific size. These potential drawbacks and world-wide pressure to refrain from the unnecessary killing of live animals subsequently led the scientists to develop chemical-based alternatives to the mouse bioassay for TTX detection and quantification. In addition, the mouse assay can neither provide any information on toxin composition, nor distinguish TTX from other neurotoxins such as paralytic shellfish poison (PSP). A few marine animals have been found to contain both TTX and PSP simultaneously. Many detection methods for TTX have been developed. A few methods including the mouse bioassay, high performance liquid chromatography (HPLC) and liquid chromatography-mass spectrometry (LC-MS) are typically used to qualitatively and quantitatively detect TTX, but other methods including gas-chromatography-mass spectrometry (GC-MS), infrared (IR) spectrometry and nuclear magnetic resonance (^{1}H-NMR) spectrometry are often used to qualitatively detect TTX. Among them, HPLC and LC-MS are the most powerful and sensitive tool for qualitatively and quantitatively detecting TTX. In addition these methods, TTX can also be identified by thin-layer chromatography (TLC) or electrophoresis. Though these methods are not instrumental analysis, these methods are simpler and more practical. In this section, an attempt has been made to review the current information and the recent progress on instrumental analysis of TTX.

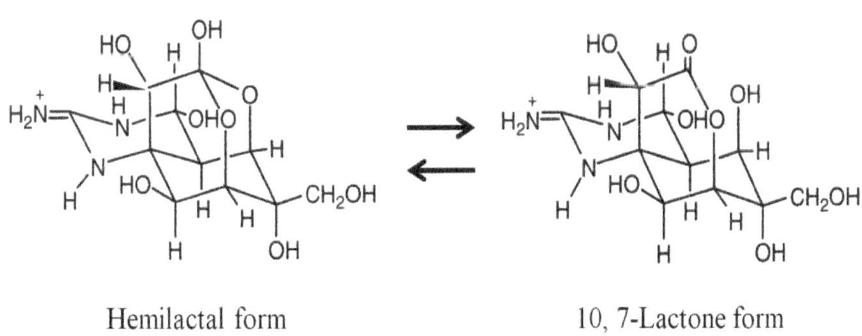

Figure 1. The tautomer of TTX

	Animals		Part
1	Platyhelminthes	Flatworms	
	Turbellaria	*Planocera* spp.	Whole body
2	Nemertinea	Ribbon worms	Whole body
		Lineus fuscoviridis	Whole body
		Tubulanus punctatus	Whole body
		Cerebratulus lacteus	Whole body
		Cephalothrix linearis	Whole body
3	Mollusca	*Charonia sauliae*	Digestive gland
	Gastropoda	*Babylonia japonica*	Digestive gland
		Tutufa lissostoma	Digestive gland
		Zeuxis siquijorensis	Digestive gland
		Niotha clathrata	Digestive gland
		Natica lineata	Whole body
		Rapana spp.	Digestive gland
		Cymatium echo	Digestive gland
		Pugilina ternotona	Digestive gland
	Cephalopoda	*Hapalochlaena maculosa*	Postsalivary gland
4	Annelida	*Pseudopotamilla occelata*	Whole body
	Polychaeta	*Lepidonotus helotypus*	Whole body
		Halosydna brevisetosa	Whole body
		Harmothoe imbricata	Whole body
5	Arthropoda	*Atergatis floridus*	Whole body
		Zosimus aeneus	Whole body
		Carcinoscorpius rotundicauda	Egg
6	Chaetognatha	Arrowworms	
		Eukrohonia hamata	Head
		Parasagitta spp.	Head
		Flaccisagitta spp.	Head
7	Echinodermata	Starfish	
		Astropecten polyacanthus	Whole body
		Astropecten latespinosus	Whole body
		Astropecten scoparius	Whole body
8	Vertebrate	*Takifugu* spp.	Skin, liver, ovary
	Pisces	*Yongeichthys criniger*	Skin, viscera, gonad
	Amphibia	*Taricha* spp.	Skin, egg, ovary, muscle, blood
		Notophthalmus spp.	Skin, egg, ovary
		Cynops spp.	Skin, egg, ovary, muscle, blood
		Triturus spp.	Skin, egg, ovary, muscle, blood
		Ambystoma sp.	Skin, egg, ovary, muscle
		Paramesotriton sp.	Skin, egg, ovary, muscle
		Polypedates sp.	Skin
		Atelopus spp.	Skin
		Colostethus spp.	Skin
9	Red Clacareous alga	*Jania* spp.	Whole body
10	Dinoflagellate	*Alexandrium tamarense*	Whole body

Table 1. Distribution of TTX in animals

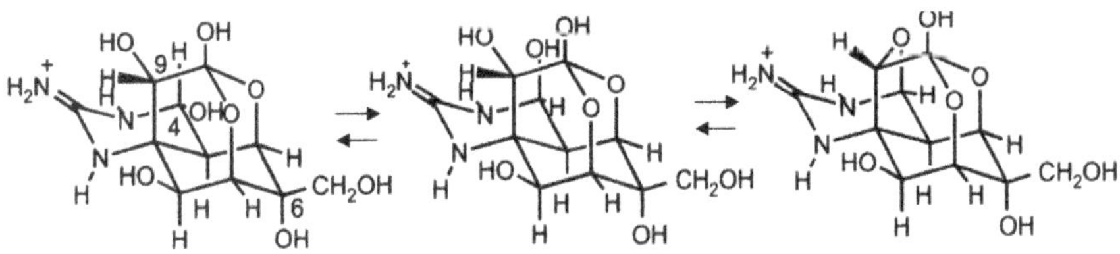

TTX	4-*epi*TTX	4, 9-anhydroTTX
(MH^+ *m/z* 320)	(MH^+ *m/z* 320)	(MH^+ *m/z* 302)

Figure 2. Chemically equilibrium of TTX, 4-*epi*TTX and 4, 9-anhydroTTX

(A) Hemilactal type

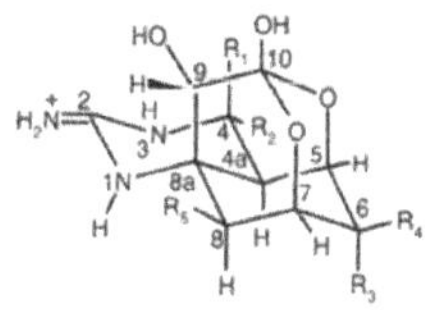

	R_1	R_2	R_3	R_4	R_5
TTX	H	OH	OH	CH_2OH	OH
4-*epi*TTX	OH	H	OH	CH_2OH	OH
6-*epi*TTX	H	OH	CH_2OH	OH	OH
11-*deoxy*TTX	H	OH	OH	CH_3	OH
6-*epi*-11-*deoxy*TTX	OH	H	OH	CH_3	OH
TTX-8-*O*-hemisuccinate	H	OH	OH	CH_2OH	$OOC(CH_2)_2COO^-$
Chiriquitoxin	H	OH	OH	*R* *S* $CH(OH)CH(NH_3^+)COO^-$	OH
11-*nor*TTX-6(*S*)-ol	H	OH	OH	H	OH
11-*nor*TTX-6(*R*)-ol	H	OH	H	OH	OH
11-*nor*TTX-6,6-diol	H	OH	OH	OH	OH
11-*oxo*TTX	H	OH	OH	$CH(OH)_2$	OH
TTX-11-carboxylic acid	H	OH	OH	COO^-	OH

(B) Lactone type

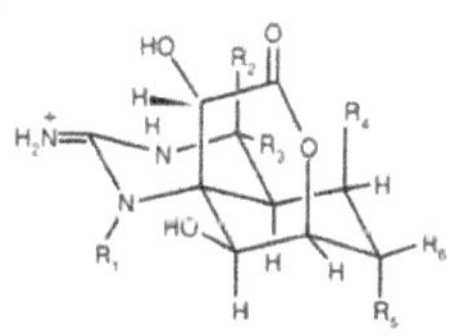

	R_1	R_2	R_3	R_4	R_5	R_6
TTX (lactone)	H	H	OH	OH	OH	CH_2OH
6-*epi*TTX (lactone)	H	H	OH	OH	CH_2OH	OH
11-*deoxy*TTX (lactone)	H	H	OH	OH	OH	CH_3
11-*nor*TTX-6(*S*)-ol (lactone)	H	H	OH	OH	OH	H
11-*nor*TTX-6(*R*)-ol (lactone)	H	H	OH	OH	H	OH
11-*nor*TTX-6,6-diol (lactone)	H	H	OH	OH	OH	OH
5-*deoxy*TTX	H	H	OH	H	OH	CH_2OH
5,11-*dideoxy*TTX	H	H	OH	H	OH	CH_3
6-*epi*-5,11-*dideoxy*TTX	H	OH	H	H	OH	CH_3
1-hydroxy-5,11-*dideoxy*TTX	OH	H	OH	H	OH	CH_3
5,6,11-*trideoxy*TTX	H	H	OH	H	H	CH_3
4-*epi*-5,6,11-*trideoxy*TTX	H	OH	H	H	H	CH_3

(C) 4,9-Anhydro type

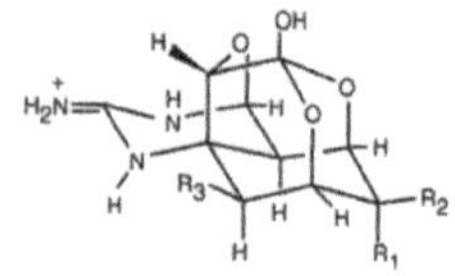

	R_1	R_2	R_3
4,9-Anhydro TTX	OH	CH_2OH	OH
4,9-Anhydro-6-*epi*-TTX	CH_2OH	OH	OH
4,9-Anhydro-11-*deoxy*TTX	OH	CH_3	OH
4,9-Anhydro-TTX-8-*O-hemisuccinate*	OH	CH_2OH	$OOC(CH_2)_2COO^-$
4,9-Anhydro-TTX-11-*O-hemisuccinate*	OH	$CH_2OCO(CH_2)_2COO^-$	OH

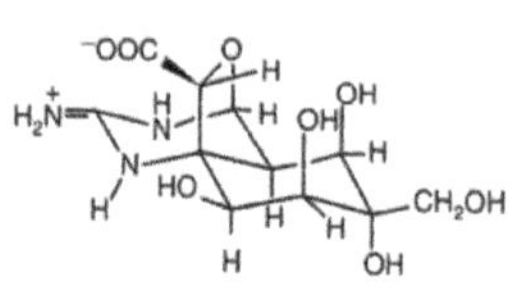

Tetrodonic acid (tetrodoic acid)

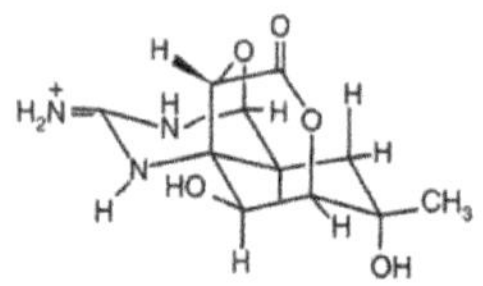

4,9-Anhydro-5,11-*dideoxy*TTX

(A) Hemilactal type, (B) Lacton type, (C) 4,9-Anhydro type

Figure 3. The structure of three types of TTX analogues

2. Isolation of TTX crystals by column chromatography

Of the TTX-bearing animals, our specimens of ribbon worms ("himomushi" in Japanese) adherent to the cultured oyster *Crassostrea gigas* hanging onto floating culture rafts, were found to be extremely toxic and to contain tetrodotoxin, during surveillance of the toxicity of various marine fouling organisms in Hiroshima Bay, Hiroshima Prefecture, which is one of the largest oyster culture areas in Japan [23]. In these analyses, the toxicity was examined on each ribbon worm specimen by the standard bioassay method for TTX. Ribbon worm specimens were collected in Hiroshima Bay between November and May from 1998 to 2005 approximately every two weeks during the harvest time for cultured oysters. A total of 764 specimens were collected, and assayed for toxicity. All specimens that were assayed throughout the season covered found to be toxic, and the toxicity scores ranged from 169 to 25,590 MU/g (Figures 4 and 5). The ratio of strongly toxic (more than 1,000 MU/g) specimens to the total number of specimens was 80%. Furthermore, the percentage of specimens possessing toxicity scores higher than 2,000 MU/g to the total was high (48%). The highest toxicity detected was 25,590 MU/g from a specimen collected on June25 (1999). The total toxicity for this sample was approximately calculated to be 5,631 MU, which is approximately equivalent to half of the minimum lethal dose of TTX in humans, which is reported to be 10,000 MU. The specimens of ribbon worms (390 g) obtained during the survey were semi-defrosted and homogenized with three volumes of 1% AcOH in 80% MeOH for 3 min, then centrifuged. This operation was repeated two more times. The supernatants (total toxicity; 2,897,000 MU) were combined, concentrated under reduced pressure, and defatted by shaking gently with approximately the same volume of chloroform several times. The aqueous layer (2,750,000 MU) was applied to an activated charcoal column and the adsorbed toxin was eluted with 1% AcOH in 20% EtOH after washing the column with dist. H_2O. The water eluate (Fr.I) and the eluate with 1% AcOH in 20% EtOH (Fr.II) was isolated (Figure 6). The main toxic fraction (Fr.II) was evaporated to dryness *in vacuo*. The resulting residue (total toxicity 2,433,000 MU; specific toxicity 99 MU/mg) was dissolved in a small amount of water, and the pH was adjusted to 5.5 with 1N NaOH. This solution was applied to a Bio-Gel P-2 column (ϕ 3.5 × 100cm). The column was washed with 3000 ml of water and then developed with 2,000 ml of 0.03 M AcOH. The toxicity was detected exclusively in the 0.03 M AcOH fraction. This fraction was concentrated to dryness under reduced pressure and the residue (3,300 MU/mg) was dissolved in a small volume of water. The resulting solution was chromatographed on a Bio-Rex 70 (H^+ form, ϕ1.0 x 100cm) column using a linear gradient of H_2O and 0.03 M AcOH at the flow rate of 0.5 ml /min. The toxic fractions were monitored for TTX via the mouse bioassay and high performance liquid chromatography (HPLC) as described later. The main toxic fractions (fr. 85–100; Fr.I) and minor fractions (fr.50-84) were obtained, and rechromatographed in the same manner (Figure 7). The toxic fraction (Fr.I) thus obtained were freeze-dried, and then dissolved in 0.5 ml of 1% AcOH. Approximately 2.0 ml of MeOH and 5.0 ml of diethylether were added to this solution, and the mixture was stored in the refrigerator overnight. During storage, stratified plate-like crystals appeared (Figure. 8). The crystals were isolated by decantation, and recrystallized by method same as that described above. Approximately, Bio-Gel P-2 column chromatography was very effective, as the specific toxicity sharply increased from 99 to 3,300 MU/mg. After recrystallization, the specific toxicity of this toxin ultimately increased to 3,520 MU/mg. From the combined homogenates with the toxicity of roughly 7,400 MU/g, approximately 25 mg of the stratified plate-like crystalline TTX was obtained. Generally, the ribbon worm has a simple structure. Since pure crystals of TTX could be obtained from this highly toxic ribbon worm efficiently by the above-described series of chromatographies, the ribbon worm is a promising source of TTX for use as a reagent in the fields of medicine and pharmacology. Previously, authentic specimens of TTX were typically prepared from pufferfish ovaries for use as reference standards, as reported in [4].

(A)

(B)

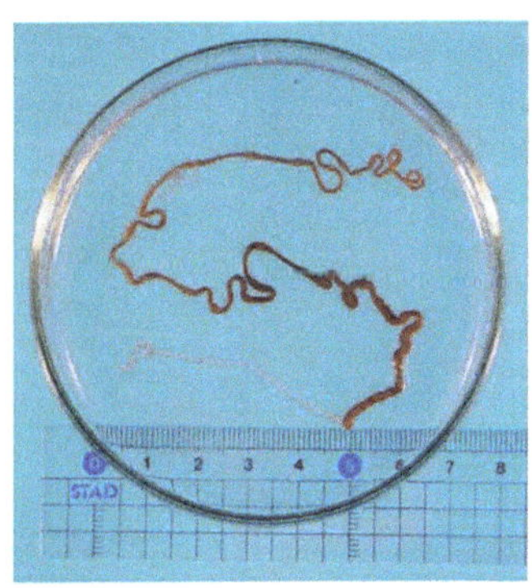

(A): Arrow indicates the ribbon worm (scale bar = 10 cm) on the surface of the cultured oyster shell.
(B): Ribbon worm removed from the cultured oyster shell

Figure 4. Ribbon worms *Cephalothrix* sp. ("himomushi" in Japanese) adherent to the cultured oyster *Crassostrea gigas* hanging onto floating culture rafts in Hiroshima Bay.

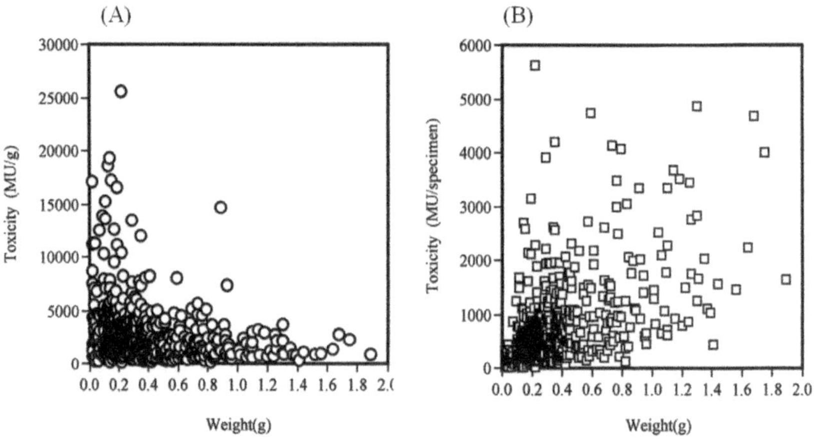

(A): Relationship between toxicity (MU/g) and body weight (g)
(B): Relationship between toxicity (MU/specimen) and body weight (g)

Figure 5. Toxicity of ribbon worm *Cephalothrix* sp. from Hiroshima Bay (1998-2005)

3. HPLC – Fluorescence detection

Rapid progress in TTX research, especially in intoxication mechanism of TTX-bearing organisms, is due to recent advancements in instrumental analysis. In particular, postcolumn-HPLC fluorescence detection (HPLC-FLD) methods expected to replace the conventional mouse bioassay, have been explored by many researchers for both qualitative and quantitative analysis of TTX and its analogs. HPLC techniques allow the separation and sensitive detection of individual TTX and its analogs irrespective of their number and group. Therefore, HPLC methods have opened up a new dimension in TTX analysis. However, the results obtained have to be comparable to those of the mouse bioassay. Additionally, accurate HPLC determination of the various TTX components in the samples is a necessity. Using these methods, the toxic principles produced peaks identical to those of authentic TTX and its derivatives. The HPLC- FLD method utilizes a computer controlled by a high pressure pump with a syringe loading sample injector or an autosampler system, a stainless steel column, a reaction pump for delivering reagents, and a fluoromonitor and chromato-recorder for calculation of the peak area. In this method, a strong alkali treatment is applied to TTX which produces a fluorescent compound with excitation and emission wavelengths of 384 and 505 nm, respectively. In this system, first, toxins are separated from the contaminants by a buffer solution on a reversed-phase column packed with C18 resin with an ion-pair reagent (sodium 1-heptanesulfonate; HSA). Then, the isolated toxins are mixed with NaOH, which converts them into fluorescent compounds that are then passed through a stainless steel tube (ϕ 0.25mm × 100cm) placed in an oven. Eventually, when the fluorescent compounds are passed through a fluoromonitor equipped with a lamp, the retention time of the toxin and fluorescence intensity are recorded. The treated toxins are identified by comparing their retention times with those of authentic TTXs. For the quantitative analysis by HPLC, the detection limit of TTX is approximately 0.03μg, which is satisfactory for practical applications. To date, several continuous improvements have been made to detect TTX and its analogs under different HPLC conditions, and a number of advances in understanding the biochemistry of TTXs are the outcomes of these developments. Briefly, a few promising methodologies are described as follows. In the early 1980's, a fluorometric continuous TTX analyzer was constructed by combining HPLC and a post-column reaction with NaOH to monitor potentially harmful puffer toxins [24]. In this system, the toxin was first separated from contaminants on a column composed of a weak cation exchange gel with a 0.06 M citrate buffer solution (pH 4.0), and toxin concentrations above 8 MU/g were detected. However, because of the poor performance of the original system in separating and detecting TTX analogs, an improved analyzer was later constructed. Using HPLC-FLD, naturally occurring TTX analogs, 4-*epi*TTX, 4,9-anhydroTTX [2,3,4,25] 6-*epi*TTX [26], 11-deoxyTTX [26], 11-oxoTTX [27], 11-*nor*TTX-6(*R*)-ol [28], 11-*nor*TTX-6(*S*)-ol [29], 1-hydroxy-5,11-dideoxyTTX [30], 5,6,11-trideoxyTTX [31], 5-deoxyTTX [32], 4,9-anhydro-6-*epi*TTX, 4-*epi*-11-deoxy-TTX, and 4,9-anhydro-11-deoxyTTX isolated from puffer and newt specimens and/or frogs [33]. The separation of TTX and 6-*epi*TTX is considered as a major achievement for this improved analyzer. In addition to this, attempts were made to apply a post-column fluorescent-HPLC system for quantitative assay of TTX and its analog 6-*epi*TTX in newts from southern Germany [34]. A reversed-phase ion-pairing HPLC method, in which HAS is used as a counter ion has also been the system preferred by many researchers for fastest and most efficient analysis of TTX and its analogs. In this method, the detection reagent for TTX and related substances does not react with any PSP component if present in the contaminant sample. A reversed - phase HPLC system (Table 2) with slight modification in the method proposed by of Nagashima *et al.*[35] is commonly used to analyze toxin compositions of extremely toxic Japanese ribbon worms, the xanthid crab *Demania cultripes* from Cebu Island, Philippines(Figure 6,7 and 9) and TTX-producing bacteria[23, 36, 37].

HPLC control system	JASCO - BORWIN/HSS-2000	
Column	LiChroCART 250-4 (Merck) (LiChrospher 100 RP-18e, 5μm)	Column size: 4 x 250 mm
Column temperature	30°C	CO-2067 plus (JASCO)
Mobile phase	60mM ammomium phosphate buffer (pH5.0) containing 10mM HAS and 2% CH_3CN	PU-2080 plus (JASCO) 0.5 ml/min
Reagent	3 M NaOH	MINICHEMI PUMP SP-D-2502 (Nihonseimitsukagaku) 0.5 ml/min
Reaction temperature	110°C	860 CO (JASCO)
Detection	Excitation 384nm, emission 505nm	JASCO FP-2025 plus

Table 2. Operating conditions of HPLC system for the analysis of TTX

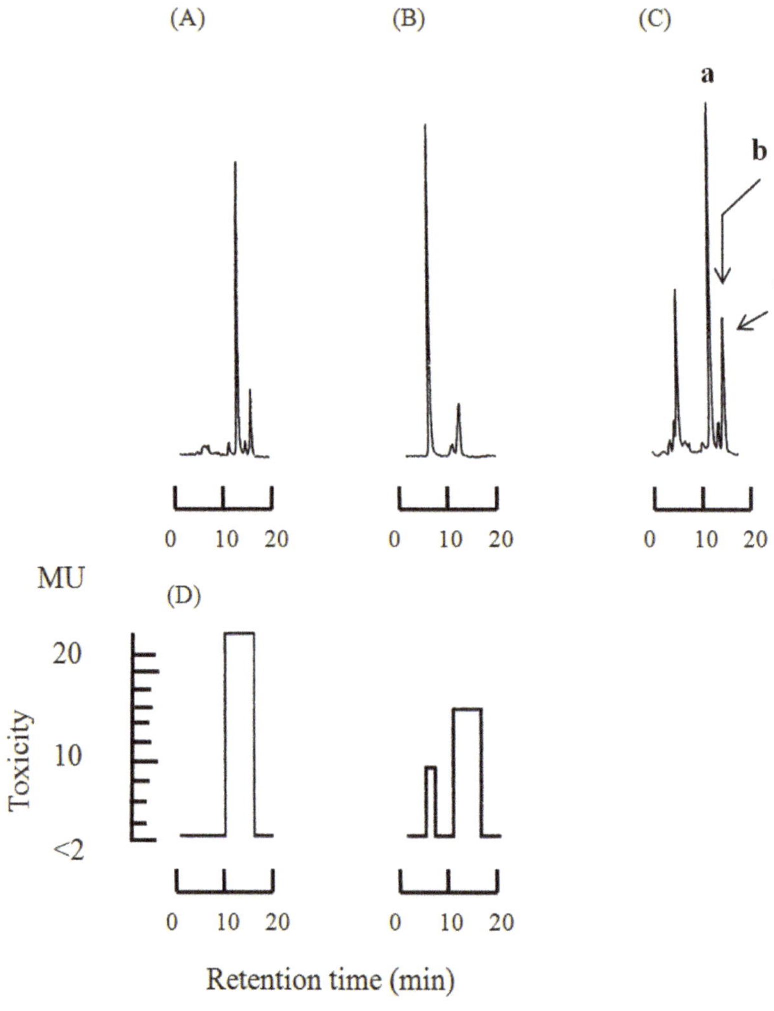

(A): Fraction II bound on activated charcoal column
(B): Fraction I unbound on activated charcoal column
(C): TTX standards; TDA (tetrodonic acid), TTX (tetrodotoxin), 4-*epi*TTX (4*epi*tetrodotoxin), 4, 9-anhyTTX (4,9-anhydrotetrodotoxin)
*HPLC-FLD: high performance liquid chromatography-fluorescence detection (HPLC-FLD)

Figure 6. HPLC-FLD * patterns (top) of fractions from the toxins contained in the ribbon worm *Cephalothrix* sp. in an activated charcoal column chromatography. The bottom patterns represent the distribution of toxicity in HPLC chromatograms, as estimated by mouse bioassay.

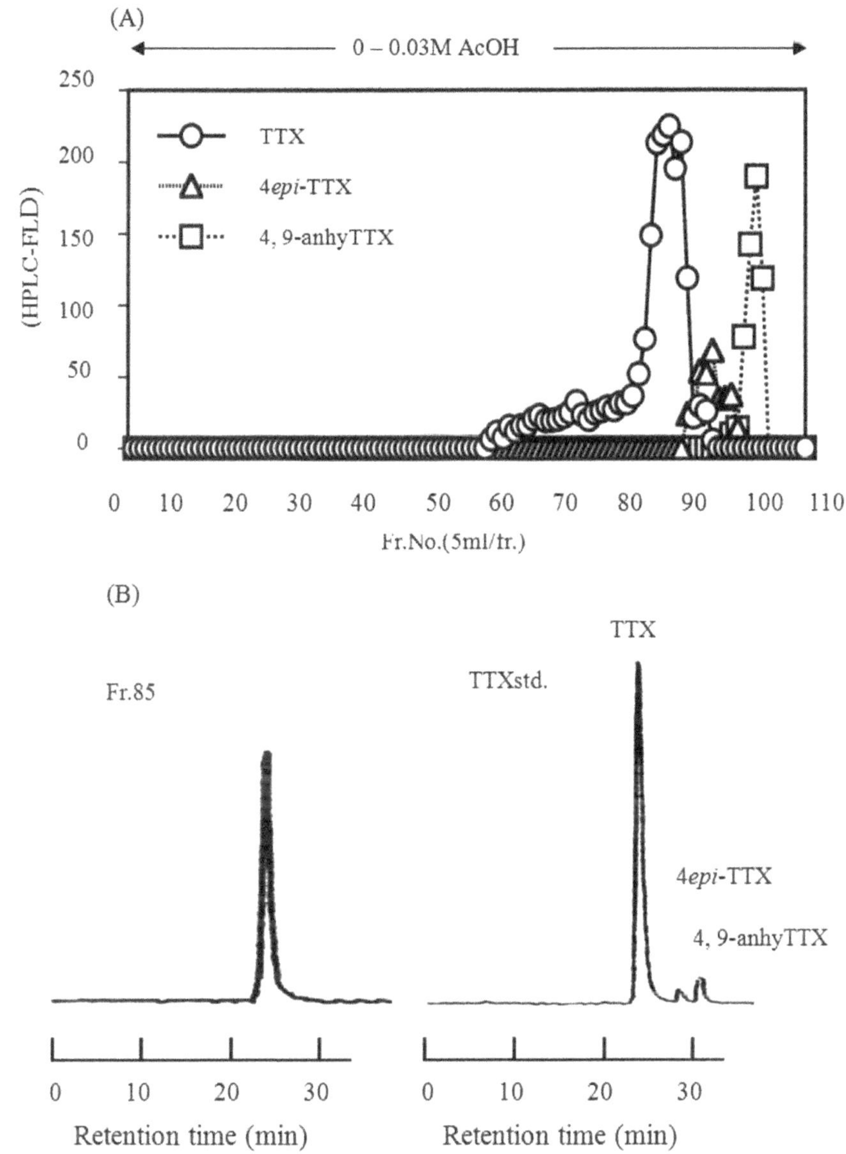

(A): Elution diagram
(B): Toxin compositions contained in fraction fr. 85 by HPLC-FLD analysis
fr.85(left), TTX standards(right)

Figure 7. Elution profile of the ribbon worm *Cephalothrix* sp. toxin from a Bio-Rex 70 column with linear gradient from 0 to 0.03M AcOH

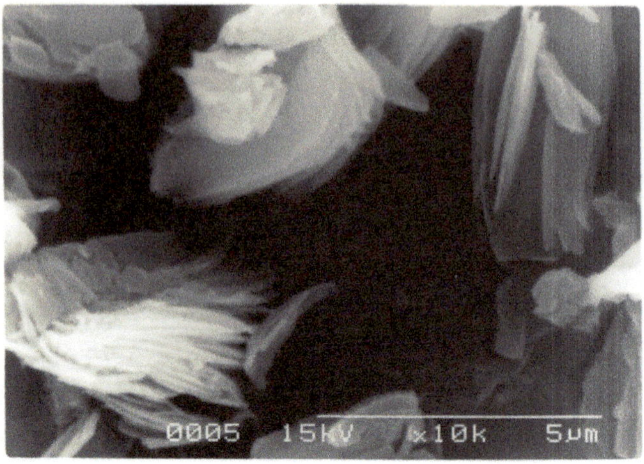

scale bar = 5 µm

Figure 8. Scanning electron micrograph of crystalline toxin isolated from the ribbon worm *Cephalothrix* sp. from Hiroshima Bay.

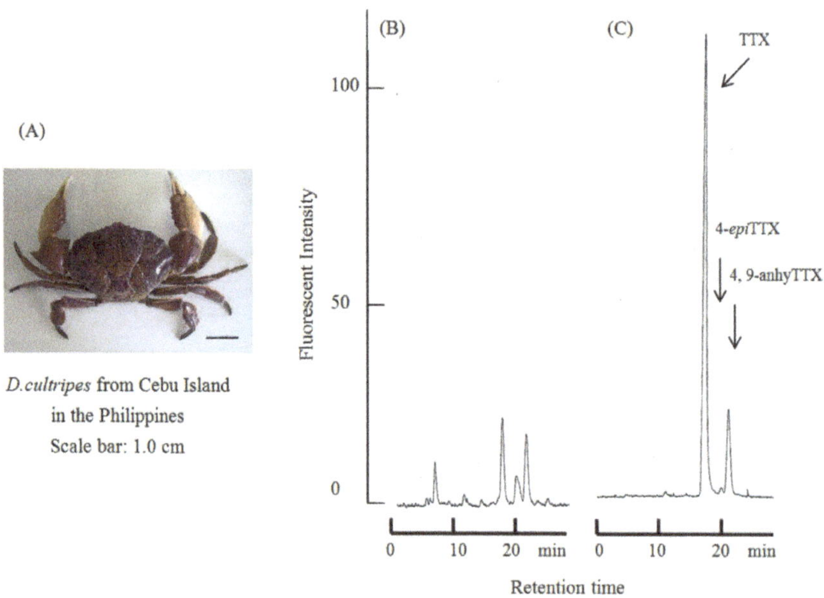

(A): *Demania cultripes*

(B): HPLC-FID analysis of the toxin contained in the viscera of *D.cultripes*

(C): TTX standards

Figure 9. HPLC-FLD analysis of TTX in the viscera of toxic crab *Demania cultripes* from Cebu Island, in the Philippines

4. Mass spectrometry

4.1 Gas-Chromatography-Mass Spectrometry

Gas chromatography (GC) and mass spectrometry (MS) form an effective combination for chemical analysis. GC-MS analysis is an indirect method to detect TTX in a crude extract which is difficult to purify in other advanced analysis methods [33]. In this method, TTX and its derivatives are dissolved in 2 ml of 3 M NaOH and heated in a boiling water bath for 30 min. After cooling to room temperature, the alkaline solution of decomposed compounds is adjusted to pH 4.0 with 1N HCl and the resulting mixture is chromatographed on a Sep-Pak C18 cartridge (Waters). After washing with H_2O first and then 10% MeOH, 100% MeOH fraction were collected and evaporated to dryness *in vacuo*. To the resulting residue, a mixture of *N*, *O*-bis acetamide, trimethylchlorosilane and pyridine (2: 1: 1) is added to generate trimethylsilyl (TMS) "C9-base" compounds (Figure 10). The derivatives are then placed in a Hewlett Packard gas chromatograph (HP-5890-II) equipped with a mass spectrometer (AutoSpec, Micromass Inc., UK). A column (ϕ 0.25 mm × 250 cm) of UB-5 (GL Sci., Japan) is used, and the column temperature is increased from 180 to 250°C at the rate of 5 or 8°C/min. The flow rate of inlet helium carrier gas is maintained at 20 ml/min. The ionizing voltage is generally maintained at 70 eV with the ion source temperature at 200°C. Scanning was performed in the mass range of *m/z* 40–600 at 3s intervals. The total ion chromatogram (TIC) and the fragment ion chromatogram (FIC) were selectively monitored. The TMS derivative of 2-amino-6-hydroxymethyl-8-hydroxyquinazoline (C9 base), prepared using gastropods and ribbon worms from Hiroshima Bay by the procedure described previously, was analyzed by GC-MS [18, 38, 39]. Recently, the isolation and characterization of bacteria from the copepod *Pseudocaligus fugu* ectoparasitie on the panther puffer *Takifugu pardalis* with an emphasis on TTX was reported [40]. The mass spectrum of the peak showed typical ions at *m/z* = 407 and 392, which correspond to M^+ and $(M\text{-}CH_3)^+$ of C9-base-$(TMS)_3$, respectively. Sharp fragment ions appear at *m/z* = 407 (parent peak), 392 (base peak) and 376, indicating the presence of quinazoline skeleton in the toxin (Figure 11). It is noteworthy that each peak of selected ion monitored at *m/z* = 376, 392 and 407 appears at the same retention time. In the selected ion-monitored mass chromatogram of the TMS derivatives of alkali-hydrolyzed from crystals prepared from ribbon worm in Hiroshima Bay, mass fragment ion peaks at *m/z* 376, 392 and 407, which are characteristic of the quinazoline skeleton (C9 base), appeared at retention times (8.33 and 8.34 min.), almost the same as those from the TMS-C9 base derived from authentic TTX (Figure 12). Screening of tetrodotoxin in pufferfish using GC-MS was reported [41]. Sensitive analysis of TTX in human plasma by solid-phase extraction and GC-MS was reported [42].

4.2. Fast atom bombardment mass spectrometry

Fast atom bombardment mass spectrometry (FAB-MS) is a direct method for the qualitative confirmation of TTX. The analysis was performed on a JEOL JMX DX-300 mass spectrometer [43]. Xenon is used to provide the primary beam of atoms, the acceleration voltage of the primary ion being 3 kV. Scanning is repeated within a mass range of *m/z* = 100-500. In this analysis, approximately 0.1 mg of TTX and glycerol are placed as the matrix on the sample

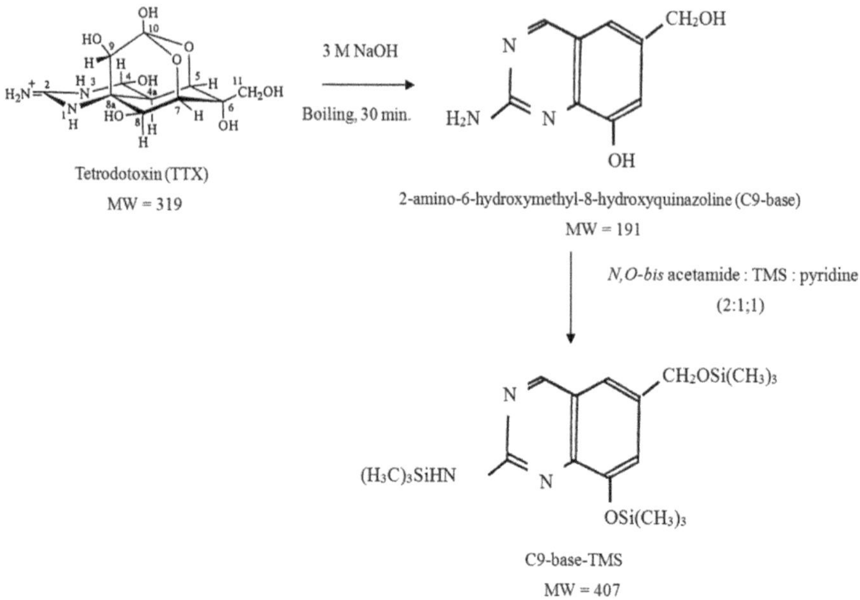

Figure 10. Reaction pathways from TTX to C9-base-TMS

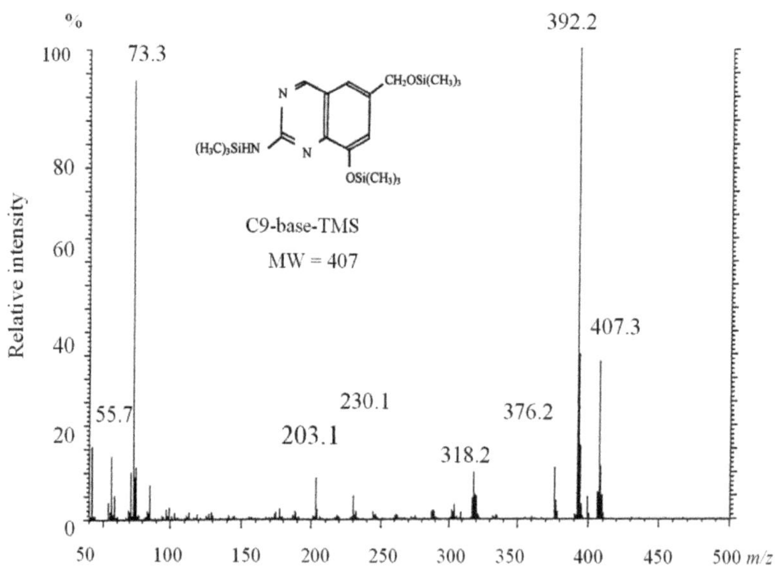

Figure 11. Mass spectrum of the trimethylsilyl (TMS) derivatives of alkali-hydrolyzed toxin from the ribbon worm *Cephalothrix* sp. from Hiroshima Bay.

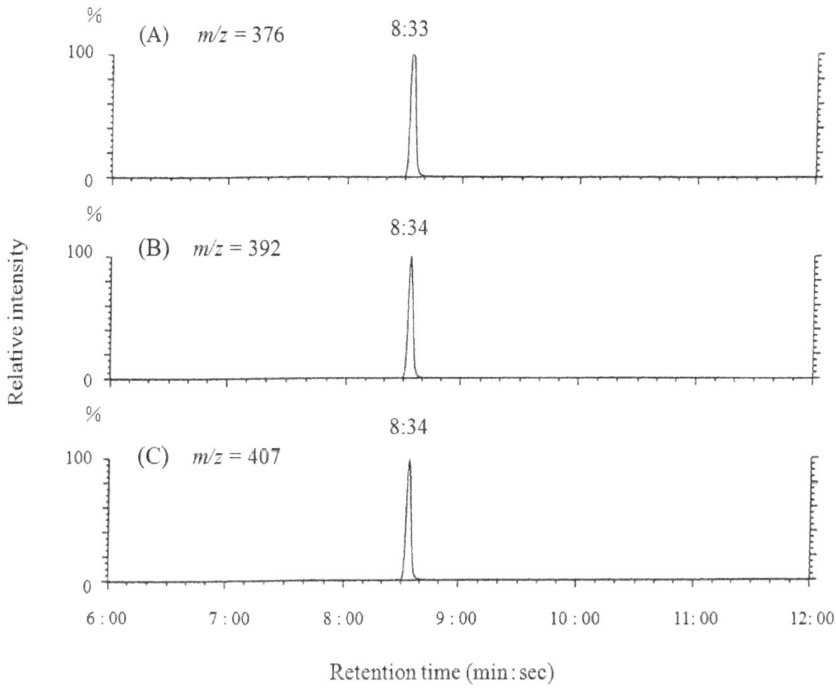

(A) *m/z* = 392 (B) *m/z* = 407 (C) *m/z* = 376

Figure 12. Selected ion-monitored (SIM) mass chromatograms of the trimethylsilyl (TMS) derivatives of alkali-hydrolyzed toxin from the ribbon worm *Cephalothrix* sp. from Hiroshima Bay.

stage of the mass spectrometer, mixed well, and placed in the ion chamber of the spectrometer. Then, both positive and negative mass spectra of TTX are measured. TTX shows $(M+H)^+$ and $(M+H-H_2O)^+$ ion peaks at *m/z* 320 and 302, respectively, in the positive mass spectrum, and an $(M-H)^-$ peak at *m/z* 318 in the negative spectrum (Figure 13). Secondary ion mass spectrometry (SIMS) performed with a Hitachi M-80B mass spectrometer gave essentially the same result as that obtained by FABMS. An extensively purified sample is required for the successful application of this method. Nagashima *et al.* developed a method to detect TTX by TLC-FAB-MS, in which the limit for detection TTX was approximately 0.1 μgTTX. [44]. TTX was also detected clearly by cellulose acetate membrane electrophoresis/FAB-MS, along with selected ion-monitored chromatograms of a mixture of TTX, anhydroTTX, and tetrodonic acid (TDA).

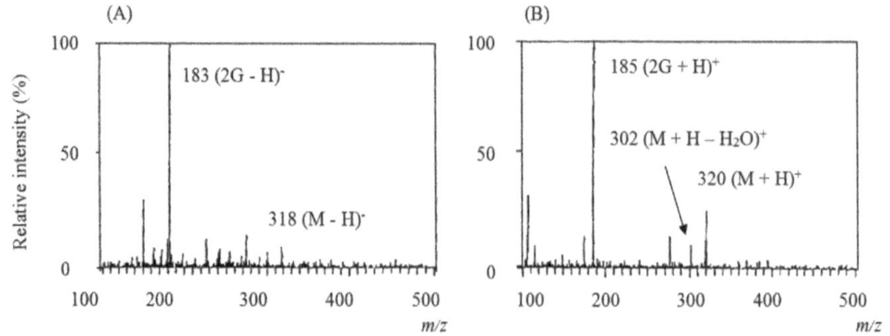

Figure 13. Positive (right) and negative (left) FAB mass spectra of TTX [41].

4.3. Liquid chromatography mass spectrometry

Liquid chromatography-mass spectrometry (LC-MS) is developed to detect TTX with considerable accuracy [45]. The major disadvantage of LC-FLD is the large difference in the structure-dependent fluorescence intensities of the analogs. In particular, the fluorescence intensities of 5-deoxyTTX and 11-deoxyTTX are approximately 1/20 and less than 1/100 of that of TTX, respectively, while those of 6-*epi*TTX and 11-norTTX-6(*R*)-ol are approximately 20-fold and 10-fold greater than that of TTX, respectively [46]. LC-MS could solve this problem, if sufficient separation and high ionization intensities could be achieved. In this method, combined HPLC-MS is performed using a Hitachi M-1000 system coupled to a mass spectrometer. The HPLC system is equipped with an ODS (ϕ 1.5 × 150 mm) column. MeOH or acetonitrile (50%, flow rate 70 µl/min) is used as the mobile phase. The effluent from the column is split to provide flow to the ion-spray interface. Brackish water puffer toxins were analyzed by LC-MS [10, 11]. An example of LC-MS of a toxin purified from ribbon worms from Hiroshima Bay is shown in Figure 14. In the MS, a protonated molecular ion peak (M+H)+ appeared at m/z = 320 showing a molecular weight for the toxin(319) in good accordance with that of TTX. Tsuruda *et al.*, detected TTX, 4-*epi*TTX, 4, 9-anhydroTTX, 6-*epi*TTX and 4, 9-anhydro-6-*epi*TTX from toxin secreted by newts *Cynops pyrrhogaster* on being subjected to "handling stimulus" [47].

4.4. Electrospray ionization – Time of flight – Mass spectrometry

Electrospray ionization time of flight mass spectrometry (ESI-TOF-MS) is applicable to many fields including the analysis of proteins, natural extracts, synthetic mixtures and medical drugs. ESI-TOF-MS is a valuable technique for identification of TTX, although it is not widely used to date in marine toxin determinations. In this analysis, a portion of purified TTX (less than 0.05 mg) is dissolved in a small amount of 1% AcOH, and the resulting solution is added to 50% aqueous MeOH. ESI-TOF-MS is run on a Micromass Q-TOF mass spectrometer. TTX in a tree frog *Polypedates* sp. extract has been successfully evaluated by ESI-TOF-MS analysis [48]. As shown in the spectrum of the toxin (Figure 15), the protonated molecular ion peak (M + H)+ appeared at *m/z* = 320.1103, suggesting the molecular weight of the toxin to be 319.1025 which agrees well with that of authentic TTX ($C_{11}H_{17}N_3O_8$ = 319.1016).

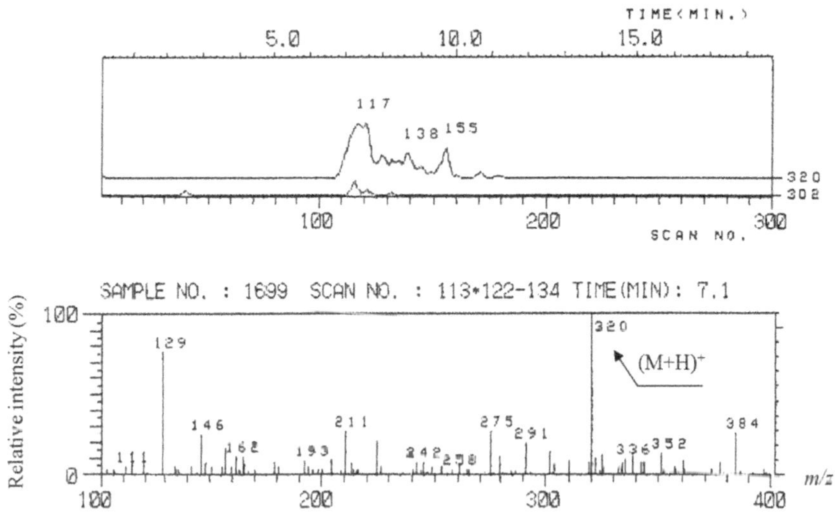

(A): Mass chromatogram of the ribbon worm toxin
(B): Mass spectrum of the ribbon worm toxin

Figure 14. LC-MS of the toxin from the ribbon worm *Cephalothrix* sp. from Hiroshima Bay.

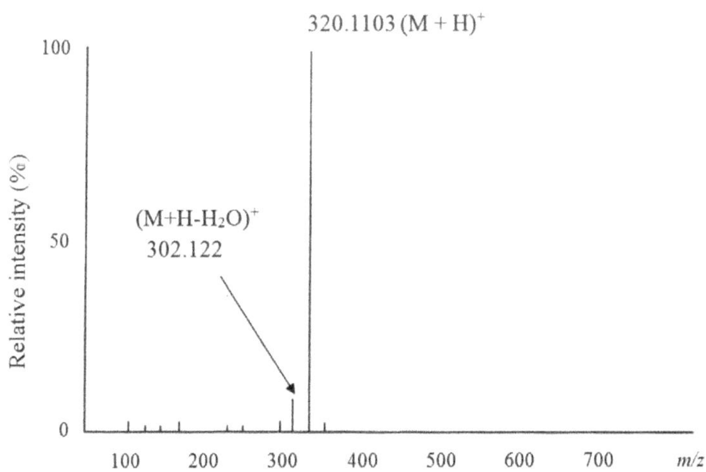

Figure 15. ESI-TOF/MS analysis of a frog *Polypedates* sp. toxin [46].

5. Infrared (IR) spectrometry

IR spectrometry is the analytical technique for the determination of functional groups in TTX. Although the IR spectrum is presumed to be complex, it is a helpful tool to identify TTX. IR-spectra of KBr pellet were acquired using IR spectrophotometer, which was used by Onoue *et al.* for determination the IR spectrum of pufferfish toxin [49]. On the other hand, Tsuda *et al.* reported the IR spectrum of a TTX-HCl salt by the "Nujol" method [2]. Here we introduce another method as mentioned below. A part of TTX crystals purified from the specimens of ribbon worms were placed on a small KBr plate, and the IR spectrum was acquired using a FT-IR spectrometer (Perkin Elmer, Spectrum 2000) equipped with FT-IR microscope. As shown in Figure 16, absorption bands at 3353, 3235, 1666, 1612 and 1076 cm^{-1} were observed in the spectrum of this crystal. This spectrum was indistinguishable from that of TTX reported previously, showing characteristic absorptions for the functional groups OH, guanidium, and COO^- [2]. The absorption near 2400 cm^{-1} was derived from the existence of CO_2 in the air. The absorption around 1800 cm^{-1} and in the range of 3,600 – 4,000 cm^{-1} was derived from H_2O in the air. Although the spectrum appears to be complex, it is a helpful tool for identification of TTX.

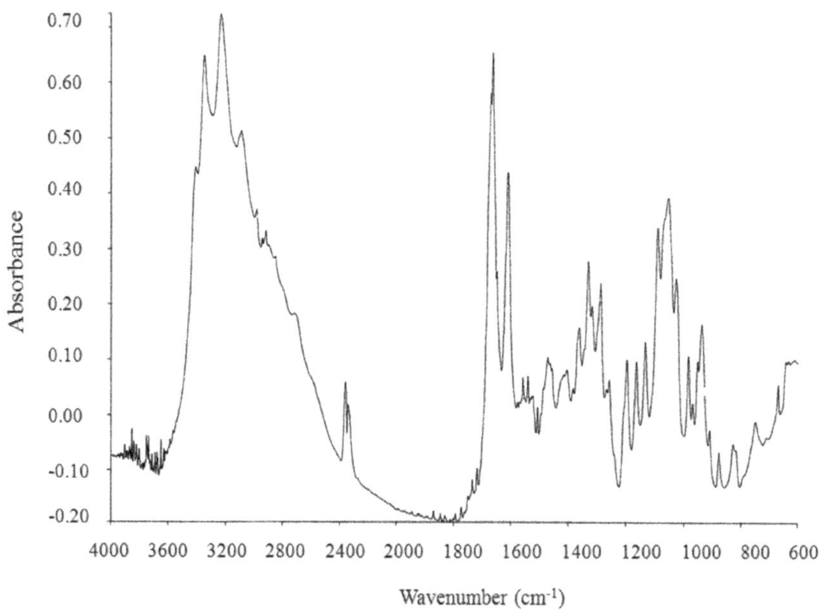

IR spectrum was taken on a FT-IR spectrometer (Perkin Elmer, Spectrum 2000) equipped with FT-IR microscope.

Figure 16. IR spectrum of the toxin isolated from the ribbon worm *Cephalothrix* sp. from Hiroshima Bay.

6. Ultraviolet (UV) spectroscopy

In UV spectroscopy, TTX is generally determined by irradiating a crude toxin with UV light. A small amount of TTX is dissolved in 2 ml of 2 M NaOH and heated in a boiling water bath for 45 min. After cooling to room temperature, the UV spectrum of the solution is examined for characteristic absorptions, associated with C9-base, 2-amino-6-hydroxymethyl-8-hydroxyquinazoline, possibly formed from TTX and/or related substances, if present. In the analysis, the UVspectrum of the alkali decomposed compounds of TTX appears as a shoulder at near 276 nm, indicating the formation of C9-base specific to TTX or related substances (Figure 17). Saito *et al.* used this method in experiments analyzing TTX and its derivatives in toxic and nontoxic pufferfish [50].

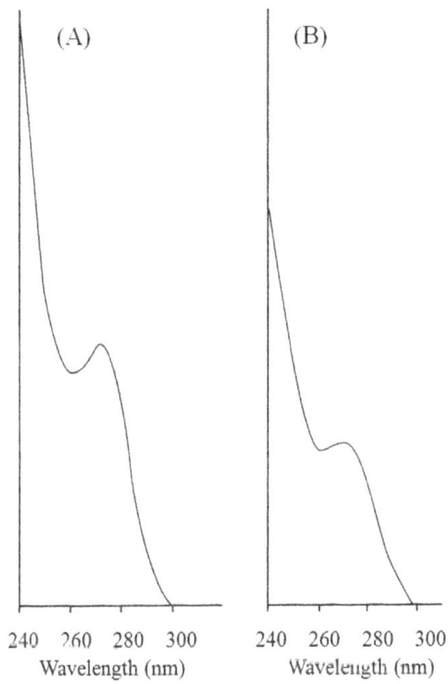

(A): TTX fraction from non toxic species of pufferfish *L. wheeleri*
(B): Authentic TTX

Figure 17. UV absorption spectra of the alkaline hydrolyzates of TTX fraction from non toxic species of pufferfish ("shirosabafugu" in Japanese) *Lagocephalus wheeleri* [48].

7. Proton nuclear magnetic resonance (^{1}H-NMR) spectrometry

^{1}H-NMR has played an important role as a complementary method to determine the absolute configuration of TTX. To date, many derivatives of TTX have been isolated, and their ^{1}H-NMR data have been reported by various investigators. In a typical ^{1}H-NMR analysis, 5 mg of TTX crystals have been dissolved in 0.5 ml of 1% CD_3COOD in D_2O, and placed in a test tube. Figure 18 shows the ^{1}H-NMR spectrum obtained with a 500 MHz JEOL JNM-500 spectrometer, using the methyl group protons of acetone as the internal standard [39]. The ^{1}H-NMR spectrum exhibited a singlet at 2.20 ppm (CH_3COCH_3), a doublet centered at 2.33 ppm (J =10.0 Hz), a large proton peak at 4.76 ppm (HDO) and a doublet centered at 5.48 ppm (J=10.0 Hz). The pair of doublets around 2.33 and 5.48 ppm, which are the hallmarks of TTX and are assigned to H-4a and H-4, respectively, have been confirmed to be coupled with each other by double irradiation (Figure 19). These results agree well with the corresponding data of TTX. The signals at 4.24, 4.06, 4.28, 3.94, 4.00 and 4.02 ppm are assigned toH-5, H-7, H-8, H-9 and H-11, respectively (Figure 20). A toxin isolated from the horseshoe crab *Carcinoscorpius rotundicauda* from Bangladesh was analyzed by HPLC-FLD, TLC, electrophoresis and ^{1}H-NMR, and was identified as TTX [51]. Identification of a neurotoxin from the blue-ringed octopus, brackish water pufferfish, marine pufferfish and so on as TTX via this method were reported [9, 10, 11, 17].

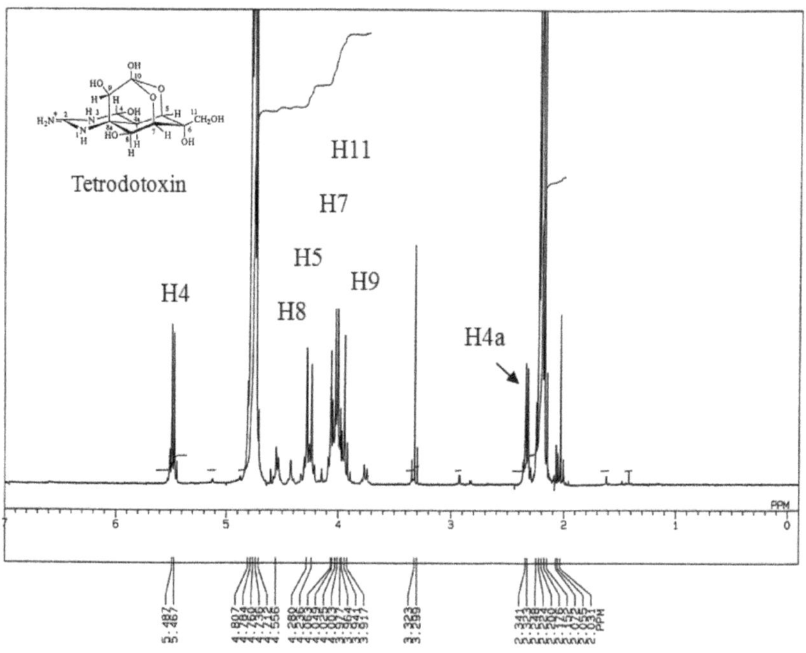

Five milligrams of the ribbon worm toxin was dissolved in 0.5ml of 1% CD_3COOD in D_2O and measured for ^{1}H-NMR spectrum measured on a JEOL JNM-500NMR spectrometer, using acetone as the internal standard.

Figure 18. ^{1}H-NMR spectrum of the toxin isolated from the ribbon worm *Cephalothrix* sp. from Hiroshima Bay.

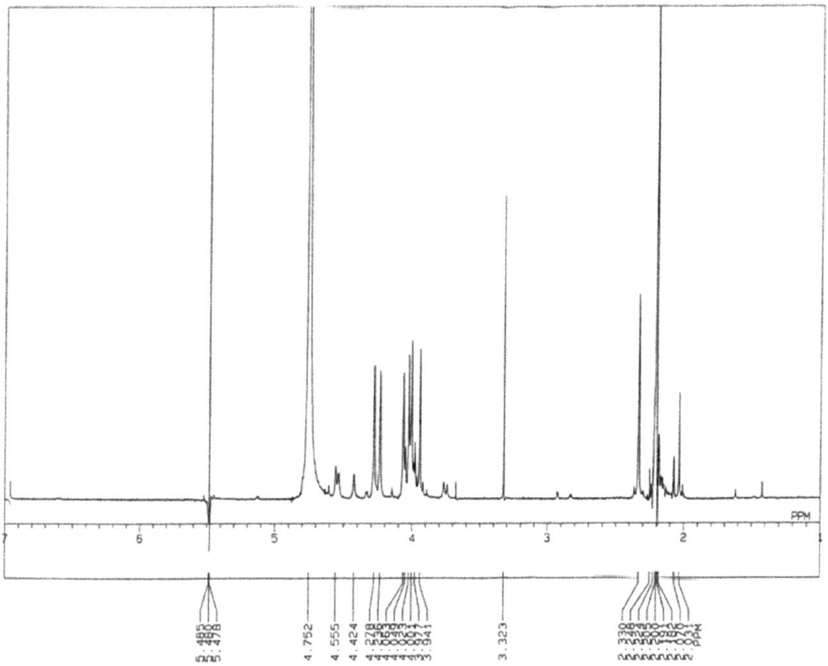

Five milligrams of the ribbon worm toxin was dissolved in 0.5ml of 1% CD_3COOD in D_2O and measured for ^1H-NMR spectrum measured on a JEOL JNM-500NMR spectrometer, using acetone as the internal standard.

Figure 19. ^1H-NMR spectrum of the toxin isolated from the ribbon worm *Cephalothrix* sp. from Hiroshima Bay by means of irradiation at C_4-H.

H	Ribbon worm toxin
4a	2.33
4	5.48
5	4.24
7	4.06
8	4.28
9	3.94
11	4 4.02

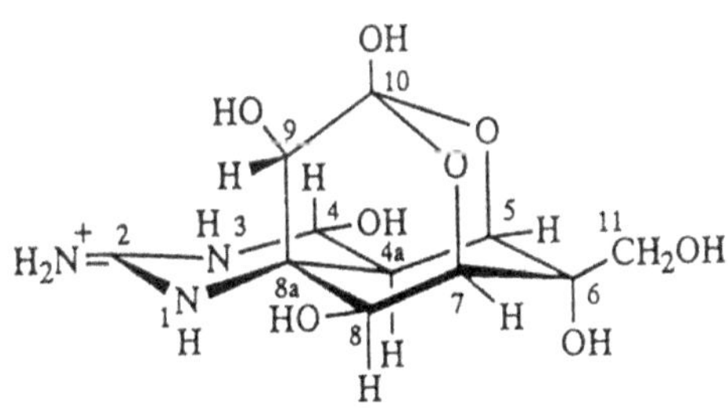

Figure 20. ^1H-NMR spectral data of the toxin isolated from the ribbon worm *Cephalothrix* sp. from Hiroshima Bay, along with the structure of TTX.

8. Thin-layer chromatography

TLC is a very commonly used technique in synthetic chemistry for identifying compounds, determining their purity. In TLC analysis, TTX is spotted onto a silica gel 60 precoated plate (Merck). The plate is developed in two different solvent systems of pyridine-ethyl acetate-AcOH -water (15:5:3:4) and 3-BuOH-AcOH-water (2:1:1) in a sealed chamber. The solvent rises by capillary action and an ascending chromatographic separation is obtained. The plate is then sprayed with 10% KOH followed by heating at 100°C for 10 minutes. The toxin is visualized as a yellow fluorescent spot under UV light (365nm). In TLC analysis, the Rf values of TTX are around 0.71(0.65) and 0.50, respectively [10, 51]. It is also possible to detect TTX on the TLC plate using the Weber reagent that gives pink spot of the toxin. The detection limit is about 2 μg of TTX (10 MU). TLC is a useful technique in those laboratories where HPLC and other costly analytical systems are not available.

9. Electrophoresis

Electrophoresis is a relatively simple and rapid method with high resolution detection of polar compounds like TTX [10, 11, 51]. When 1 μl of TTX (10 MU, corresponding to 2 μg) is applied onto a 5 x 18 cm cellulose acetate membrane (Chemetron, Milano), the ion molecules of TTX move toward the cathode with a mobility (Rm) clearly smaller than that of authentic of STX. The analysis is performed for 30 minutes in an electrolytic buffer solution of 0.08 M Tris-HCl (pH8.7), under the influence of an applied electric field with a constant current of 0.8 mA/cm width. The toxin is visualized in the same manner as described for TLC.

10. Capillary isotachophoresis

Capillary isotachophoresis proved to be a very effective technique for the analyses of organic acids, carbohydrates, drugs and amino acids. It is a rapid and accurate detection technique for TTX [52]. It is performed using a cationic system, as TTX exists as cation under acidic and neutral conditions. Conditions for capillary isotachophoresis composed of a leading electrolyte of 5 mM potassium acetate (pH6.0), containing 0.2% Triton X-100 and 0.5 volume of dioxane, and a terminating electrolyte of 10mM β-alanine adjusted to pH 4.5 with acetic acid. When TTX is applied to isotachophoretic analyzer (Shimadzu IR-2A) equipped with a potential gradient 0.32, it is eventually monitored by the detector. The quantitative detection limit by this method is about 0.25μg of TTX. It was possible to quantify TTX content of contaminated puffer extracts without any pretreatment.

11. Conclusion

In an attempt to protect consumers from TTX-intoxication, the mouse bioassay has historically been the most universally applied tool to determine the toxicity level in monitoring programs. This bioassay, however, shows low precision and requires a continuous supply of mice of a specific size. These potential drawbacks and world-wide pressure to refrain from the unnecessary killing of live animals subsequently led scientists to develop alternative chemical methods to the mouse bioassay for TTX detection and quantification. TTX levels in pufferfish are normally estimated using the mouse bioassay. However, this assay and other techniques such as TLC, electrophoresis, LC, spectrophotometry, and the enzyme-linked immunosorbent assay (ELISA) pose ethical concerns, are not specific and lack sensitivity and precision at low concentrations. HPLC-FLID and LC-MS/GC-MS are sensitive techniques for identification of TTX. However, due to the complexity of sample matrices and insolubility of TTX in organic solvents, HPLC-FLD and LC-MS (or LC-MS/MS) are more preferred methods than GC-MS. MS spectrometry is a powerful technique that also has an important future for the analysis of marine toxins. In addition to high sensitivity and selectivity, MS can provide structural information useful for the confirmation of toxin identity and the identification of new toxins. The drawback of LC-MS and LC-MS/MS analyses is that they involve the use of expensive instruments, which require higher maintenance compared to GC-MS. Nevertheless, for routine analysis of TTXs, HPLC-FID and LC-MS (LC-MS/MS) are expected to replace the conventional mouse bioassay.

Author details

Manabu Asakawa* and Keisuke Miyazawa
Department of Bioresource Science and Technology, Department of Bioresource Science and Technology, Graduate School of Biosphere Science, Hiroshima University, Kagamiyama, Higashi-Hiroshima, Hiroshima, Japan

Yasuo Shida
Clean Energy Research Center, University of Yamanashi, Takeda, Kofu, Yamanashi, Japan

Tamao Noguchi
Tokyo Healthcare University, Setagaya, Tokyo, Japan

Acknowledgement

This work was supported by a Grant-in-Aid for Scientific Research (C) (2) (No.11660205), (C)(No.18580204), (C)(No.20580220) and (B)(No.20380110) from Japan Society for the Promotion of Science. We are grateful to Dr.Hiroshi Kajihara (Hokkaido University) for his helpful advice on ribbon worms.

12. References

[1] Yokoo, A. (1950) Chemical studies on pufferfish toxin (3) – separation of spheroidine. *Nippon KagakuZasshi*, 71:590-592.

[2] Tsuda, K., Ikuma, S., Kawamura, M., Tachikawa, R., Sakai, K., Tamura, C., Amakasu, D. (1964) Tetrodotoxin. VII. On the structures of tetrodotoxin and its derivatives. *Chem.Pharm. Bull.* 12:1357-1374.

[3] Woodward, R.B. (1964) The structure of tetrodotoxin. *Pure Appl. Chem.* 9:49-74.

* Corresponding Author

[4] Goto, T., Kishi, Y., Takahashi, S., Hirata, Y. (1965) Tetrodotoxin. *Tetrahedron* 21:2059-2088.

[5] Narahashi, T. (2001) Pharmacology of tetrodotoxin. *J. Toxicol.-Toxin Rev.* 20:67-84

[6] Noguchi, T., Ebesu, J.S.M. (2001) Puffer poisoning: epidemiology and treatment. *J. Toxicol.-Toxin Rev.* 20: 1-10.

[7] Hashimoto, Y. (1979) Marine Toxins and Other Bioactive Marine Metabolites. Japan Scientific Societies Press, Tokyo, pp.70-91.

[8] Khora, S.S., Isa, J., Yasumoto, T. (1991) Toxicity of puffers from Okinawa, Japan, *Nippon Suisan Gakkaishi*, 57:163-167.

[9] Mahmud, Y., Tanu, M.B, Takatani, T., Asayama, E., Arakawa. O., Noguchi, T. (2001) *Chelonodon patoca*, a highly toxic marine puffer in Japan. *J. Nat Toxins*, 10:69-74.

[10] Mahmud, Y., Yamamori, K., Noguchi, T. (1999) Occurrence of TTX in a brackishwater puffer "Midorifugu" *Tetraodon nigroviridis*, collected from Thailand. *J.Food Hyg. Soc.*, 40:363-367.

[11] Mahmud, Y., Yamamori, K., Noguchi, T. (1999) Toxicity and tetrodotoxin as the toxic principle of a brackishwater puffer *Tetraodon steindachneri*, collected from Thailand, *J.Food Hyg. Soc.*, 40:391-395.

[12] Hwang, D.F, Kao, C.Y., Yang, H.C., Jeng, S.S., Noguchi, T., Hashimoto, K. (1992) Toxicity of puffer in Taiwan, *Nippon Suisan Gakkaishi*, 58:1541-1547.

[13] Mahmud, Y., Tanu, M.B., Noguchi, T. (1999) First occurrence of a food poisoning incident due to ingestion of *Takifugu oblongus*, along with a toxicological report on three marine puffer species in Bangladesh. *J.Food Hyg. Soc.*, 40:473-480.

[14] Mosher,H.S., Fuhrman,F.A., Buchwald,H.G., Fischer, H.G. (1964) Tarichatoxin - tetrodotoxin, a potent neurotoxin. *Science* 144: 1100-1110.

[15] Noguchi, T., Kao, H., Hashimoto, Y. (1971) Toxicity of the goby, *Gobius criniger*. *Bull. Jpn. Soc.Sci.Fish., 37:642*-647.

[16] Kim, Y.H., Brown,G.B., Mosher, H.S.,Furman, F.A. (1975) Tetrodotoxin : occurrence in Atelopid frogs of Costa Rica. *Science* 189:151-152.

[17] Sheumack, D.D., Howden, M.E.H., Spence, I., Quinn, R.J. (1978) Maculotoxin: a neurotoxin from the venom glands of the octopus *Hapalochlaena maculosa* identified as tetrodotoxin. *Science* 199:188-189.

[18] Narita,H., Noguchi, T., Maruyama,J, Ueda,Y., Hashimoto, K., Watanabe, Y., Hida,K. (1981) Occurrence of tetrodotoxin in a trumpet shell "boshubora" , *Charonia sauliae. Bull. Japan. Soc. Sci. Fish.*, 47:935-941.

[19] Noguchi,T., Maruyama,J., Ueda,Y., Hashimoto,K. Harada,T. (1981) Occurrence of tetrodotoxin in the Japanese ivory shell *Babylonia japonica. Bull. Jpn. Soc.Sci.Fish.*, 47:909-913.

[20] Noguchi, T., Maruyama, J., Narita, H., Hashimoto, K. (1984) Occurrence of tetrodotoxin in the gastropod mollusk *Tutufa lissostoma* (frog shell). *Toxicon* 22:219-226.

[21] Noguchi, T. Narita, H. Maruyama, J. Hashimoto, K. (1982) Tetrodotoxin in the starfish *Astropecten polyacanthus*, in association with toxification of a trumpet shell, "boshubora", *Charonia sauliae. Bull. Japan. Soc. Sci. Fish.* 48, 1173-1177.

[22] Kawabata, T., 1978. In: Ministry of Health and Welfare of Japan, (Ed.), Assay Method for Tetrodotoxin, Food Hygiene Examination Manual, vol. 2. Japan Food Hygiene Association, Tokyo, pp. 232–239.

[23] Asakawa, M., Toyoshima, T., Shida, Y., Noguchi, T., Miyazawa, K.(2000) Paralytic toxins in a ribbon worm *Cephalothrix* species (Nemertean) adherent to cultured oysters in HiroshimaBay, Hiroshima Prefecture, Japan. *Toxicon* 38, 763–773.

[24] Yasumoto, T., Nakamura, M., Oshima, Y., Takahata, J. (1982) Construction of a continuous tetrodotoxin analyzer. *Bull.Jpn.Soc.Sci.Fish.*, 48:1481-1483.

[25] Nakamura, M., Yasumoto, T. (1985) Tetrodotoxin derivatives in puffer fish. *Toxicon* 23:271-276.

[26] Yasumoto, T., Yotsu, M., Murata, M., Naoki, H. (1988) New tetrodotoxin analogues from the newt *Cynops ensicauda*. *J.Am.Chem. Soc.*, 110:2344-2345.

[27] Khora, S. S., Yasumoto, T. (1989) Isolation of 11-oxotetrodotoxin from the puffer *Arothron nigropunctatus*. *Tetrahedron Lett.* 30: 4393-4394.

[28] Endo, A., Khora, S.S., Murata, M., Naoki, H., Yasumoto, T. (1988) Isolation of 11-nortetrodotoxin-6(R)-ol and other tetrodotoxin derivatives from the puffer *Fugu niphobles*. *Tetrahedron Lett.* 29: 4127-4128.

[29] Yotsu, M., Hayashi,Y., Khora, S.S., Sato, S., Yasumoto, T. (1992) Isolation and structural assignment of 11-nortetrodotoxin-6(S)-ol from the puffer *Arothron nigropunctatus*. *Biosci. Biotech. Biochem.* 56: 370-371.

[30] Kotaki, Y., Shimizu, Y. (1993) 1-hydoxy-5, 11-dideoxytetrodotoxin, the first N-hydroxy and ring-deoxy derivative of tetrodotoxin found in the newt *Taricha granulosa*. *J.Am.Chem.Soc.* 115: 827-830.

[31] Yotsu-Yamashita, M., Yamagishi, Y. Yasumoto, T. (1995) 5, 6, 11-trideoxytetrodotoxin from the puffer fish *Fugu poecilonotus*. *Tetrahedron Lett.* 36: 9329-9332.

[32] Yotsu-Yamashita, M., Schimmele, B., Yasumoto, T. (1999) Isolation and structural assignment of 5-deoxytetrodotoxin from the puffer fish *Fugu poecilonotus*. *Biosci. Biotech. Biochem.* 63: 961-963.

[33] Yotsu, M., Endo, A., Yasumoto, T. (1989) An improved tetrodotoxin analyzer. *Agric. Biol. Chem.*, 53: 893-895.

[34] Yotsu-Yamashita, M., Meds, D., Kwet, A., Schneider, M. (2007). Tetrodotoxin and its analog 6-*epi*tetrodotoxin in newts (*Triturus* spp.; Urodela, Salamandridae) from southern Germany. *Toxicon* 50: 306-309.

[35] Nagashima, Y., Maruyama, J., Noguchi, T., Hashimoto, K. (1987) Analysis of paralytic shellfish poison and tetrodotoxin by ion-pairing high performance liquid chromatography. *Nippon Suisan Gakkaishi*, 53:819-823.

[36] Asakawa,M., Delan,G.G., Tsuruda,S., Shimomura,M., Shida,Y., Taniyama,S., Quilantang, M.B., Shindo,J. (2010) Toxicity assessment of the xanthid crab *Demania cultripes* from Cebu Island, Philippines. *J.Toxicology.*, Article ID172367.

[37] Noguchi, T., Jeon, J.K., Arakawa,O., Sugita,H., Deguchi,Y., Shida, Y., Hashimoto, K. (1986) Occurrence of tetrodotoxin and anhydrotetrodotoxin in *Vibrio* sp. isolated from the intestines of xanthid crab, *Atergatis floridus*. *J.Biochemistry*, 99:311-314.

[38] Narita, H., Noguchi, T., Maruyama,J., Nara, M., Hashimoto, K. (1984) Occurrence of a tetrodotoxin-associated substances in a gastropod, "Hanamushirogai" *Zeuxis siquijorensis*. *Bull.Jpn.Soc.Sci.Fish*, 50:85-88.

[39] Asakawa, M., Toyoshima,T., Ito, K., Bessho, K., Yamaguchi, C., Tsunetsugu, S., Shida, Y., Kajihara, H., Mawatari, S. F., Noguchi, T., Miyazawa, K.(2003) Paralytic toxicity in the ribbon worm Cephalothrix species (Nemertea) in Hiroshima Bay, Hiroshima Prefecture, Japan and the isolation of tetrodotoxin as a main component of its toxins. *Toxicon*, 41:747-753.

[40] Maran,B.A,V., Iwamoto, E., Okuda, J., Matsuda, S., Taniyama, S., Shida,Y., Asakawa, M., Ohtsuka, S., Nakai, T., Boxshall G.A. (2007) Isolation and characterization of bacteria from the copepod *Pseudocaligus fugu* ectoparasitic on the panther puffer T*akifugu pardalis* with the emphasis on TTX. *Toxicon*, 50:779-790.

[41] Man,C.N., Noor, N.M., Harn, G.L., Lajis, R. Mohamad, S. (2010) Screening of tetrodotoxin in puffers using gas chromatography-mass spectrometry. *J.Chrom.A*, 1217:7455-7459.

[42] Kurono,S., Hattori,H., Suzuki,O., Yamada,T., Seno, H. (2001) Sensitive analysis of tetrodotoxin in human plasma by solid-phase extraction and gas chromatography/mass spectrometry. *Anal.Letts.*, 34:2439-2446.

[43] Noguchi, T., Ali, A.E., Arakawa, O., Miyazawa, K., Kanoh, S., Shida, Y., Nishio, S., Hashimoto, K. (1991) Tetrodonic acid-like substance: a possible precursor of tetrodotoxin. *Toxicon*, 29:845-855.

[44] Nagashima,Y., Nishio, S., Noguchi, T., Arakawa,O., Kanoh,S., Hashimoto,K. (1988) Dtection of tetrodotoxin by thin-layer chromatography/fast atom bombardment mass spectrometry. *Anal.Biochem*. 175:258-262.

[45] Shida,Y., Arakawa, O., Onoue, Y. and Noguchi,T. (1998) LC/MS of marine toxin-1. In "Proceedings of the 46th Annual Conference on Mass Spectrometry,"

[46] Shoji, Y., Yamashita-Yotsu, M., Miyazawa, T., Yasumoyto,T. (2001) Electrospray ionization mass spectrometry of tetrodotoxin and its analogs: liquid chromatography/mass spectrometry, tandem mass spectrometry, and liquid chromatography/tandem mass spectrometry. *Analy. Bioch.*, 290: 10-17.

[47] Tsuruda,K., Arakawa, O., Kawatsu, K., Hamano, Y.,Takatani, T. and Noguchi, T. (2002) Secretory glands of tetrodotoxin in the skin of the Japanese newt *Cynops pyrrhogaster*. *Toxicon* 40:131-136.

[48] Tanu, M.B., Mahmud, Y., Tsuruda,K., Arakawa,O., Noguchi, T. (2001) Occurrence of tetrodotoxin in the skin of a rhacophoridid frog *Polypedates* sp. from Bangladesh. *Toxicon*, 39:937-941.

[49] Onoue,Y., Noguchi, T., Hashimoto,K. (1984) Tetrodotoxin determination method. *In* "Seafood Toxin" (E.P.Ragelis, ed.) pp.345-355. American Chemical Society, Washington, DC.

[50] Saito, T., Kanoh,S., Noguchi,T., Shida,Y., Hashimoto, K. (1987) Occurrence of anhydrotetrodotoxin in several nontoxic species of puffer. *In* "Progress in venom and toxin research" (P.Gopalakrishnakone and C.K.Tan, ed.) pp.314-324. National University of Singapore, Singapore.

[51] Tanu, M.B., Noguchi, T. (1999) Tetrodotoxin as a toxic principal in the horseshoe crab *Carcinoscorpius rotundicauda* collected from Bangladesh. *J.Food Hyg. Soc.*, 40:426-430.

[52] Shimada, K., Ohtsuru, M., Yamaguchi, T., Nigota, K. (1983) Determination of tetrodotoxin by capillary isotachophoresis. *J.Food Sci.*, 48:665-667.

Chapter 10

Chromatographic Analysis of Nitrogen Utilization and Transport in Arbuscular Mycorrhizal Fungal Symbiosis

Jin HaiRu and Jiang Xiangyan

Additional information is available at the end of the chapter

1. Introduction

The combination of chromatographic methods and mass spectrometry (MS) techniques is very useful in analysis of metabolites or components of complex biological samples. Chromatography has become the most important technique for separation and analyses, whereas MS is one of the most effective analytical methods used today for the determination of element concentrations, especially in the trace range; for the structural studies of organic and bioorganic compounds as well as isotopic analysis, due to its very high sensitivity, low detection limits and very small sample volumes needed[1-2]. Using chromatography and MS to separate and measure the concentration of amino acids has been well documented. Bengtsson[3] was the first to introduce a micro-method for the analysis of free amino acids in natural waters by gas chromatography (GC); the technique includes removal of interfering organic substances by chloroform extraction and purification of amino acids by cation exchange. Later, the procedure for analysis of amino acids from protein acid hydrolyzates as their *tert*-butyl dimethylsilyl derivatives by gas chromatography and mass spectrometry has been developed[4-9] and applied in biological matrix analysis[10-13]. Johansen et al.[12] and Rolin et al.[13] studied the nitrogen metabolism of external hyphae of the AM fungus using measurement of *tert*-butyldimethylsilyl(tBDMS)-derivatized amino acids levels by gas chromatography and mass spectrometry. Arbuscular mycorrhizal (AM) fungus is the oldest obligate symbiont,[14] benefiting the host plants by taking up N, P and other macronutrients, trace elements, and water from the soil. Among many components important for plant nutrition, nitrogen is often the most limiting but the AM fungi can improve the nitrogen (N) levels of their hosts.[15-17] The extraradical hyphae of the fungi effectively acquire nitrate (NO_3^-)[18-20] , ammonium (NH_4^+)[12,21-23], and amino acids[24-27] from the external medium. However, for a long time, it has been unclear in what form nitrogen is translocated along the hyphae (extraradical mycelium, ERM) of the fungus to the fungal structures within roots (intraradical mycelium, IRM), and how it is transferred across the mycorrhizal interface to the plant. To follow the uptake, assimilation and transfer of nitrogen in the arbuscular mycorrhizal symbiosis, we added isotopically labeled substrates to *in vitro* arbuscular mycorrhizal cultures of carrot (*Daucus carota* L.) roots colonized by *G. intraradices.*[28-30] When grown in divided Petri plates, this model mycorrhiza excludes other microorganisms and prevents diffusion of nonvolatile solutes between the compartments. This model system shows normal life cycle and development of fungal morphology.

Following N uptake, its incorporation into amino acids via the glutamine synthetase/glutamine 2-oxoglutarate amidotransferase (GS/GOGAT) cycle has been observed in AM fungi. Using chromatographic separation and mass spectrometry analysis of amino acids of the hyphae, Johnsen et al.[12] have proved that *G. intraradices* grown in a medium containing $^{15}NH_4^+$ generated abundant free AAs in the ERM. Among these amino acids, ^{15}N- labeled glutamate (Glu), glutamine (Gln), asparagine (Asn), aspartate, and alanine were predominant. Jin et al.[28] have confirmed the mechanism of N transport to the host plant via the AM fungi proposed by Bago et al.[31] (Fig.1). When mycorrhizae of *G. intraradices* and Ri T-DNA-transformed carrot roots were grown in two-compartment Petri dishes, containing $^{15}NH_4Cl$ synthetic medium in the fungal compartment, the measurement of amino acids N-Methyl-N-(tert-butyldimethylsilyl)trifluoroacetamide (MTBSTFA) derivatives with GC-MS revealed that all free amino acids of the ERM were ^{15}N-labeled to a high level. Among these amino acids, arginine (Arg) was the most abundant (over 90% of the total ^{15}N in the free amino acids).This was confirmed by analysis of phenyl isothiocyanate (PITC)-derivatized amino acids with high performance liquid chromatography (HPLC). When the [U-^{13}C] Arg (MTBSTFA derivative with an *m/z* of 448) was used as the labeled substrate in the fungal compartment, the isotopomer analysis with GC-MS confirmed that this amino acid was intact both in the mycorrhizal root tissue (*m/z* of 448) and in the ERM. This result suggests that Arg is either taken up or synthesized by the ERM and transported intact to the IRM; the conclusion is also consistent with the outcome of other [U-^{13}C/U-^{15}N] Arg labeling experiments.[30] This finding indicates that the AM fungi take up N and incorporate it into Arg, and this organic form of N is then transported to IRM (Fig.1).

When either [U-^{13}C] or [U-^{15}N/U-^{13}C]-labeled Arg was added to the fungal compartment to follow up Arg catabolism in AM fungi (Fig. 1), GC-MS[30] analysis of Arg isotopomer found either [U-^{13}C] or [U-^{15}N/U-^{13}C] ornithine(Orn) in the mycorrhizal root tissues. This result demonstrates that once Arg is translocated to the potential N-limited sites in the mycelium of AM fungi, it is degraded into Orn and urea. Although N released from Arg degradation is transferred to the host plant, it has been shown (using $^{13}C_{1,2}$-acetate labeling in fungal compartment and subsequent mass spectrometry analysis of amino acids of mycorrhizal root) that the Arg-originated C is not incorporated into the host C pool and remains in the IRM[29]. Analysis of ^{15}N-labeled Glu and Gln isotopomers using GC-MS has demonstrated that, following Arg degradation, Orn is recycled to Glu and Gln, which serve as C donors. Small amounts of ^{15}N label have been found in Glu and Gln (in spite of their low levels) after [guanido-$^{15}N_2$] Arg translocation from the ERM to the mycorrhizal roots, whereas other amino acids have displayed only negligible ^{15}N labeling[30]. Thus, the Orn cycle seems to be an efficient path for Arg biosynthesis, nitrogen transfer to the host, and carbon recycling in AM symbiosis.

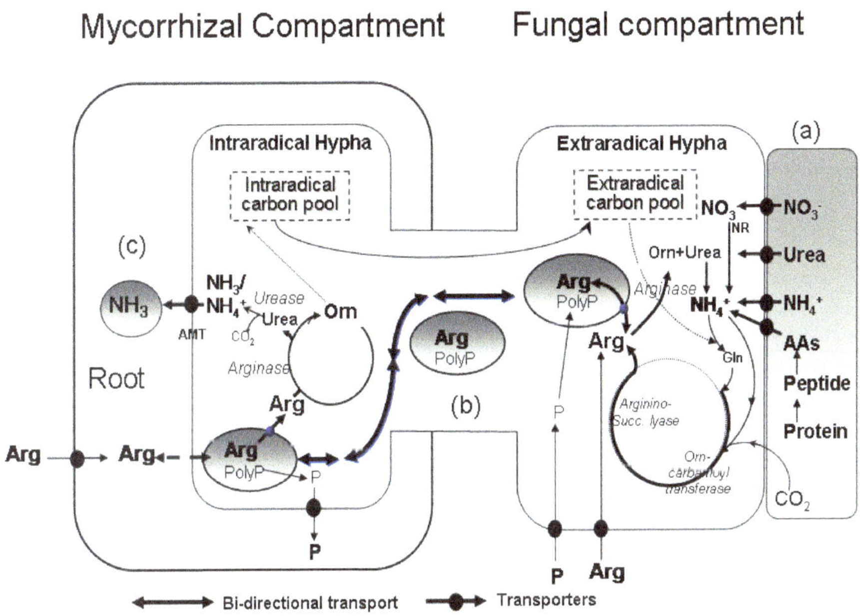

Figure 1. A model of N uptake, translocation, degradation and transfer in the AM fungal symbiotic system. Various forms of N sources (NO_3^-, NH_4^+, amino acids, peptides, and proteins) are taken up, assimilated, and incorporated into arginine (Arg) by the extraradical mycelium (ERM). The accumulated Arg as well as polyphosphate (PolyP) is then bi-directionally translocated along the coenocytic fungal hyphae from the ERM to the intraradical mycelium (IRM) or from the mycorrhizal compartment tissue to the ERM. Arg is catabolized through the catabolic arm of the urea cycle in the IRM, releasing NH_3/NH_4^+ in the arbuscules. The NH_4^+ ion is deprotonated prior to its transport across the plant membrane by AMT protein and released in its uncharged NH_3 form into the plant cytoplasm. Figure modified from Govindarajulul et al. (2005).

In addition, the results of MS analysis of ^{15}N-labeled amino acids isotopomers have indicated that germinating spores break down stored N (Arg or proteins) during pre-symbiotic growth.[32] HPLC analysis of amino acids derivatives using PITC[33] revealed that the germinating AM spores combine the C skeletons originating from the degradation of the internally stored lipids with the N released from the stored N compounds (Arg or proteins) for *de novo* biosynthesis of free amino acids, mostly producing serine and glycine. This is consistent with a large flux in the glyoxylate cycle and the utilization of lipids as a C source for carbohydrate and amino acid biosynthesis reported by Lammers at al.[34] Although exogenous N is not required for pre-symbiotic growth, it can be used for *de novo* biosynthesis of amino acids. MS analysis of ^{15}N enrichment of amino acids after ^{15}N labeling experiment showed that NH_4^+ and urea are assimilated more rapidly than NO_3^- and exogenous amino acids. *De novo* biosynthesis of free amino acids in the AM spores was increased greatly after the uptake of exogenous NH_4^+, urea, and NO_3^-. In cases of a low C:N ratio (no exogenous glucose), the measurements of PITC-derivatized AAs with HPLC showed that Asn was the predominant amino acid in the AM spores. These results suggest that during spore germination, the main carbon source for amino acids biosynthesis is derived mostly from the degradation of stored lipids and the glyoxylate cycle. In contrast, HPLC analysis of PITC-amino acids derivatives has revealed that at a high C:N ratio (available exogenous glucose) Arg is the main amino acid produced and incorporated into the proteins of germinating AM spores[33]. This is consistent with the report of Tisserant et al [35] showing that the transcripts coding for the enzymes of Arg biosynthesis are highly expressed in germinating spores of AM fungus *G. intraradices*.

To summarize, chromatographic separation and analysis of $^{15}N/^{13}C$-labeled amino acids have determined in what form nitrogen is taken up and assimilated, and clarified the mechanisms of Arg transport and degradation in AM fungal symbiosis. In the following sections, we will discuss in some detail the application of chromatographic methods in the studies of nitrogen metabolism and transport in AM fungal system.

2. Cleanup and separation of amino acids using ion-exchange chromatography

Sample preparation is often required to improve the analysis, eliminate interference and increase sensitivity.[1] This is necessary when a sample cannot be directly analyzed or when the analysis generates poor results. Most cleanup and concentration techniques are based on separations. Separation techniques have a number of characteristics such as fractionation capacity, load capacity, adaptability to analyte volatility, type of selectivity, speed and convenience, static and dynamic procedures. The type of selectivity depends on physicochemical properties of the components in the sample. For example, differences in boiling point allow separation by distillation and are important in gas chromatographic separation, and acid/base dissociation constant is important in ion-exchange chromatographic analysis. Chromatographic separations are used both for sample preparation and core analytical separation and measurements. To check nitrogen utilization and amino acids metabolism in AM fungal symbiotic system, free amino acids from cultured tissues can be extracted with a mixture of methanol/chloroform/water, then separated on a cation exchange column (DOWEX 50 *4-200, hydrogen form) and recovered after freeze-drying [28-30].

In the method published by Jin at al., after 3-week culture of mycorrhizal roots,[28] the extraradical mycelium (ERM) and mycorrhizal root tissues were recovered on a 38-μm sieve, rinsed with deionized water and lyophilized. The lyophilized mycorrhizal roots and ERM were ground in a mortar with a pinch of acid-washed sand and extracted with a mixture of methanol/ chloroform/water (12 : 5 : 3, v/v/v), which recovered 30–35% more amino acids than extraction with NH_4HCO_3 buffer (pH 8 with 0.2% NaN_3) or 80% ethanol.[13]

Methylene chloride and water were added to the extraction solution to facilitate the separation of chloroform and the methanol–water phases. The methanol–water phase containing the amino acids was collected and evaporated in a rotary evaporator at 50°C Bengtsson & Odham[3] have pointed out that losses of amino acids during evaporation prior to derivatization are negligible, and, using a radioactive amino acid tracer, demonstrated low losses of amino acids co-precipitated with carbonates and hydroxides. Losses from nutrient rich samples were further reduced by acidifying the sample before evaporation. During evaporation, Maillard reaction can be avoided by keeping the temperature at 50°C and reducing evaporation pressure. A direct cation exchange has been shown to be inadequate in obtaining a sufficiently pure solution for derivative formation. However, it has been demonstrated that extraction of the aqueous sample with chloroform prior to ion exchange efficiently removes interfering organic substances without detectable losses of amino acids.[12]

Most amino acids are not soluble in nonpolar solvents and are soluble in water;[36] they display amphoteric properties (caused by COOH and NH_2 groups), and many exist as zwitterions in the form R-CH(NH_3^+)-COO^-. In acidic solutions, the amino groups are at least partly protonated whereas the ionization of the carboxyl group is very low. For the cation exchange, strong acidification is therefore necessary to convert the monoamino-acids completely to the univalent cation form. For example, at pH 2.5, 35% of Phe, 66% of Thr, and 100% of the diamino-acids are in the cationic form. In the micro-method procedure published by Bengtsson[3], the residues containing the amino acids were dissolved in 2 ml of 0.01 M HCl and loaded onto a cation exchange column, previously washed with 2 M NH_4OH, deionized H_2O and 2 M HCl, and followed by a wash with deionized H_2O until the effluent was neutral. The neutral compounds, principally carbohydrates, were washed off the column with 5 ml of water, and the free amino acids (except cysteine(Cys) and methionine(Met), whose recoveries were low), were eluted with 2 ml of 1 M NH_4OH. Sulfur-containing amino acids are partly oxidized during the ion-exchange procedure or derivatization, therefore, this method is not suitable for recovery and purification of Cys and cystine. Nevertheless, Myung et al.[37] have developed a method employing SPME (solid-phase micro-extraction) technique and GC–MS to determine homocysteine (Hcy), Cys and Met levels in aqueous samples. This method provides a new approach to the studies of S uptake and transfer in AM symbiosis.

3. Determination of amino acid concentrations with high performance liquid chromatography (HPLC)

Since amino acids are non-volatile compounds and most of them show low UV absorbance, they have been commonly analyzed by liquid chromatography (LC) methods with pre-column or post-column derivatization using UV chromophore or fluorophore reagents. The use of HPLC analysis is extremely common because this technique has no specific analyte volatility or thermal stability restrictions.[38-39] Derivatization can make the analysis more sensitive, gives a linear detection response and avoids specific interference. The common approach to the preparation of derivatized samples for HPLC and GC analysis is to replace the active hydrogens to form a desired physical property. However, in HPLC, the elimination of all active hydrogens from the analyte is not usually necessary. Some derivatizations requiring only the attachment of a chromophore or fluorophore group to the analyte use one of the functional groups (such as phenyl isothiocyanate, O-phthalaldehyde or 9-fluorenylmethyl chloroformate) which react only at the amino group. Other derivatizations of amino acids involve both NH_2 and COOH groups; for example, when isothiocyanates are used to form thiohydantoins. The quantification of derivatized amino acids, such as phenylthiocarbamoyl (PTC) derivatives, is commonly used prior to their analysis by HPLC. Phenyl isothiocyanate (PITC) reacts both with the primary and secondary amino-groups, at room temperature, within 5-20 min; PITC-amino acids are very stable in dried samples, and their elution and detection only requires a binary gradient pump and a UV detector.[39] The disadvantages of the method are its limited sensitivity (because of the lack of fluorogenic derivatives) and the need for removal of excess reagent. Therefore, in our experiments to determine the levels of free amino acids in the ERM and mycorrhizal root tissues, we used a Waters Pico-Tag amino acid analyzer (HPLC), employing the Pico-Tag method. As in the study by Endres & Mercier[40], the amount of each amino acid was measured by high-performance liquid chromatography (HPLC) of the PITC derivatives. The extracted amino acids were dissolved in 0.1 M HCl and vacuum-dried in a Pico-Tag workstation, then ethanol/ water/triethylamine mixture was added and evaporated by vacuum-drying. 20 ml of ethanol/ water/triethylamine/ phenylisothiocyanate (7 : 1 : 1 : 1) was added to derivatize the amino acids at 23°C for 20 min. The samples were then dried under vacuum and re-dissolved in 100 μl of Pico-Tag sample diluent. 20 μl of each sample was loaded onto a reverse-phase C18 column (3.9 mm ID X 150 mm long) using a Waters 510 autosampler. An eluent gradient consisting of 38 ml Pico-Tag Eluent A (0.05 M sodium acetate) and Eluent B (0.1 M sodium acetate/acetonitrile/methanol (46:44:10) was used as mobile phase. The flow rate was 1.0 ml min^{-1}, with the proportion of Eluent B rising from 0–100%. The elution was monitored at 254 nm with a Waters 486 tunable absorbance detector. The concentrations of the amino acids were calculated by comparing the integrated peak area with those for standard amino acids at known concentrations using Waters MILLENIUM software (Waters Chromatography Division). The threshold for detection of amino acids in standard solutions was 30 pM of each amino acid per assay, corresponding to <10 nmol g^{-1} of dry weight of tissue.

As shown in Table1, our HPLC analysis of free amino acid levels reveals that Arg is by far the most abundant fungal amino acid (between 50 and 200mM depending on developmental stage), representing *c.* 90% of the total free amino acids in the ERM. Arg levels are also substantially higher in colonized than in un-colonized roots (54.2 ± 19.3% versus 10.9± 4.8% of free amino acids). Johansen *et al.*[12] have observed, without reporting absolute levels, that Arg is the dominant free amino acid in extraradical mycelium of *Glomus claroideum*. However, they have not measured Arg levels in *G. intraradices* because of the problems with derivatization and decomposition of the silated product.

Free amino acids	Concentration in mycorrhizal root compartment tissue (nmol mg^{-1} d.wt)		Concentration in extraradical mycelium (nmol mg^{-1} d.wt)	
	1wks	3wks	1wks	3wks
Aspartate	0.40±0.01	6.87±0.56	0.52±0.13	12.26±1.87
Glutamate	3.09±1.91	7.35±0.74	6.44±2.81	22.20±2.34
Asparagine	21.01±12.52	4.80±0.23	25.58±1.24	17.12±1.58
Glutamine	14.05±5.49	8.3±0.38	11.62±3.86	6.66±0.87
Serine	1.54±0.83	3.12±0.21	17.51±6.10	10.95±1.32
Glycine	/	/	12.93±3.35	4.37±0.58
Arginine	3.64±2.50	9.85±0.59	167.22±32.95	227.93±5.8
Threonine	0.58±0.31	1.84±0.11	/	/
Alanine	1.43±0.90	1.2±0.08	8.39±1.89	2.81±0.34
Proline	0.22±0.12	0.80±0.23	1.57±0.33	1.53±0.23
Tyrosine	0.38±0.21	0.74±0.12	1.67±0.53	2.27±0.67
Valine	0.91±0.50	1.44±1.43	1.73±0.37	2.2±0.11
Methionine	0.76±0.70	/	9.485±1.31	/
Cysteine	/	/	/	/
Isoleucine	0.647±0.39	0.64±0.07	1.14±0.311	0.69±0.09
Leucine	0.587±0.33	0.94±0.03	1.537±0.31	0.77±0.06
Phenylalanine	0.262±0.16	0.37±0.12	1.027±0.33	0.50±0.11
Lysine	0.19±0.10	0.92±0.06	1.63±1.17	3.03±0.56
Ornithine	3.94±1.22	3.67±1.66	40.43±34.46	16.01±3.63

* Mean ± standard deviation.

Table 1. Concentration of free amino acids, established using HPLC, in the root compartment tissue and extraradical mycelium of AM fungus *G. intraradices* after culturing for 1 or 3 weeks in two-compartment Petri dishes, with$^{15}NH_4Cl$ labeling in fungal compartment .*

4. Identification of derivatized amino acids and their isotopomer analysis with gas chromatography-mass spectrometry

Derivatization involves reactions with one or more reagents to change the chemical nature of the analyte to make it more suitable for analysis. As chemical reactions,[41] derivatizations are efficient chemical processes between the analyte and the reagent, such as reactions forming acyl, alkyl or aryl derivatives, silylation reactions, adding to carbon-hetero multiple bonds, formation of cyclic compounds, etc. These reactions result in a replacement of active hydrogens in an analyte in functional groups such as OH, COOH, SH, NH, CONH.

The purpose of derivatization varies depending on the analyte, the matrix of the sample, and the analytical method to be applied.[41] Some derivatizations are used in the sample cleanup or concentration process. Much more frequently, they are done to change the analyte properties for the chromatographic separation, to achieve better thermal stability, better detectability and improve separation in GC analysis. In GC-MS analytical technique, derivatization may help in spectra identification. For HPLC using liquid mobile phase, derivatization is performed mainly to increase detectability and improve the separation.

The trimethylsilyl (TMS) derivatives are obtained in one-step derivatization procedure, whereas almost all other derivatives are formed in two or more reaction steps. The derivatization with the formation of silylated derivatives is applied to replace the active hydrogens in an analyte in groups such as OH, SH, NH, CONH, POH, and SOH. The purpose of silylation is to reduce the polarity of the analyte, increase its stability and improve detectability. Although the TMS derivatives are by far the most commonly used for analytic purposes, TBDMS is used when compounds more resistant to hydrolysis are required. The thermal stability of TBDMS derivatives is better than that of TMS derivatives. The TBDMS derivatives give reproducible results in amino acid analysis. Gehrke[36] pointed out that the best foundation for a successful amino acid analysis by GC is (a) reproducible and quantitative conversion of amino acids to suitable derivatives; and (b) separation and quantitative elution of the derivatives from the chromatographic column. For satisfactory analysis of amino acids by GC, a complete derivatization is essential. In my own experiments, free amino acids samples were derivatized with MTBSTFA containing 1% N-methyl-N-t-butyldimethylchlorosilane; such derivatized Arg and ornithine (Orn) are shown in Fig.2. [30]

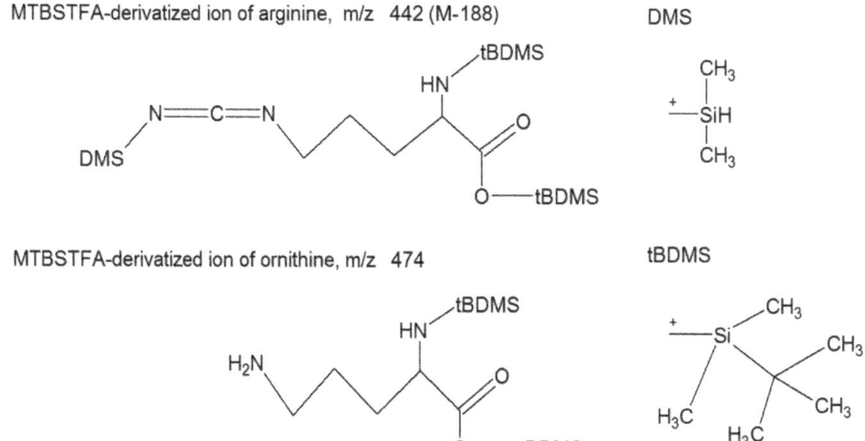

Figure 2. Molecular structure of ions of N-methyl-N-(t-butyldimethylsilyl)trifluoroacetamide (MTBSTFA)-derivatized Arg and Orn. This figure gives the structure of the ion remaining after a 70 eV impact on the MTBSTFA derivative of Arg and Orn. It demonstrates that derivatized Arg has an *m/z* of 442, which is the M-188 (molecular ion minus 188) fragment arising from losing the guanido Ns and a t-butyl group. This fragmentation is different from that of Orn, which only loses a t-butyl group (M-57) in its fully derivatized form.[30]

MTBSTFA has been reported as a very powerful tBDMS silyl donor capable of tert-butyldimethylsilylating active protic functions (hydroxyl, amino, carboxylic and thiol

amino acid was determined by measuring the M-57 ions which result from the loss of a t-butyl group (i.e., [M - $C(CH_3)_3$]) from the molecules of MTBSTFA derivatives, except for Arg whose m-188 ion was used. The Arg ion examined had a *m/z* of 442, which corresponds to M-188 (molecular ion minus 188) fragment arising from the loss of one guanido nitrogen together with a tBDMS and DMS group from the tetra-substituted *tert*-butyldimethylsilyl(tBDMS)-derivatized Arg (Fig. 2). When using [guanido-2-^{15}N] Arg, we observed an ion at an *m/z* of 443 (M-189, molecular ion minus 189). This isotopomer corresponds to the derivatized [guanido-2-^{15}N]Arg because the ion loses one of the guanido nitrogens by fragmentation at 70 eV.

5. Analysis of ^{15}N-labeled amino acid isotopomers with gas chromatography-mass spectrometry

Unlabeled, derivatized Arg yields an ion at an *m/z* of 442, which corresponds to a molecular fragment containing three ^{14}N atoms. Thus, the maximum number of ^{15}N atoms detected is three, resulting in a mass isomer distributions of M, M+1, M+2 and M+3. These were used to calculate the isotopic enrichment in each amino acid after correction for natural isotopic contents by comparison with the mass isomer distributions measured for unlabeled standards. As shown in Fig.1, to test whether Arg is translocated from the ERM to the IRM,

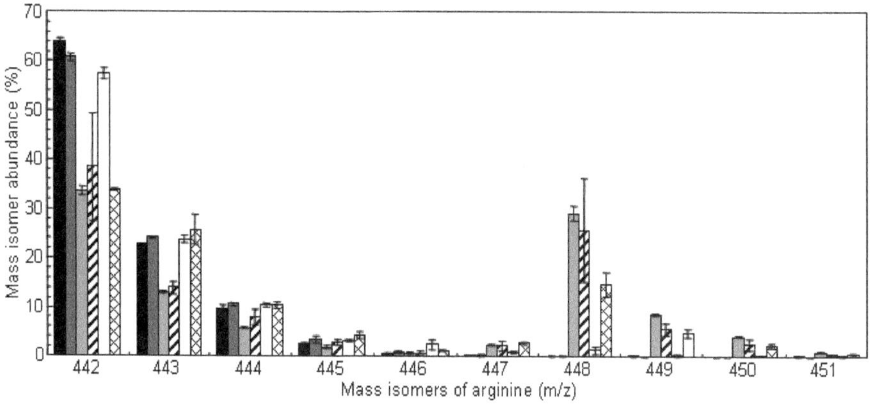

Figure 4. Labeled arginine after addition of 2 mM $^{13}C_{U6}$ arginine to the ERM compartment for 6 weeks. Mass isomer distributions were measured by mass spectrometry after extraction of free amino acids or hydrolysis of extracted soluble protein followed by derivatization (see methods): black bars, unlabeled arginine standard showing the natural abundance mass isomer distribution; dark grey bars, arginine extracted from unlabeled mycorrhizal root tissue; medium grey bars, arginine extracted from ERM after labeling; hatched bars, arginine extracted from mycorrhizal roots after labeling; white bars, arginine from soluble protein of mycorrhizal roots after labeling; checkered bars, arginine from soluble protein of un-colonized roots after exposure to $^{13}C_{U6}$ arginine (positive control, showing that if arginine is available to the root tissue, it is detectable in root protein).

$^{13}C_{U6}$ arginine was added to the ERM. After 6 weeks, MS analysis of MTBSTFA-derivatized Arg isotopomer revealed that 34% of the free Arg in the ERM and 33% of the free Arg in the colonized roots showed $^{13}C_{U6}$ labeling (M+6, Fig. 4). The mass spectra showed that the free Arg molecules in the colonized roots are either completely unlabeled (natural abundance mass isomer distribution) or labeled in all six carbon positions, thus indicating that Arg is transported intact from ERM to IRM.

6. Conclusion

Chromatographic and mass spectrometry analysis of amino acids, in combination with isotopic tracing, shows that various forms of N sources (NO_3^-, NH_4^+, amino acids) are taken up, assimilated, and incorporated into Arg by the ERM. The accumulated Arg as well as polyP is then bidirectionally translocated along the coenocytic fungal hyphae from the ERM to the IRM or from the mycorrhizal compartment tissue to the ERM. Arg is catabolized through catabolic arm of the urea cycle (utilizing arginase and urease activities) in the IRM, releasing NH_3/NH_4^+ in the arbuscules. The NH_3/NH_4^+ acquired by the plant is either transported into adjacent cells or immediately incorporated into AAs, as shown in Fig.1 (modified from Govindarajulul et al.[29]).

However, although Ala, Gly, Val, Leu, Ileu, Pro, Ser, Thr, Phe, Asn, Asp, Glu, Orn, Gln, Arg and tyrosine(Tyr) are detected, GC-MS of samples from AM fungal tissues performed after MTBSTFA-derivatized amino acids cleaned up on a cation exchange column (DOWEX 50 *4-200, hydrogen form) does not detect some of S-containing amino acids and basic amino acids lysine (Lys) and His. Some amino acids cannot be quantified with chromatographic MS due to non-linear response. Nevertheless, analysis of PITC-derivatized amino acids with HPLC shows excellent linear relationship between the molar concentrations of amino acids and peak areas in the chromatogram, and thereby can be used for effective quantification. Although many techniques are already in use in this field, we will need some novel methods, yet to be developed, to achieve a simultaneous measurement and identification of all free amino acids in biological tissues.

Author details

Jin HaiRu and Jiang Xiangyan
College of Chemistry and Life Sciences, Zhejiang Normal University, Jinhua, China

Abbreviations

GC: gas chromatography; GLC: gas liquid chromatography; MS: mass spectrometry; HPLC: high performance liquid chromatography; MTBSTFA: N-methyl-N-(t-butyldimethylsilyl) trifluoroacetamide; tBDMS: *tert*-butyldimethylsilyl; TMS: trimethylsilyl; PTC: phenylthiocarbamoyl; PITC: phenyl isothiocyanate; SPME: solid-phase microextraction.

Ala: alanine; Gly : glycine; Val: valine; Leu: leucine; Ile: isoleucine; Pro: proline; Ser: serine; Thr: threonine; Phe: phenylalanine;Asp: aspartate; Glu: glutamate;Orn: ornithine; Asn:

asparagine;Gln: glutamine; Arg: arginine;Tyr: tyrosine; Lys: lysine; Cys: cysteine; Met: methionine, His: histidine; Hcy: homocysteine.

AM: arbuscular mycorrhizae; ERM: extraradical mycelium; IRM: intraradical mycelium; GS/GOGAT: glutamine synthetase/ glutamine 2-oxoglutarate amidotransferase.

7. References

[1] S.C. Moldoveanu, V. David. Overview on sample preparation for chromatography. In Sample preparation in Chromatography. Journal of Chromatography—volume 65. First edition, Elsevier, (2002) ,pp137-163.

[2] J. Sabine Becker. Inorganic Mass Spectrometry - Principles and Applications.Wiley, (2007),pp1-6.

[3] G. Bengtsson, G. Odham. A micromethod for the analysis of free amino acids by gas chromatograph and its application to biological systems. Anal. Biochem. 92 (1979) 426–443.

[4] T.P. Mawhinney, R.S. Robin Robinett, A. Atalay, M.A. Madson. Analysis of amino acids as their tert-butyldimethylsilyl derivatives by gas-liquid chromatography and mass spectrometry. J. Chromatogr. 358 (1986) 231–242.

[5] H. J. Chaves Das Neves, A. M. P. Vasconcelos. Capillary gas chromatography of amino acids, including asparagine and glutamine: Sensitive gas chromatographic-mass spectrometric and selected ion monitoring gas chromatographic-mass spectrometric detection of the N,O(S)-tert.-butyldimethylsilyl derivatives. J. Chromatogr. 392 (1987)249-258.

[6] R.J. Early., J.R. Thompson., G.W. Sedgwick, J.M. Kelly., R J. Christopherson. Capillary gas chromatographic analysis of amino acids in blood and protein hydrolysates as tert.-butyldimethylsilyl derivatives. J. Chromatogr. 416 (1987) 15-23

[7] C.J. Biermann, C.M. Kinoshita, J.A. Marlett, R.D. Steele. Analysis of amino acids as tert.-butyldimethylsilyl derivatives by gas chromatography. J. Chromatogr. 351 (1986) 330-334.

[8] C.J. Goh, K.G. Craven., J.R. Lepock, E.B. Dumbroff. Analysis of all protein amino acids as their tert.-butyldimethylsilyl derivatives by gas-liquid chromatography. Anal. Biochem. 163 (1987) 175-181.

[9] M. Dauner, U. Sauer. GC-MS analysis of amino acids rapidly provides rich information for isotopomer balancing. Biotech. Prog. 16 (2000) 642-649.

[10] J. Persson, T. A. Näsholm. A GC-MS method for determination of amino acid uptake by plants. Physiol. Planta. 113 (2001) 352-358.

[11] K.R. Kim, J.H. Kim, E. Cheong, C. Jeong. Gas chromatographic amino acid profiling of wine samples for pattern recognition. J. Chromatogr. A.722 (1996) 303-309.

[12] A, Johansen, R.D. Finlay, P.A. Olsson. Nitrogen metabolism of the external hyphae of the arbuscular mycorrhizal fungus *Glomus intraradices.* New Phytol. 133 (1996) 705–712.

[13] D. Rolin, P.E. Pfeffer, D.D. Douds, H.M. Farrell, Y. Shachar-hill. Arbuscular mycorrhizal symbiosis and phosphorus nutrition: effects on amino acidfs production and turnover in leek. Symbiosis.30 (2001) 1-14.
[14] S.E. Smith, D.J. Read. Mycorrhizal Symbiosis. Third edition. Academic Press,(2008), pp13-90.
[15] X.M. Cheng, K. Baumgartner. Arbuscular mycorrhizal fungi-mediated nitrogen transfer from vineyard cover crops to grapevines. Biol. Fert Soils. 40 (2004) 406–412.
[16] S.M. Meding, R.J. Zasoski. Hyphal-mediated transfer of nitrate, arsenic, cesium, rubidium, and strontium between arbuscular mycorrhizal forbs and grasses from a California oak woodland. Soil Biol. Biochem. 40(1) (2008) 126–134.
[17] J.W. McFarland, R.W. Ruess, K. Kielland. Cross-ecosystem comparisons of in situ plant uptake of amino acid-N and NH_4^+. Ecosystems. 13 (2010) 177–193.
[18] R. Tobar, R. Azcón, J.M. Barea. Improved nitrogen uptake and transport from ^{15}N labelled nitrate by external hyphae of arbuscular mycorrhiza under water-stressed conditions. New Phytol. 126 (1994) 119–122.
[19] B. Bago, H. Vierheilig, Y. Piché, C. Azcón-Aguilar. Nitrate depletion and pH changes induced by the extraradical mycelium of the arbuscular mycorrhizal fungus *Glomus intraradices* grown in monoxenic culture. New Phytol. 133 (1996) 273–280.
[20] R. Azcón, R. Rodriguez, E. Amora-Lazcano, E. Ambrosano. Uptake and metabolism of nitrate in mycorrhizal plants as affected by water availability and N concentration in soil. Eur J. Soil Sci. 59 (2008) 131–138.
[21] H.J. Hawkins, A. Johansen, E. George. Uptake and transport of organic and inorganic nitrogen by arbuscular mycorrhizal fungi. Plant Soil. 226 (2000) 275–285.
[22] M. Breuninger, C.G. Trujillo, E. Serrano. Different nitrogen sources modulate activity but not expression of glutamine synthetase in arbuscular mycorrhizal fungi. Fungal Genet. Biol. 41 (2004) 542–552
[23] J.P. Toussaint, M. St-Arnaud, C. Charest. Nitrogen transfer and assimilation between the arbuscular mycorrhizal fungus *Glomus intraradices* Schenck & Smith and Ri T-DNA roots of *Daucus carota* L. in an in vitro compartmented system. Can J Microbiol. 50 (2004) 251–260.
[24] R. Azcón, J.M. Ruiz-Lozano, R. Rodriguez. Differential contribution of arbuscular mycorrhizal fungi to plant nitrate uptake of ^{15}N under increasing N supply to the soil. Can J. Bot. 79 (2001) 1175–1180.
[25] Y. Shachar-Hill, D.B. Rolin, P.E. Pfeffer. Uptake and transfer of N to the host by an arbuscular mycorrhizal (AM) fungus. Plant Physiol. 114 (Suppl) (1997) 39.
[26] M.M. Vazquez, J.M. Barea, R. Azcon. Impact of soil nitrogen concentration on *Glomus* spp.- Sinorhizobium interactions as affecting growth, nitrate reductase activity and protein content of *Medicago sativa*. Biol. Fert. Soils. 34 (2001) 57–63.
[27] A. Hodge, A.H. Fitter. Substantial nitrogen acquisition by arbuscular mycorrhizal fungi from organic material has implications for N cycling. PNAS, 107(31) (2010) 13754–13759.
[28] H.R. Jin, P.E. Pfeffer, Y. Shachar-Hill, D.D. Douds, E. Piotrowski, PJ. Lammers. The uptake,metabolism,transport and transfer of nitrogen in an arbuscular mycorrhizal symbiosis. New Phytol. 168 (2005) 687-696.

[29] M. Govindarajulu, P.E .Pfeffer, H.R. Jin, J. Abubaker, D.D .Douds, J.W.B. Allen, H. Bücking, P.J. Lammers, Y. Shachar-Hill. Nitrogen transfer in the arbuscular mycorrhizal symbiosis. Nature 435 (2005) 819–823.

[30] H.R. Jin. Arginine bi-directional translocation and breakdown into ornithine along the arbuscular mycorrhizal mycelium. Sci in China Series C: Life Sci. 52(4) (2009) 381-389.

[31] B. Bago, Y. Shachar-Hill, P.E. Pfeffer. Could the urea cycle be translocating nitrogen in the arbuscular mycorrhizal symbiosis? New Phytol. 149 (2001) 4–8.

[32] M. Guether, B. Neuhäuser, R. Balestrini. A mycorrhizal-specific ammonium transporter from *lotus japonicus* acquires nitrogen released by arbuscular mycorrhizal fungi. Plant Physiol. 150 (2009) 73–83.

[33] H.R. Jin, D.H. Jiang, P.H. Zhang. Effect of carbon and nitrogen availability on the metabolism of amino acids in the germinating spores of arbuscular mycorrhizal fungi. Pedosphere 21（4）(2011) 432-442.

[34] P.J. Lammers, J. Jun, J. Abubaker. The glyoxylate cycle in an arbuscular mycorrhizal fungus. Carbon flux and gene expression. Plant Physiol. 127 (2001) 1287–1298.

[35] E. Tisserant1, A. Kohler1, P. Dozolme-Seddas. The transcriptome of the arbuscular mycorrhizal fungus *Glomus intraradices* (DAOM 197198) reveals functional tradeoffs in an obligate symbiont. New Phytol. 193 (2012) 755–769.

[36] C.W. Gehrke. Quantitation of Amino Acids by Gas-Liquid Chromatography. In: Quantitation of Amino Acids and Amines by Chromatograph methods and protocols. Journal of Chromatography, vol. 70. Edited by Ibolya Molnár-Perl. First edition, Elsevier, (2005),pp39-97.

[37] S.W. Myung, M. Kim, H.K. Min, E.A. Yoo , K.R. Kim. Determination of homocysteine and its related compounds by solid-phase microextraction-gas chromatography-mass spectrometry. J. Chromatogr. B. 727 (1999) 1–8.

[38] I. Molnár-Perl. Quantitation of amino acids and amines in the same matrix by high-performance liquid chromatography, either simultaneously or separately. J. Chromatogr. A. 987 (2003) 291-309.

[39] I. Molár-perl. HPLC of Amino Acids as Phenylthiocarbamoyl Derivatives. In:Quantitation of Amino Acids and Amines by Chromatograph methods and protocols. Journal of Chromatography, vol. 70. Edited by Ibolya Molnár-Perl. First edition, Elsevier, (2005), pp137-162.

[40] H. Endres, A. Mercier. A. Amino acid uptake and profile in bromeliads with different habits cultivated *in vitro*. Plant Physiol. Biochem. 41 (2003) 181–187.

[41] S.C. Moldoveanu, V. David. Purpose of derivatization in chromatographic analysis. In Sample preparation in Chromatography. Journal of Chromatography—volume 65. First edition, Elsevier, (2002) ,pp473-524.

[42] T.P. Mawhinney, M.A Madson. N-Methyl-N-(tert-butyldimethylsilyl) trifluoroacetamide and related N-tert- butyl dimethyl -silyl amides as protective silyl donors. J. Org. Chem., 47 (1982) 3336-3339.

[43] S.L. MacKenzie, D. Tenaschuk. Gas—liquid chromatographic assay for asparagine and glutamine. J. *Chromatogr.* 322 (1985) 228-235.

Chapter 11

Chromatography as the Major Tool in the Identification and the Structure-Function Relationship Study of Amylolytic Enzymes from Saccharomycopsis Fibuligera R64

Wangsa T. Ismaya, Khomaini Hasan, Toto Subroto,
Dessy Natalia and Soetijoso Soemitro

Additional information is available at the end of the chapter

1. Introduction

Chromatography, at both preparative and analytical levels, has been a key experimental technique in the study of proteins, primarily liquid chromatography in the separation, purification, and analysis. Recent developments in the study of proteins lean towards simplification and miniaturization, thus chromatography becomes less involved and explored. For example, development of protein tags and their associated affinity matrices enables purification of a protein in one step. This chapter describes the identification and structure-function relation study of amylolytic enzymes from *Saccharomycopsis fibuligera* R64, where liquid chromatography was used as the major tool, not only in the purification but also biochemical analysis of the enzymes.

2. *S. fibuligera* R64 secretes amylolytic enzyme

S. fibuligera is a food-borne yeast that is widely used in the production of rice- or cassava-based fermented food, i.e. *Tape* in Indonesia and other Southeast Asia countries (1). The yeast, in combination with *Saccharomyces cerevisiae* or *Zymomonas mobilis*, has been employed in the production of ethanol using cassava starch as the starting material (2), where the starch is firstly degraded into simple sugars prior to (bio-) ethanol. Bioethanol has been promoted as a renewable energy replacing fossil fuels and at the moment is already used as an additive. As the demand for renewable energy grows, *S. fibuligera* emerges as an attractive workhorse for the bio-ethanol production.

Degradation of starch into sugars is performed by amylolytic enzymes, such as α-amylase, glucoamylase, β-amylase, isoamylase, pullulanase, exo(1-4)-α-D-glucanase, α-D-glycosidase, and cyclomaltodextrin-D-glucotransferase (Fig. 1) (3). *S. fibuligera* secretes amylolytic enzymes, almost exclusively α-amylase and glucoamylase. α-Amylase acts as an endo-enzyme, cleaving 1,4-α-glycosidic bond at random positions to result in liquefaction of starch. Glucoamylase, on the other hands, is an exo-enzyme that cleaves 1,4-α-glycosidic bond only at the non-reducing end to result in saccharification (4). Thereby upon the combined action of α-amylase and glucoamylase, starch is degraded into maltose, maltotriose, or dextrin, and subsequently hydrolyzed to glucose.

The use of amylolytic enzymes for the ethanol production in the course of renewable energy requires an ability to act on raw starch, allowing the use of biomass as the starting material. The ability of *S. fibuligera* α- and glucoamylase to degrade raw starch has been reported (4, 5). Interestingly, only 10% of amylolytic enzyme-secreting organisms are capable of producing raw starch degrading α-amylase (6, 7). Since starch degradation begins with α-amylase action to produce simpler sugars, raw-starch acting α-amylase is highly desired. This situation strengthens the position of *S. fibuligera* for bioethanol production, being a raw starch degrading enzyme producer.

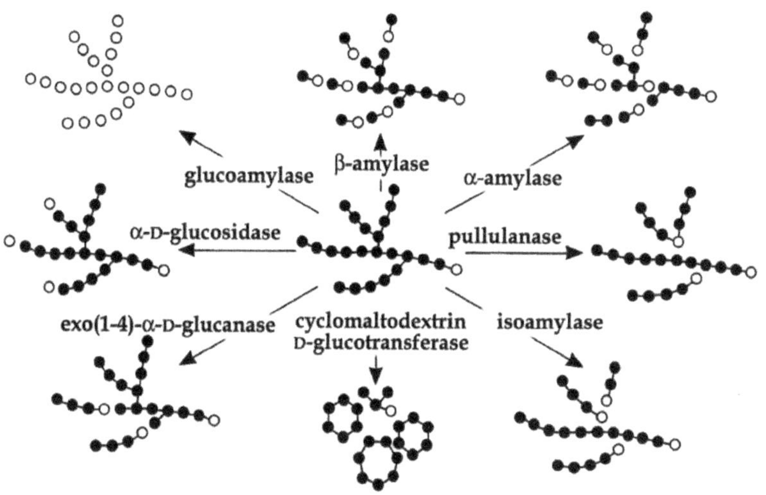

Figure 1. Starch degradation by amylolytic enzymes (3). The open and black coloured circles represent the reducing and non-reducing sugars, respectively. Note that the cleavage occurs at the reducing sugar.

α-Amylase is commonly used in food, beverage, paper industries (8), in textile industry and as additive in detergents (9, 10), for renewable energy (11, 12) and medical purpose (13). Glucoamylase has been the most important enzyme in food industry, mainly in the production of sugar or ethanol (14). Glucoamylase is normally employed in combination with amylolytic enzymes that are able to act on more complex polysaccharides, such as α-amylase and pullulanase (14). Recently, amylolytic enzymes are employed in the production of lactic acid and ethanol by-product by lactic acid bacteria (LAB) (15), demonstrating that the application of amylolytic enzymes continues to expand.

Interestingly, numbers and characteristics of amylolytic enzymes secreted by *S. fibuligera* vary from one strain to another. For example, strain IFO 0111 secretes only glucoamylase whilst strain KZ secretes α-glucosidase along with α-amylase and glucoamylase (4). Further, glucoamylases from strain IFO 0111 and HUT 7212 demonstrate raw starch degradation whereas that from the strain KZ does not. Glucoamylase from strain KZ, however, demonstrates better thermal stability (4). Fascinating to find these variations in their properties despite of their near identical acid sequences (16). Strain KJ-1, an *S. fibuligera* strain from Indonesia, is also reported to secrete only glucoamylase and its partial amino acid sequence (residues 28-47) is identical to that of glucoamylase from strain KZ (1). This observation indicates differences in expression and secretion of amylolytic enzymes between *S. fibuligera* strains. Nevertheless, the practicality of *S. fibuligera* in producing amylolytic enzymes applauds the proposal to sequence its whole genomic DNA (2).

Of 136 isolates screened from various places in Indonesia, *S. fibuligera* strain R64 was selected for demonstrating the highest amylolytic activities (α-amylase and glucoamylase). Strain R64 is able to degrade raw starch and its amylolytic enzymes demonstrate potent thermal stability. This finding makes the amylolytic enzymes from strain R64 attractive.

Extra-cellular amylolytic enzyme production by strain R64 is relatively simple, using a medium that contains of 1% sago starch and 1% yeast extract. The enzyme was harvested after 4-5 days of cultivation in a one-litre bioreactor, with constant aeration rate 1 vvm, volumetric oxygen transfer coefficient (k_{La}) 1.53 per hour, agitation speed 100 rpm, 30°C, and pH 7.0. Under this condition, enzyme activity observed on starch degradation was 1320 U/ml (Table 1). The production was also easily reproduced at a laboratory scale by means of Erlenmeyer flask (5). Thus, strain R64 offers a simple but convenient production scheme.

3. Isolation of the amylolytic enzyme complex

The amylolytic enzyme complex from strain R64, consisting of α-amylase (AMY) and glucoamylase (GLL1), is secreted into the production medium. Thus, the enzyme complex was harvested by simply cold-centrifugation (~4°C) at 6000 *g* for 30 minutes, to remove the yeast cells. The enzyme complex in the supernatant, which was designated as the crude extract, was subjected to a diafiltration system (Millipore Minitan II, Tangential Flow Filtration system, Merck Pte Ltd, Singapore) over a 10-kDa cut off membrane disc-plate, at a flow rate of 10-20 ml per minute, at room temperature. The enzyme complex was recovered in the retentate. Diafiltration step tremendously reduced the size of the sample, which is advantageous because the subsequent step to capture the amylolytic enzyme complex in the crude extract was precipitation with ammonium sulfate by 0-100%, on ice (~4°C). These sequential procedures demonstrate a straightforward way to isolate extra-cellular proteins that was accomplished within one-day operation, which is important for protein works due to, for example, possible degradation by proteases.

4. Purification of AMY and GLL1 in chromatography columns

Since the final step in the isolation procedure in the pilot experiment was precipitation with high concentration of ammonium sulfate, subsequent purification on a size-exclusion chromatography (SEC) column appears to be the most appropriate approach because it also functions as a desalting procedure. SEC column requires minimum size of sample upon application (recommended less than 3-5% of the column volume), which can easily be overcome by dissolving the protein precipitate from ammonium sulfate precipitation in a small volume. Another option is purification using an anion or cation exchanger chromatography (abbreviated as AEX and CEX, respectively) columns. However, this either type of column requires the removal of salts prior to sample application. This requirement can be countered by either diluting the sample until the conductivity of the sample solution is low or by dialysis against a buffer that contains low salt concentration.

The ammonium sulfate precipitate containing AMY and GLL1 was immediately dissolved in a small portion (3-5 ml) of 25 mM Tris-HCl buffer, pH 8.0. Unfortunately, AMY and GLL1 were not separated in an SEC column as suggested from the enzyme activity assays. In a sodium dodecyl-sulfate polyacrylamide gel electrophoresis (SDS PAGE) analysis, AMY and GLL1 was not separated and appeared as one protein band with a molecular mass of 55±2 kDa. Further, purification in a DEAE-cellulose AEX column has also failed to separate the two enzymes. This finding indicated a similarity in the distribution of charge on the surface of AMY and GLL1. In addition, there was a recommendation to avoid the use of carbohydrate- based column material for the purification of proteins with an affinity towards carbohydrate, because of the possible interaction between the protein and the supporting column matrix (17). Thus, both SEC and AEX columns were obliterated from the separation of AMY from GLL1.

As purification strategies based on either protein size or charge failed to separate AMY from GLL1, exploiting differences of their protein surface hydrophobicity profiles emerged as an alternative. This strategy was tested on a hydrophobic interaction chromatography (HIC) column, where the purification proteins is based on the hydrophobic character of the surface of the proteins (18). In an HIC column, proteins in solution are conditioned with high salt concentration, which enforces interaction between proteins and the hydrophobic resin. Separation of AMY from GLL1 in an HIC column is lucrative because it can be carried out immediately after precipitation with high concentration of ammonium sulfate. Thus, the use of the HIC column is complementary to the developed isolation procedure.

4.1. Separation of AMY and GLL1 in an HIC column

The ammonium sulfate precipitated protein was dissolved in 25 mM Tris-HCl buffer, pH 8.0, containing 25% ammonium sulfate (w/v). This protein solution was applied to a butyl-Toyopearl 650M HIC column (Tosoh Bioscience Corp., Tokyo, Japan), which had previously been equilibrated with the same buffer. The column was then eluted with 25 mM Tris-HCl buffer, pH 8.0, containing a decreasing concentration of ammonium sulfate (25-0% of 5%

decrement, w/v). The separation profile of AMY and GLL1 in the HIC column is presented in Fig. 2. AMY was eluted at the ammonium sulfate concentration of 15% (w/v) whilst GLL1 was at 5% (w/v). This result demonstrated that AMY and GLL1 were successfully separated based on their hydrophobicity. Moreover, the elution profile suggests that the surface character of GLL1 is more hydrophobic than AMY.

Purification of proteins in HIC column is influenced by the pH of the solution where hydrophobic interaction is stronger at lower pH (19). This means, less hydrophobic proteins are bound stronger onto the HIC matrix at lower pH value. This phenomenon may explain different separation profile of AMY and GLL1 observed upon purification in an HIC column using phosphate-citrate buffer system at pH 5.8 (5), where α-amylase activity was detected in the GLL1 fraction. Although this hypothesis is yet to be proven, the separation at neutral to basic pH (above 7.0) is recommended for optimum operation.

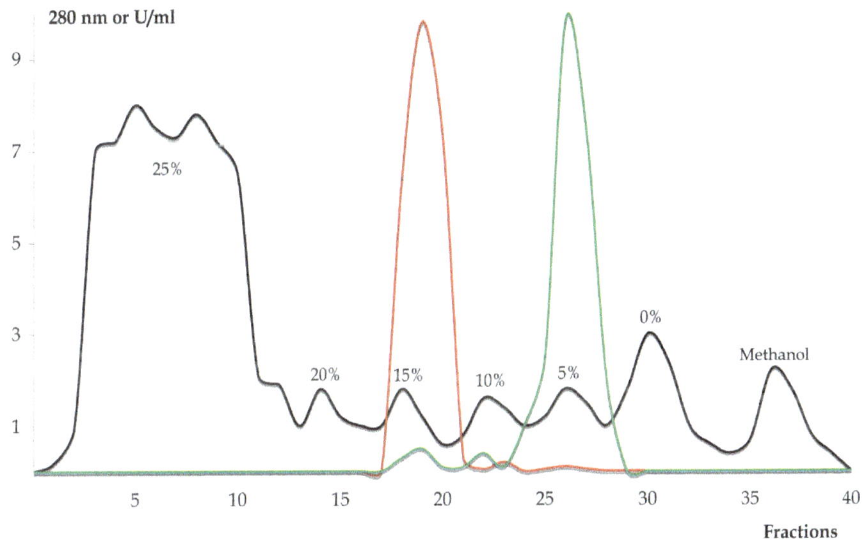

Figure 2. Separation profile of AMY and GLL1 in the butyl-Toyopearl 650M HIC column. The black, red, and green lines represent the protein absorbance at λ 280 nm, α-amylase activity (X 500 U/ml), and glucoamylase activity (X 10 U/ml), respectively. α-Amylase activity was based on starch degradation (20) whilst glucoamylase was on reducing sugar (21).

The use of HIC column to separate *S. fibuligera* α-amylase from glucoamylase was reported previously with an amylolytic enzyme preparation from strain KZ (22). That attempt was performed using a Spheron 300 LC HIC column, following successful separation of hog's pancreas α-amylase from its amylolytic enzyme complex (23). However, unlike butyl-Toyopearl column where proteins are captured by the butyl ligand, hydrophobic interactions in a Spheron column occur between the proteins and the hydrophobic backbone of the matrix, similar to separation of proteins in a reversed-phase high performance liquid chromatography (RP-HPLC) column. In the Spheron 300 LC column, α-amylase (strain KZ) was eluted at an ammonium sulfate concentration of 10% whilst glucoamylase was at 5%. The separation profile, however, is strikingly similar to that of AMY and GLL1 with butyl-Toyopearl column. However, separation of AMY from GLL1 in the butyl-Toyopearl column has better resolution, suggesting variations in the surface properties of *S. fibuligera* α-amylase and glucoamylase from different strains, or the butyl-Toyopearl resin serves for better separation because there is no interaction between the proteins and the supporting matrix. Nevertheless, these results emphasize the power of HIC to separate α-amylase from glucoamylase.

4.2. Subsequent purification of AMY or GLL1 in chromatography columns

Intended for their characterization, AMY and GLL1 were independently collected and subjected to subsequent purification with DEAE-Toyopearl 650M AEX column. An SDS PAGE analysis showed that these purification steps resulted in pure AMY (5) and GLL1 (16). The final purification scheme of AMY is summarized in Table 1. Further, whilst the presence of two and three types of α-amylase and glucoamylase, respectively, were reported upon purification of these amylolytic enzymes from strain KZ by a Mono-Q anion exchanger column (22), AMY and GLL1 appeared to be the only amylolytic enzyme species from strain R64. Nevertheless, additional analysis was performed to confirm that the isolation of purified AMY and GLL1 was unanimous.

	Volume (ml)	Total activity (Units)	Total protein (mg)	Specific activity (U/mg)	Yield (%)
Crude	2000	2640000	16500	160	100
25% AS	2200	2244000	13046	172	85
Butyl Toyopearl	148	1082400	753	1438	41
95% AS	35	765600	213	3595	29
DEAE-Toyopearl	60	580800	100	5808	22

Table 1. The final purification scheme of AMY. AS stands for ammonium sulfate.

4.3. Identification of AMY and GLL1 using RP-HPLC for protein

Nowadays, identification of proteins usually employs techniques with sophisticated and delicate instrumentation e.g. peptide-mass finger print mass spectrometry (PMF-MS) (24). Unfortunately, this technique is rather pricey and furthermore, requires convenient access to the protein database and to the amino acid sequence of the protein. Particularly in the lack of the latter, which is a very common situation in the early stage of protein works, the separation of AMY from GLL1 was confirmed by a more simple and robust technique, exploiting the use of an RP-HPLC system (25). The analysis using RP-HPLC is based on the fragmentation profile of a protein after proteolytic digestion. Proteolytic digestion of AMY was expected to result in fragmentation profile that differs to that of GLL1, as reflected from their chromatographic profiles.

The analysis of AMY and GLL1 by RP-HPLC was carried out essentially following Soedjanaatmadja and co-workers on the identification of pseudo-hevein from the latex of rubber tree (26). Firstly, purified AMY was incubated for four hours at 37°C with chymotrypsin (EC. 3.4.21.1), at an AMY to chymotrypsin mass ratio of 100:1, in 200 mM ammonium bicarbonate buffer, pH 8.0. The reaction was stopped by an addition of 10 mM hydrochloric acid, to lower the pH of the solution. Preparation of GLL1 sample was done in the same way, including the substrate to chymotrypsin mass ratio.

The reaction mixture was immediately applied to an RP-HPLC nucleosil 10 C18 column (30 x 0.45 cm) and the separation was performed for 60 minutes, at a flow rate of 1 ml/min, using 0-70% acetonitrile gradient in 0.1% trifluoroacetic acid (TFA). Elution of fragments was monitored at λ 214 nm, which is specific for detection of peptide bonds. The fragments are eluted according to their hydrophobicity, where more hydrophobic fragments are retained longer in the RP-HPLC column.

Chymotrypsin cleavage takes place on peptide bonds at the C-terminal part of preferably tyrosine, phenylalanine, tryptophan, and leucine residues, and with (much) less extent of methionine, valine, isoleucine, histidine, glycine and alanine residues (27). Due to this broad specificity, the cleavage may result in many fragments varying in size and hydrophobicity, as observed in the chromatographic profiles of proteolytically digested AMY and GLL1 (Fig. 3).

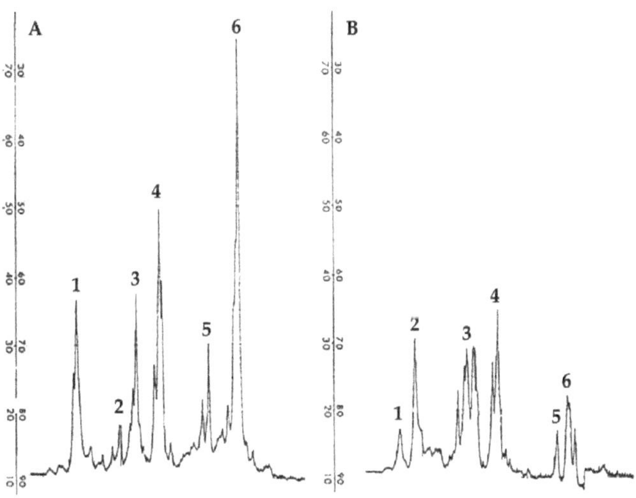

Figure 3. Chromatographic profile of fragments from proteolytic digestion AMY (A) or GLL1 (B) as monitored at λ 214 nm.

The fragmentation profile of AMY was significantly different from that of GLL1. Six sets of resolved peaks were recovered in both cases (Fig. 3) but their retention time, intensities, and peak distribution were different. Intensity of the peaks suggested much less materials were recovered from GLL1 digested samples than from AMY. This situation is likely contributed from highly hydrophobic or negative charged fragments that were not eluted from the RP-HPLC column due to their poor solubility in the solvent used (28). The amino acid sequences of AMY and GLL1 (see 3.5) showed that the charged amino acid distribution in the amino acid sequence of GLL1 (100 charged amino acid residues out of 494, ~20.2%) is higher than that of AMY (86 out of 468, ~18.4%), thus AMY contains more hydrophobic and non-charged amino acids. Unfortunately, correlation between the result from RP-HPLC and hydrophobic amino acid distribution cannot firmly be established because the nature of chymotrypsin digestion was unclear. The results may indeed indicate that hydrophobic amino acids in GLL1 are likely more clustered than in AMY, resulting in highly hydrophobic peptide fragments in GLL1. This hypothesis may be related to the structure of GLL1 (Fig. 8), which consists of one globular molecule that opposes two separable domains of AMY. However, possibility for the existence of highly negative charged peptide fragments from GLL1 can also not be excluded. Nevertheless, of the eluted fragments (Fig. 3), majority of AMY fragments are localized at peaks 5 and 6 whilst GLL1 are at peaks 3 and 4, suggesting different fragmentation had occurred. Thus, the RP-HPLC profiles of AMY and GLL1 indicated successful separation of the two enzymes.

4.4. Analysis of starch hydrolysis products to discriminate AMY and GLL1

Glucoamylase hydrolyzes starch at the non-reducing end of amylose or amylopectin to result in glucose therefore the enzyme activity assay is normally based on the release of reducing sugar (21, 29). Unfortunately, α-amylase random digestion of starch may also result in reducing sugar i.e. maltose (30) therefore detection of glucoamylase activity in an α-amylase preparation can be anticipated. Glucoamylase can also, at lesser extent, act on 1,6-α-glycosidic bond of amylopectin (31), although its efficiency diverse greatly depending on the source organisms. Hydrolysis of the 1,6-α- bond may result in a less integrated starch molecule, hampering the formation of iodine-starch complex, which is the basis of standard α-amylase activity assay (20). Therefore, cross-detection of the two enzyme activity assays is inevitable. This phenomenon was notorious upon separation of AMY from GLL1 at pH 5.8 (5), where α-amylase activity was detected in GLL1 fractions.

One approach to discriminate α-amylase and glucoamylase is *via* the evaluation of their hydrolysis product. α-Amylase action results in various kind of oligomeric sugars (32) whilst glucoamylase hydrolysis product is glucose (monomeric). Based on this principle, successful separation of GLL1 from AMY was assessed through their hydrolysis products, as analyzed with Thin Layer Chromatography (TLC) (16), using purified GLL1 that was obtained from the purification procedure carried out at pH 5.8.

Purified GLL1 was incubated with soluble starch substrate at 37°C and samples were taken after 5, 10, 15 and 45 minutes of incubation. Each sample was then applied onto a silica gel 60 TLC plate (20 cm x 20 cm, Merck, Darmstadt, Germany) using capillary glass tube. The plate was then placed in a TLC separation chamber that had been equilibrated with the mobile phase, which was the mixture of butanol : ethanol : water (at a ratio of 5:5:3, v/v/v). The plate was retrieved from the TLC separation chamber after the mobile phase migration reached three quarters of the length of the plate and then immediately short-submersed (few seconds) in a mixture of water : methanol : sulphuric acid (at a ratio of 45:45:10, v/v/v). The plate was then heated at 120°C for 15 minutes on a hot plate for visualization of the sugars.

The TLC profile (Fig. 4) shows that the product of starch hydrolysis by GLL1 was solely monosaccharide i.e. glucose (G1). The intensity of the G1 spot was increasing from 5 to 45 minutes of incubation, indicating that glucose was produced over time. The spots at the sample application points were from the un-hydrolyzed soluble starch substrate. Apparently, most of the starch molecules were hydrolyzed within 15-45 minutes of incubation. Most importantly, no higher oligomeric sugar forms were detected, even during the first 5-10 minutes of incubation, suggesting the absence of α-amylase. This result provided the undisputed proof that GLL1 had successfully been separated from AMY despite of α-amylase activity was detected in the GLL1 fraction. Thus, the detected α-amylase activity was not originated from AMY.

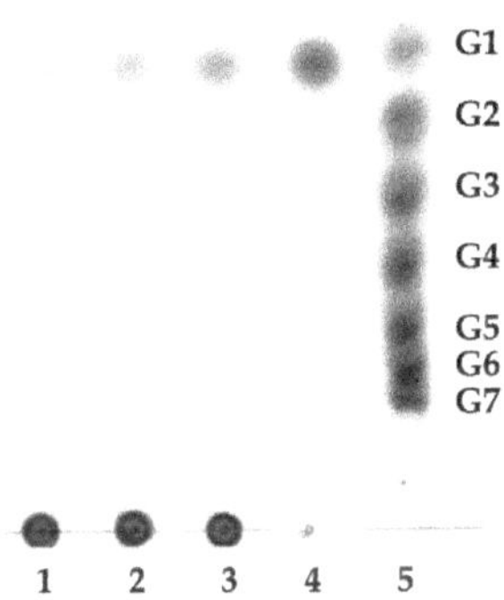

Figure 4. Analysis of starch hydrolysis products of GLL1 on a TLC plate (16). Lane 1-4 is hydrolysis products after 5, 10, 15, and 45 minutes, respectively. Lane 5 is the mono- (G1), di-(G2), and oligo-saccharide (G3-G7) markers.

4.5. Properties of AMY and GLL1

Since the purity of AMY and GLL1 was firmly established, each enzyme could now be characterized. AMY was found active at a pH range of 5.0-7.5 and a temperature range of 30-60°C, with an optimum at 5.5 and 50°C, respectively (5). GLL1 was also found active a pH range of 4.6-6.8 and a temperature range of 30-65°C, with an optimum at 5.6-6.4 and 50°C, respectively (16). These results suggest that the two enzymes are active at similar conditions. Moreover, their characters are similar to most other *S. fibuligera* amylases.

Recently, the amino acid sequences of AMY (GenBank accession code HQ172905) and GLL1 (HQ415729) were successfully elucidated as derived from their chromosomal DNA (16, 33).

The amino acid sequence of AMY is similar to that of published earlier by Itoh and coworkers (34), having mutations at six amino acid residues (Asp>Asn153, Ile>Val159, Ser>Asn190, Ser>Xaa216, Asp>Asn239, and Ser>Thr295). The six deviating residues in AMY comprise an additional predicted glycosylation site (Asn153), which according to a structural modelling, is highly plausible because it resides in a long surface loop (33).

The amino acid sequence of GLL1 is similar to glucoamylases from strain HUT 7212 (GLU) and KZ (GLA) (16). GLU and GLA share high sequence identity (4), with only seven amino acid residues different. However, these seven residues are responsible for differences in their characteristics, where GLA exhibits potential thermal stability and GLU has higher affinity towards substrate. The amino acid sequence of GLL1 also differs to both GLU and GLA at precisely those particular seven residues, which four are being identical to GLU and three to GLA. Interestingly, these mutations result in GLL1 to adopt the potential thermal stability and higher affinity towards substrate. Thus, GLL1 behaves as a hybrid of GLU and GLA.

The calculated theoretical pI values (35) of AMY and GLL1 (based on the amino acid sequences of their mature protein sequences: AMY sequence starts at residue Glu27 of the full-length protein containing signal and pre-peptides as encoded by *AMY* gene whilst GLL1 sequence starts at residue Ala1 as reported in the data base) were 4.4 and 4.3, respectively. The isoelectric point (pI) value of AMY was confirmed experimentally, being 4.6±0.2. This finding supports the early observation that AEX or CEX columns are unable to separate the two enzymes. Similarly, the calculated theoretical molecular masses of AMY and GLL1 were also similar, being 51.7 kDa and 54.6 kDa, respectively. Both values were confirmed experimentally, being 54±2 kDa for AMY and 56±2 kDa for GLL1. Additional ±2 kDa of AMY and GLL1 masses is likely contributed from glycosylation. This result also confirms the inability of SEC column to separate them.

The activity of AMY is diminished in the presence of ethylene diamine tetra acetate (EDTA), a chelating agent. Inactivation of α-amylase by chelating agent is well known, as the enzyme requires calcium ion for its activity and integrity (36). This inactivation by chelating agent was not observed with GLL1. However, the activities of both AMY and GLL1 were increased in the presence of calcium or magnesium ions (data not shown). Further, AMY demonstrated lower activity in phosphate-citrate buffer. This phenomenon may arise from the citrate, which is reported to interact with calcium ion (37). The activity of AMY decreased concomitantly with the increase of the citrate buffer concentration but was fully recovered upon back-titration with calcium chloride (data not shown). Based on this observation, citrate buffers (1, 5) should only be used with considerable reservation and the use of tris buffer is recommended. This recommendation is in line with the necessity to perform purification at basic pH for HIC. In addition, this finding may also explain higher glucoamylase activity detected in AMY fractions when the purification was carried out using phosphate citrate buffer, pH 5.8 (5). Since the buffer does not influence the glucoamylase activity as it does to AMY, the disparity between the two enzyme activities was much less pronounced in comparison to the one presented in Fig. 2.

The hydrophobicity profile of AMY and GLL1 was analyzed based on the amino acid sequence of the mature enzymes (38), using an online analysis service at ExPASy (http://web.expasy.org/protscale/) (35). Another handful online service is also available at http://www.vivo.colostate.edu/molkit/hydropathy/index.html. The window size values for the frame normalization were scanned from 3 to 21 and compared to determine the significance of the regions that represent the hydrophobic character. The profile at window frame 9 (recommended for hydrophilic protein) and 19 (for hydrophobic protein) are presented in Fig. 5. The prediction points were linearly weighted with 100% relativity at the window edges. The score for hydrophobicity is ranged from -4.5 (hydrophilic) to 4.5 (hydrophobic), where the curve above the midpoint (zero) is interpreted as regions with hydrophobic character. As shown in Fig. 5, regions with hydrophobic character in AMY are at the residues 50-75, 210-230, 310-330, 375-410, and 425-440, whilst for GLL1 are at 225-230, 375-380, 425-445, and 475-480. This *ab initio* results suggest that AMY has more fragments with hydrophobic character, which may also be due to the presence of two separable domains in AMY that each has its own hydrophobic core. However, this result may not be corresponded with RP-HPLC profile as the fragmentation of AMY and GLL1 upon digestion with chymotrypsin occurs randomly, where more hydrophobic fragments could be produced from GLL1.

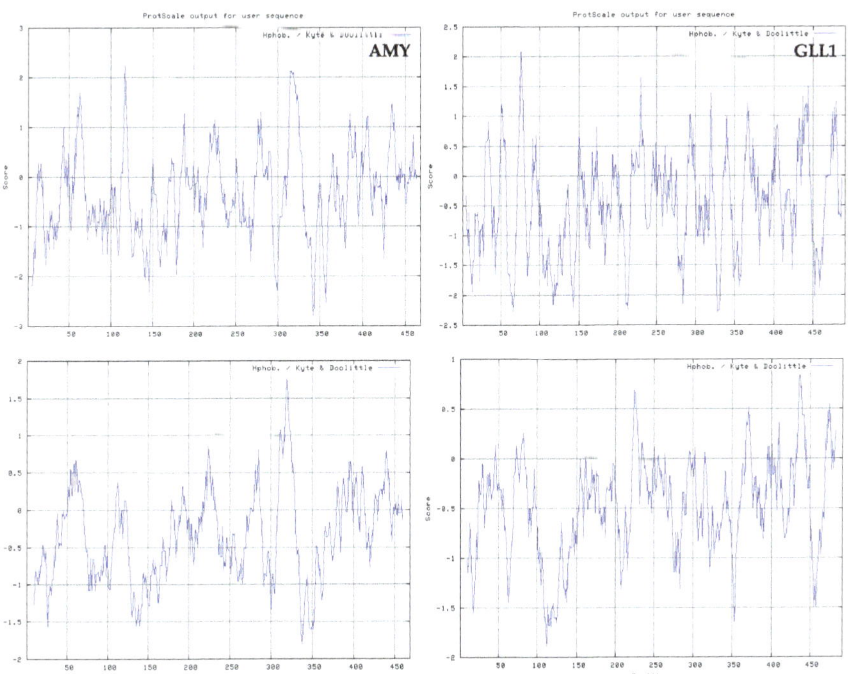

Figure 5. Hydrophobicity profile of AMY and GLL1 as calculated according to their amino acid sequences (38). The upper graphs are produced with window frame of 9 whilst the lower was with that of 19.

Another computational study was performed using the on-line hydrophobic cluster analysis (HCA) program (39). This approach has previously been done to compare the hydrophobic clusters in α-amylases (40). The sequence of AMY and GLL1 were submitted to the drawcha server (http://bioserv.impmc.jussieu.fr/hca-form.html) and the resulted profiles were analyzed manually following Gaboriaud *et al.* (41). The profiles were manually mismatched with the structures of AMY (*in silico*, generated from its homolog, see section 5) and GLU, correspondingly. AMY was predicted to have 34 hydrophobic clusters whilst GLL1 was 17. However, only 3 hydrophobic clusters of AMY are exposed on the protein surface as oppose to 7 of GLL1. The analysis suggests the presence of more hydrophobic patches on the surface of GLL1 than on that of AMY. Thus the result corresponded with the experimental finding with HIC column.

5. AMY recombinant behaviour on purification in chromatography columns

The biochemical characteristics of AMY and GLL1 are to be improved to meet specific conditions for application, such as resistance to high temperature, chemical inactivation, and proteases (42). This can be achieved by engineering at both gene and protein levels, which require convenient access to the genetic information and protein structure. Although the genes encoding for AMY or GLL1 are successfully elucidated, commencing an educated and directed genetic engineering entails the structure of the enzymes. The amino acid sequence of GLL1 is nearly identical to that of GLU (43), which its structure has been reported (in complex with amylases inhibitor acarbose, PDB accession code 2F6D). Therefore, structural study of GLL1 was performed employing the structure of GLU. The structure of *S. fibuligera* α-amylase is, on the other hands, not available. The structural study of fungal α-amylase has been employing the structure of the enzyme from *Aspergillus niger* (PDB code 7TAA) (44), which shares only 35% homology to AMY (5). Although the use of that structure is amenable, it might not be able to describe the details of characteristics of AMY. Therefore, elucidating the structure of AMY became a priority. Leading to this aim, heterelogous expression of AMY in *Escherichia coli* was attempted.

Overexpression of soluble and functional AMY in *E. coli*, with a His_6-tag at its N-terminus for easy purification on a Nickel-agarose affinity column, was unsuccessful. This situation was not improved after change of the vectors that harbour the AMY gene and of *E. coli* strains, and manipulation of overproduction conditions. No amylolytic activity or protein band at the expected molecular mass upon an SDS PAGE analysis was detected in the cell lysate. Overexpression of proteins from higher organisms in bacterial system often results in the formation of inclusion bodies because of the lack of post-translational machinery (45). As α-amylase is a glycoprotein (46), which requires post-translational modification for its expression, this lack of that machinery was an obvious suspect. Actually, was AMY overexpressed successfully in *E. coli* as soluble protein, attempt leading to structure determination would be easier because post-translational modifications are reported to hamper protein crystallization (47), the initial step for structure determination by means of

X-ray crystallography. Nevertheless, based on the result of overexpression, the attempt for heterelogous expression of AMY in *E. coli* was withdrawn for the use of expression system from higher organism i.e. yeast *S. cerevisiae.*

S. cerevisiae has been a popular choice for heterelogous expression of proteins (48) and tested to express α-amylase and glucoamylase from *S. fibuligera* strain HUT 7212 (49, 50). *S. cerevisiae* expression system for expression of *S. fibuligera* α-amylase and glucoamylase has also been improved, i.e. the enzymes expression was drastically increased under the control of the *S. cerevisiae* constitutive phosphoglycerate kinase (PGK) promoter (51). Motivated from that success, AMY was overexpressed in *S. cerevisiae* INVSc1, using yEP-secretex vector and galactose as the inducer. However, although AMY recombinant demonstrated similar activity to AMY, its molecular mass is substantially higher, being 67±2 kDa. Unfortunately, no such information was provided from the expression of *S. fibuligera* α-amylase from strains HUT 7212 (50) and Eksteen (51). Since no polypeptide chain or protein tag was added, different glycosylation profile was likely the reason for the increase in the molecular mass of the recombinant protein. The finding of a plausible additional glycosylation site in the amino acid sequence of AMY applauds this proposal.

5.1. AMY recombinant behavior upon purification in AEX columns

Likewise AMY, AMY recombinant was secreted into overproduction medium therefore it was harvested by cold centrifugation at 6000 *g* for 30 minutes to remove the yeast cells. The enzyme was captured from the medium by fractionation with ammonium sulfate at a saturation degree of 0-100%. The ammonium sulfate protein precipitate was dissolved in 50 mM Tris-HCl buffer, pH 8.0 (~5 ml) and then dialyzed overnight against one litre of that respective buffer in the cold room (~4°C).

The dialyzed AMY recombinant was then loaded onto a resource-Q AEX column (GE Healthcare Europe GmbH, Diegem, Belgium), which had been equilibrated with that respective buffer. Purification was performed in a cold cabinet (~7°C) using a fast protein liquid chromatography (FPLC) Äkta system (GE Healthcare Europe GmbH, Diegem, Belgium) and monitored on-line with the Unicorn program. The enzyme was recovered from the column upon an elution with an increasing gradient of sodium chloride 0-1 M. As the control, purified AMY was also subjected to purification with the same column.

Purification of AMY in the resource-Q column showed that minor contaminants were still present in the purified sample (Fig. 6). The major contaminant has, however, higher absorbance at λ 260 nm, suggesting that it may not be protein. This contaminant appeared yellowish in colour, which may be originated from the overproduction medium component that was co-purified in HIC, AEX, and SEC columns. Further, the elution profile of AMY recombinant was very similar to that of AMY, except for an additional large protein peak upon sample application and washing step prior to the sodium chloride salt gradient. These additional peaks unmistakably are originated from other proteins and components of overproduction medium because the AMY recombinant sample applied was not purified prior to this column. AMY recombinant was not pre-purified with the HIC because no glucoamylase was co-produced. Nevertheless, similarity of their elution profiles in the resource-Q AEX column suggests that AMY and AMY recombinant share similar surface charge distribution.

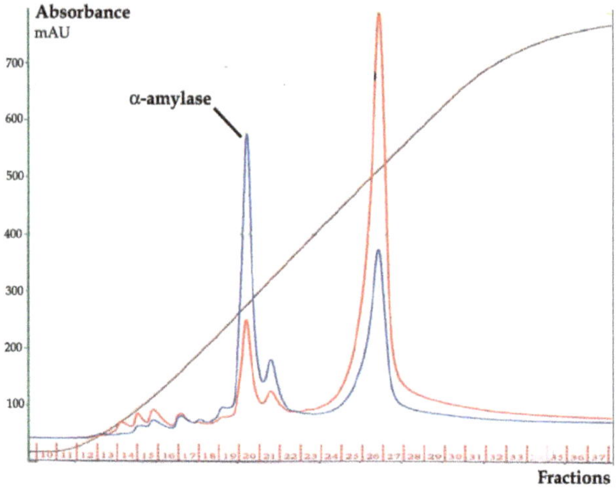

Figure 6. Elution profile of AMY on a resource-Q AEX column. The blue, red, and brown lines represent absorption at λ 280 nm (protein), at λ 260 nm, and the conductivity (mS/cm).

Further, AMY recombinant was subjected to purification in DEAE-52 cellulose AEX column (Whatman Nederland BV, s'Hertogenbosch, The Netherlands) following the purification of *S. fibuligera* α-amylase strain HUT7212 (46). Although DEAE cellulose have failed to separate AMY from GLL1 in the previous attempt, the binding of AMY to this AEX resin at purification conditions similar to that of Matsui's was observed. Thus, AMY recombinant should produce a comparable elution profile, as demonstrated previously during its purification on the resource-Q AEX column.

Supernatant from cold centrifugation was mixed with DEAE-52 cellulose AEX resin that had been equilibrated with 50 mM sodium acetate buffer, pH 5.2. After 1.5 hours of incubation at 4°C, the resin was allowed to settle and the unbound proteins were carefully decanted. The resin was washed three times with and then suspended in the respective buffer. The DEAE-52 cellulose AEX suspension was poured into an empty chromatographic column and eluted with the respective buffer. The proteins were recovered from the column by a sodium chloride salt gradient 0-1 M.

Surprisingly, AMY recombinant was not bound to DEAE-52 cellulose AEX resin, as judged from both SDS PAGE analysis and enzymatic assay. Reflecting back to purification on the resource-Q AEX column at pH 8.0, AMY and AMY recombinant were eluted at 30-35% of B solution (1 M sodium chloride, thus 300-350 mM), suggesting that the enzymes were not strongly bound to the AEX matrix. Thus, to observe no interaction between AMY recombinant and AEX resin at pH 5.2 is logical, although the pH is still higher than the pI of the enzyme. However, AMY demonstrated binding to DEAE-52 cellulose matrix in that condition, suggesting the recombinant protein has different protein surface character.

5.2. AMY recombinant behavior upon purification in sugar affinity columns

Another approach to purify AMY recombinant is the use of α-, β-, or γ-cyclodextrin (CD) columns, which was reported previously for the purification of amylolytic enzymes (52). CD is cyclic polymer of D-glycopyranosyl that is linked by α-D-(1-4) glycosidic bond and has no non-reducing or reducing ends. The α-, β-, or γ- variant of cyclodextrin matrices differs only on the number of glucose residue that builds up the dextrin ring, being six, seven, and eight, respectively (53). The protein target is bound to the interior of the CD molecule *via* hydrophobic interactions, since this interior part is less hydrophilic than its surroundings. From this perspective, CD columns have similar function to HIC column. CD also interacts with the raw starch affinity but not active sites of α-amylases (53), unlike pullulanase or cyclodextrin glycosyltranferase. Interestingly, the binding to CD column also occurs to the raw starch degrading but not adsorbing α-amylases. These reports showed that the purification of amylolytic enzymes on a CD column is based on their affinity towards substrate like matrix material.

The use of CD affinity columns is lucrative because the purification can directly be carried out after the harvesting step. After cold centrifugation, the supernatant that contained AMY recombinant was directly loaded onto α-, β-, or γ- CD affinity columns, which had already been equilibrated with 10 mM sodium acetate buffer, pH 5.5. After an extensive elution with the same buffer to remove unbound proteins, the column was eluted with 1% (w/v) α-, β-, or γ- CD in 10 mM sodium acetate buffer, pH 5.5, respective to the type of the column. Samples taken during the elution were subjected to an analysis with SDS PAGE. As the control, this purification procedure was also applied to the purified AMY.

Purification of AMY on the CD columns showed an equally strong interaction with both α- and β-CD matrices variant but weakly to γ-CD, as judged from the amount of AMY eluted from the respective CD column (Fig. 7). The result is in agreement with the reported use of CD column to separate α- from β amylase from the amylolytic complex in higher plants (52). Surprisingly, AMY recombinant was not bound onto any of the CD matrices. This result set further doubt on the surface character of AMY recombinant as being different to AMY.

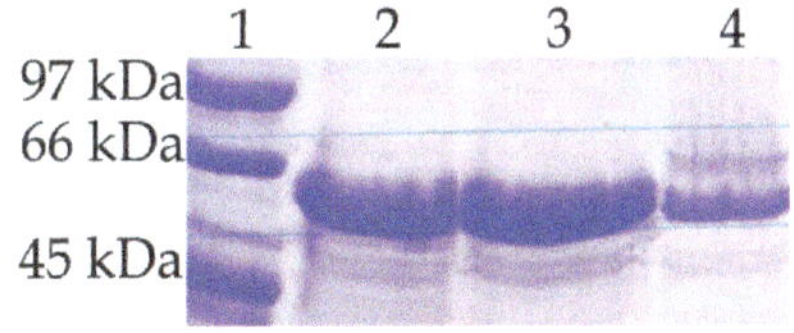

Figure 7. Elution of AMY from the α- (lane 2), β- (lane 3), and γ- (lane 4) CD columns. In lane 1 is the molecular mass marker.

Excessive amount of contaminants from the production medium was one of the suspects for this altered AMY recombinant behaviour upon purification with AEX DEAE-52 cellulose or CD columns, but the result from the latter challenged that possibility. Instead, different glycosylation pattern emerged as the prime suspect because glycosylation has indeed been reported as the main drawback for heterelogous expression of protein in *S. cerevisiae* (48). Hypermannosylation was detected upon the overexpression of *A. niger* glucoamylase in *S. cerevisiae* (54). Change of glycosylation profile may have negative impact as it can demolish enzyme activity or even change protein function (48). Furthermore, different glycosylation pattern may explain the observed higher molecular mass of AMY recombinant. Unfortunately, comparable information from similar works on the expression of *S. fibuligera* α-amylase from strains HUT 7212 (50) and Eksteen (51) was also not available.

Overexpression of amylolytic enzymes in *S. cerevisiae* is very attractive because it leads to direct conversion of starch to ethanol (11), for generation of renewable energy. Heterelogous expression of AMY in yeast *S. cerevisiae* has been successful to produce AMY recombinant, in term of enzyme activity. This AMY recombinant could be employed in further work leading to production of bio-ethanol. However, the purification profile of AMY recombinant differs substantially to AMY and the cause for this problem is yet to disclose. Therefore, the use of AMY is still preferred for the structure-function study.

6. Structure function study of AMY in the absence of the structure

The domain organization of AMY was studied by means of limited proteolysis by trypsin-TPCK, which has specific activity to cleave lysine and arginine residues. This procedure was employed to study the domain organization of bacterial cellulase (55). The fragments recovered from the limited proteolytic digestion of AMY were separated in a SEC column and subjected onto functional analysis (5).

Benefiting from the availability of amino acid sequences of AMY and GLL1, structural study *in silico* was performed (33). The structural model for AMY was prepared using the online model building program from the EMBL-EBI (56). Briefly, the amino acid sequence of AMY derived from its DNA (GenBank ID: HQ172905) was submitted to the online program SWISS-MODEL (57) using the structure of Taka-amylase from *A. niger* (PDB code 7TAA) as the template. The resulted model (7TAAmt) was evaluated using the program MolProbity (58), and manually assessed using the program *COOT* (59). The graphic representation was prepared with the program PyMOL (60), equipped with the programs PDB2PQR (61) and APBS (62) for the calculation of the surface charge distribution. As for GLL1, the structure of GLU was adopted because of near identical amino acid sequence.

The protein surface of AMY and GLL1 is mainly composed by negatively charged amino acid residues (Fig. 8). AMY does have more hydrophobic residues but they are concentrated at the interface between A/B to C domains. Discounting these domain interface hydrophobic residues in AMY, GLL1 has more hydrophobic patches on its surface. The large hydrophobic patch on the surface of AMY (Fig. 8, top right) is interrupted by positively charged residues (lysine) and putative additional glycosylation site, which increases the overall hydrophilicity of that hydrophobic patch. The surface representation suggests that the surface profile of GLL1 is more hydrophobic, thus the *in silico* study supports results from the purification in HIC column. The overall negative charged residues on the protein surface might also explain the need for basic pH for the buffers used in the purification as well as the low pI of the two enzymes.

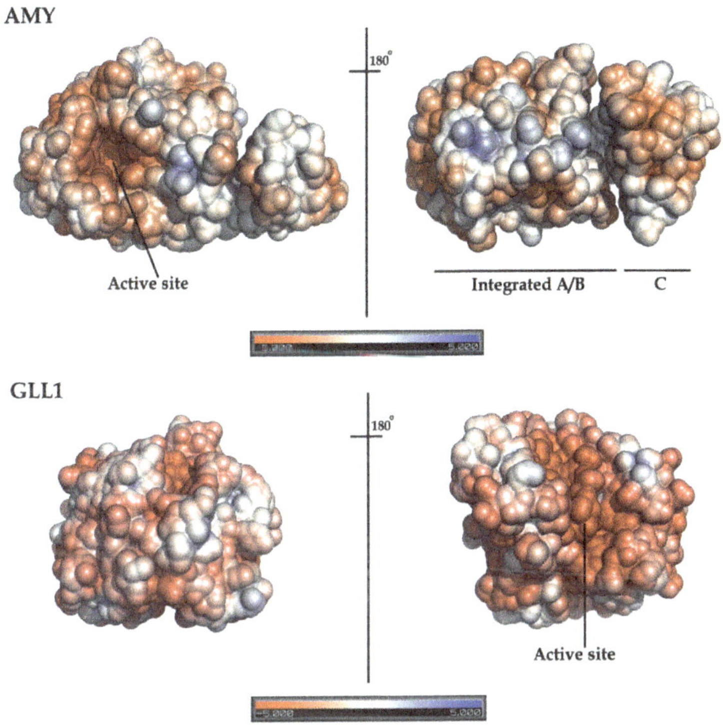

Figure 8. Surface charge distribution representation of AMY and GLL1. The red and blue colour represents negatively and positively charged amino acid residues, whilst the whitish are neutral or hydrophobic amino acids. The C domain of AMY is shifted away for clarity.

6.1. Proteolytic digestion of AMY

Purified AMY was incubated with trypsin treated with tosyl phenylalanyl chloromethyl ketone (TPCK) for 72 hours at 37°C at a substrate to protease mol ratio of 15:1. The reaction was carried out in 25 mM Tris-HCl buffer, pH 8.3 containing 20 mM calcium chloride. The reaction mixture was then transferred to -20°C for storage prior to further analysis in SDS PAGE analysis, or immediately applied to a sephadex G50 SEC column for the enzyme's domain separation (3).

Digestion of AMY by trypsin-TPCK resulted in two fragments with molecular masses of ~39 kDa (f39) and ~10 kDa (f10) (5), as judged from an SDS PAGE analysis. Based on the size of the fragments and proteolytic cleavage prediction according to its amino acid sequence, the f39 is designated as the N-terminal domain whilst f10 as the C-terminal. α-Amylases structure comprises of an $(\alpha/\beta)_8$-TIM barrel structural motif that is built up from the N-terminal part (domain A and B) and of the C-terminal part (domain C) (63). These two major domains are linked by a long surface loop. The integrated domain A/B is assigned as the catalytic domain whilst domain C is postulated as the starch-binding domain. As the two major domains of AMY were apparently separated upon proteolytic digestion, the functioning of f39 and f10 were evaluated.

6.2. Separation of f39 and f10 in an SEC column

The separation of f39 and f10 was performed in a Sephadex G50 SEC column (∅1.3 x 50 cm, bed volume ~48 ml) with gravity flow, in 20 mM phosphate citrate buffer, pH 5.8. Fractions of 5 ml were collected and the protein elution profile was measured by absorbance at λ 280 nm (UV-160, Shimadzu Corp., Tokyo, Japan). Only the collected protein absorption peak fractions were used for further analysis. As the control, purified AMY was also applied to the column and eluted with the same conditions for the proteolytically digested sample. The amount of AMY applied was also kept similar to that of used in the proteolytic reaction for fair comparison.

Two distinct protein peaks were recovered from the proteolytic digestion reaction mixture (Fig. 9) as oppose to one peak from the purified AMY. The use of SEC column at a 50 kDa cut off allowed a clear separation because AMY was eluted right at the end of the void volume retention whilst digested AMY was eluted after the void volume. Trypsin (~23 kDa) was not detected because its amount was very small (out numbered by f39 and f10, having an AMY to trypsin mass ratio of 34:1). Should trypsin be detected, it may contribute to a small increase of absorbance at fraction 7 of the digested AMY. Fractions 3 (of AMY), 4 and 8 (of f39 and f10, respectively) were selected for further analysis.

Only AMY and f39 demonstrated α-amylase activity, confirming the assignment of f39 function as the catalytic domain. However, the K_M value of f39 suggested lower affinity towards starch substrate. Interestingly, lower f39 K_M value was not followed by the decrease of the k_{cat} value of the reaction, suggesting that the catalytic efficiency was not disturbed (5). These observations further supported the assignment of f39 as the catalytic domain as well as suggested the function of f10 in substrate binding. Furthermore, the half-life time (IC50) value of f39 upon incubation at 50°C was also lower than that of AMY (5). This finding suggested that the absence of f10 also resulted in lower stability of f39. In conclusion, these findings served as an evidence for the proposed function of f10 to house the substrate recognition site (63) and to maintain thermostability (64) of α-amylase. This finding assigned AMY as a raw starch degrading but not-adsorbing enzyme. Ultimately, these results were confirmed by the independent group working on AMY homolog from *S. fibuligera* strain KZ (65).

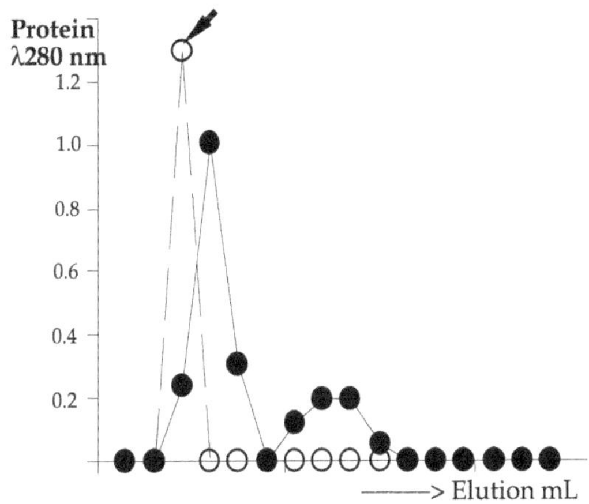

Figure 9. Separation of the f39 and f10 using a Sephadex G50 SEC column (5). The open circle represents absorption profile of AMY (54 kDa) whilst the closed circle is f39 and f10. The arrow marks the peak of AMY.

Further, AMY was pre-treated under various conditions that resulted in denatured and partially denatured enzymes prior to proteolysis. Similar experiments were also carried out using a chemically modified AMY (33). The results were employed to assess the domain organization and assignment of AMY as well as to predict the precise location of trypsin cleavage and the nature of the catalytic domain. These results are being reported elsewhere (33).

7. Conclusion

Amylolytic enzymes from *S. fibuligera* R64 (AMY and GLL1) were successfully separated and identified using chromatography as the key tool. The two enzymes have a different protein surface hydrophobicity profile and their fragmentation profiles provided undisputed proof for their separation. Their assignment was confirmed by the analysis of the products from enzymatic hydrolysis. Further, the domain organization and functioning of AMY has been explored, which led to the structural study of AMY in the absence of its structure. Thereby, this chapter demonstrates how the results from the chromatographic analysis of AMY and GLL1 are complementary to the structural study of the enzymes.

Author details

Wangsa T. Ismaya
Department of Chemistry, Faculty of Mathematics and Natural Sciences, Universitas Padjadjaran, Jatinangor, Sumedang, Indonesia
Institute for Bioengineering and Bioscience, School of Chemistry and Biochemistry, Georgia Institute of Technology, IBB Petit Building, Atlanta, Georgia, USA

Khomaini Hasan
Department of Chemistry, Faculty of Mathematics and Natural Sciences, Universitas Padjadjaran, Jatinangor, Sumedang, Indonesia
Loschmidt Laboratories, Department of Experimental Biology and Research Centre for Toxic Compound in the Environment, Faculty of Science, Masaryk University, Brno, Czech Republic

Toto Subroto and Soetijoso Soemitro
Department of Chemistry, Faculty of Mathematics and Natural Sciences, Universitas Padjadjaran, Jatinangor, Sumedang, Indonesia

Dessy Natalia
Biochemistry Research Division, Faculty of Mathematics and Natural Sciences, Institut Teknologi Bandung, Bandung, Indonesia

Acknowledgement

This paper is dedicated to Prof. Oei Ban Liang (1930-2010), the founder of the Inter-University Centre for Biotechnology, Bandung Institute of Technology, where most of the works presented in this chapter were carried out.

The work has been supported by Indonesian Ministry of Research and Technology (RUT 1993-1996 to S.S), Indonesian Ministry of National Education (HB 2000-2003 to W.T.I), Royal Netherlands Academy of Arts and Science (KNAW SPIN mobility program 2004 to S.S), Indonesian Toray Science Foundation (STRG 2006 to K.H.), and Institut Teknologi Bandung (RU-ITB 2007-present to D.N.). We thank Prof. J.J. Beintema (Rijks*universiteit* Groningen) for valuable advice and discussion.

Abbreviations

AEX, anion exchange chromatography; AMY, α-amylase from *S. fibuligera* R64; CD, cyclodextrin; CEX, cation exchange chromatography; DEAE, diethyl amino ethyl; DNA, deoxyribose nucleic acid; EDTA, ethylene diamine tetra acetate; FPLC, fast protein liquid chromatography; GLA, glucoamylase from *S. fibuligera* KZ; GLL1, glucoamylase from *S. fibuligera* R64; GLU, glucoamylase from *S. fibuligera* HUT7212; HCA, hydrophobic cluster analysis; HIC, hydrophobic interaction chromatography; LC, liquid chromatography; PDB, Protein Data Bank; PGK, phosphoglycerate kinase; RP-HPLC, reversed-phase high performance liquid chromatography; SDS PAGE, sodium dodecyl-sulfate polyacrylamide gel electrophoresis; SEC, size-exclusion chromatography; TFA, trifluoroacetic acid; TLC, thin layer chromatography; TPCK, tosyl phenylalanyl chloromethyl ketone.

8. References

[1] Sukara E, Kumagai H, Yamamoto K. Study on amyloglucosidase of a newly isolated *Saccharomycopsis* sp. TJ-1 from the Indonesian fermented food (tape). *Annal Bogor*. 1998; 5 (2):77-83.

[2] Zhenming C, Chi Z, Liu G, Wang F, Ju L, Zhang T. *Saccharomycopsis fibuligera* and its applications in biotechnology. *Biotechnol Adv.* 2009; 27:423-431.

[3] Kennedy JF, Cabalda VM, White CA. Enzymic starch utilization and genetic engineering. *Trends Biotech.* 1988; 6 (8):184-189.

[4] Hostinova E. Amylolytic enzymes produced by the yeast *Saccharomycopsis fibuligera. Biologia.* 2002; 57 (11):247-251.

[5] Hasan K, Ismaya WT, Kardi I, Andiyana Y, Kusumawidjaya S, Ishmayana S, et al. Proteolysis of alpha-amylase from *Saccharomycopsis fibuligera*: characterization of digestion products. *Biologia.* 2008; 63 (6):1044-1050.

[6] Rodriguez-Sanoja R, Oviedo N, Sanchez S. Microbial starch-binding domain. *Curr Opin Microbiol.* 2005; 8:260-267.

[7] Mahovic M, Janecek S. Starch-binding domains in the post-genome era. *Cell Mol Life Sci.* 2006; 63:2710-2724.

[8] van der Maarel MJEC, van der Veen B, Uitdehaag JCM, Leemhuis H, Dijkhuizen L. Properties and applications of starch-converting enzymes of the alpha-amylase family. *J Biotechnol.* 2002; 94:137-155.

[9] Nielsen JE, Borchert TV. Protein engineering of bacterial alpha-amylase. *Biochim Biophys Acta.* 2000; 1543:253-274.

[10] Araujo R, Casal M, Cavaco-Paulo A. Application of enzymes for textile fibres processing. *Biocatal Biotransform.* 2008; 26 (5):332-349.

[11] Shigechi H, Koh J, Fujita Y, Matsumoto T, Bito Y, Ueda M, et al. Direct production of ethanol from raw starch corn starch via fermentation by use of a novel surface-engineered yeast strain codisplaying glucoamylase and alpha-amylase. *Appl Environ Microbiol.* 2004; 70:5037-5040.

[12] Sun H, Zhao P, Ge X, Xia Y, Hao Z, Liu J, et al. Recent advances in microbial raw starch degrading enzymes. *Appl Biochem Biotechnol.* 2010; 160 (4):988-1003.

[13] McCue PP, Shetty K. A role for amylase and peroxidase-linked polymerization in phenolic antioxidant mobilization in dark-germination soybean and implications for health. *Process Biochem.* 2004; 39:1785-1791.

[14] Kumar P, Satyanarayana T. Microbial glucoamylases: characteristics and applications. *Crit Rev Biotechnol.* 2009; 29 (3):225–255.

[15] Reddy G, Altaf M, Naveena BJ, Venkateshwar M, Kumar EV. Amylolytic bacterial lactic acid fermentation - A review. *Biotechnol Adv.* 2008; 26 (1):22-34.

[16] Natalia D, Vidilaseris K, Satrimafitrah P, Purkan, Ismaya WT, Permentier H, et al. Biochemical characterization of a glucoamylase from *Saccharomycopsis fibuligera* R64. *Biologia.* 2011; 66 (1):27-32.

[17] Lawson CL, van Montfort R, Strokopytov B, Rozeboom HJ, Kalk KH, de Vries GE, et al. Glycosyltransferase from *Bacillus circulans* strain 251 in a maltose-dependent crystal form. *J Mol Biol.* 1994; 236:590-600.

[18] Alois J. Chromatographic media for bioseparation. *J Chromatogr A.* 2005; 1065 (1):3-12.

[19] Halperin G, Breitenbach M, Tauber-Finkelstein M, Shaltiel S. Hydrophobic chromatography on homologous series of alkylagaroses : A comparison of charged and electrically neutral column materials. *J Chromatogr A.* 1981; 215 (0):211-228.

[20] Fuwa H. A new method for microdetermination of amylase activity by the use of amylose as the substrate. *J Biochem (Tokyo)*. 1954; 41:583-603.
[21] Somogyi M. Notes on sugar determination. *J Biol Chem*. 1952; 195:19-23.
[22] Gasperik J, Kovac L, Minarikova O. Purification and characterization of the amylolytic enzymes of *Saccharomycopsis fibuligera*. *Int J Biochem*. 1991; 23 (1):21-25.
[23] Strop P, Mikes F, Chytilova Z. Hydrophobic interaction chromatography of proteins and peptides on spheron P-300. *J Chromatogr A*. 1978; 156 (2):239-254.
[24] Webster J, Oxley D. Peptide Mass Fingerprinting. Chemical genomics: reviews and protocols. Heidelberg, Germany: Springer; 2005. p. 227-240.
[25] Rittinghaus K, Franzen KH. HPLC in protein analysis: an alternative to gel-filtration and gel-electrophoresis. *Fresen J Anal Chem*. 1980; 301 (2):144-144.
[26] Soedjanaatmadja UMS, Hofsteenge J, Jeronimus-Stratingh CM, Bruins AP, Beintema JJ. Demonstration by mass spectrometry that pseudo-hevein and hevein have ragged C-terminal sequences. *Biochim Biophys Acta - Prot Struct Mol Enzymol*. 1994; 1209 (1):144-148.
[27] Walter A. Chymotrypsin: Molecular and catalytic properties. *Clin Biochem*. 1986; 19 (6):317-322.
[28] Tarr GE, Crabb JW. Reverse-Phase High-Performance Liquid Chromatography of hydrophobic proteins and fragments thereof. *Anal Biochem*. 1983; 131:99-107.
[29] Fourier E. Colorimetric quantification of carbohydrate. *Curr Protoc Food Anal Chem*. 2001:E1.1.3-E.1.1.4.
[30] Ramesh MV, Lonsane BK. End product profiles of starch hydrolysis by bacterial alpha-amylases at different temperature and pH values. *Biotech Lett*. 1989; 11 (9):649-652.
[31] Joutsjoki VV, Parkkinen EEM, Torkkeli TK. A novel glucoamylase preparation for grain mash saccharification. *Biotechnol Lett*. 1993; 15 (3):277-282.
[32] Saelim K, Dissara Y, H-Kittikun A. Saccharification of cassava starch by *Saccharomycopsis fibuligera* YCY1 isolated from Loog-Pang (rice cake starter). *Songklanakarin J Sci Technol*. 2008; 30 ((supp. 1)):65-71.
[33] Ismaya WT, Hasan K, Kardi I, Zainuri A, Rahmawaty RI, Permanahadi S, et al. Chemical modification of *Saccharomycopsis fibuligera* R64 alpha-amylase to improve its stability against thermal, chelator, and proteolytic inactivation. *Appl Biochem Biotech*. submitted.
[34] Itoh T, Yamashita I, Fukui S. Nucleotide-sequence of the alpha-amylase gene (Alp1) in the yeast *Saccharomycopsis fibuligera*. *FEBS Lett*. 1987; 219 (2):339-342.
[35] Gasteiger E, Hoogland C, Gattiker A, Duvaud S, Wilkins MR, Appel RD, et al. Protein identification and analysis tools on the ExPASy server. In: Walker JM, editor. The proteomics protocols handbook. New York: Humana Press; 2005. p. 571-607.
[36] Janecek S, Balaz S. Alpha-amylases and approaches leading to their enhanced stability. *FEBS Lett*. 1992; 304:1-3.
[37] Davies CW, Hoyle BE. 842. The Interaction of calcium ions with some phosphate and citrate buffers. *J Chem Soc*. 1953:4134-4136.
[38] Kyte J, Doolittle RF. A simple method for displaying the hydropathic character of a protein. *J Mol Biol*. 1982; 157 (1):105-132.

[39] Callebaut I, Labesse G, Durand P, Poupon A, Canard L, Chomilier J, et al. Deciphering protein sequence information through hydrophobic cluster analysis (HCA): current status and perspectives. *Cell Mol Life Sci.* 1997; 53:621-645.
[40] Raimbaud E, Bullion A, Perez S, Henrissat B. Hydrophobic cluster analysis of the primary sequences of α-amylases. *Int J Biol Macromol.* 1989; 11 (August):217-225.
[41] Gaboriaud C, Bissery V, Benchetrit T, Mornon JP. Hydrophobic cluster analysis: an efficient new way to compare and analyse amino acid sequences. *FEBS Lett.* 1987; 224 (1):149-155.
[42] Ulmer KM. Protein Engineering. *Science.* 1983; 219 (4585):666-671.
[43] Ševčík J, Hostinová E, Solovicová A, Gašperík J, Dauter Z, Wilson KS. Structure of the complex of a yeast glucoamylase with acarbose reveals the presence of a raw starch binding site on the catalytic domain. *FEBS J.* 2006; 273 (10):2161-2171.
[44] Matsui I, Yoneda S, Ishikawa K, Miyairi S, Fukui S, Umeyama H, et al. Roles of the aromatic residues conserved in the active center of *Saccharomycopsis fibuligera* alpha-amylase for transglycosylation and hydrolysis activity. *Biochem.* 1994; 33:189-196.
[45] Sahdev S, Khattar S, Saini K. Production of active eukaryotic proteins through bacterial expression systems: a review of the existing biotechnology strategies. *Mol Cell Biochem.* 2008; 307 (1):249-264.
[46] Matsui I, Matsui E, Ishikawa K, Miyairi S, Honda K. The enzymatic and molecular characteristics of *Saccharomycopsis* α-amylase secreted from *Saccharomyces cerevisiae Agric Biol Chem.* 1990; 54 (8):2009-2015.
[47] Qiu X, Janson CA. Improving protein crystallizability by modifications and engineering. In: Bergfors TE, editor. Protein Crystallization. 2nd ed. La Jolla, CA, USA: International University Line; 2009. p.
[48] Romanos MA, Scorer CA, Clare JJ. Foreign gene expression in yeast: a review. *Yeast.* 1992; 8:423-488.
[49] Yamashita I, Itoh T, Fukui S. Cloning and expression of the *Saccharomycopsis fibuligera* glucoamylase gene in *Saccharomyces cerevisiae. Appl Microbiol Biotechnol.* 1985; 23 (2):130-133.
[50] Yamashita I, Itoh T, Fukui S. Cloning and expression of the *Saccharomycopsis fibuligera* alpha-amylase gene in *Saccharomyces cerevisiae. Agric Biol Chem.* 1985; 10:3089-3091.
[51] Eksteen JM, van Rensburg P, Cordero Otero RR, Pretorius IS. Starch fermentation by recombinant *Saccharomyces cerevisiae* strains expressing the α amylase and glucoamylase genes from *lipomyces kononenkoae* and *Saccharomycopsis fibuligera. Biotechnol Bioeng.* 2003; 84 (6):639-646.
[52] Subbaramiah K, Sharma R. Affinity purification of amylases on cyclodextrin-sepharose columns. *Stärke.* 1989; 41 (9):357-359.
[53] Hamilton LM, Kelly CT, Fogarty WM. Review: cyclodextrins and their interaction with amylolytic enzymes. *Enzyme Microb Tech.* 2000; 26 (8):561-567.
[54] Sauer Jr, Sigurskjold BW, Christensen U, Frandsen TP, Mirgorodskaya E, Harrison M, et al. Glucoamylase: structure/function relationships, and protein engineering. *Biochim Biophys Acta - Prot Struct Mol Enzymol.* 2000; 1543 (2):275-293.

[55] Gilkes NR, Kilburn DG, Miller RC, Warren RAJ. Structural and functional-analysis of a bacterial cellulase by proteolysis. *J Biol Chem*. 1989; 264 (30):17802-17808.

[56] Goujon M, McWilliam H, Li W, Valentin F, Squizzato S, Paern J, et al. A new bioinformatics analysis tools framework at EMBL-EBI. *Nuc Acids Res*. 2010; 38:W695-9.

[57] Arnold K, Bordoli L, Kopp J, Schwede T. The SWISS-MODEL workspace: a web-based environment for protein structure homology modelling. *Bioinformatics*. 2006; 22:195-201.

[58] Chen VB, Arendall III WB, Headd JJ, Keedy DA, Immormino RM, Kapral GJ, et al. MolProbity: all-atom structure validation for macromolecular crystallography. *Acta Crystallogr D*. 2010; 66:12-21.

[59] Emsley P, Lohkamp B, Scott W, Cowtan K. Features and development of Coot. *Acta Crystallogr D*. 2010; 66:486-501.

[60] DeLano WL. *The PyMOL molecular graphics system, Delano Scientific LLC, Palo Alto, CA - USA*. 2008.

[61] Dolinsky TJ, Nielsen JE, McCammon JA, Baker NA. PDB2PQR: an automated pipeline for the setup, execution, and analysis of Poisson-Boltzmann electrostatics calculations. *Nuc Acids Res*. 2004; 32:W665-W667.

[62] Baker NA, Sept D, Joseph S, Holst MJ, McCammon JA. Electrostatics of nanosystems: application to microtubules and the ribosome. *Proc Natl Acad Sci USA*. 2001; 98:10037-10041.

[63] Janecek S, Sevcik J. The evolution of starch binding domain. *FEBS Lett*. 1999; 456:119-125.

[64] Iefuji H, Chino M, Kato M, Limura Y. Raw-starch digesting and thermostable alpha-amylase from yeast *Cryptococcus* sp. S2: purification, characterization, cloning, and sequencing. *Biochem J*. 1996; 318:989-996.

[65] Hostinová E, Janeček Š, Gašperík J. Gene sequence, bioinformatics and enzymatic characterization of α-amylase from *Saccharomycopsis fibuligera* KZ. *Prot J*. 2010; 29 (5):355-364.

Chapter 12

Purification of Azurin from *Pseudomonas Aeuroginosa*

Sankar Ramachandran, Moganavelli Singh and Mahitosh Mandal

Additional information is available at the end of the chapter

1. Introduction

The use of microorganisms and their products as possible therapeutic agents in the control of cancer begins at the latter part of the nineteenth century. The search of new drugs from microbial sources against infectious disease has been augmented when Alexander Fleming (1928) discovered penicillin [1]. The secondary metabolites from microorganisms play a vital role in developing antibiotics and chemotherapeutics [2, 3]. Several researchers have reported various anticancer molecules from different microbial sources [4]. Even though chemotherapy is efficient in enhancing patient survival with primary tumors continue to have deprived prognosis. The rapid advances in the field of antibiotics have inspired new hope that the search among biological systems will disclose a chemical agent which will exert a destructive effect upon neoplastic growth without seriously affecting normal cells. Using live or attenuated pathogenic bacteria or its metabolites in treatment of cancer excretes toxic effects among patients. Azurin, a redox protein recently fascinated biomedical researcher's immense interest as an anti cancer therapeutic agent which enters human breast cancer cells and induces apoptosis without any adverse effects in cancer patients [5]. Azurin, a secondary metabolite derived from bacterial species especially from *P. aeruginosa* function as a donor in terminal electron transfer process [6]. Azurin also termed as blue small copper proteins highly stable in nature. The presence of copper ion in the polypeptide chain contributes to the azurin stability [6-8]. Azurin reported as a potential anticancer protein against breast cancer cell lines, evoked the researchers of novel methods for enhanced synthesis of azurin has initialized. *P. aeruginosa* a common gram negative opportunistic pathogenic bacterium found naturally [9]. They are considered as facultative anaerobic grow in partial or total oxygen depletion cultural conditions. This organism can achieve anaerobic growth with nitrate as a terminal electron acceptor. *P. aeruginosa* secretes a variety of pigments, including pyocyanin, pyoverdine and pyorubin.

Previous researchers [10] adopted genetic engineering techniques and other bacterial species for purification of azurin. This study is concerned of enhanced azurin synthesis from different strains of *P. aeruginosa* with lucid homogeneity by customized methods. The growth of different *P. aeruginosa* MTCC strains 1934, 741, 2453, and 1942 for the synthesis of azurin were scrutinized for enhanced azurin synthesis. The enhanced azurin synthesis from *P. aeruginosa* strains was improved by the $CuSO_4$ and KNO_3 containing media under facultative anaerobic condition. The purification of azurin had been performed by ion-exchange and gel-filtration chromatography. High yield was reported in *P. aeruginosa* 2453 strain than other strains.

2. Materials and methods

2.1. Chemicals and reagents

Growth medium constituents were of analytical grade obtained from Hi-Media laboratories, India. The buffer ingredients were purchased from Merck Chemicals Ltd, India. Sephadex G-25, G-75, diethyl amino ethyl cellulose (DEAE) cellulose and carboxy methyl (CM) cellulose were all obtained from Sigma-Aldrich, USA. The 3, 5-dimethoxy, 4-hydroxy cinnamic acid otherwise called as sinnapinic acid a matrix-assisted laser desorption/ionization-time of flight (MALDI-ToF) was also acquired from Sigma-Aldrich. Protein concentrations were measured by Lowry's method with bovine serum albumin as standard. Standard dialysis bag with 3 kDa cutoff was purchased from Sigma-Aldrich. Powder of 3-(4,5-Dimethylthiazol-2-yl)-2,5-diphenyltetrazolium (MTT), propidium iodide (PI) and dimethylsulfoxide (DMSO) solution were procured from Sigma Aldrich, India. Cell culture media and other constituents of media are purchased from Hi-media Laboratories Ltd, India. Fetal bovine serum was obtained from Invitrogen Life Technologies, USA.

2.2. Cultivation of P. aeruginosa MTCC 2453

A freeze dried culture of *P. aeruginosa* strains MTCC2453,741,1934,1942 was obtained from the Microbial type culture collection center, Chandigarh, India and was grown in a medium containing 7g yeast extract, l0g peptone, 20g of KNO3, 6.4g of KH_2PO_4, 3.6g of Na_2HPO_4 (anhydrous), 2.5g of NaCl, 5μg/ml of CuSO4 per liter. The initial pH was adjusted to 6.5 with NaOH. The strains were maintained on nutrient broth with 50% glycerol concentration and stored at -70°C for further study.

Inoculums' of *P. aeruginosa* strains was prepared by inoculating a loopful of colonies in individual 100 ml conical flasks with the exact constituents of the above prescribed media and incubated at 37°C, in stirring mode at 100/rev for 21 hours. These inoculums was used to seed the bulk 500ml x 4 sterile medium in 2000 liter conical flasks (separate conical flask used for all four strains) which was also kept at stirred mode (100/rev) for 21 hours at 37°C [11].

2.3. Impact of copper sulphate and potassium nitrate on culture medium

All P. *aeruginosa* strains MTCC2453, 741, 1934, and 1942 are inoculated separately in a sterile medium. Impact of copper sulphate and potassium nitrate in azurin synthesis were observed by adding different concentration of copper sulphate (1μg/ml- 5μg/ml) and potassium nitrate (5μg/l – 20μg/l) at different flasks for each concentration distinctly. The azurin protein optimization and quantification was studied in UV spectrometer (Perkin Elmer, Massachusetts, USA) at 595 $_{nm}$ by Bradford's method. The azurin synthesized from P. aeruginosa MTCC2453 is significantly higher than other strains [11, 12].

2.4. Extraction of cellular protein

After 21 hrs incubation, cells were harvested by centrifugation method at 13200 g for 15-20 minutes by using ultra centrifuge (Eppendorf, Hamburg, Germany). Cell pellets was collected and suspended in the appropriate volume of 0.02M potassium phosphate buffer pH 7 with protease inhibitor and kept in the ice basket for sonication. Cells were sheared by Ultra sonicator (Cole Parmer, USA) of approximately 100 ml batches of cell suspension. All batches were sonicated for 1-2 minutes at 100W. After sonication the samples was stirred vigorously and centrifuged at 10000g for 20 minutes which removes cell wall debris. The green-brown crude supernatant was stored. Resuspended the precipitate in same buffer, stirred it vigorously and centrifuged as before and the supernatant were stored with the previous extracts [11, 12].

2.5. Ammonium sulfate precipitation of proteins

The Crude (supernatant) was saturated to 45% (277g/l) by slowly adding ammonium sulfate salt at 4^{0}C for precipitation, kept it for overnight [5,6]. After precipitation the solution was centrifuged at 20,000g for 25 minutes [6,]. Collected the yellow supernatant saturated again to 95% by adding (NH4)2SO4 (372g/l) slowly and kept at 4^{0}C for overnight. The overnight precipitated solution was centrifuged at 23000g for 45 minutes. Pale supernatant was discarded. Precipitate (contains azurin) were collected and resuspended in 0.02M Potassium Phosphate buffer pH 7[11, 12].

2.6. Dialysis of the supernatant

Azurin suspended in 0.02M potassium phosphate buffer pH 7 was dialysed by standard dialyses bag purchased from Sigma-Aldrich, (Kolkata, India) having 3 kDa molecular weight cut off at 4^{0}C for 20 hours on the same buffer for overnight with continuous gentle stirring. Dialysis was done until the solution attains its buffer pH. The solutions were kept at 4^{0}C after dialysis for further purification [11, 12].

2.6.1. Purification of Azurin on Ion – Exchange chromatography

2.6.1.1. DEAE cellulose treatment

Dialysate (contains azurin) were initially treated with DEAE. 100 ml slurry of DEAE cellulose equilibrated in 0.02M potassium phosphate buffer pH 7 were treated with the dialysate and stirred for 20-30 min at 4^0C.The suspension was centrifuged at 10,000g for 15 min. Azurin does not adsorb in the gel remains in supernatant but most of the unwanted proteins like yellow flavo proteins are removed. Supernatant was collected. DEAE cellulose precipitate was resuspended in the same buffer and again centrifuged at 10,000g for 15 minutes to remove all unattached proteins [11-13].

The supernatant after DEAE treatment was saturated to 100% (766g/l) with (NH4)2SO4 at 4^0C for overnight for precipitation. After saturation, precipitates are mixed gently and kept for centrifugation at 10000g for 10 min. supernatant was collected for dialysis at 4^0C for overnight with gentle stirring with the same before. Dialysis was continued till the solution pH attains its buffer pH [11-13].

2.6.2. *Purification of Azurin on gel-filtration chromatography*

2.6.2.1. *Chromatography on Sephadex G-25*

Sephadex G-25 beads were equilibrated in the 0.02M potassium phosphate buffer pH 7 [Parr S R et al 1976] for overnight, and tightly packed in 3cm x 25cm length glass column without any bubbles. The column was initially washed with 0.02M potassium phosphate buffer for twenty volumes of the gel packed. The Flow rate was adjusted to one minute per ml. slowly the dialysate (after DEAE treatment) was added with the eluent buffer 0.02M potassium phosphate buffer pH 7 on the column. Thirty fractions were collected at one minute interval [12-14].

2.6.2.2. *Chromatography on Sephadex G-75*

Sephadex G-75 beads in powder form are equilibrated in 0.01M Tris/Hcl buffer pH 7.5 for overnight. After equilibration the beads were tightly packed in a 3cm x 45cm glass column. The column was washed with same equilibrating buffer for fifty volumes of the column value. After washing with buffer one ml of the sample (fraction (a) collected from the G-25) were passaged and eluted with the same equilibrating buffer. Seventy five fractions were collected at 1 ml/6minutes flow rate [12-14].

2.6.3. *Purification of Azurin on ion – Exchange (anionic) chromatography*

2.6.3.1. *Chromatography on CM cellulose*

The CM cellulose beads from Sigma–Aldrich (Kolkata, India) were equilibrated for overnight in the ammonium acetate buffer pH 3.9 adjusting the pH by 0.05M acetic acid with 2M NH_3.After swelling, the beads were packed in a 5cm x 15cm glass column and washed for ten times of the column volume. Gently one ml of the sample added (Fraction (e) collected from G-75) over the top of the column and left it for 5-10 minutes to bind the protein inside the beads. After 10 minutes the column was eluted with ammonium acetate buffer pH 4.65 [12-14].

2.6.4. *Characterization of Azurin (purified from P. aeruginosa MTCC2453)*

2.6.4.1. *Molecular weight determination by matrix-assisted laser desorption/ionization time of flight (MALDI-ToF)*

The most successful method to analyze biopolymers such as, proteins, peptides, sugars and large organic molecules which are tend to be fragile and fragment when ionized by more conventional ionization methods [15]. The Fractions collected from G-25, G-75 and CM cellulose were performed MALDI for molecular weight determination. Two micro liter from each fraction of the chromatography was added with 20µl of 3, 5-Dimethoxy, 4-Hydroxy cinnamic acid otherwise called as sinnapinic acid (Sigma-Aldrich. Kolkata, India). Tiny spots were made on silver plate and kept for drying for 4-6 hours to drain the water molecules. Further spots were dried with a vacuum drier to make a crystalline molecule. After drying the samples were placed in the MALDI-ToF chamber (Voyager De pro, applied systems Illinois, USA) for analysis by using nitrogen laser at 337 nm.

2.6.5. *Purification profile of Azurin ((synthesized from P. aeruginosa MTCC2453) by SDS-PAGE*

Five ml of 12% resolving gel contains 1ml distilled water, 30% acryl amide, 1.5M Tris (pH 8.8), 10% SDS, 10% APS and 0.002µl TEMED for polymerization was casted in the glass slab without any bubbles and kept it for 10-15 minutes. After polymerization of the resolving gel, 3ml of stacking gel (4%) were loaded over the resolving gel which contains 0.68 ml distilled water, 30% acryl amide, 1M Tris (pH 6.8) 10% SDS,10% APS, and 0.001ml TEMED. After casting the gel, proteins purified from different chromatography were loaded with bromophenol (molecular weight marker dye) at different lanes for profiling the protein purification process.

Glass slab gel were kept in the electrophoresis tank with tank buffer (196 mM glycine, 0.1%SDS, 50mM Tris-Hcl pH 8.3 made by diluting a 10x stock solution). This setup was connected with power pack initially in 80mV to 100 mV. After running the gel up to its anode end, was removed and stained with 0.2% coomassie brilliant blue for overnight. Destained with destaining solution (45: 45: 10 – methanol: water: acetic acid) which destains the comassie blue until it reveals the bands. The bands (figure 4) were observed under UV transilluminator (Biorad, PA, USA) [16].

2.6.6. *FTIR analysis*

Infrared spectroscopy experiments were performed using a Nexus 870 (Thermo Nicolet Corporation, Madison, USA) spectrometer equipped with a potassium bromide (KBr) beam splitter and DTGS (deuterated triglycine sulfate) detector in the range of 3,000-4000 cm^{-1}. We recorded 32 scans per spectrum at a 2 cm^{-1} resolution for 100 µl of azurin liquid samples in 0.02 M PBS buffer (pH 7.0). We kept the same buffer as a background medium and performed all measurements at room temperature. We corrected spectra for the moisture and carbon dioxide in the optical path. The curves were deconvoulted and imported into Omnic's peak fit software (Thermo scientific, Illinois, USA) and a Gaussian curve fitting performed [17].

3. Results

3.1. Growth of P. aeruginosa strains

The inoculated growth of *P. aeruginosa* MTCC strains 2453, 741, 1934 and 1942 under facultative anaerobic conditions, yields total dry cell protein in the range of 150- 170 g/l medium (Fig. 2.1a). We observed *P. aeruginosa* 2453 produces lesser amount of cellular proteins than other strains. The quality assay was performed after incubation for contamination of any other unwanted organisms. A unique green colour colony in nutrient agar medium was observed and hence we confirmed it as *P. aeruginosa* colonies (Fig 1.).

3.2. Effect of copper sulphate and Nitrate in azurin synthesis

Earlier studies showed that azurin production by different bacterial strains were similar to the azurin produced by *P. aeruginosa* MTCC 2453 but with more yield than previous procedures. Four strains were tested for high yield of azurin productions were *P. aeruginosa* 2453, 741, 1942, and 1934. We observed a significant increase in the yield of azurin secreted by *P. aeruginosa* 2453 than genetically engineered strains and other strains. This remarkable increase in the yield of azurin was obtained by addition of $CuSO_4$ and KNO_3 in the medium with specific facultative anaerobic cultural conditions. In contrast to earlier studies, adding both $CuSO_4$ (4-5 μg/ml) and KNO_3 (0.02 μg/ml) in the medium under facultative anaerobic conditions generate high amount of azurin (Figure 2.), rather adding either $CuSO_4$ or KNO_3 (Table 1.).

(a) (b)

Figure 1. (a) Bacterial culture medium incorporated with $CuSO_4$ and KNO_3 (b) Green colour colonies a unique characteristic of *P. aeruginosa* colonies

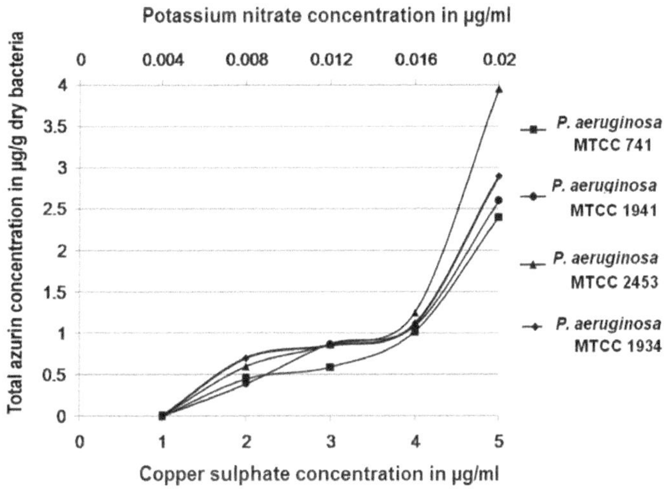

Figure 2. Quantification of azurin synthesis by different strains of *P. aeruginosa* MTCC 741, 1934, 1942 and 2453 and impact of $CuSO_4$ and KNO_3: 1-5 µg/ml range of $CuSO_4$ concentration with 0.004-0.02 µg/ml of KNO_3 was added in the culture medium to study the impact of azurin synthesis.

Purification step	*P. aeruginosa* MTCC 1934	*P. aeruginosa* MTCC 741	*P. aeruginosa* MTCC 2453	*P. aeruginosa* MTCC 1942
Total dry cell yield in g/l medium	1650	1780	**156 0**	1590
Protein concentration after 45/95 % (NH4)2SO4 precipitation (g/l)	1520/1410	1460/1550	**1105/1250**	1205/1300
Protein concentration after DEAE treatment in g/l medium	560	610	**460**	490
Protein concentration after G-25 treatment g/l.	440	485	**315**	384
Protein concentration after G-75 treatment g/l.	320	350	**295**	302
Total Azurin synthesis in mg/g dry bacteria. (CM cellulose)	2.9	2.4	**3.95**	2.6

Table 1. Azurin yield from different strains (*P. aeruginosa* MTCC 741, 1934, 1942 and 2453) in addition of 5 µg/ ml $CuSO_4$ and 0.02 µg/ml KNO_3 in the culture medium

3.3. Chromatography methods for azurin Purification

DEAE and G-25 are gel filtration columns which remove positively and negatively charged proteins respectively. The unwanted flavo proteins and positively charged proteins were removed during DEAE chromatography. The collected fractions from G-25 were quantified for protein concentration in the UV-Spectrophotometer at 280_{nm} wavelength. Azurin and other proteins more than 5 kDa were eluted immediately after void volume is plotted as graph (Figure 3.).

Peak (a) from G-25 was loaded on G-75 for further purification. The G-75 fractions were quantified for protein concentration in the UV-Spectrophotometer at 280nm wavelength. Azurin a 14 kDa protein will elute after binding in to the beads when the elution buffer elutes it. Thus, azurin and some other proteins will elute very lately, was confirmed from the OD values of the spectrometer, when plotted as graph (Figure 4). The azurin will form a thick band when passages through CM cellulose column which was eluted by ammonium acetate buffer pH 4.65. Ten fractions were collected and absorbed under UV spectrometer at 280nm wavelengths for azurin concentration (Figure 5.).

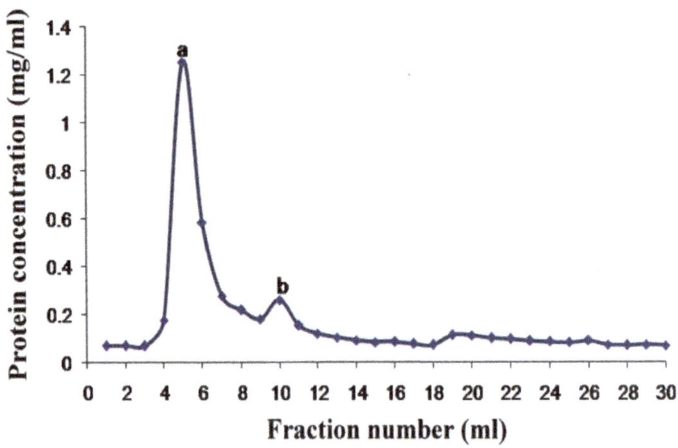

Figure 3. Elution on sephadex G-25: The active fraction from DEAE was loaded on G-25 column. Upon thirty fractions only peak (a) collected for further purification.

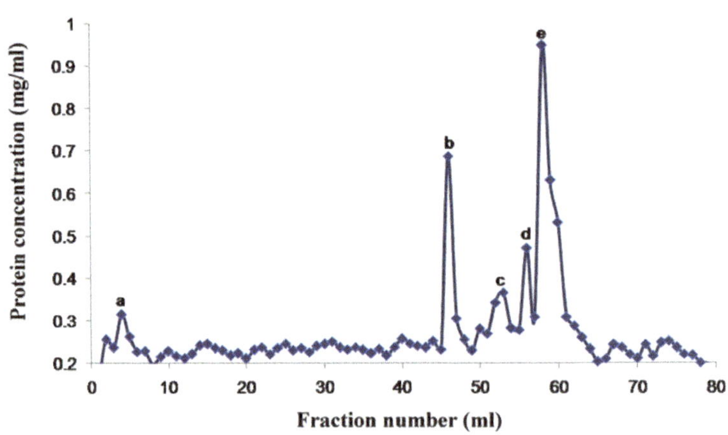

Figure 4. Elution on sephadex G-75: Fraction (a) collected from the G-25 was loaded with eluent buffer (PBS) to elute bounded proteins. Seventy five fractions were collected at 1 ml/6minutes flow rate. Peak (e) collected for further purification.

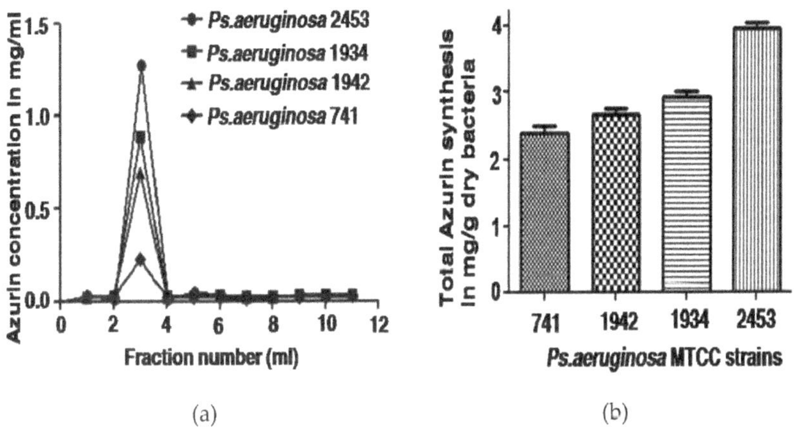

Figure 5. Azurin purified from various strains of *Pseudomonas aeruginosa*: **a** Azurin forms thick bands in CM cellulose column chromatography during their purification process. Later this was eluted by ammonium acetate buffer pH 4.65. All *Pseudomonas aeruginosa* strains, particularly *Pseudomonas aeruginosa* 2453 strain shows more significant amount of azurin production. **b** the mean of the azurin production by various *Pseudomonas.aeruginosa* strains: The production of azurin was enhanced by the copper sulphate (5μg/ml) and KNO_3 (0.02 μg/ml) containing media under facultative anaerobic condition. The bar graph shows *Pseudomonas aeruginosa* 2453 secrets more azurin than any other strains like *Pseudomonas aeruginosa* 741, 1942, 1934 strains tested.

3.4. Characterization of Azurin (Purified from P. aeruginosa MTCC2453)

In this study we profiled our purification process at every step by MALDI-ToF (Figure 6.) and SDS-PAGE (Figure 7.) and to confirm the azurin presence in our experiments. Cellular proteins loaded in lane 2 of SDS-PAGE reveals whole cell proteins of *P. aeruginosa* MTCC 2453. Fraction (a) collected from G-25 gel filtrations were loaded in lane 3; it shows proteins above 5 kDa in SDS-PAGE which was also confirmed in MALDI-ToF results. Most unwanted proteins were deduced during G-75 gel filtration. Our SDS and MALDI results shows fraction (e) from G-75 contains azurin (14 kDa). The azurin was again purified and concentrated in CM cellulose.

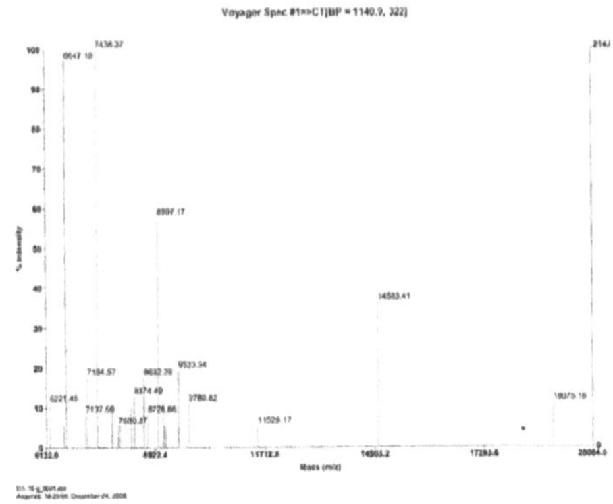

(a)

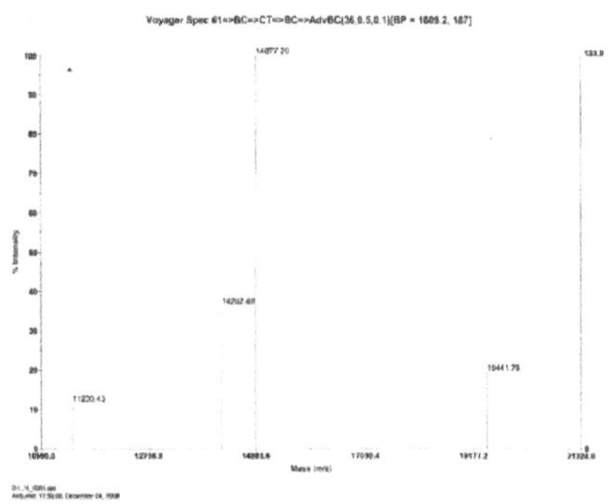

(b)

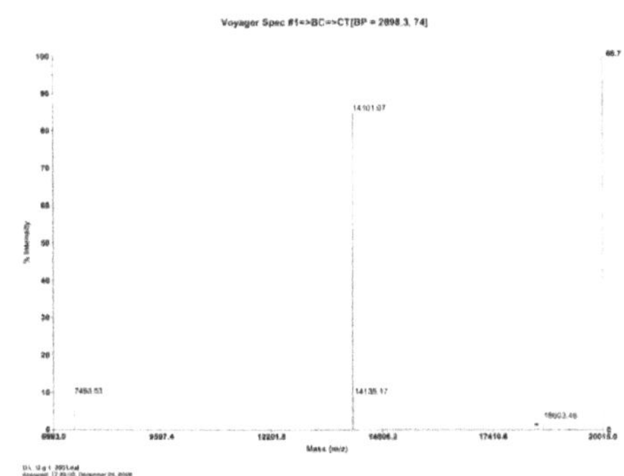

(c)

Figure 6. (a) Protein purification was assayed at each step of chromatography. Peak (a) from G-25 was analyzed in MALDI-ToF using Nitrogen laser at 337 nm, confirming the 14 kDa molecular weight of azurin.

(b) Protein purification was assayed at each step of chromatography. Peak (a) from G-25 was analyzed in MALDI-ToF using Nitrogen laser at 337 nm, confirming the 14 kDa molecular weight of azurin.
(c) Peak (e) from G-75 was analyzed in MALDI-ToF

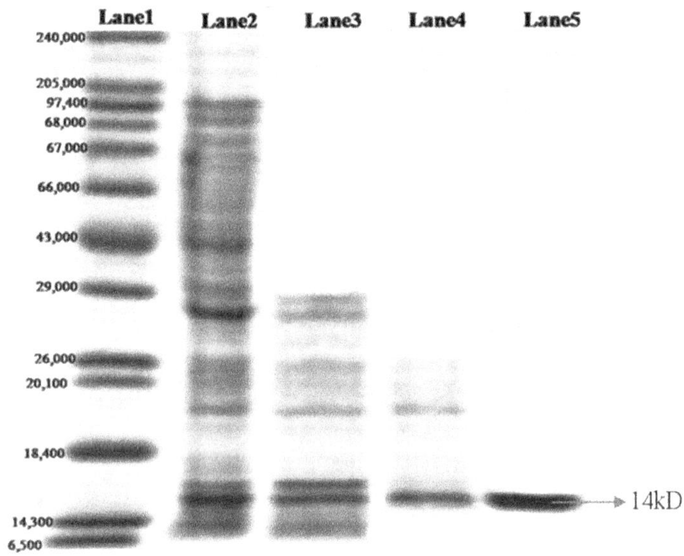

Figure 7. Protein purification profile further was confirmed by SDS-PAGE analysis: Lane 1: Molecular weight markers 6.5-240 kDa (Bangalore Gene, India), Lane 2: Total cellular proteins, Lane 3: G-25 Fraction [peak (a)], Lane 4: G-75 fraction [peak (e)], Lane 5: CM cellulose purified azurin.

3.5. FTIR analysis

The functional groups of azurin were studied using FTIR spectrum. The presence of the amide I band was indicated by the peak around 1650 cm^{-1} region, which arises primarily because of the stretching vibration of the main chain of carbonyl groups in the protein backbone coupled with the in-plane N-H bending and C-N stretching modes. Furthermore, the presence of an amide band around 1650 cm^{-1} signifies α-helix secondary structure of azurin. Azurin synthesized from all strains showed a significant shift in the amide I band with one another, indicating differences in their helix secondary structure of azurin. The most prominent among all strains is *P. aeruginosa* 2453 which showed peak around 1646.936 whereas, others showed peak around 1642.269, 1639.446, 1637.873 for *P. aeruginosa* 741, 1942, 1934 respectively (Fig. 8.). The peaks at 3695 and 3251 cm^{-1} are the amide A and B bands, respectively, which arise from a Fermi resonance between the first overtone of amide and the N-H stretching vibrations. The 1495 cm^{-1} peak refers to the amide II band, which arises because of the C-N stretching as well as the C-N-H bending motions. The 1352 peak is the amide III band, which arises predominantly because of the in-phase combination of N-H in plane bending and C-N stretching vibrations.

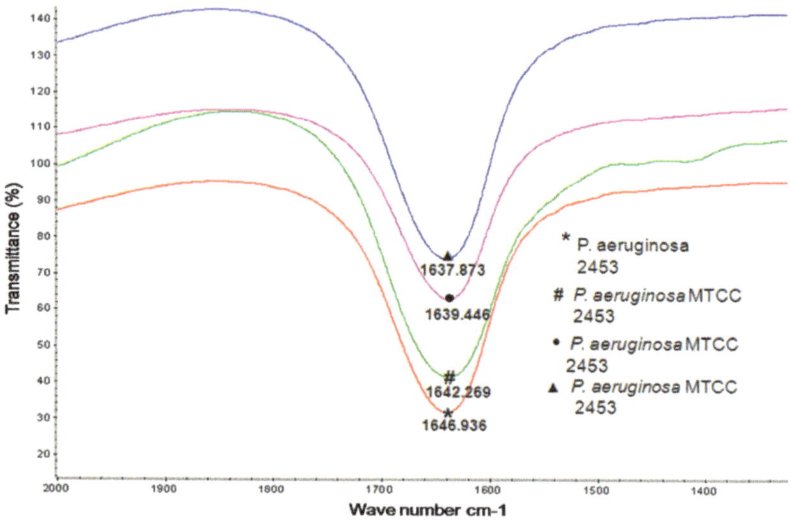

Figure 8. FTIR analysis showed peak around 1646.936 in *P. aeruginosa* 2453 whereas, others showed peak around 1642.269, 1639.446, 1637.873 for *P. aeruginosa* 741, 1942, 1934 respectively.

4. Discussion

Azurin production from *P. aeruginosa* MTCC 2453 was enhanced when 5µg/ml of copper sulphate and the potassium nitrate (0.02µg/ml) was added. At each step of the purification process protease inhibitor was added to the protein sample for inhibition of protein lyses. During dialysis most of the lower (approx) proteins up to 3 kDa molecular weight are pierced out. The retained proteins (above 3 kDa molecular weight) were washed in sephadex G-25 which serves as a desalting column and also has 3-5 kDa molecular weight fractionation range. High molecular weight more than 5 kDa proteins were eluted immediately after void volume which was revealed in SDS-PAGE and MALDI-ToF spectrometer.

The fraction collected from G-25 containing only more than 5 kDa proteins were passed through on G-75 which has 5-80 kDa fractionation range. The higher proteins above 80 kDa molecular weight elute after void volume; the remaining proteins between 6-70 kDa were bounded within the beads later eluted by the buffer. The fraction which showed 14 kDa molecular weight by analyzing in MALDI-ToF spectrometer for all the fractions (MALDI-ToF results not shown for all fractions which showed peak) were collected and again purified in CM cellulose chromatography. The fraction which showed peak in CM cellulose was again observed in MALDI-ToF spectrometer to confirm the presence of 14 kDa molecular weight of Azurin.

Our idea of adding copper in the culture medium was not only for the enhanced azurin synthesis, but to reveal the differences of azurin's stability in the secondary structure for all

P. aeruginosa strains. The FTIR investigation showed azurin has C=O (protein backbone) stretching, which is the unique nature of the amide I band. The presence of the amide band at 1650 cm^{-1} signifies the α-helix secondary structure of azurin. The significant shift among four strains synthesized azurin implies that there was a difference in their secondary structure which may be due to their physiological or genetic variations among strains. The impact of the differences in the secondary structure of azurin synthesized from all four strains tested, were also reflected in the apoptosis generation of all strains.

Abbreviations

MALDI-Matrix-Assisted Laser Desorption/Ionization, SDS-PAGE-Sodium Dodecyl Sulfate Polyacrylamide Gel Electrophoresis, FTIR- Fourier Transform Infrared Spectroscopy, $CuSO_4$ – copper sulphate, KNO_3 – Potassium nitrate, MTT -3-(4,5-Dimethylthiazol-2-yl)-2,5-diphenyltetrazolium, PI-Propidium Iodide, DMSO-dimethylsulfoxide, MALDI-ToF-Matrix-Assisted Laser Desorption/Ionization-Time of Flight, MTCC- Microbial Type Culture Collection center, CM-carboxymethyl, DEAE-Diethylaminoethyl Cellulose

Author details

Sankar Ramachandran[1,2,*], Moganavelli Singh[2] and Mahitosh Mandal[1]
[1]*School of Medical Science and Technology,*
Indian Institute of Technology, Kharagpur, West Bengal, India
[2]*Department of Biochemistry, School of Life sciences,*
University of KwaZulu-Natal, Westville, Durban, South Africa

5. References

[1] Fleming A. On the antibacterial action of cultures of a penicillium with special reference to their use in the isolation of *B. influenzae*. British Journal Experimental Pathology 1929;10 226-236.

[2] Ambler RP, Brown LH. The Amino Acid Sequence of Pseudomonas fluorescens azurin. Journal of Biochemistry 1967;104 784-825.

[3] Chakrabarty AM. Microorganisms and Cancer:Quest for a Therapy. Journal of bacteriology 2003;185 2683-2686.

[4] Apiyo D, Wittung-Stafshede P. Presence of the cofactor speeds up folding of *Desulfovibrio desulfuricans* flavodoxin. Protein Science 2002;11 1129-1135.

[5] Yamada T, Goto M, Punj V, Zaborina O, Chen ML, Kimbara K, Majumdar D, Cunningham, E, Das Gupta TK, Chakrabarty AM. Bacterial redox protein azurin, tumor suppressor protein p53, and regression of cancer. PNAS 2002;99 14098-14103.

[6] Pozdnyakova I, Guidry J, Wittung-Stafshede P. Copper stabilizes azurin by decreasing the unfolding rate. Archives of Biochemistry and Biophysics 2001;390 146-148.

* Corresponding Author

[7] Pozdnyakova I, Guidry J, Wittung-Stafshede P. Studies of *Pseudomonas aeruginosa* azurin mutants: cavities in β-barrel do not affect refolding speed. Biophysical Journal 2002;82 2645-2651.

[8] Pozdnyakova I, Wittung-Stafshede P. Copper binding before polypeptide folding speeds up formation of active (holo) *Pseudomonas aeruginosa* azurin. Biochemistry 2001;40 13728-13733.

[9] Sankar R, Mandal M. Inhibitory effect of azurin synthesized from *P. aeruginosa* MTCC 2453 against Dalton's lymphoma ascites model. Biomedicine & Pharmacotherapy 2011; 65 461-466.

[10] Goto M, Yamada T, Kimara K, Horner J, Newcomb M, Gupta TK, Chakarborty AM. Induction of apoptosis in macrophages by *Pseudomonas aeruginosa* azurin: tumor suppressor protein p53 and reactive oxygen species, but not redox activity, as critical elements in cytotoxicity. Biochemistry (Mosc) 2003;68 958-962.

[11] Parr SR, Barber D, Greenwood C, Phillips BW, Melling J. A Purification Procedure for the Soluble Cytochrome Oxidase and some other Respiratory Proteins from *Pseudomonas aeruginosa*. Biochemical Journal 1976;157 423-430.

[12] Sutherland IW. The Production of Azurin and Similar Proteinsm. Archly fiir Mikrobiologie 1966;54 350-357.

[13] Sankar R, Mandal M. Copper sulphate enhanced synthesis of azurin from *Pseudomonas aeruginosa* MTCC 2453, Research & Reviews: A Journal of Microbiology & Virology 2011;1 1-4.

[14] Sankar R, Mandal M. Azurin synthesis from Pseudomonas aeruginosa MTCC 2453, properties, induction of reactive oxygen species, and p53 stimulated apoptosis in breast carcinoma cells. Journal of Cancer Science and Therapeutics 2011;3(5) 104-111.

[15] Fukuo T, Kubota N, Kataoka K, Nakai M, Suzuki S, Arakawa R. Matrix-assisted laser desorption/ionization and electro spray ionization mass spectrometry analysis of blue copper proteins azurin and Mavicyanin. Rapid Communication in Mass Spectrometry 1998;12 1967-1971.

[16] Weber K, Osborn M. The Reliability of molecular weight determinations by dodecyl sulfate polyacrylamide gel electrophoresis. The Journal of Biological Chemistry 1969;244 4406-4412.

[17] Surewicz WK, Szabo A G, Mantsch H H. Conformational properties of azurin in solution as determined from resolution-enhanced Fourier-transform infrared spectra. European Journal of Biochemistry 1987;167 519-523.

Chapter 13

Isolation and Purification of Sperm Immobilizing/Agglutinating Factors from Bacteria and Their Corresponding Receptors from Human Spermatozoa

Vijay Prabha and Siftjit Kaur

Additional information is available at the end of the chapter

1. Introduction

For successful fertilization, motility is the most obvious and most essential sperm function and has been repeatedly shown to be predictive of fertilization in vitro [1]. Several studies have shown that the motility characteristics of spermatozoa are of the utmost importance for the men's fertility [2]. Spermatozoa dysfunction is the single most important cause of infertility. A decrease in spermatozoa motility with time is universal phenomenon. This reduction differs from species to species and also among individuals of the same species, as in the human male. Most investigators agree that the majority of spermatozoa cease to move within the first 24 hours. The survival of spermatozoa after ejaculation is dependent on the environmental conditions under which they are kept. In the female genital tract they may remain active for several days [3], but their activity is of much shorter duration if they remain in the seminal fluid outside the body.

Male genital infections are relevant cause in the etiology of infertility due to abnormalities in sperm quality [4,5], affecting spermatozoal motility. The comparison of semen characteristics between infected and non-infected men show that motile spermatozoa are lower when the microorganisms are present in the semen [6]. It appears that bacteria have a direct effect on sperm motility with negative consequences in fertility. Among bacterial species that interact with spermatozoa are well-known causative pathogens of genitourinary infections such as *Escherichia coli, Ureaplasma urealyticum, Mycoplasma hominis, Chlamydia trachomatis* [7]. Of the various bacteria, *E. coli* is the most extensively studied microorganism in relation to infertility as a result of interaction with spermatozoa [8]. It is also the primary bacteria associated with prostatitis and epididymitis [9]. Several authors describe sperm agglutination and immobilization by *E. coli* [10,11]. Paulson & Polakoski [12] investigated the mechanism of how *E. coli* immobilizes spermatozoa and they reported a factor, apparently excreted by the bacteria which immobilizes spermatozoa without agglutinating it. However, Diemer et al. [13] reported that *E. coli* inhibits sperm motility by directly adhering to and agglutinating spermatozoa. Rapidity and extent of sperm–*E. coli* agglutination indicated strong adhesive forces. Bartoov et al. [8] proposed that mannose plays a critical role in adherence of *E. coli* to sperm. Although, a number of studies have evaluated the ability of *E. coli* to affect sperm motility by adherence, agglutination and dialyzable factors, however, none have identified the exact mechanism of interaction between spermatozoa and bacteria.

In addition to *E. coli, Staphylococcus aureus,* the predominant flora in infertile men, has also been reported to cause a significant decrease in sperm motility [14]. Emokpae et al. [15] while studying the contribution of seminal tract infection to sperm density, asthenozoospermia and teratozoospermia, observed *S. aureus* as the causative organism accounting for 68.2% of seminal infections. Most practitioners dismiss this infection as mere contamination, which is assumed to have no significance. Semen that passes through the genital tract is routinely contaminated with staphylococci. However, since the prevalence of abnormal sperm indexes is high, it was suggested that *S. aureus* infection should be treated and no longer ignored when managing male factor infertility [15]. *S. aureus* is known to produce various toxins and enzymes that may be exerting damaging effect on human sperm, but its mechanism of action also needs further investigation. Therefore, the present work was undertaken to study the mechanism of immobilization/agglutination of spermatozoa by *S. aureus* and *E. coli*.

2. Microorganisms

The bacterial isolates *Staphylococcus aureus* either showing immobilization or agglutination of human spermatozoa, were isolated from the cervices of women suffering from unexplained infertility, attending the Department of Obstetrics and Gynecology, Government Multi Speciality Hospital, Sector-16, Chandigarh, India. The isolates of *Escherichia coli* were isolated from semen samples of infertile males attending the infertility clinic at the Department of Urology, Post Graduate Institute of Medical Education and Research (PGIMER), Chandigarh.

3. Extraction and purification of sperm agglutinating factor (SAF) from *S. aureus*

Sperm agglutinating factor was extracted and purified from 72 h old cell culture of *S. aureus* by the method standardized in the laboratory [16]. The cell culture was centrifuged at 10,000 g for 20 minutes and the clear supernatant and the cells obtained were washed twice with sterile phosphate buffered saline (PBS) (50 mM, pH 7.2) and resuspended in the same buffer. When washed cells and cell free supernatant were studied for their sperm agglutinating property, the results showed that only washed cells were able to agglutinate the spermatozoa whereas culture supernatant failed to do so indicating that the agglutination of sperm might be associated with bacterial cells and not their metabolites. Pretreatment of *S. aureus* by sonication produced bacterial fragments that were unable to agglutinate sperm. Centrifugation of *S. aureus* fragments at 10,000 g for 5 min did not eliminate sperm agglutinating elements from the solution, indicating the sperm agglutinating factor to be present in sonicated supernatant. Based on its sperm agglutinating activity, the bioactive molecule from the sonicated supernatant was purified by ammonium sulphate precipitation, gel permeation chromatography and ion exchange chromatography. The sonicated supernatant was fractionated with ammonium sulphate so as to get 20, 40, 60 80 and 100% saturation. The flasks were kept at 4°C overnight and next day the precipitates were collected by centrifugation at 10,000 g for 15 min at 4°C and were redissolved in minimum amount of PBS. SAF was precipitated by ammonium sulphate at 40% saturation. The protein was dialyzed in preactivated dialysis bags against PBS under cold conditions and concentrated against polyethylene glycol (PEG) 6000 at 4°C. Further purification was done using Sephadex G-200 and DEAE cellulose column chromatography. The fractions containing approximately 1mg protein were applied on Sephadex G-200 (Pharmacia Fine Chemicals, Uppsala), column (2 cm X 31 cm) equilibrated and eluted with PBS. The aspirator bottle with PBS was joined with a fine capillary tube to maintain constant head pressure and allowed to run for 24 h. Fractions of 3 ml each were collected and each fraction was read at 280 nm on U.V. spectrophotometer. The fractions showing spermagglutinating activity i.e. fraction (4-7) with a peak value in fraction 5 were pooled and concentrated against PEG 6000 (Figure 1a) and applied to DEAE cellulose (Hi Media Laboratories Ltd., Mumbai, India) column. A column of 2 cm x 15 cm was made with activated slurry of DEAE cellulose, an anion exchanger with a pressure of about 1 cm water/cm of height of gel bed. The column was washed with PBS until the ion exchanger reached ionic equilibrium with the starting buffer. The aspirator bottle with PBS was joined with a fine capillary tube to maintain a constant head pressure and allowed to run. Final elution was done with PBS containing 0.05, 0.1, 0.2, 0.4 and 0.6 M NaCl. Fractions of 4 ml each were collected and read at 280 nm on U.V. spectrophotometer (Hitachi U-2900). All fractions were again checked for agglutination of spermatozoa. The fractions showing agglutination of sperm (35-39) were pooled and concentrated (Figure 1b). The purified pooled fractions were dialyzed, lyophilized and then subjected to protein estimation. The purification status of SAF was checked by gel electrophoresis (PAGE). Molecular weight estimation was done by sodium dodecyl sulphate polyacrylamide gel electrophoresis (SDS-PAGE) using a 12% resolving gel containing SDS [17]. 10μl of the pooled and concentrated sample obtained from DEAE cellulose column was mixed with loading dye (Bromophenol blue: glycerol) in the ratio of 1:50. The sample was applied to the gel and subjected to a current of 15 mA and run initially at 50V and thereafter at 100V. The gel was visualized by staining with Coomassie Brilliant Blue R-250. The molecular weight of SAF was estimated to be approximately, 65 kDa (Figure 1c).

4. Extraction and purification of sperm immobilization factor (SIF) from *S. aureus*

For extraction of SIF, *S. aureus* was grown in Brain Heart Infusion (BHI) broth for 72 h at 37°C under shaking conditions (150 rpm). The cell culture was subjected to centrifugation at

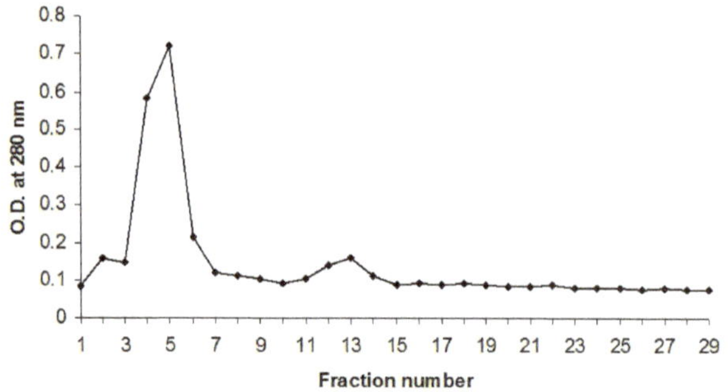

1a

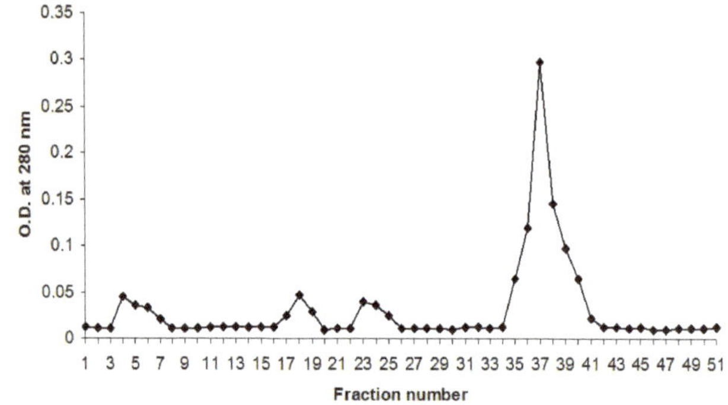

1b

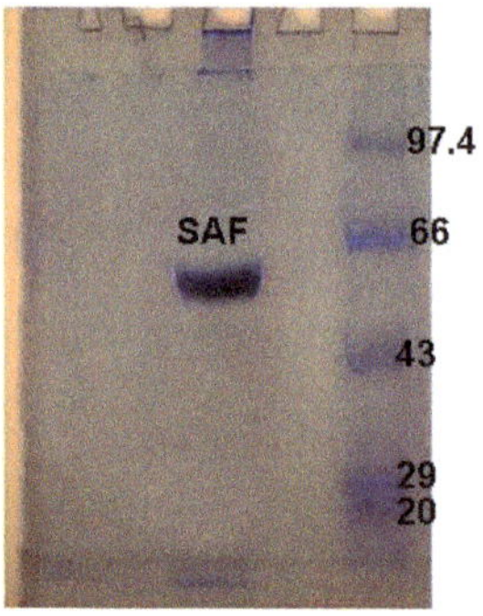

1c

Figure 1. **(a)** Elution pattern of SAF from *S. aureus* after gel filtration through Sephadex G-200 column showing the presence of SAF in fractions 4-7 with a peak value in fraction 5. **(b)** Elution pattern of SAF obtained after DEAE cellulose column showing sperm agglutination activity in fractions 35-39 with peak value in fraction 37. **(c)** SDS-PAGE of purified SAF, with **Lane1** containing DEAE cellulose purified and concentrated fraction. Molecular weight approximately, 65 kDa and **Lane2** containing Standard protein markers.

10,000 g for 30 min at 4°C. Both the resultant cell pellet and culture supernatant were checked for sperm immobilization activity. As the activity resided with cell supernatant it was subjected to ammonium sulphate precipitation so as to get 20, 40, 60, 80 and 100% saturation. SIF could be precipitated at 60-80% saturation. These precipitated proteins containing the bioactive molecule were dialyzed extensively against PBS and applied to Sephadex G-100 column (2 cm x 31 cm) equilibrated and eluted with PBS. The head pressure maintained to achieve a flow rate of 10 ml/h. Fractions of 3 ml each were collected, and the absorbance of each fraction was read at 280 nm. Fractions (12–18) showing the immobilization activity were pooled and concentrated against PEG 6000 at 4°C (Figure 2a). These fractions were then applied onto DEAE cellulose column. The column was equilibrated with PBS and head pressure was maintained to achieve a flow rate of 60 ml/h. Final elution was done with 0.05, 0.1, 0.2, 0.4 and 0.6 M NaCl dissolved in PBS. Fractions of 4 ml each were collected, and the absorbance was read at 280 nm. The fractions (47–52) that caused immobilization of spermatozoa were pooled and concentrated (Figure 2b). The fractions concentrated after DEAE cellulose chromatography were dialyzed, lyophilized, and subjected to PAGE analysis. The presence of a single band suggested apparent homogeneity of the protein. This purified fraction was then analyzed by SDS-PAGE against standard molecular weight markers (Figure 2c). The molecular weight of SIF was estimated to be ~20 kDa [18].

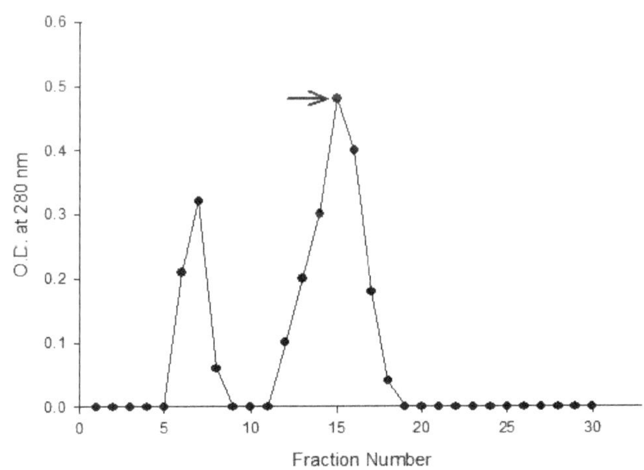

2a

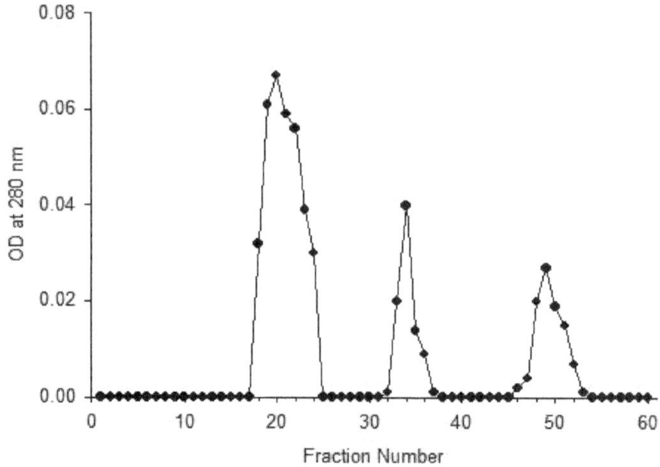

2b

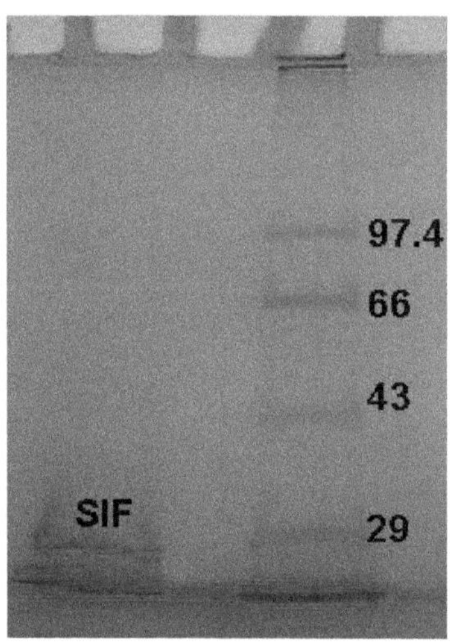

2c

Figure 2. **(a)** Elution pattern of SIF from *S. aureus* after gel filtration through Sephadex G-100 column showing the presence of SIF in fractions 12-18 with a peak value in fraction 15. **(b)** Elution pattern of SIF from *S. aureus* obtained after DEAE cellulose column showing sperm immobilization activity in fractions 47-52 with peak value in fraction **49.** **(c)** SDS-PAGE of purified SIF, with **Lane1** containing DEAE cellulose purified and concentrated fraction. Molecular weight approximately, 20 kDa and **Lane 2** containing standard protein markers.

5. Extraction and purification of SAF from *E. coli*

48 h old culture of *E. coli* grown in Luria broth was centrifuged at 10,000 g for 10 min at 4ºC. Cell free supernatant was prepared by passing the supernatant through a 0.22 μm Millipore filter. The cell pellet was washed twice with sterile PBS. Both the cell free supernatant and the washed cells were checked for spermagglutinating activity by incubating with semen samples. As the washed cells showed spermagglutinating activity, therefore further studies were carried out with washed cells. Extraction of sperm ligand from washed cells was done by salt treatment. The washed cells of *E. coli* (1000 ml, 48 h old cell culture) were incubated with 1, 2, 3, 4 and 5 M solution of NaCl under shake conditions (150 rpm) at 37ºC for different time intervals 2, 4, 8, 12 and 24 h, separately. The cells were centrifuged at 10,000 g for 30 min. The resulting cell pellet and supernatant (which was dialyzed against double distilled water overnight at 4ºC and passed through the UM05 Amicon filter) were analyzed for sperm agglutinating activity. As pellet did not show sperm agglutinating activity, further work was carried out with the supernatant. SAF could be efficiently extracted by treatment of cell pellet with 3 M NaCl for 12 h under shaking conditions (150 rpm) at 37ºC. Purification of crude sperm ligand consisted of filtration of dialyzed and concentrated fraction through a Sephadex G-200 column, equilibrated and eluted with PBS. Fractions of 3 ml each were collected and read at 280 nm on U.V. spectrophotometer. Fractions showing sperm agglutinating activity i.e. 6-9, with a peak in fraction 7 (Figure 3a) were pooled and concentrated using polyethylene glycol (PEG) 6000 at 4ºC. These fractions were applied to DEAE cellulose column. Final elution was done with 0.05, 0.1, 0.2, 0.4 and 0.6 M NaCl dissolved in PBS. Fractions of 4 ml each were collected and read at 280 nm on U.V. spectrophotometer. Most of the SAF could be eluted with PBS containing 0.4 M NaCl (fractions 46-49, Figure 3b). Fractions showing agglutinating activity were further pooled, dialyzed and concentrated. The protein content at each step was assayed by the method of Lowry et al. [19] against a standard curve calibrated with bovine serum albumin. PAGE of purified sperm ligand was carried out to check the purification status. Molecular weight estimation by SDS-PAGE, using the standard molecular weight markers showed the presence of ~71 kDa protein band (Figure 3c) [20].

6. Extraction and purification of SIF from *E. coli*

The isolate of *E. coli* showing immobilization of spermatozoa was grown in BHI broth under shaking conditions (150 rpm) at 37ºC for 72 h. The culture was centrifuged at 10,000 g for 15 min at 4ºC and cell-free supernatant was prepared by passing the supernatant through a 0.22 μm Millipore filter. The supernatant was then subjected to ammonium sulphate precipitation so as to get 20, 40, 60, 80, and 100% saturation. The precipitates so obtained were dissolved in a minimum amount of PBS. The precipitated protein was dialyzed against PBS under cold conditions and checked for sperm immobilization. SIF could be saturated at 60-80% of ammonium sulphate. Further purification of the factor was done by gel filtration through a Sephadex G-100 column (2 cm x 31 cm) equilibrated, and eluted with PBS.

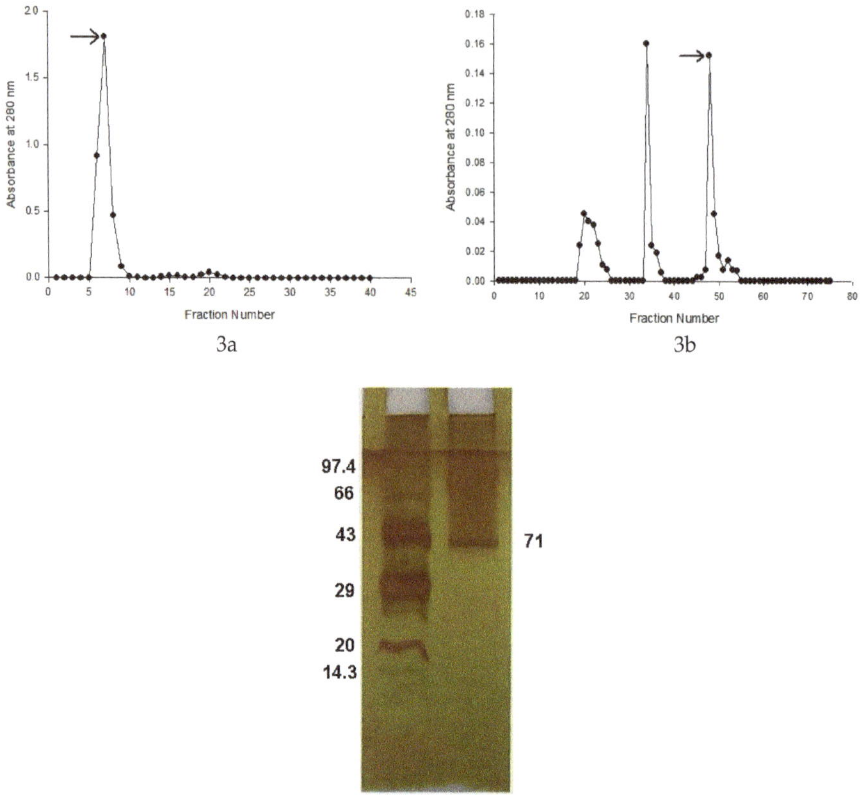

3c

Figure 3. **(a)** Elution pattern of SAF obtained after Sephadex G-200 gel filtration of dialyzed and filtered supernatant containing sperm ligand on *E. coli*, showing the presence of ligand in fractions 6-9 with peak value in fraction 7 (arrow refers to sperm ligand on *E. coli*). **(b)** DEAE cellulose chromatography of G-200 pooled and PEG concentrated fractions revealed the presence of three peaks but the agglutinating activity was present only in the fractions 46-49 with peak value in fraction 48 (arrow refers to sperm ligand on *E. coli*). **(c)** SDS-PAGE of purified sperm ligand, with **Lane1** containing Standard protein marker and **Lane2** containing DEAE cellulose purified and concentrated fraction. Molecular weight approximately, 71 kDa.

Fractions of 3 ml each were collected and each fraction was read at 280 nm on U.V. spectrophotometer. Fractions 8-14, showing the immobilization of spermatozoa were pooled and concentrated using PEG 6000 under cold conditions (Figure 4a). These fractions were applied on to a DEAE cellulose column. First of all, 80 ml of elution buffer, PBS (50 mM pH 7.2) was allowed to run down the column. Final elution was done with 0.05, 0.1, 0.2, 0.4, and 0.6 M NaCl dissolved in PBS (50 mM, pH 7.2). Fractions of 4 ml each were collected and read at 280 nm on U.V. spectrophotometer. The fractions showing immobilization of spermatozoa i.e. 3-7 were again pooled and concentrated (Figure 4b).

To verify the purification status of all the preparations, PAGE was carried out (10% resolving and 5% stacking gel). Estimation of molecular weight by SDS-PAGE, showed that SIF had a molecular weight of ~56 kDa (Figure 4c), compared to standard protein markers [21].

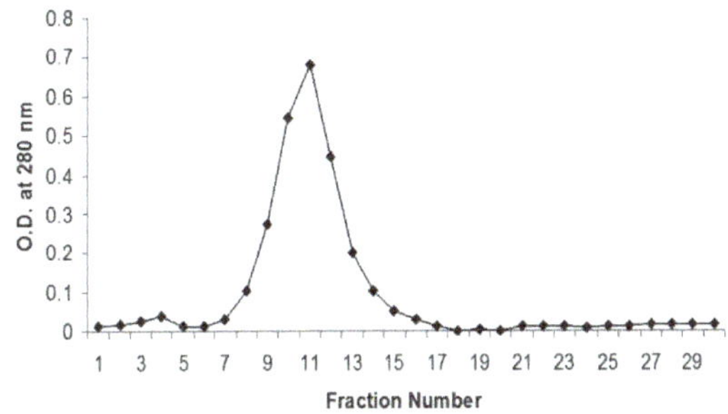

4a

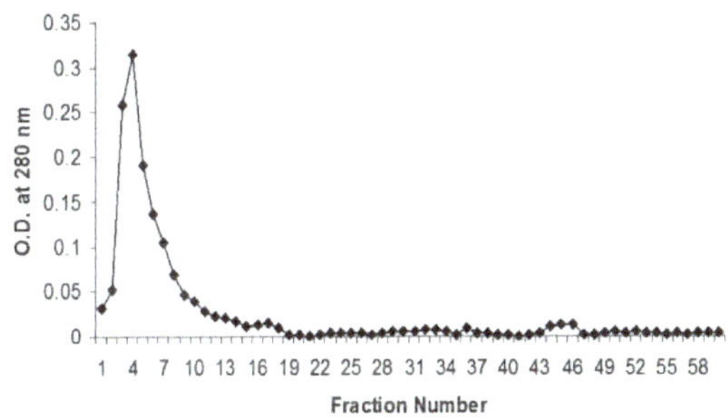

4b

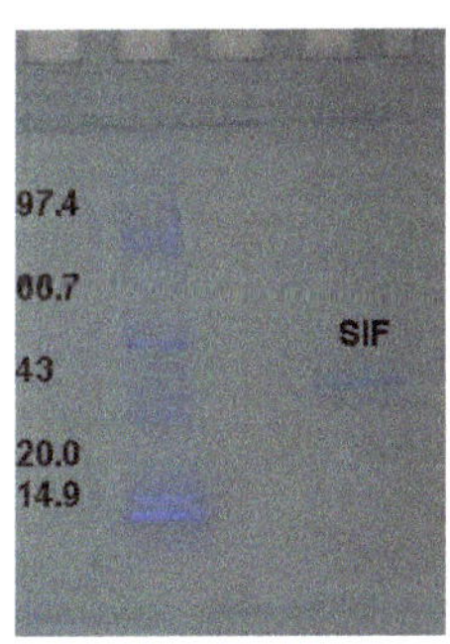

4c

Figure 4. (**a**) Elution pattern of SIF from *E. coli* on Sephadex G-100 column. (**b**) Elution pattern of SIF from *E. coli* on DEAE cellulose column. (**c**) SDS-PAGE of purified SIF, with **Lane1** containing Standard protein marker and **Lane2** containing DEAE cellulose purified and concentrated fraction. Molecular weight approximately, 56 kDa.

7. Isolation and purification of receptor from human spermatozoa for SAF isolated from *S. aureus*

Spermatozoa were washed twice with PBS and resuspended in the same buffer. Washed spermatozoa were treated with 1, 2, 3 and 4 M NaCl for different intervals to optimize the best concentration and time combination for receptor extraction. The treated mixture was centrifuged at 10,000 g for 20 min. The pellet and supernatant were examined for blockage of agglutination activity. Receptor was efficiently extracted by 3 M NaCl when incubated for 12 h at 37°C while shaking at 150 rpm. Crude receptor was further dialyzed extensively against PBS under cold conditions and then concentrated using PEG 6000. Receptor was purified by filtration through a 2 x 31 cm Sephadex G-200 column, equilibrated and eluted with PBS. Head pressure was maintained to achieve a flow rate of 10 ml/h. Fractions (3 ml) were collected and each was read at 280 nm using a U.V. spectrophotometer. Fractions were analyzed for the presence of receptor by blocking the agglutination of spermatozoa by SAF. Fractions (12-17) representing receptor were pooled, concentrated against PEG 6000 at 4°C (Figure 5a). One-dimensional SDS-PAGE was done using a vertical slab gel apparatus with stacking gel containing 5% polyacrylamide and resolving gel containing 12% polyacrylamide. The protein sample was diluted 1:1 with reducing sample buffer and heated for 5 min at 100°C. The sample was loaded and subjected to electrophoresis at 50 V for 15 min, followed by 150 V for 45 min. The gel was stained with Coomassie Brilliant Blue R-250 and calibrated using molecular weight markers. The molecular weight of the receptor was estimated to be approximately, 57kDa (Figure 5b) [22].

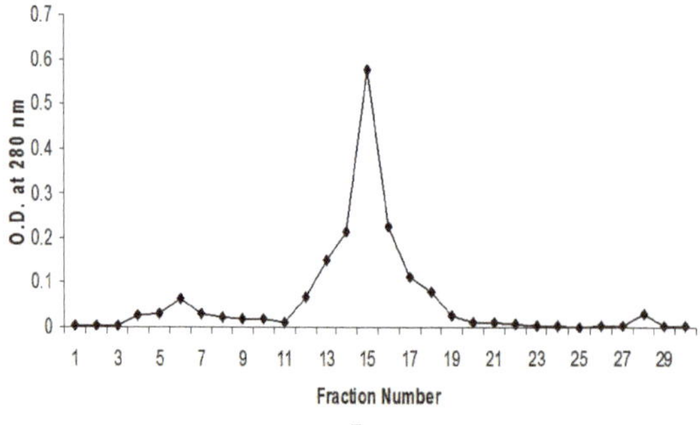

5a

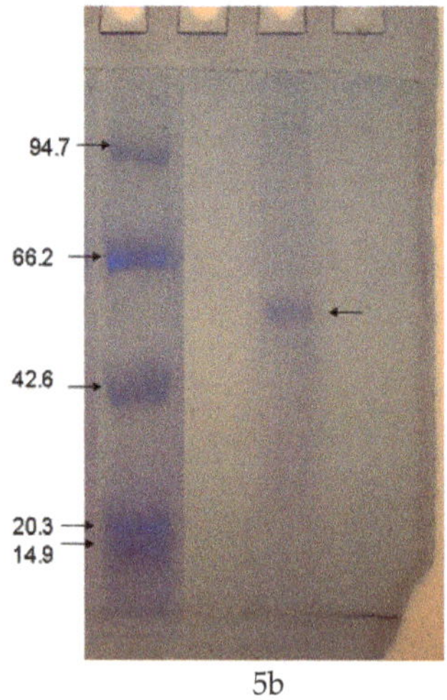

5b

Figure 5. **(a)** Elution pattern of receptor from human sperm on Sephadex G-200 column. **(b)** SDS-PAGE (molecular weight determination) **Lane 1**: Standard proteins markers, **Lane 2**: Purified receptor protein.

8. Isolation and purification of receptor for SIF isolated from *S. aureus*

Salt extraction of the receptor from human spermatozoa was done by treating the washed sperm sample with 1, 2, and 3 M NaCl for different time intervals i.e. 2, 4, 8, 12, 24 and 48 h. Purification of the receptor was further done by gel filtration through a Sephadex G-100 column (2 cm × 31 cm) equilibrated and eluted with PBS (50 mM, pH 7.2). Fractions of 3 ml each were collected and absorbance read at 280 nm to determine the SIF receptor concentration. Fractions showing the blockage of immobilization of spermatozoa (2-5, with a peak in fraction 3) were pooled and concentrated using PEG 6000 under cold conditions (Figure 6a). To check the purification status and biological activity of pooled and concentrated fractions, PAGE was carried out. Molecular weight of the purified receptor as estimated using SDS-PAGE was found to be approximately, 62kDa (Figure 6b) [23].

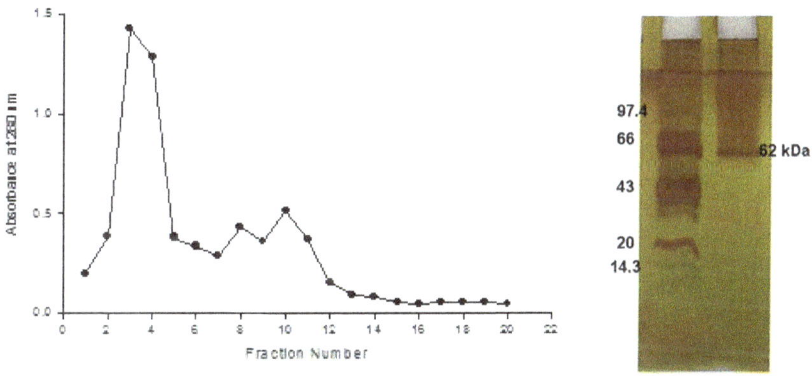

6a 6b

Figure 6. (a) Elution pattern of receptor from human sperm on Sephadex G-100 column. **(b)** SDS-PAGE (molecular weight determination) **Lane 1**: Standard proteins markers, **Lane 2**: Purified receptor protein

9. Isolation and purification of receptor for SAF isolated from *E. coli*

The corresponding receptor for SAF was isolated from human spermatozoa using the ligand as a tool. The procedure involved the use of salt solution for the extraction of molecules bound to the cell surface. The spermatozoa were washed twice with PBS and then resuspended in same buffer. They were then treated with different concentrations of NaCl i.e. 1, 2, 3 and 4 M for different time intervals and incubated at 37°C under shaking conditions (150 rpm). The salt treated sample was centrifuged at 1500 g for 15 min. Cell debris was suspended in minimum amount of PBS. Both cell debris and supernatant were dialyzed against PBS at 4°C overnight, concentrated against PEG 6000 and checked for blocking of agglutination induced by SAF. Results showed that receptor for SAF from *E. coli* could be efficiently extracted by 2 M NaCl when incubated for 18 h. Purification of the receptor was further carried out by filtration through a Sephadex G-200 column (1.5 cm X 31 cm) equilibrated and eluted with PBS. The head pressure was maintained to achieve a flow rate of 35 ml/h. Fractions of 3 ml each were collected and each fraction was read at 280 nm. Fractions showing blockage of agglutination activity were pooled and concentrated with PEG 6000 at 4°C (Figure 7a). The homogeneity of the preparations was checked by PAGE and the molecular weight of the purified receptor was estimated by SDS-PAGE. The molecular weight of the receptor for SAF from *E. coli* was approximately, 125 kDa (Figure 7b) [24].

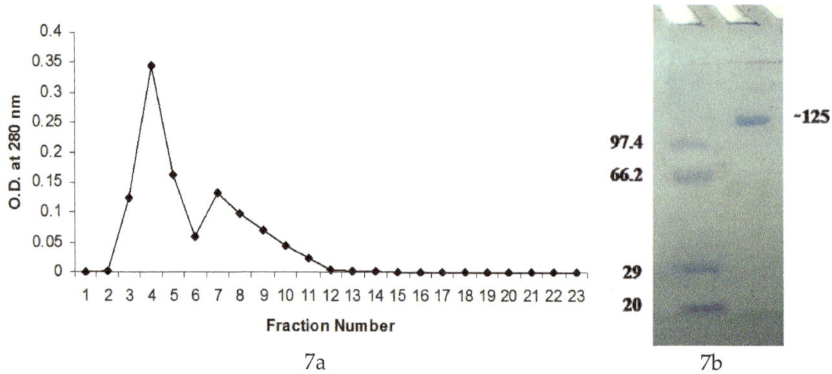

Figure 7. **(a)** Elution pattern of receptor from human sperm on Sephadex G-200 column. **(b)** SDS-PAGE (molecular weight determination) **Lane 1**: Standard proteins markers, **Lane 2**: Purified receptor protein

10. Isolation and purification of receptor for SIF isolated from *E. coli*

The spermatozoa were washed twice with PBS (50 mM, pH 7.2) and then resuspended in same buffer. The washed spermatozoa were treated with different concentrations of NaCl i.e.1, 2, 3 and 4 M for different time intervals in order to optimize the best concentration and time combination for receptor isolation. These were incubated at 37°C under shaking conditions (150 rpm). The salt treated sample was centrifuged at 1500 g for 15 min. Cell debris was suspended in minimum amount of PBS. Both the cell debris and supernatant after dialysis and concentration (against PEG 6000) were checked for blocking of immobilization induced by SIF. Preliminary results showed that SIF receptor could be efficiently extracted by 2 M NaCl from spermatozoa when incubated for 20 h at 37°C under shaking conditions.

Purification of the receptor was further done by filtration through a Sephadex G-200 column (1.5 cm X 31 cm) equilibrated and eluted with PBS. The head pressure was maintained to achieve a flow rate of 35 ml/h. Fractions of 3 ml each were collected and each fraction was read at 280 nm (U.V. spectrophotometer). The column chromatographic pattern showed that the receptor could be eluted in the fractions 3-4 which could block the SIF induced sperm immobilization (Figure 8a). These fractions were pooled and concentrated with polyethylene glycol (PEG 6000) at 4°C and were applied to ion exchange column. First of all, 80 ml of elution buffer (PBS) was allowed to run down the column. Final elution was done with 0.05, 0.1, 0.2, 0.4 and 0.6 M NaCl dissolved in PBS. Fractions of 4 ml each were collected and read at 280 nm on U.V. spectrophotometer. The fractions showing blockage of sperm immobilization i.e. 2-4 were pooled and concentrated (Figure 8b). Molecular weight of the purified SIF receptor was estimated by SDS-PAGE. The purified SIF receptor was first denatured and then loaded onto the gel along with the standard protein markers. After the gel was run, silver staining was done and molecular weight was estimated to be ~113 kDa (Figure 8c) [25].

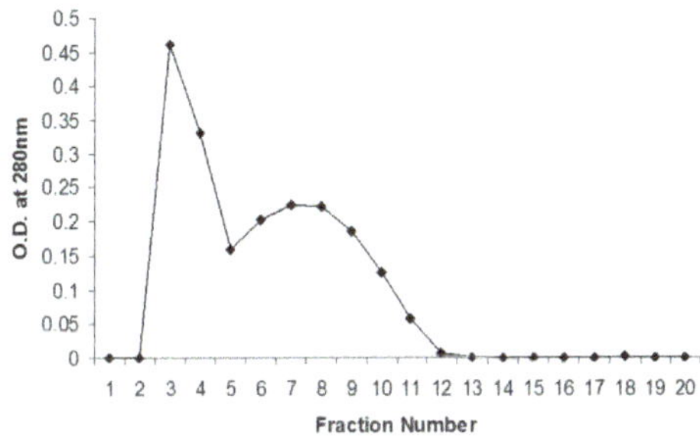

8a

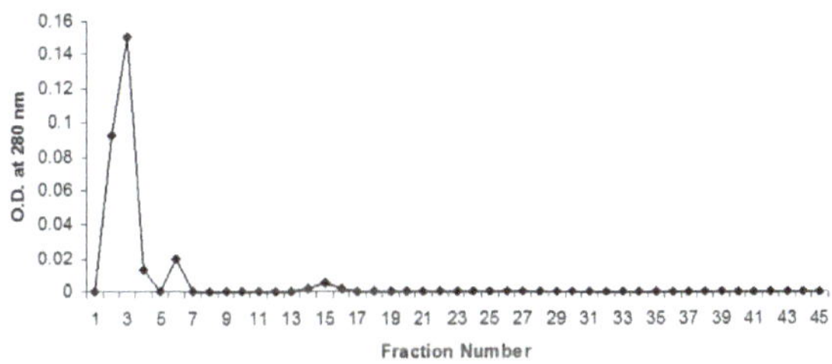

8b

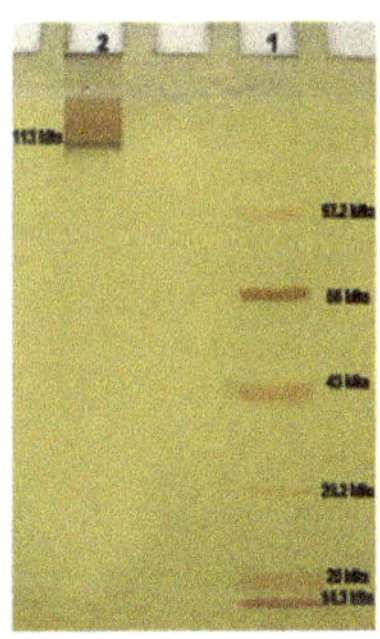

8c

Figure 8. **(a)** Elution pattern of receptor from human sperm on Sephadex G-200 column. **(b)** Elution pattern of receptor from human sperm on DEAE cellulose column. **(c)** SDS-PAGE (molecular weight determination) **Lane 1**: Standard proteins markers, **Lane 2**: Purified receptor protein

11. Conclusion

This study identifies receptor-ligand interaction between bacteria and spermatozoa that results in sperm immobilization/agglutination.

12. Further research

Understanding bacteria-spermatozoa interactions at receptor-ligand level may hold potential for infertility treatment.

13. Abbreviations

~	:	approximately
%	:	percentage
°C	:	degree celcius
BHI	:	brain heart infusion
DEAE	:	Diethylaminoethyl
h	:	hour
i.e.	:	id est (that is)
kDa	:	kilodalton
µg	:	microgram
µl	:	microlitre
µm	:	micrometre
mA	:	milliampere
min	:	minutes
ml	:	millilitre
M	:	molar
mM	:	millimolar
mm	:	millimetre
cm	:	centimetre
nm	:	nanometre
mg	:	milligram
NaCl	:	sodium chloride
O.D.	:	optical density
PAGE	:	polyacrylamide gel electrophoresis
PBS	:	phosphate buffered saline
PEG	:	polyethylene glycol
rpm	:	revolutions per minute
s	:	seconds
SDS	:	sodium dodecyl sulphate
SAF	:	sperm agglutinating factor
SIF	:	sperm immobilization factor
U.V.	:	ultraviolet
V	:	volts

Author details

Vijay Prabha and Siftjit Kaur
Department of Microbiology, Panjab University, Chandigarh, India

Acknowledgement

This work was supported by funds from University Grants Commission, New Delhi and Indian Council of Medical Research, New Delhi, India.

14. References

[1] Bonde JP, Ernst E, Jensen TK, Hjollund NH, Kolstad H, Henriksen TB, Scheike T, Giwercman A, Olsen J, Skakkebaek NE (1998) Relation between semen quality and fertility: a population-based study of 430 first-pregnancy planners. Lancet 352: 1172-1177.

[2] Karpuz V, Gokturk A, Meral Koyuturk M (2007) The effects of sperm morphology and motility on the outcome of intracytoplasmic sperm injection. Marmara med. j. 20: 92-97.

[3] Hafez ESE (1980) Physiology of spermatozoal insemination in the female reproductive tract. In: Emperiare JC, Audebent A, Hafez ESE editors. Homologous artificial Insemination. The Hague, Mastinus Nijhoff Publishers .pp.328-334.

[4] Golshani M, Taheri S, Eslami G, Suleimani Rehber AA, Fallah F, Goudarzi H (2006) Genital tract infection in asymptomatic infertile men and its effect on semen quality. Iranian j. publ. health 35: 81-84.

[5] Pellati D, Mylonakis I, Bertoloni G, Fiore C, Andrisani A, Ambrosini G, Armanini D (2008) Genital tract infections and infertility. Eur. j. obstet. gynecol. reprod. biol. 140: 3-11.

[6] Menkveld R, Wong WY, Lombard CJ, Wetzels AMM, Thomas CMJ, Merkus HMWM, Steegers-Thuenissen RPM (2001) Semen parameters, including WHO and strict criteria morphology, in a fertile and subfertile population: an effort towards standardization of in-vivo thresholds. Hum. reprod. 16: 1165-1171.

[7] Cunningham KA, Beagley KW (2008) Male genital tract chlamydial infection: implications for pathology and infertility. Biol. reprod. 79:180-189.

[8] Bartoov B, Ozbonfil D, Maayan MC, Ohad E, Nitzan Y (1991) Virulence characteristics of male genital tract *Escherichia coli* isolated from semen of suspected infertile men. Andrologia 23: 387-394.

[9] Comhaire F, Verschraegen G, Vermeulen L (1980) Diagnosis of accessory gland infection and its possible role in male infertility. Int. j. androl. 3: 32-45.

[10] Huwe P, Diemer T, Ludwig M, Liu J, Schiefer HG, Weidner W (1998) Influence of different uropathogenic microorganisms on human sperm motility parameters in an *in vitro* experiment. Andrologia 301: 55-59.

[11] Khalili MB, Sharifi-Yazdi MK (2001) The effect of bacterial infection on the quality of human's spermatozoa. Iranian j. publ. health 30: 119-122.

[12] Paulson JD, Polakoski KL (1977) Isolation of a spermatozoal immobilization factor from *Escherichia coli* filtrates. Fertil. steril. 28: 182-185.

[13] Diemer T, Weidner W, Michelmann HW, Schiefer HG, Rovan E, Mayer F (1996) Influence of *Escherichia coli* on motility parameters of human spermatozoa in vitro. Int. j. androl. 19: 271-277.

[14] Fowler JE Jr, Mariano M (1984) Difficulties in quantitating the contribution of methral bacteria to prostatic fluid and seminal fluid cultures. J. urol. 132:471-473.

[15] Emokpae NA, Uadia PO, Sadiq NM (2009) Contribution of bacterial infection to male infertility in Nigerians. OJHAS 8:6. Available at: http://openmed.nic.in/3277/01/2009-1-6.pdf

[16] Ohri M, Prabha V (2005) Isolation of sperm agglutinating factor from *Staphylococcus aureus* isolated from a woman with unexplained infertility. Fertil. steril. 84: 1539-1541.

[17] Laemmeli UK (1970) Cleavage of structural proteins during the assembly of the head of bacteriophage T4. Nature 227: 680-685.

[18] Prabha V, Gupta T, Kaur S, Kaur N, Kala S, Singh A (2009) Isolation of a spermatozoal immobilization factor from *Staphylococcus aureus* filtrates. Can. j. microbiol. 77: 874-878.

[19] Lowry OH, Rosebrough Al, Randall RJ (1951) Protein measurement with Folin-Phenol reagents. J. biol. chem. 193: 265-275.

[20] Prabha V, Thakur N, Kaur S, Kaur N, Singh A, Kala S (2009) Agglutination of human spermatozoa due to human semen culture bacterial isolates bearing sperm ligand. Am. j. biomed. sci. 1: 126-132.

[21] Prabha V, Sandhu R, Kaur S, Kaur K, Sarwal A, Mavuduru RS, Singh SK (2010) Mechanism of sperm immobilization by *Escherichia coli.* Adv. urol. 2010:1-6. doi: 10.1155/2010/240268.

[22] Kaur S, Prabha V, Sarwal A (2010) Receptor mediated agglutination of human spermatozoa by spermagglutinating factor isolated from *Staphylococcus aureus.* J. urol. 184: 2586-2590.

[23] Prabha V, Chaudhary N, Kaur S (2011) Molecular mimciry between bacteria and spermatozoa. J. urol.186: 2442-2447.

[24] Kaur K, Kaur S, Rishi P, Singh SK, Prabha V (2012) Evidence for the occurrence of receptor in sperm for spermagglutinating factor isolated from *Escherichia coli.* Gynecol. endocrinol.-int. j. gynecol. obstet. 34: 207-209.

[25] Kumar V, Prabha V, Kaur S, Kaur K, Singh SK (2011). Receptor dependent immobilization of spermatozoa by sperm immobilization factor isolated from *Escherichia coli:* Proof of evidence. Int. j. urol. 18: 597–603.

Chapter 14

Characterization of Apple Pectin – A Chromatographic Approach

Maria Helene Giovanetti Canteri, Alessandro Nogueira,
Carmen Lúcia de Oliveira Petkowicz and Gilvan Wosiacki

Additional information is available at the end of the chapter

1. Introduction

The apple tree belongs to the *Plantae* kingdom, *Magnoliophyta* class, *Rosales* order, *Rosaceae* family, *Pomoidae* sub-family and *Malus* genus. Originating in the mountains of Asia, the apple (*Malus domestica* Borkh) is a temperate fruit with an ancient history [1].

The center of the origin of the *Malus* cultivars is understood to be in Asia Minor, in the Caucasus, located between the Caspian and the Black Seas, the Indian Himalayas, Pakistan and Western China[2], all areas that had access to the Silk Road where marketable products were transported from the east to Europe since antiquity. This is a series of trade routes that crossed Eurasia by land and by stretches of sea or river for over 2,000 years until 1,500 BC, and which enabled the movement of important materials that promoted the exchange of ideas and innovations between different cultural groups. Over the centuries, gunpowder, the compass, the printing press, silk, mathematics and ceramics all migrated along this road, as well as stringed instruments. Resources, information and news were all spread in this fashion among many cultures for such a long period that is often difficult to identify the source of many traditions.

Whatever may have been the mechanism of transport of varieties of apples it is certain that animals, birds, cattle and bats, among others, played an important role [2]. The caravans, with their humans and camels, ate the fruit and left the seeds in fertile areas along the highway since the Neolithic period and many cultivars were perfectly defined in the Middle East around 4000 BC. These regions are characterized by a temperate climate, with low winter temperatures - important for the winter hibernation of apple trees. Commercial apples today are descendants of species from Asia, including the Gala, Golden Delicious and Fuji.

2. Geographical distribution of apple agroindustries

The cultivation of apples is found in diverse temperate climates around the world, in both hemispheres but with a greater distribution in the northern hemisphere. In the first decade of the twenty-first century, in a situation with many global changes taking place, the five biggest producers of fruit are those known by the acronym BRICS (Brazil, Russia, India, China and South Africa).

The worldwide production of apples of 68 million metric tons, contains 40% that is relative to the main producing country, China, which is characterized by having a monovarietal culture based on the Fuji variety [3]. The area under cultivation in China has been stable over the past three years, with about 2,000 ha and a yield of 16 ton/ha, and the country still imports 58,000 tons which brings the domestic market up to a total of 32,058 million tons. The consumption of fresh fruits is 25.6 million tons, with a per capita index of 19.2 kg, and 1,460,000 tons of these high quality apples are exported, leaving 4,998 million tons for industrial use, of which 87.5% are destined for use as concentrated juice [4]. These statistics place China in a prominent position in the apple agroindustry. This means that just as China had increased its production to become the largest producer as forecast [5] many other countries had decreased their production to justify decline in production in the rest of the world of -2%. The production of the European Economic Community, including the UK, demonstrates stability of production and points to a fall of 5% between 2002 and 2011, demonstrating how the largest producers such as Poland and Italy (> 2,000,000 tons), France (> 1,000,000 tons) and Germany (900,000 tons), which together account for 70% of this activity of apple production available for both the common market and for export [6]. The United States, the second largest producer, accounted for 6.2% of the volume of apples produced in 2008. In South America, the main producers are, in order, Chile, Argentina and Brazil, although none of the three produced more than 2% of the world total in 2008. In that year, production in Brazil accounted for 1.6% of the world total, positioning the country as 12th in the ranking of countries by apple production [7].

There are thousands of domesticated and commercial species of apples that act as a producer or pollinator in commercial orchards and they are ranked periodically in relation to their commercial importance. The top ten apple varieties (excluding those produced in China) in descending order are Golden Delicious, Delicious, Gala, Granny Smith, Fuji, Jonagold, Idared, Jonathan, Braeburn and McIntosh [8].

The major advances in the cultivation of apples are listed by [9] and include the development of new varieties and disposal of others in the process of applying emerging research processes. Granny Smith, Gala, Fuji, Jonagold, Pink Lady and Empire are all types of apples that have excelled more effectively than expected. A new feature has been the practice of treating all the varieties in a similar way as a means of maintaining the unit price, since the producers are moving towards producing predetermined and defined amounts of fruit. This homogenous practice spread to Australia, but in Europe there are already 10 sets of varieties and in the United States there are at least seven such sets. The dwarfing of apples is regarded as an improvement as well as the optimization of planting density systems, with the solution of problems of overheating solved by setting branches at angles, which demonstrates the practical aspects of the solutions. The development of growth regulators and genomics shows the dependence on current scientific development [9].

3. Principal products from apples

In the apple processing industry the principal products are apple juice and fermented products such as cider and vinegar. In some countries (France, Spain, Switzerland and England), genetically improved fruit are used with special physical and chemical characteristics in order to standardize products, which are called 'industrial apples'. On the other hand, the production of apples generates a discarding process both at the time of harvest and in the classification process for the fresh market. These discarded apples can vary from between 10 to 30% of total production [10].

These discarded fruits are taken for processing to extract the juice or must. In Brazilian companies, besides the mixture of cultivars, there are fruits in different stages of ripening (from pre-maturation to senescence) and with possible phytopathological defects due to high temperatures (25-35°C) at harvest time. This occurs in summer and fruit for processing is stored externally, as in Europe, but the temperature at harvesting time in France is 7-15ºC [11].

Various technologies can be used in the extraction of apple juice. The system of continuous press is the most used, but in small companies discontinuous vertical hydraulic presses are also used. The addition of pectinolytic enzymes and maceration may increase the extraction yield either by pressing or centrifuging. However, the use of enzymes should not alter the physical-chemical parameters established by international legislation or the standards of each company that will use the juice or must [10].

In the process of extracting the juice or must a by-product is produced, which is known as apple pomace. This residue can represent 20-35% of the production. The yield of pomace that is generated and its composition will depend upon the cultivars, their stage of conservation and the technology used in processing.

3.1. Apple pomace

Traditional stabilization technology takes into account that apple pomace is an agro-industrial residue with a view to the fact that its disposal creates environmental problems of high cost. However, apple pomace is an interesting material that has attracted considerable attention as a potential source of dietary sugar, fiber, phenolic compounds and pectin.

These products can be used for many purposes in the pharmaceutical, cosmetics and food industries. The commercial production of apples in Brazil, based on only two cultivars, was designed to meet highly demanding customer requirements both with respect to quality and the retail prices of table fruit, and more recently the industrial product, either as juice or in a fermented form. Industrial apple pomace has in its composition all the constituents of the fruit in varying amounts, more or less, and is the residue from the pressing of the grated mass of unfermented juice, cider, wines, brandies and distillates or vinegars 12.

In Brazil, 70% of apples produced are sold for fresh consumption, while 30% are considered 'industrial' fruit. One third of this fraction is considered to contain fruit of low quality, which must be discarded or used for the production of distillates such as alcohol or vinegar and the remaining two thirds are fruits that can be used for the processing of apple juice 13. On the global scene, in the 1980s, with reference to industrial fruit, 75% of the products consisted of apple juice or must and 25% was apple pomace, when 8 kg of raw material was required to produce the amount of 6 kg of apple juice necessary to make1 kg of concentrated juice. In 1996, the yield had risen by 85% due to the use of enzymes in the preparation of the must. Since that time, the yield has increased slightly depending on the variety of the fruit and the degree of maturity, which means that currently 7 kg of apples are required for 1 kg of concentrate. But it is possible to increase this by 1% with the use of the most up-to-date enzymes [14].

The drying of apple pomace seems to be the most economic technological approach to stabilize the product because it dramatically reduces the volume of water and makes transport cheaper. The yield of dried pomace at 60°C is about 50.0 g kg^{-1} in 10 hours, or 5% of the raw material. The appearance of the dried pomace is dependent on the adiabatic drying temperature. From 50 to 60°C, darkening enzymatic reactions are stimulated [15], while from 90 to 100°C Maillard reactions occur, with products appearing darker than those obtained in the range of 70 to 80°C. However, if the criterion of stopping the process is the time at which the temperature of the pomace starts to rise, that temperature will never rise to values higher than 52°C and thus the final temperature tends to be homogeneous.

The instability of apple pomace is related to its physical-chemical composition and the presence of some enzymes that are activated after the disintegration of plant tissues [16,17,14]. Apple pomace is composed of water (76.3%), soluble solids (23.7%) and is obtained from the epi-mesocarp (95.5%), seeds (4.1%) and stems (1.1 %). It has an average humidity of 80% and 14% of total soluble solids including glucose, fructose and sucrose. Its composition is related to the cultivar and processing [17]. The fiber content varies from 11.6 to 44.5%, and includes cellulose (12.0 to 23.2%), lignin (6.4 to 19.0%), pectin (3.5 to 18%,) and hemicellulose (5.0 to 6.2%). The average dietary fibers (35.8%) and sugars (54.4%) make up 91.2% of pomace and the other components are proteins, lipids and ash [18]. The chromatic characteristics L = 51.8, a = 5.4 and b = 18.2 have been determined in samples of apple pomace [19].

The use of apple pomace as a potential source of nutrients for the production of glucosidase by *Aspergillus foetidus* was suggested by [20]. Ten years later [21] suggested its use for other technological purposes such as the recovery of phenolic compounds. Apple pomace is also recommended for biotechnological applications such as ethanol production [22], flavorings, citric acid, pectin, enzymes and molds for the extraction of dietary fiber and mineral coal [23].

The apple pectin that is derived from the extraction of apple pomace is dark brown in color compared to orange pectin. Studies are being carried out regarding the potential use of apple pomace in the recovery of native phenolic compounds associated with darkening, for use as antioxidants, resulting in a lighter color of the pectin obtained. These points to the growing trend of industry to find alternatives that promote the 'recycling' of waste, with maximum recovery and greater commercial exploitation of components previously considered as by-products.

3.2. Pectin

Pectin refers to a family of polysaccharides and oligosaccharides, which have common characteristics but are extremely diverse in their fine structure. The pectic skeleton is primarily a homopolymer of galacturonic acid bound in $\alpha(1\rightarrow4)$, with varying degrees of methyl esterified carboxyl groups 24;25. Pectin must consist of at least 65% galacturonic acid, according to the FAO (UN Food and Agriculture Organization) and EU (European Union) 25.

Pectin is one of the most important substances found in apples because it provides approximately 10% of daily fiber requirements. Pectin is a soluble fiber that is not absorbed by the intestine, i.e. the fibers are not degraded by digestive juices, but they increase the volume of fecal material, assist in the proper functioning of the intestine, retain water and various residual substances, facilitate the elimination of toxins along with stools, promote the protection of the intestinal mucosa and help in the treatment of diarrhoea. Pectin is also highly recommended for diabetics because it reduces the absorption of glucose. The daily consumption of approximately 2 small apples provides the required dose of pectin. Pectin also assists in reducing bad cholesterol because it forms a fiber barrier in the intestinal wall preventing the absorption of cholesterol and other fats.

In the industrial sector, pectic polysaccharides promote increased viscosity and act as a protective and stabilizing colloid in foods and beverages such as jams and jellies, fruit preparations for yoghurts, concentrated fruit juices and drinks, milk and fruit-based desserts, gelled dairy products, confectionery, and dairy products that are directly acidified or fermented. Other properties include the prevention of flotation in fruit preparations, stability of bread products, protein stabilization, softness in texture, increase in volume and the control of syneresis 25.

3.2.1. History of pectin

The first citation relating to pectin is found in an English article from 1750 about the preparation of apple jelly [26]. The process of extracting liquid pectin was recorded in 1908 in Germany and the process spread rapidly to the United States, where a patent was obtained in 1913 by Douglas (U.S. Patent no. 1,082.682) [27]. As for commercial production, in the 1930s Hermann Herbsthreith discovered the potential use and application of apple pomace, a hitherto discarded by-product of the production of fruit juice [28].

The content of pectic substances varies depending on the botanical source of plant material. There are four by-products of the agricultural and food industries that are high in pectic substances (content over 15% on a dry basis): pomace from apples, citric albedo, sugar beet pulp and /sunflower rinds 29.

The cell wall, a dynamic compartment of plants, can be divided into two layers called 'primary' and 'secondary'. The primary cell wall can be classified into: [1] type I, mainly composed of cellulose, xyloglucans, pectin and extensin, generally present in dicots and some monocots (non-comelinoides) and [2] type II, comprising mainly cellulose, glucoarabinoxylans and phenolic compounds with a lower proportion of pectin, found in Poaceae and in most monocotyledonous plants. The matrix of pectin controls, among other properties, porosity. The middle lamella consists of pectin molecules that are joined by cross-linked chains with homogalacturonan layers subsequently deposited in pectin of opposing cells 30.

Most of the pectin used by the food industry originates from such raw materials and is extracted under conditions of low acidity and high temperature, resulting in chains that are primarily homogalacturonan 25.

The ratio of raw materials and solvent in the extraction of pectin can be adjusted in order to separate the solid and liquid phases, the filtration of the extract and the costs of water evaporation in the process. Thus, it is possible to control the extraction of pectin to optimize its potential use 24; 31.

3.2.2. Structure and composition of pectin

In 1934, citrus pectins were recognized as linear chains of galacturonic acid and since then it has been found that pectin is a highly complex molecule. The challenge of recent times has been to accommodate all the available information in a single structural model [25]. Pectin are formed by seventeen different monosaccharide, arranged in distinct polysaccharides from more than twenty different connections that form a network when joined together [25; 30] and they are grouped into different types of chains consisting of uronic acids, hexoses, pentoses and methyl pentoses. Several structural units may be replaced by methanol, acetic acid and phenolic acids. Sugars can exist in furanic or pyranic forms and with different anomers (α or β) with various types of linkages between monomers such as α(1→4), α(1→5), β(1→3) e β(1→4) e β(1→6) β(1→6) [32]. Recently there has been progress in the understanding of the very complex fine structure of pectic polymers.

Homogalacturonan (HG) is the most abundant pectic polysaccharide in the cell wall, equivalent to about 60-65% of the total pectin [25;33]. It presents units of α-D- galacturonic acid in 1→4 links in a linear pattern. The carboxyl groups are partly methyl-esterified. The chains may be, depending on the plant source, partially O-acetylated at C-3 or C-2 [34; 33; 30].

Rhamnogalacturonan I (RG-I) has a chain represented by the disaccharide [→4-α-D-GalA-(1→2)-α-L-Rha-(1→]$_n$ [34; 25; 33; 30]. In summary, a variety of different glucan chains (mainly arabinan and galactan) are linked to rhamnose units. The chain length may vary considerably and the composition of RG-I sugars may be highly heterogeneous [34]. RG-I represents 20-35% of pectin, with a high degree of cell specialization and expression depending on the development, the type and number of simple sugars and oligosaccharides attached to this chain. The reason for this level of variation in RG-I is not known, but it suggests a functional diversity.

Rhamnogalacturonan II (RG-II) is the most structurally complex segment and comprises 10% of pectin. This structure, highly conserved in most plant species, consists of a homogalacturonan skeleton of approximately eight (probably more) monomeric units, containing side chains of up to 12 different types of sugars, some highly peculiar such as apiose, aceric acid, Docosahexaenoic acid (DHA) and 3-Deoxy-D-manno-oct-2-ulosonic acid (KDO). RG-II usually exists in cell walls as cross-linked dimers by a borate diol ester apiosil between units in the side chain [34; 25].

Xilogalacturonana (XGA) is a homogalacturonan substituted with xylose linked to position 3. The degree of control of xylose may vary between 25% (watermelon) to 75% (apple). This xylose may be additionally substituted in 0-4 in conjunction with another xylose in β, which is more prevalent in reproductive tissues such as fruits and seeds [25;33].

Arabinogalactan I (ARA-I) is composed of skeleton β -D-Gal*p*; residues of α-L-Ara*f* may be linked to galactosyl units in position 3. Arabinogalactan II (ARA-II) is primarily associated with proteins (3-8%), also called arabinogalactan-proteins (AGPs). AG-II is composed of a β-D-Gal*p* 1→3 skeleton, containing short chains [α-L Ara*f* (1→6) β-D-Galp (1→6)]$_n$, where n = 1, 2 or 3. The protein part is rich in proline, hydroxyproline, alanine, serine and threonine [25]. Arabinan (ARA) is composed of an α-L-Ara*ff* skeleton in 1→5 links, where there may be side chains of α-L-Ara*f* (1→3) [35]. There is also another chain, not shown schematically, which is apiogalacturonan (API), HG substituted in O-2 or O-3 with D-apio*f* the 2-O-or D-3 apio*f*. It is present in aquatic monocots such as *Lemna* [33].

4. The application of chromatographic techniques

The properties and application of pectin are affected by several parameters related to pectin structure, including the composition, presence and distribution of side chains, degree of methyl-esterification (DE), degree of acetylation (DA), molar mass, and the charge distribution along their backbone. Differently from proteins, whose structure is defined in relation to a template, pectin are a mixture of heterogeneous polymers and preparations that contain pectins that have been isolated by chemical and enzymatic treatments likely to cleave covalent bonds. The conditions used for pectin extraction can also release other cell wall polymers. Preparative chromatography can be used to separate homogeneous fractions of pectic polysaccharides from a mixture.

Determining the fine structure of pectins is a challenging task which encompasses the determination of molar mass, monosaccharide composition, configuration and ring size of monosacchrides, identification of glycosidic linkages, sequence of glycosyl units in the polymer, determination of DA and DE, as well as the distribution of the acetyl and methyl esters groups along the chain. A combination of chemical, spectroscopic and chromatographic methods is commonly used to fully characterize pectic polysaccharides. In recent years, X-ray diffraction, circular dichroism, light scattering, electron and atomic force microscopy and theoretical approaches has also been used in order to examine the fine structure and interactions of pectins. Although chromatography is primarily a separation technique, chromatography techniques are powerful tools to study the composition, DE, DA and molar mass of pectins [36].

5. Determination of molar mass by HPSEC technique with MALLS detector and RI

Size exclusion chromatography (also called gel permeation chromatography) has been widely used to analyze the molar mass distribution of polymers. The eluted fractions can be examined by different detection techniques such as: colorimetric methods, refractive index, ultraviolet, light scattering and viscosity measurements.

Size exclusion chromatography has been used to determine the molar mass of polysaccharides. In many instances, a calibration curve, previously obtained by using dextran molar mass standards, is used for this purpose. However, in size exclusion chromatography, molecules in solution are separated by their size rather than by their molar mass. Since the size is related to the conformation of the polysaccharide in a given solvent, the use of dextrans as molar mass standards is highly restricted. Experimentally, a problem in this approach is the absence of commercial molar mass standards for the different types of polysaccharides, including pectins.

The use of high performance size exclusion chromatography coupled to a differential refractometer (RI) and a multi-angle laser light scattering (MALLS) detector allows the study of the molar mass distribution of pectins as well as the determination of the absolute molar mass. While RI gives a signal proportional to the concentration, MALLS response increases with the concentration and molar mass. The MALLS instrument enables the determination of absolute molar masses of polymers from below 1000 g/mol to hundreds of millions without a calibration curve with reference standards. The molar mass is estimated directly from the angular dependence of scattered light intensity as a function of concentration, as formulated by light scattering theory. However, the molar mass calculation is only possible when the eluted fraction is detected by both RI and MALLS. An accurate determination of the refractive index increment with concentration, dn/dc, is required [37; 38]. Figure 1 displays a unimodal distribution of molar mass for a polysaccharide with M_W 1.39 × 10^6 g/mol, determined by HPSEC-MALLS.

In addition to the average molar mass from HPSEC analyses coupled to light scattering detection it is also possible to obtain: a) the ratio M_W/M_N which is used as an index of polydispersity; b) the radius of gyration; c) the differential molar mass distribution plot which shows how much material (differential weight fraction) is; distribution for the same polysaccharide extracted using three different times of; extraction); d) the cumulative distribution of molar mass curve which gives for each molar mass, the weight fraction of material having molar mass less than the given molar mass (Figure 3 shows the cumulative distribution of molar mass for a same polysaccharide extracted using two different times of extraction); e) the log-log plot of root mean square radius of gyration as a function of molar mass which gives information about the molecular conformation. Theoretical slopes of 0.33, 0.50 and 1.0 occur for spheres, random coils in theta solvents, and rigid rods, respectively. Usually, most real random coils have slopes in the range 0.55-0.6 [37]; f) using a viscometer connected in series with MALLS and RI in the HPSEC system, the intrinsic viscosity can be determined [40].

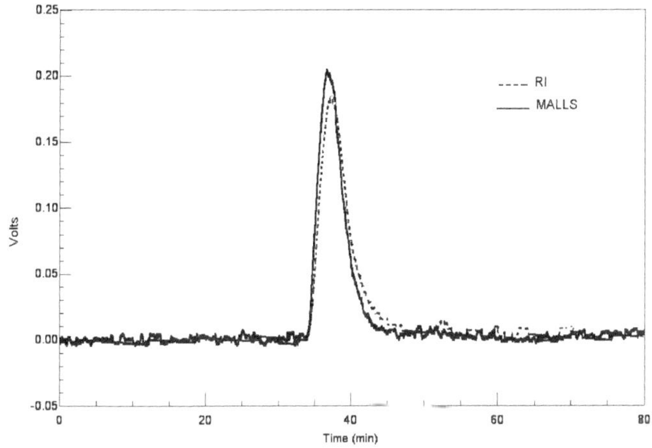

Figure 1. HPSEC elution profile of a polysaccharide isolated from seed [39].

A comparison between MALLS and LALLS detector in HPSEC has been shown to give results in very good agreement [40]

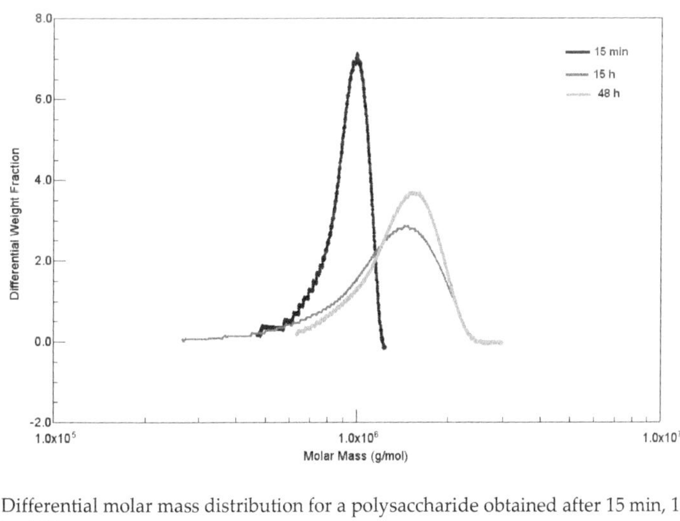

Figure 2. Differential molar mass distribution for a polysaccharide obtained after 15 min, 15 h, and 48 h of extraction [39].

HPSEC-MALLS/RI was used to compare the molar mass distribution of pectins obtained from the dried pomace of eleven apple cultivars [41]. Although the pectins were obtained under the same extraction conditions, some differences were identified among the fractions. According to the results, pectins from different variety of apple can display differences in their molar mass profiles as depicted in Figure 4 (for clarity only RI detector is shown).

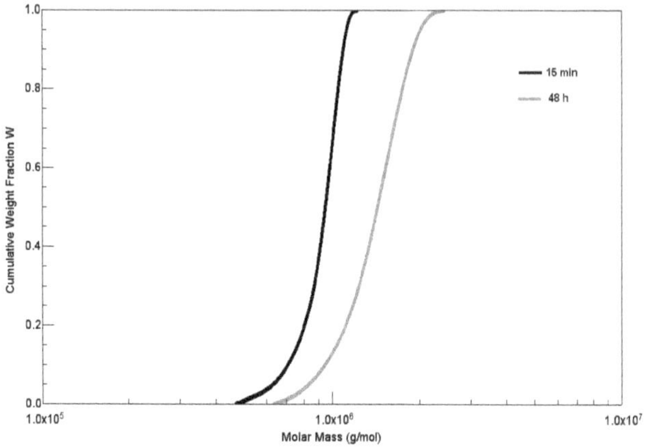

Figure 3. Cumulative molar mass distribution for a polysaccharide obtained after 15 min and 48 h of extraction [39].

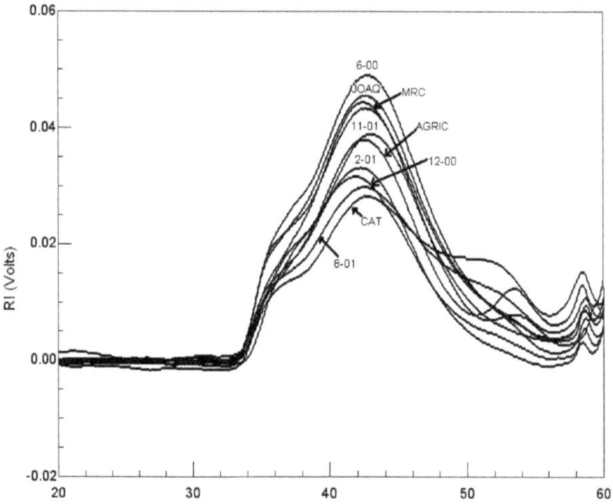

Figure 4. HPSEC elution profiles of pectins obtained from the dried pomace of eleven apple cultivars (columns: Waters Ultrahydrogel 2000/500/250/120 connected in series; eluent: 0.1M $NaNO_2$ solution, containing NaN_3 (0.5 g/l); flow: 1.5 mg/ml) [41].

6. Monosaccharide composition

Information on the composition of pectins can be obtained by different chromatographic methods. Regardless of the method used to determine the monosaccharide composition, the pectin under study needs to be previously hydrolyzed in order to release its monomeric units. It is well known that the glycosidic linkages with acidic monosaccharides are more resistant to hydrolysis. For pectins, the different sensitivities to acid hydrolysis are: GalA-GalA > GalA-Rha > Rha-GalA > neutral sugar-neutral sugar 42. The chosen conditions must be a compromise between maximum hydrolysis with minimal degradation.

Although the more simple chromatographic methods, such as paper chromatography (PC) and thin layer chromatography (TLC), can give only qualitative information about the composition of pectins, they can be useful to check the adequacy of the conditions of hydrolysis.

7. Determination of monosaccharide composition by GC

Gas-liquid chromatography (GC) is widely used for the analysis of carbohydrate. GC is used in the separation and analysis of complex mixtures of many components that can be vaporized without decomposition. For monosaccharides released after total hydrolysis, derivatization is necessary due to polar groups of carbohydrates which make them nonvolatile. Derivatization methods consist of the substitution of the polar groups of monosaccharides in order to increase their volatility. Many types of derivatives can be employed in the analysis of monosaccharide composition by GC, acetylated and silylated derivatives being the most popular. The advantages of acetylated derivatives include the presence of a single peak for each derivatized monosaccharide and their high stability. The derivatization reaction involves reduction with sodium borohydride followed by the acetylation itself. However, in the acetylation only the hydroxyl groups are derivatized. Complete quantification and identification of individual neutral and acidic sugars using acetylated derivatives can be accomplished by an additional step, which includes the carbodiimide-activated reduction of the carboxyl groups of uronic acids in order to give the corresponding neutral sugars [43]. Alternatively, GC can be used just to quantify the neutral monosaccharides, while titration or colorimetric method can be used to evaluate the amount of acidic units. This approach was chosen to study the composition of pectins from the dried pomace of eleven apple cultivars (Table 1) [41].

In GC analysis, compounds that have similar properties often have the same retention times. Sometimes extraneous background peaks can be a problem for the identification and quantification of monosaccharides present in minor amounts in complex mixtures. Gas chromatography associated to mass spectroscopy (GC-MS) can be used to overcome this difficulty. GC-MS combines two techniques to form a single method of analyzing mixtures of chemicals. For a pectic polymer, gas chromatography separates the monosaccharides derivatives present in the mixture and mass spectroscopy characterizes each of the derivatives individually by their mass fragments. MS has the advantages of high selectivity, specificity, and sensitivity 43.

Cultivar	Fraction (%)								
	AUA	MeO	Rha	Fuc	Ara	Xyl	Man	Gal	Glc
1	46.90	5.88	0.65	0.12	3.33	1.31	2.23	2.51	37.3
2	51.20	6.49	0.85	0.31	4.70	3.21	0.05	4.67	28.9
3	50.90	6.34	0.82	0.32	4.70	2.46	4.10	2.94	27.5
4	51.60	6.67	0.01	0.01	3.48	3.55	9.55	5.11	24.7
5	51.60	6.70	0.96	0.23	4.23	2.50	2.99	2.37	28.5
6	51.30	6.66	1.35	0.54	5.05	3.90	4.31	3.46	23.4
7	53.40	6.93	1.08	0.33	4.88	3.00	3.25	4.38	22.8
8	50.30	6.35	2.09	0.74	8.54	3.78	3.95	3.14	21.2
9	50.40	6.30	0.34	0.00	4.79	3.86	3.43	4.31	25.4
10	48.70	6.06	0.91	0.20	3.76	2.00	2.78	3.27	32.4
11	52.90	6.72	1.08	0.35	5.03	2.90	3.96	4.68	22.4
Average	50.80	6.46	0.92	0.29	4.77	2.95	3.69	3.71	26.8
Standard Deviation	1.72	0.30	0.51	0.21	1.33	0.79	2.17	0.91	4.60
Coefficient of Variation (%)	3.39	4.62	55.57	72.00	27.86	26.82	58.92	24.44	17.21

Source: Sato et al., (2011)[41].

Table 1. Monosaccharide composition of pectins from the dried pomace of eleven apple cultivars. Neutral sugars were determined as alditol acetates by GC using a DB-210 capillary column (0.25 mm internal diameter x 30 m), film thickness 0.25 µm and flame ionization detector. Contents of uronic acids (AUA) were determined by titration.

When carboxyl-reduction of glycosyluronic acid is performed the use of sodium borodeuteride ($NaBD_4$) and GC-MS is preferred. The use of NaBD4 provides an easily identified tag by MS analysis that allows the quantitative determination of the content of uronic acid as its corresponding neutral sugar [43].

GC-MS has also been used to identify and quantify unusual sugars present in RG-II. This approach is very useful, since commercial standards are not available for all unusual monosaccharides present in RG-II. The contents of DHA, KDO, aceric acid, 2-methyl-xylose and 2-methyl-fucose have been determined by GC-MS of their trimethylsilyl-esters *O*-methyl glycosides after acidic methanolysis and derivatization [43; 44].

8. Determination of monosaccharide composition by HPLC and HPAEC-PAD

High performance liquid chromatography (HPLC) is characterized by using high pressure to force chemical compounds to pass the column containing a stationary phase. Since monosaccharides released after total hydrolysis do not present enough volatility to be analyzed by GC, HPLC should be the most useful technique for the determination of monosaccharide composition. Monosaccharides cannot be detected by absorption due to the lack of chromophore in the molecular structure and are usually detected by refraction index detector. However, refraction index detector is sensitive to eluent composition and sample matrix, making the application of the gradient elution method limited and not allowing the complete separation of all monosaccharides [43].

Acid-catalysed or enzymatic hydrolysis followed by HPLC analysis to determine the GalA content in pectins has been proposed by different authors. When enzymes that release and specifically degrade pectins are used, the method can be useful for analysis of GalA in pectin samples as well as in the raw material that can be used for pectin extraction [45].

In recent years, high performance anion exchange chromatography coupled with pulsed amperometric detection (HPAEC-PAD) had been used to directly analyze the contents of carbohydrates. The development of this new technology had readily solved the above-mentioned problems. HPAEC-PAD can be used for simultaneous determination of monosaccharides and uronic acids. The method had been applied for the separation of GalA, GlcA, Rha, Fuc, Ara, Xyl, Man, Gal and Glc. Monosaccharides are weak acids with p*Ka* above 11. HPAEC uses NaOH as an eluent making sugars in their anionic form. In general, the elution sequence of monosaccharide is related to their p*Ka* value, and in the same conditions, the time of elution increases with the value of p*Ka* [46].

9. The use of chromatographic techniques for determining DA and DE

DE is usually determined, together with galacturonic acid content, by titration. However, the presence of acetyl can contribute to an overestimation of DE, since the acetic acid released during the saponification will be titrated. Chromatographic methods can be used for determination of DE and DA. An HPLC method for simultaneous determination of DE and DA was proposed by Voragen et al (1986) [47] and modified by Levigne et al (2002) [31]. In the former, an Aminex HPX-87H column is used while the second method uses a C18 column. The pectin is saponified and precipitated followed by methanol and acetic acid separation from the supernatant by HPLC and quantification by refractometry using an internal standard. The method allows accurate determination in a single run and within a short time, the DE and DA using low amounts of pectins (~ 5mg). This method presents some advantages when compared to that using FT-IR for DE calculation. FT-IR method requires pectins standards with different DEs in order to construct a calibration curve and the presence of acetyl can contribute to an overestimation of the DE.

Huisman et al. (2004) [48] proposed a method for DE determination using head-space GC. The approach is similar to that used in the HPLC method. Head-space GC is used for the quantification of methanol released from pectin by saponification. A lower amount of pectin is needed (~2mg) and the chromatogram shows a symmetrically shaped methanol peak which is very easy to integrate. However, the method is only applicable for the determination of DE and another approach has to be used in order to estimate DA.

10. Chromatographic techniques as an aid in the study of the fine structure of pectic polysaccharides

Chemical and enzymatic hydrolysis has been used to produce fragments of pectic polysaccharides in order to study the fine structure of pectins. The differences in the lability of glycosidic linkages to acid hydrolysis (see section 3.2) allowed the homogalacturonan region to be isolated and its minimal length estimated. Under controlled conditions, acid hydrolysis of RG-I gives a fraction of intermediate molar mass that is rich in GalA and Rha. The neutral side chains are quickly hydrolysed to form low molar mass oligosaccharides [49; 42].

RG-II was first isolated from cell walls using enzymatic treatment. The structure of RG-II was investigated using a combination of enzymatic and controlled acid hydrolysis. Using chemical fragmentation, four oligosaccharides from side chains were obtained [42].

The products from chemical and enzymatic hydrolysis are further fractionated by chromatographic techniques and structurally characterized. Usually, ion-exchange and size exclusion chromatography are used to fractionate the segments of pectic chains produced in the partial hydrolysis. The recovery of oligosaccharides with low degree of polymerization greatly facilitates the structural studies. Elucidation of the sequence of glycosyl units with the methods current available is only possible for oligosaccharides. This approach allowed the identification of xylogalacturonan in the modified hairy region of apple pectin [50].

One of the steps in the structural analysis of pectins is the determination of the glycoside linkages. Methylation analysis is one of the main tools to determine glycosydic linkages of polysaccharides and their oligomers. In this procedure, the free hydroxyl groups of pectins or their oligosaccharides are methylated. Then, the partially methylated material is hydrolysed and submitted to reactions to afford volatile derivatives. The products from this procedure are separated and analyzed using the chromatographic method of GC-MS [51].

11. Final considerations

Due to the need in recent years to seek alternatives that meet sustainable development, polysaccharides have emerged as source of sustainable products for fuel, food, materials and medicine. In this context, pectin is found in all higher plants and can be obtained from renewable sources. Although nowadays, commercial pectin is usually derived from citrus peel or apple pomace, other by-products from agroindustry can be used for pectin isolation. However, pectin preparations from non-traditional sources with a set of properties for specific applications require complete structural elucidation of the polymers. Chemical and enzymatic modification of pectins can be used to obtain new functionalities and again, an understanding of pectin fine structure–function relationships is required. Considerable progress has been made in the elucidation of the fine structure of pectin mainly due to the progress of chromatographic and spectroscopic methods; however a better understand is necessary in order to be able to design pectins for specific industrial or biological applications either by chemical modification or genetic engineering. At the moment, apple pectin remains a well suited option for different applications.

List of abbreviation

FAO	Food and Agriculture Organization
HG	Homogalacturonan
RG-I	Rhamnogalacturonan I

RG-II	Rhamnogalacturonan II
DHA	Docosahexaenoic acid
KDO	3-Deoxy-D-manno-oct-2-ulosonic acid
XGA	Xilogalacturonana
ARA-I	Arabinogalactan I
ARA-II	Arabinogalactan II
API	Apiogalacturonan
DE	Degree of methyl-esterification
DA	Degree of acetylation
RI	Differential refractometer
LALLS	Low angle laser light scattering
MALLS	Multi-angle laser light scattering
HPSEC	High Pressure Size Exclusion Chromatography
PC	Paper chromatography
GC	Gas-liquid chromatography
TLC	Thin layer chromatography
AUA	Uronic acids
GC-MS	Gas chromatography associated to mass spectroscopy
HPLC	High performance liquid chromatography
HPAEC-PAD	High performance anion exchange chromatography coupled with pulsed amperometric detection
FT-IR	Fourier transform infrared spectroscopy

Author details

Maria Helene Giovanetti Canteri
Federal Technological University of Paraná, Ponta Grossa, Paraná, Brazil

Alessandro Nogueira and Gilvan Wosiacki*
State University of Ponta Grossa, Av. General Carlos Cavalcanti, Ponta Grossa, Paraná, Brazil

Carmen Lúcia de Oliveira Petkowicz
Federal University of Paraná, Curitiba, Paraná, Brazil

12. References

[1] Denardi F, Camilo AP. Novas cultivares de macieira: proposta de nova composição de pomares com polinizadoras/produtoras. Revista Agropecuária Catarinense, 1997; 10(2): 25-30.

[2] Juniper BE, Watkins R, Harris SA. The origin of the apple. In: Eucarpia Symposium on Fruit Breeding and Genetics, XIII, Poland, Warsaw, 11-15 Sept. 2011. Available in eucarpia2011.woiak.sggw.pl.

* Corresponding Author

[3] World Fruit Market Analysis. Available in : www.e_belrose.com/ArticleAppleIndustryKeyFacts2009.html.7/22/2/12.
[4] Apfelproduktion China. Flüssiges Obst, 2011; 77(2), booklet FRUCHTWELTEN Märkte International. Confructa medien GmbH, 2011.
[5] China now world leader in apple production. Chronica Horticulturae. 2003; 43(2): 21.
[6] Apple production by country from 2002 till 2011. Fruit Processing. 2011; 21 (5), booklet World of Fruit Processing. Markets International. Confructa medien GmbH, 2011.
[7] Apfel production nach Ländern; südlich Hemisphäre. Flüssiges Obst. 2010; 77(11): booklet Fruchtwelten Märkte International. Confructa medien GmbH, 2010.
[8] O'Rourke D, Janic J, Sansavini S. World apple cultivar dynamics, horticultural science news. Plant Species Project. Chronica Horticulturae. 2003; 43(3): 10-14.
[9] Robinson T. Advances in apple culture worldwide. Revista Brasileira de Fruticultura. 2011; 23(1):1-10.
[10] Nogueira A, Wosiacki G. Apple cider fermentation. In: Hui YH, Evranuz ÖE. Editors. Handbook of Plant-Based Fermented Food and Beverage Technology, 2nd ed, CRC Press, p. 206-235, 2012.
[11] Wosiacki G, Nogueira A. Suco de maçã. In: Venturini Filho, WG editor. Bebidas Não-alcoólicas. Ciência e Tecnologia. São Paulo, Edgard Blücher, p.269-302, 2010.
[12] Smock RM, Neubert AM. Apples and apples products. New York: Interscience, 1950.
[13] Wosiacki G, Nogueira A, Silva NCC, Denardi F, Camilo AP. Apple varieties growing in subtropical áreas. The situation in Santa Catarina – Brazil. Fruit Processing.2002; 12(1): 19-28.
[14] Issenhuth F. The new generation of Panzym mash. Fruit Processing. 2008; 18(5): 254-55.
[15] Sataque EY, Wosiacki G. Caracterização da polifenoloxidase de maçã (variedade Gala). Arquivos de Biologia e Tecnologia. 1987;30(2):287-99.
[16] Endreβ HU. High quality resulting from product integrated environment protection - PIUS. Fruit Processing. 2000;10(7):273-76.
[17] Kennedy M, List D, Lu Y, Foo LY, Newman RJ, Sims IM, Bain PJS, Hamilton B, Fenton G. Apple pomace and products derived from apple pomace: uses, composition and analysis. In: Linskens G, Jackson JF editors. Modern methods of plant analyses: analysis of plant waste materials. Berlin, Heidelberg, Springer Verlag, vol.20, p.5-119, 1999.
[18] Carson KJ, Collins JL, Penfield MP. Unrefined, dried apple pomace as a potential food ingredient. Journal of Food Science. 1994; 59(6): 1213-15.
[19] Shuda ML, Baskaran V, Leelavathi K. Apple pomace as a source of dietary fiber and polypolyphenols and its effect on the rheological characteristics and cake making. Food Chemistry. 2007; 104(2): 686-92.
[20] Hang YD, Woodams EE. Apple pomace: a potential substrate for production of β-glucosidase by *Aspergillus foetidus*. Lebensmittel-Wissenschaft Technologie. 1994; 27(6): 587-89.
[21] Schieber A, Hilt P, Berardini N, Carle R. Recovery of Pectin and Polyphenolics from Apple Pomace and Mango Peels. In: Waldron K, Faulds C, Smith A. editors. Total food 2004, exploiting co-products minimising waste. Norwich, Institute of Food Research, p.144-49, 2004.

[22] Paganini C, Nogueira A, Silva NCC, Wosiacki G. Aproveitamento de bagaço de maçã para a produção de álcool e obtenção de fibras alimentares. Ciência e Agrotecnologia. 2055; 29(6): 1231-38.

[23] Tsurumi R, Shiraishi S, Ando Y, Yanagida M, Takeda K. Production of flavour compounds from apple pomace. Journal of the Japanese Society for Food Science and Technology. 2001; 48 (8): 564-69.

[24] Voragen GJ, Pilnik W, Thibault JF, Axelos MAV, Renard CMGC. "Pectins", in: Food polysaccharides and their applications, cap. 10, Stephen A. M. (ed.), Marcel Dekker Inc., New York (1995).

[25] Voragen AGJ, Coenen GJ., Verhoef RP, Schols HA. Pectin, a versatile polysaccharide present in plant cell walls. Structural Chemistry, 2009; 20 (2), p. 263-275.

[26] Kertesz ZI. The pectic substances. New York: Interscience, 1951.

[27] Pectinas. Ação e utilização nos alimentos. Available from www.insumos.com/aditivos_e_ingredientes/matérias/365.pdf. 22/3/2012

[28] Canteri MHG, Moreno LR, Wosiacki G, Scheer AP. Pectin: from raw material to the final product. Polimeros, 2012, 22(2) 1-13.

[29] Thibault JF. Les substances pectiques. In: Les polymeres vegetaux. Monties B. Editor, Paris: Gaulthier-Vilars, 1980.

[30] Vincken Jp, Scholz Ha, Oomen Rjfj, Mccann Mc, Ulvskov P, Voragen Agj, Visser Rgs. If homogalacturonan were a side chain of rhamnogalacturonan I. Implications for cell wall architecture. Plant Physiol. 2003; 132 (4): 1781-1791.

[31] Levigne S, Thomas M, Ralet MC, Quemener B, Thibault JF. Determination of the degrees of methylation and acetylation of pectins using a C18 column and internal standards. Food Hydrocolloids. 2002; 16 (6) : 547-50.

[32] Thibault JF, Saulnier L, Axelos MAV, Renard CMGC. Determination of the degrees of methylation and acetylation of pectin using ac18column and internal standards. Bull. Soc. Bot. Fr. 1991; 314 (6):319. Available in
http://www.sciencedirect.com/science/article/pii/s0268005x02000152 21/7/2012

[33] Mohnen D. Pectin Structure and Synthesis. *Curr. Opin. Plant Biology* 2008 (11) :266-27.

[34] Willats WGT, Knox JP, Mikkelsen JD. Pectin: new insights into and old polymers are starting to gel. 2006; 17(3): 97(3) 97-104...

[35] Thakur BR, Singh RK, Handa AK. Chemistry and Uses of Pectin--A Review. Crit. Rev. Food Sc. Nut. 1997; 37(1):47-73.

[36] Pérez S, Mazeau K, Du Penhoat CH The three-dimensional structures of the pectic polysaccharides. Plant Physiology and Biochemistry. 2000; 38(1/2):37-55.

[37] Wyatt P. Light scattering and the absolute characterization of macromolecules. Analytica Chimica Acta. 1993; 272(1):1-40.

[38] Knobloch JE, Shaklee PN. Absolute molecular weight distribution of low-molecular-weight heparins by size-exclusion chromatography with multiangle laser light scattering detection. Analytical Biochemistry. 1997; 245(2):231-241.

[39] Kai K, Petkowicz CLO. Influence of extraction conditions on properties of seed xyloglucan. International Journal of Biological Macromolecules. 2010; 46(2):223–28.

[40] Reed WF. Data evaluation for unified multi-detector size exclusion chromatography-molecular mass, viscosity and radius of gyration distributions. Macromolecular Chemistry and Physics. 1995; 196(5):1539-75.

[41] Sato MF, Rigoni DC, Canteri MHG, Petkowicz CLOP, Nogueira A, Wosiacki G. Chemical and instrumental characterization of pectin from dried pomace of eleven apple cultivars. Acta Scientiarum. Agronomy. 2011; 33(3):383-89.

[42] Novosel'skaya LL, Voropaeva NL, Semenova LN, Rashidova SS. Trends in the science and applications of pectins. Chemistry of Natural Compounds. 2000; 36(1): 1-10.

[43] Biermann CJ, Mcginnis GD. Analysis of carbohydrates by GLC and MS. Florida: CRC press, Boca Raton, 2000.

[44] Ayestarán B, Guadalupe Z, León D. Quantification of major grape polysaccharides (*Tempranillo* v.) released by maceration enzymes during the fermentation process. Analytica Chimica Acta. 2004; 513 (1): 29-39.

[45] Garna H, Mabon N, Nott K, Wathelet B, Paquot M. Kinetic of the hydrolysis of pectin galacturonic acid chains and quantification by ionic chromatography. Food Chemistry. 2006; 96:477-84.

[46] Lina L, Ping Z, Yaq C, Shifen M. High-performance anion exchange chromatography with pulsed amperometric detection for simultaneous determination of monosaccharides and uronic acids. Chinese Journal of Analytical Chemistry. 2006; 34: 1371-74.

[47] Voragen GJ, Schols HA, Pilnik W. Determination of the degree of methylation and acetylation of pectins by HPLC. Food Hydrocolloids. 1986; 1 (1): 65-70.

[48] Huismann MMH, Oosterveld A, Schols HA. Fast determination of the degree of methyl esterification of pectins by head-space GC. Food Hydrocolloids. 2004; 18(4): 665-68.

[49] Renard CMGC, Crépeau M-J, Thibault J-F. Structure of the repeating units in the rhamnogalacturonic backbone of apple, beet and citrus pectins. Carbohydrate Research. 1995; 275 (1): 155-65.

[50] Schols HA, Bakx EJ, Schipper D, Voragen AGJ. A xylogalacturonanan subunit present in the modified hairy regions of apple pectin. Carbohydrate Research. 1995; 279(1): 265-279.

[51] Ralet M-C, Lerouge P, Quéméner B. Mass spectrometry for pectin structure analysis. Carbohydrate Research. 2009; 344 (14):1798–1807.

Chapter 15

The Use of Solvation Models in Gas Chromatography

Fabrice Mutelet

Additional information is available at the end of the chapter

1. Introduction

Gas chromatography is widely used for determining thermodynamic properties of pure substances or solvent properties of binary mixtures. From retention data, the solute activity coefficient at infinite dilution, the gas–liquid partition coefficient and others thermodynamic properties of mixing can be easily obtained. Using these parameters and appropriate models allows understanding of the intermolecular interactions responsible for solvation in the stationary phase [1-2]. The solvation parameter model is now well established as a useful tool for obtaining quantitative structure–property relationships for chemical, biomedical and environmental processes. The model correlates a free-energy related property of a system to six free-energy derived descriptors describing molecular properties. The ultimate goal is to establish a suitable quantitative structure–property relationship (QSPR) to facilitate the prediction of further system properties for compounds lacking experimental values. Two broad strategies are generally employed in QSPR studies. The first approach is based on theoretical descriptors. The advantage of using the QSPR approach based on theoretical descriptors is that all of the necessary parameters for prediction can be calculated purely from the three-dimensional representation of the molecular structure of each of the compounds of the mixtures, including mixtures of chemically diverse compounds [3-4]. The main weakness of this approach is that the selected descriptors may be difficult to understand and the models may lack obvious chemical significance. The second approach is based on descriptors determined using experimental technique such as gas chromatography.

In this review, we will present the different possibility of using chromatographic methods to facilitate the rapid and convenient measurement of the LSER model presentsed in reference [5]. An application of the LSER model on ionic liquids will be presented. The last paragraph will be focused on the use of the LSER model coupled to a group contribution model for the estimation of the partition coefficient of organic compounds in ionic liquids.

2. Solvation models used in gas chromatography

2.1. Linear solvation energy relationship

In numerous solvation models, the partition of a solute between the gas phase and a solvent (or stationary phase) may be described by a cavity theory of solution [6-7] in which the solvation process is divided into three steps. In the step 1, a cavity of suitable size to accommodate the solute is created in the solvent. This process is endoergic because work is required to disrupt solvent–solvent interactions. In the second step, the solvent molecules around the cavity are reorganized from their original positions to the positions they will adopt when the solute is in equilibrium with the solvent. Of course, these positions are not fixed but are averages of solvent positions. The Gibbs energy change for such reorganization is assumed to be negligible, by analogy with the melting of a solid. However, enthalpy and entropy changes in reorganizations may be large, again by analogy with the melting of a solid. In the last step, the solute is introduced into the reorganized cavity, and various solute–solvent interactions are set up, all of which are exoergic.

Several thermodynamic properties are related to partitioning between water and other phases, for example octanol (K_{ow}) or the pure compound itself (water solubility). These partitioning processes can be understood from thermodynamic concepts—like free energy, chemical potential and fugacity. The equilibrium partition constant between two phases, on a mole fraction basis, can be expressed as:

$$K_{i\,12} = \exp(-\Delta_{12}G_i \,/\, RT) \tag{1}$$

where $\Delta_{12}G_i$ is the Gibbs energy (or free energy) of transfer between the two phases, R is the general gas constant and T is the absolute temperature. The Gibbs energy sums up both the enthalpic ($\Delta_{12}H$) and entropic ($\Delta_{12}S$) effects resulting from changes in intermolecular interactions:

$$\Delta_{12}G_i = \Delta_{12}H_i - T\Delta_{12}S_i \tag{2}$$

The calculation of partitioning from structural or other descriptors therefore requires the modelling of these effects. The Gibbs energy change can also be separated into the contributions of van der Waals and polar interactions, assuming that these are additive:

$$\Delta_{12}G_i = \Delta_{12}G_i^{\,vdW} + \Delta_{12}G_i^{\,polar} \tag{3}$$

In the early 1980's, Taft & Kamlet [8-12] have developed the basic concept of linear solvation energy relationships (LSERs). They have demonstrated for thousands of chemical systems that some property which is linearly related to either a free energy of reaction, a free energy of transfer, or an activation energy can be correlated with various fundamental molecular properties of the solvents or solutes involved. Chromatographic retention and in particular logarithmic retention factors (log *k*′), logarithmic partition coefficients ($\log K_L$) are linear free energy parameters and as such one can linearly correlate these data with the molecular properties of the solutes using the LSER model [13-16].

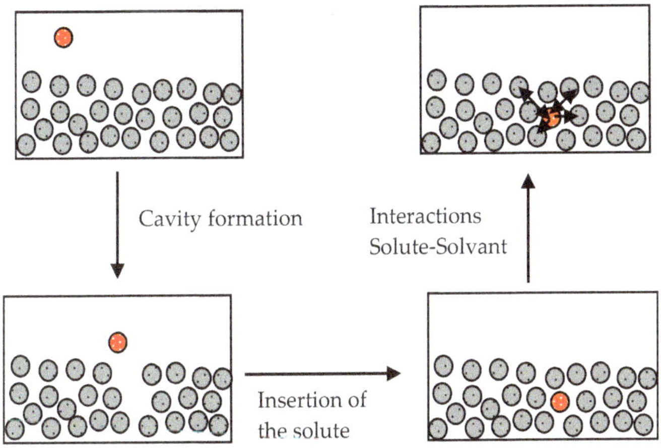

Figure 1. Model of the solvation process.

The most recent representation of the LSER model proposed by Abraham and coworkers [5, 17-19] is given by equation (4)

$$\log SP = c + r \cdot R_2 + s \cdot \pi_2^H + a \cdot \sum \alpha_2^H + b \cdot \sum \beta_2^H + l \cdot \log L^{16} \quad (4)$$

Where SP is a solvation parameter related with the free energy change such as gas-liquid partition coefficient, specific retention volume or adjusted retention time at a given temperature. The capital letters represent the solutes properties and the lower case letters the complementary properties of the stationary phase. The solute descriptors are the excess molar refraction R_2, dipolarity/ polarizability π_2^H, hydrogen bond acidity basicity, $\sum \alpha_2^H$ and $\sum \beta_2^H$, respectively, and the gas-liquid partition coefficient on n-hexadecane at 298 K, $\log L^{16}$. The solute descriptors may be determined using inverse gas chromatography or estimated using a group contribution method. A databank of descriptors for about 3000 compounds may be found in the literature [2, 20, 21]. The coefficients *c, e, s, a, b* and *l* are not simply fitting coefficients, but they reflect complementary properties of the solvent phase. These coefficients are determined by mutiple linear regression of equation (4). This model was strongly applied to characterize chemicals products, petroleum fluids.

2.2. Determination of LSER parameters of pure solutes

2.2.1. Determination of Log L^{16}

To preserve the general character of equation (4), all characteristic parameters should be carefully determined and correlation between parameters should be avoided. Experimental procedures of successive determination of LSER parameters were described in the literature [20, 22-24]. Log L^{16} characterizes the most general interactions present in every physical system and should be determined before other parameters [24]. The original values of log

L^{16} were determined from retention data of organic compounds on n-hexadecane coated packed columns at 298.2 K [5]. A number of papers proposed techniques based on the use of either capillary or packed columns in wide temperature ranges and replacing the n-hexadecane with other non-polar stationary phases. The gas liquid partition coefficient of a solute is directly related to the experimental capacity factor, k, by equation:

$$L = k\frac{V_M}{V_S} = \frac{k}{\Phi} \tag{5}$$

where V_M and V_S are the volumes of the mobile and stationary phases, respectively, and Φ is the phase ratio (V_S/V_M). Experimental determination of log L^{16} is often very difficult. Adsorption phenomena introduce an important error in determination of the capacity factor. Zhang et al. [23] determined log L^{16} with capillary columns coated with n-hexadecane and concluded that results were not influenced by adsorption in this case. However, recent results presented in reference [25] showed that this improvement is not general and that an interfacial adsorption still exists with capillary columns. Moreover, it is very difficult to obtain absolute retention data using this technique. Li et al. [24] and Abraham et al. [25] studied influence of the solute support and of the stationary phase loading on adsorption phenomena. They concluded that the high loading ratio (up to 20 %) of the stationary phase and the high temperature of the column allow to reduce adsorption. In this case, the knowledge of R_2, $\sum\alpha_2^H$ and π_2^H parameters is necessary. The number and the nature of parameters needed for calculation depend on the stationary phase used.

Serious difficulties arise when log L^{16} of non-volatile compounds is to be measured. This is due to the definition of log L^{16} itself. Indeed, direct experimental determination of log L^{16} of compounds less volatile than n-hexadecane is impossible. In the case of heavy compounds slightly more volatile than n-hexadecane the experiment is possible but difficult, especially at 298.2 K. Often it is recommended to measure retention times at higher temperatures and then extrapolate partition coefficients to the ambient temperature. In this case, the quality of results depends strongly on the extrapolation method used. The problem of the temperature dependence of retention times was often discussed in literature. A suitable extrapolation procedure was described in reference [26]. In the case of compounds less volatile than n-hexadecane several authors proposed to work with columns coated with long chain branched paraffins and to establish relationships between corresponding partition coefficients and log L^{16} [16, 26]. Defayes et al. [26] worked with apolane coated stationary phase (apolane is a $C_{87}H_{176}$ branched alkane). This column can be used at temperatures up to 550 K without a weight loss of the stationary phase. Moreover, it was shown that the effect of adsorption at the liquid-gas interface is negligible in this case [26]. However, this opinion is not generally accepted and Werckwerth et al. [16] found the influence of adsorption in the case of strongly polar compounds. The same authors observed that a linear relationship exists between gas-apolane partition coefficients, log L^{87} and log L^{16} and that the data obtained with apolane can be used to estimate the value of log L^{16}. Moreover, they demonstrated that strong correlation between both partition coefficients exists also for log

L^{87} determined at significantly higher temperature [16]. Recently, several authors investigated the use of predictive methods to estimate log L^{16} [27, 28]. This approach is particularly interesting to determine the log L^{16} of nonvolatile compounds.

2.2.2. *Determination of LogL^{16} using capillary columns*

A direct determination of the stationary phase mass is difficult in the case of capillary or Megabore columns and the use of relative methods to determine partition coefficient from retention data is often preferred. In this case, different approaches were proposed to calculate partition coefficient based on the well-established value of log Ln-hexane [23]. Corresponding equation is as follows:

$$L_X = L_{n-hexane} \frac{t_{R,X} - t_m}{t_{R,n-hexane} - t_m} \tag{6}$$

where t_R and t_m are the solute retention time and the dead time of the column respectively. Retention data of the solute X and of n-hexane should be determined at the same temperature. Available data of log Ln-hexane on apolane were determined at 312.4 K [16]. On the other hand, it was shown that partition coefficients determined at two temperatures are linearly correlated [16].

$$\log L_T^{87} = a \log L_{T'}^{16} + b \tag{7}$$

New relationship can be established between the partition coefficient of the solute X at temperature T and the partition coefficient of the n-hexane at temperature T′. This relationship is based on the observation verified with several n-paraffin stationary phases that equation (7) for one stationary phase is reduced to the following form:

$$\log L_T^{87} = \log L_{T'}^{87} + b \tag{8}$$

Equations (6) and (8) lead to relationship between the partition coefficient at temperature T of the solute X and the partition coefficient of n-hexane determined at temperature T′:

$$(L_X)_T = (L_{n-hexane})_T \times \left(\frac{t_{R,X} - t_m}{t_{R,n-hexane} - t_m} \right)_{T'} \tag{9}$$

Apolane coated capillary columns are considered as an appropriate tool to determine log L^{16} of heavy compounds. Studies of Defayes et al. [26] and Werckwerth et al. [16] provided arguments supporting this opinion. As it can be seen in Table 1, results obtained with a similar but deactivated column are in good agreement with literature. Chromatographic peaks obtained with non-polar and polar compounds were symmetric. While this method gave good results at high temperatures, the column was deactivated irreversibly within few hours. A probable explanation of this phenomenon is that the adhesion between apolane

Compounds	log $L^{16\,a}$	log L^{16exp}	Compounds	log $L^{16\,a}$	log L^{16exp}
n-Hexane	2.668	2.660	Triethylamine	3.040	2.947
n-Octane	3.677	3.670	1-Butanol	2.539	2.545
n-Nonane	4.182	4.180	1-Pentanol	3.106	3.094
n-Decane	4.685	4.686	1-Octanol	4.569	4.556
n-Undecane	5.191	5.189	1-Decanol	5.628	5.631
n-Dodecane	5.696	5.699	Butanone	2.287	2.274
n-Tridecane	6.200	6.198	2-Pentanone	2.755	2.758
n-Tetradecane	6.705	6.701	Pyridine	3.022	3.033
n-Pentadecane	7.209	7.205	Thiophene	2.819	2.809
n-Hexadecane	7.714	7.711	Naphthalene	5.161	5.162
Benzene	2.792	2.810	1-Methyl naphthalene	5.789	5.785
Toluene	3.325	3.331	2-Methyl naphthalene	5.771	5.772
Ethylbenzene	3.778	3.782	1,4-Dimethylnaphthalene	6.339	6.338
o-Xylene	3.939	3.943	1,6- Dimethylnaphthalene	6.280	6.284
m-Xylene	3.868	3.870	cis-Decahydronaphtalene		5.162
p-Xylene	3.839	3.841	Indane	4.590	4.598
1,3,5-Trimethylbenzene	4.344	4.348	Phenanthrene	7.632	7.638
1,2-Diethylbenzene	4.732	4.728			

Table 1. Partition coefficient log $L^{16\,exp.}$ at 298.2K obtained with a $C_{50}H_{102}$ packed column at 373.2K. log $L^{16\,a}$: data taken from [5,14,21,31].

and the deactivated silica does not assure the film stability at higher temperature [24, 27]. Our experience indicates that the use of commercially available apolane coated capillary columns should be limited to low temperature ranges. In the case of heavy compounds this implies very long retention times and imposes injection of samples of important volume, which induces adsorption effects. Consequently, to determine log L^{16} of heavy compounds we decided to use packed columns with long chain n-alkane stationary phase. Moreover, results obtained with a non-deactivated column indicate that retention times are influenced by adsorption phenomena. Indeed, polar solutes exhibited severely asymmetric peaks and their retention times strongly depended on the sample size. Retention times of alcohols are longer than the literature values that may indicate the presence of active sites.

2.2.3. *Determination of LogL^{16} using packed columns*

Problems of the capillary column stability encouraged us to review the possibility of application of packed columns for determining log L^{16} of non-volatile compounds. Stationary phases used were long chain n-alkanes, n-hexatriacontane and n-pentacontane. They were used at temperatures up to 320 K without significant loss of weight. The essential problem encountered with packed columns concerned adsorption effects [30, 23]. Mutelet & Rogalski [2] used teflon columns, inert and stable up to 330 K. Selecting an appropriate support material can reduce the adsorption on the surface of the support. Preliminary tests showed that the best results were obtained with the Chromosorb PAW DMCS and the

Chromosorb WHP. Both supports were loaded with 25% of n-pentacontane. The fact that with Chromosorb PAW DMCS retention times depend on the sample size and chromatographic peaks are asymmetric indicates the presence of adsorption. Moreover, retention times of alcohols are longer than expected, indicating the presence of active adsorption sites on the support surface. The Chromosorb WHP support has a lower specific area and a smaller concentration of hydroxyl groups which reduces the adsorption. Results obtained are in good agreements with literature data for most of the compounds studied and retention times depend only slightly on the sample size. However, retention times observed with polyaromatic hydrocarbons are still longer than expected. Good results were obtained by deactivating the column (Table 1) with Silyl 8, as recommended by [31]. The use of packed columns with Chromosorb WHP coated with n-alkane and deactivated with Silyl 8 made it possible to obtain a homogenous set of log L^{16} in good agreement with literature data.

2.2.4. Determination of LogL^{16} using temperature gradient method

The packed column technique can be used to measure log L^{16} data of volatile organic compounds. The reasonable limit of application of this method is the retention time of n-eicosane. Experimental difficulties make hazardous quantitative determination log L^{16} of heavier compounds. To enlarge the applicability of chromatographic methods to organic compounds less volatile than n-eicosane, a method based on the temperature gradient chromatography can be used. Recently, Donovan [32] showed that retention times of heavy organic compounds obtained in a gradient mode are linearly related to the logarithm of the vapor pressure at 298.2 K. The authors used a DB-1 megabore column at high flow rates of the gas phase. This method making it possible to reduce considerably retention time was applied to determine vapor pressures of pesticides and polyaromatic hydrocarbons. Nevertheless, the stationary phase DB-1 is slightly polar [24]. Corresponding system parameters of the poly(dimethylsiloxane) immobilized in DB-1 column were published in the reference [24]. Values determined at t= 60°C are as follows: r =0, s = 0.211, a = 0.308 and b = 0. Therefore, experimental results obtained with a DB-1 column can be used to determine log L^{16} only if LSER parameters expressing solute polarity are known. No general relationship between the reduced retention time and log L^{16} valid with all organic compounds can be obtained without the knowledge of above parameters.

However, this approach can be used to establish relationship between the reduced retention time and log L^{16} within a series of compounds. Indeed, polar parameters vary only slightly and in a regular way within a series. Moreover, certain parameters decrease strongly with rising temperature [24]. Therefore, it can be supposed that the effect of the stationary phase polarity is nearly constant within a homologous series of moderately polar compounds. Measurements performed in a gradient mode with several homologous series confirmed this hypothesis. However, linear relationship does not afford the precision required for the log L^{16} determination. It was noticed that not only the reduced retention time t_R but also the corresponding temperature T is needed to establish the appropriate function. Function log L^{16} = f(t_R,T) is linear with R=0.996 that is not enough to represent the log L^{16} with the precision required. We found that the suitable function is as follows:

$$\log L^{16} = exp\left(\frac{f(t_R)}{T}\right) \quad (10)$$

In the case of n-alkanes, function $f(t_R)$ was obtained with log L^{16} literature data of n-alkanes from n-dodecane up to n-docosane. The plot of $f(t_R)$ and values of parameters determining this function are given in Figure 1. The log L^{16} of n-alkanes up to n=38 calculated with equation (10) using gradient mode results are presented in Table 2. It is reasonable to suppose that partition coefficients of heavy n-alkanes up to approximately $C_{45}H_{92}$ can be obtained with gradient method. It should be pointed out that the present approach based on the gradient mode chromatography can be used only to determine log L^{16} within a homologous series of moderately polar compounds. The use of the present method with less polar stationary phase (recently, Li et al. [24] shown that in the case of SPB columns r = 0, a = 0, b = 0) can facilitate the study of polar compounds and perhaps obtain more general results.

Compounds	t_R	log $L^{16\,exp}$	Log $L^{16\,a,b}$
n-Undecane	3.70	5.221	5.191[a]
n-Dodecane	5.87	5.705	5.696[a]
n-Tridecane	8.18	6.195	6.200[a]
n-Tetradecane	10.60	6.707	6.705[a]
n-Pentadecane	12.98	7.221	7.209[a]
n-Hexadecane	15.15	7.707	7.714[a]
n-Heptadecane	17.42	8.237	8.218[a]
n-Octadecane	19.47	8.739	8.722[a]
n-Nonadecane	21.42	9.239	9.226[a]
n-Eicosane	23.30	9.743	9.731[a]
n-heinecosane	25.12	10.253	10.236[a]
n-Docosane	26.85	10.760	10.740[a]
n-Tricosane	28.48	11.259	11.252[a]
n-Tetracosane	30.05	11.759	11.758[a]
n-Pentacosane	31.57	12.262	12.244[b]
n-Hexacosane	33.07	12.779	12.744[b]
n-Heptacosane	34.47	13.280	13.244[b]
n-Octacosane	35.82	13.780	13.744[b]
n-Nonacosane	37.15	14.291	14.244[b]
n-Heinetriacontane	39.70	15.321	15.244[b]
n-dotriacontane	40.80	15.787	15.744[b]
n-Pentatriacontane	44.00	17.223	17.244[b]
n-Hexatriacontane	45.08	17.736	17.744[b]
n-Octatriacontane	47.00	18.684	18.744[b]

Table 2. Retention times t_R and partition coefficients log $L^{16\,exp}$ of n-alkanes determined at 298.2°C using the gradient method on DB-1 column. Log $L^{16\,a,b}$ a: -literature data [5,14,21,31]. b: -estimates calculated using the group contribution method proposed by reference [2].

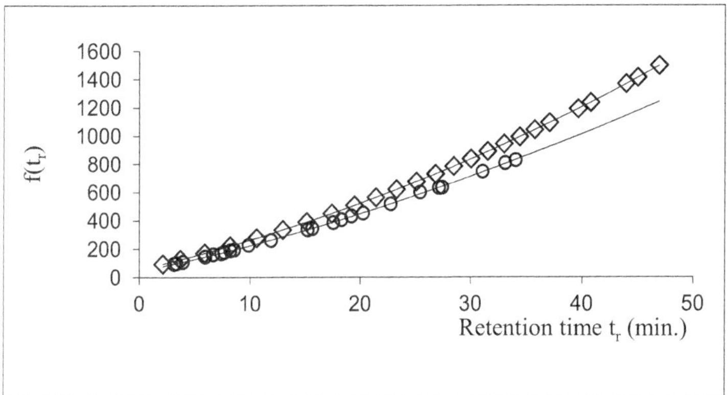

Figure 2. Plot of f(tR) function established with temperature gradient method; aromatics and polyaromatics (o), n-alkanes (◊).

2.2.5. Determination of polar LSER parameters

The excess molar refraction E is defined by the difference between the value for the solute molar refraction and the molar refraction for an alkane of the same characteristic volume:

$$R_2 = MR_X(\text{solute}) - MR_X(\text{alkane of the same } V_X) \tag{11}$$

The solute molar refraction is calculated from the following equation:

$$MR_X = \frac{10 \cdot (n^2 - 1) \cdot V_X}{(n^2 + 2)} \tag{12}$$

Where V_X is the specific volume in $cm^3.mol^{-1}/100$ and n the refractive index of the solute.

Abraham et al. [18] set out to construct scales of solute hydrogen bond acidity and hydrogen bond basicity using logK values for reaction 1 in tetrachloromethane. The authors set out logK values for a series of acids against 45 given bases. If log K values for acids in a given reference base is plotted against log K values for acids in another reference base, a series of straight lines is observed with an intersection at a magic point of -1.1 log units. It is found that:

$$\log K(\text{series of acids against reference base B}) = L_B \log K_A^H + D_B \tag{13}$$

Where L_B and D_B characterize the base and $log K_A^H$ values characterize the series of acids. The $\sum \alpha_2^H$ and $\sum \beta_2^H$ parameters are then defined by:

$$\Sigma\alpha_2^H = \left(\log K_A^H + 1.1\right)/4.636 \tag{14}$$

$$\Sigma\beta_2^H = \left(\log K_B^H + 1.1\right)/4.636 \tag{15}$$

The $\Sigma\alpha_2^H$ and $\Sigma\beta_2^H$ parameters can be determined at the same time with π_2^H parameter. Retention data (such as retention volume, partition coefficients) of the solute in three stationary phases at least of different polarity can be used to determine the $\Sigma\alpha_2^H$, $\Sigma\beta_2^H$ and π_2^H . Each stationary phase can be described by equation (16):

$$\log K_L = c + r \cdot R_2 + s \cdot \pi_2^H + a \cdot \Sigma\alpha_2^H + b \cdot \Sigma\beta_2^H + l \cdot \mathrm{Log}L^{16} \tag{16}$$

At first, the LSER parameters c, s, a, b and l of each stationary phase are determined by multiple linear regression using solutes for which R_2, $\Sigma\alpha_2^H$, $\Sigma\beta_2^H$, π_2^H and $\log L^{16}$ are known. Then, $\Sigma\alpha_2^H$, $\Sigma\beta_2^H$, π_2^H can be determined by multiple linear regression.

2.2.6. Group contribution Method for calculation of LSER parameters of organic compounds

Predictive methods allow to calculate these physico-chemical parameters which are inacessible via direct experiment. This alternative is particularly interesting in the case of log L^{16} of nonvolatile compounds. We consider that experimental methods described in the preceeding paragraph are useful for determination of LSER parameters of volatile and moderatly non-volatile compounds. Therefore, the large data bank of log L^{16} values already available in the literature can be used to establish group additivity rules and to predict log L^{16} of less volatile compounds. Havelec & Sevcik [27,28] presented a general group contribution method making it possible to calculate accurate estimates of log L^{16} of about 2000 organic compounds. The number of groups necessary to obtain good estimates of log L^{16} depends on the complexity of the molecular structure and rises in the case of polyfunctional molecules. This explains a high number of adjustable parameters used in the model [27,28]. For instance, log L^{16} of non-aromatic hydrocarbons is established with 33 parameters and 9 structural contributions. The total number of all group parameters, interactional parameters and structural contributions is of 131. The contribution of a given group is represented in the reference [27,28] with three parameters related to the structure of the molecule and to its interactions with the stationary phase. As log L^{16} is dependent on the solute vapor pressure and on the infinite dilution activity coefficient this approach is basically correct. However, molecular interactions are always related to n-hexadecane and certain parameters can be correlated. Platts et al. [20] recently proposed a new predictive method based on a careful analysis of contributions of various functional groups to establish log L^{16} and other LSER parameters. Therefore, molecular segments were defined in view to obtain good estimates of each. The log L^{16} of hydrocarbons is calculated with 9 parameters only. This method was established with 81 parameters, using a databank of 1908 compounds. A new model was proposed to calculating log L^{16} for nonvolatile organic compounds with special attention paid to heavy hydrocarbons. Data for 550 organic compounds containing mainly hydrocarbons and members of homologous series were used in regression. Basic heteroatom segments were taken into account but the polyfunctional organic compounds were not dealt with. Values of log L^{16} were taken from literature [5, 21]. To elaborate the group contribution method a simple and efficient approach was used. Accordingly, log L^{16} of the compounds X was calculated with the following expression:

$$\log L_X^{16} = \sum_i n_i \cdot c_i \quad (17)$$

where c_i is the contribution of the group "i" and ni is the number of groups "i" in the compound X.

Platts et al. [20] have developed and tested additive models for six important molecular LFER descriptors, namely, R_2, $\sum\alpha_2^H$, $\sum\beta_2^H$, π_2^H and $\log L^{16}$. Five of these six, all bar $\sum\alpha_2^H$ are calculated from a single set of 81 atom and group fragments, while $\sum\beta_2^H$ is calculated from a separate set of 51 fragments. In general, the linear fit obtained with these additive models is good, with R_2 and log L^{16} in particular giving excellent correlation. Splitting the data into training and test sets has also tested the predictive ability of such models, and is found to be almost as accurate as the full regressions. The performance of the method in calculating descriptors for "difficult" structures, ones containing intramolecular interactions such as hydrogen bonds, has been analyzed. Variations in descriptors due to such interactions are generally found to be reproduced, though inevitably some small discrepancies are found. In conclusion, this model is particularly powerful and useful for the prediction of LSER parameters of heavy and complicated molecules.

3. Application of linear solvation energy relationship on ionic liquids

The LSER model may be used to characterize the stationary phases in chromatography. In this case, a large number of solutes (between 20 and 50) for which LSER parameters R_2, $\sum\alpha_2^H$, $\sum\beta_2^H$, π_2^H and $\log L^{16}$ are known have to be injected. The LSER parameters characterizing the stationary phases (c, s, r, a, b and l) are determined by multiple linear regression. In the literature, there is a large amount of data of partition coefficients or activity coefficients measured by gas–liquid chromatography or by dilutor technique. Some system constants for various ionic liquids and classical solvents at 25 °C are summarized in Table 3. The data for the 1-ethanol-3-methylimidazolium tetrafluoroborate, 1-ethanol-3-methylimidazolium hexafluorophosphate, 1,3-dimethylimidazolium dimethylphosphate and 1-ethyl-3-methylimidazolium diethylphosphate [33], 1-Butyl-3-methylimidazolium tetrafluoroborate [34], n-Acryloyloxypropyl-N-methylimidazolium bromide and n-Methacryloyloxyhexyl-N-methylimidazolium bromide [35], 1-Propenyl-3-alkyl-imidazolium bromide [36], 1-butyl-3-methylimidazolium octyl sulfate and 1-ethyl-3-methylimidazolium tosylate [37], Triethylsulphonium bis(trifluoromethylsulfonyl)imide [38], 1-Methyl-3-ethylimidazolium bis(trifluorosulfonyl)-amide and 1.2-Dimethyl-3-ethylimidazolium bis(trifluorosulfonyl)-amide [39] were taken from the sources indicated. Poole & Poole [40] found that the system constants of LSER model for the room temperature ionic liquids fall into the range $e = -0.62$ to 0.86, $s = 1.7$–2.8, $a = 2.1$–7.3, $b = 0$–1.07, and $l = 0.35$–0.96. Compared with the scale of the polar organic solvents $e = -0.60$ to 0.82, $s = 0.54$–2.8, $a = 0.28$–5.50, $b = 0$–4.8, and $l = -0.21$ to 0.98, we can see that both scales are similar indicating that the solvation properties for the room temperature ionic liquids are classical and fit quite well into the scales developed for polar molecular solvents.

The (c + lL) term gives information on the effect of cohesion of the ionic liquids on solute transfer from the gas phase. In general, the ionic liquids are cohesive solvents; they interact weakly via nonbonding and π-electrons (*r* system constant is zero) and are not much different to other polar non-ionic liquids. The ionic liquids are roughly as dipolar/polarizable as classical solvents. The hydrogen-bond basicity of the ionic liquid (*a* system constants) are considerably larger than values obtained for non phases (0-2.1) [1]. The hydrogen-bond basicity of Ils depends on the anion grafted on the cation. ionic liquids can be slightly more hydrogen-bond basic than dimethyl sulfoxide and *N*-methylpyrrolidinone, and are weak to moderate hydrogen-bond acids, similar to the aliphatic alcohols. From Table 3 and data collected in the reference [40], we can see that the hydrogen-bond acidity of the ionic liquids depends largely on the cation and is lower for the 1,3-dialkylimidazolium salts with an alkyl group at C-2 position than 1,3-dialkylimidazolium salts.

4. Predictive models based on LSER model coupled to a group contribution method

Solvation model may be also used to set up correlation between thermodynamic properties and LSER parameters. Abraham et al. [41,42] reported mathematical correlations based on the general Abraham solvation parameter model for the gas-to-solvent, K_L, and water-to-solvent, P, partition coefficients. Recently, [43-46] modified the Abraham solvation parameter model:

$$\begin{aligned}\mathrm{Log}K_L = {} & c_{cation} + c_{anion} + (e_{cation} + e_{anion})\cdot\mathbf{E} + (s_{cation} + s_{anion})\cdot\mathbf{S} + \\ & + (a_{cation} + a_{anion})\cdot\mathbf{A} + (b_{cation} + b_{anion})\cdot\mathbf{B} + (l_{cation} + l_{anion})\cdot\mathbf{L}\end{aligned} \tag{18}$$

$$\begin{aligned}\mathrm{Log}P = {} & c'_{cation} + c'_{anion} + (e'_{cation} + e'_{anion})\cdot\mathbf{E} + (s'_{cation} + s'_{anion})\cdot\mathbf{S} \\ & + (a'_{cation} + a'_{anion})\cdot\mathbf{A} + (b'_{cation} + b'_{anion})\cdot\mathbf{B} + (v_{cation} + v_{anion})\cdot\mathbf{V}\end{aligned} \tag{19}$$

by rewriting each of the six solvent equation coefficients as a summation of their respective cation and anion contribution. The dependent variables in equations (18) and (19) are solutes descriptors as follows: **E** and **S** refer to the excess molar refraction in units of $(cm^3.mol^{-1})/10$ and dipolarity/polarizability descriptors of the solute, respectively, **A** and **B**

Ionic liquids and classical solvents	System constants					
	e	s	a	b	l	c
1-ethanol-3-methylimidazolium hexafluorophosphate	0	3.03	2.89	1.13	0.47	-1.14
1-ethanol-3-methylimidazolium tetrafluoroborate	0	3.03	3.64	0.763	0.5	-1.35
1,3-dimethylimidazolium dimethylphosphate	0.86	2.59	7.27	0	0.35	-0.61
1-ethyl-3-methylimidazolium diethylphosphate	0.26	1.97	6.9	0	0.54	-0.09
1-Butyl-3-methylimidazolium tetrafluoroborate	0.56	2.82	3.27	0.48	0.5	-0.77
n-Acryloyloxypropyl-N-methylimidazolium bromide	0	2.88	5.5	0	0.48	-1.03
n-Methacryloyloxyhexyl-N-methylimidazolium bromide	0	2.46	5.36	0	0.57	-0.87
1-Propenyl-3-methyl-imidazolium bromide	0	2.16	5.19	0	0.53	-1.86
1-Propenyl-3-octyl-imidazolium bromide	0	1.72	4.96	0	0.57	-1.6
1-Propenyl-3-decyl-imidazolium bromide	0	1.73	4.89	0	0.66	-1.58
1-Propenyl-3-dodecyl-imidazolium bromide	0	1.44	4.87	0	0.72	-1.51
1-Butyl-3-methylimidazolium octyl sulfate	0	1.47	4.05	0	0.68	-0.237
1-Ethyl-3-methylimidazolium tosylate	0.54	2.4	4.81	0.17	0.48	-0.84
n-Butylammonium thiocyanate	0.14	1.65	2.76	1.32	0.45	-0.75
1-Methyl-3-ethylimidazolium bis(trifluorosulfonyl)-amide	0.148	2.277	2.172	1.041	0.629	-0.439
1.2-Dimethyl-3-ethylimidazolium bis(trifluorosulfonyl)-amide	0.214	2.347	2.075	0.896	0.655	-0.565
Triethylsulphonium bis(trifluoromethylsulfonyl)imide	0.114	2.37	2.34	0.696	0.642	-0.803
1-Ethyl-3-methylimidazolium Trifluoroacetate	0.608	1.63	4.21	1.81	0.584	-0.918
1-Butyl-3-methylimidazolium Trifluoromethanesulfonate	0.399	2.03	3.49	0.681	0.647	-0.784
Trifluoroethanol	-0.547	1.339	2.213	3.807	0.645	-0.092
Methanol	-0.22	1.17	3.7	1.43	0.769	-1.27
Water	0.82	2.74	3.9	4.81	-0.213	0
1.2 Dichoroethane	-0.47	1.676	0.92	0.486	0.927	0.025
Dry methyl acetate	-0.447	1.675	2.625	0.213	0.874	0.129
Dry ethyl acetate	-0.352	1.316	2.891	0	0.916	0.182
Ethylene glycol	0.217	1.427	4.474	2.687	0.568	-0.898
Benzene	-0.313	1.053	0.457	0.169	1.02	0.107
2-(Cyclohexylamino)ethanesulfonate	0.07	1.57	3.67	0	0.51	-0.83

Table 3. LSER descriptors of ionic liquids determined at 313.15 K.

are measures of the solute hydrogen-bond acidity and basicity, **V** is the McGowan volume in units of ($cm^3.mol^{-1}$)/100 and **L** is the logarithm of the gas-to-hexadecane partition coefficient at 298 K. Sprunger et al. calculated equation coefficients for 8 cations and 4 anions using a database that contained 584 experimental log K_L and 571 experimental log P values. No loss in predictive accuracy was observed by separating the equation coefficients into individual cation-specific and anion-specific values. The major advantage of splitting the equation coefficients into individual cation-specific and anion-specific contributions is that one can make predictions for more ILs. In Sprunger's approach, the major advantage of splitting the equation coefficients into individual cation-specific and anion-specific contributions is that one can make predictions for more Ils. Most of the cations are alkylimidazolium. The use of this model is somewhat limited since it can not be extrapolated to alkylimidazolium ionic liquids not initially defined by the method (e.g. with long alkyl chains).

In the development of Mutelet et al. [47], the cation with its alkyl chains is splitted in different contributions: (CH_3, CH_2, N, CH_{cyclic}...). The approach allows to have a predictive model. The aim of this work was to develop a group contribution method allowing to estimate the log K_L and Log P of organic compounds in ionic liquids at 298 K. Using the LSER model proposed by Abraham, the group contribution method expresses LSER coefficients c_i, e_i, s_i, a_i, b_i and l_i of equation (19). or $c_i^{'}$, $e_i^{'}$, $s_i^{'}$, $a_i^{'}$, $b_i^{'}$ and v_i of equation (20) by:

$$\begin{aligned} \mathrm{Log}K_L &= \sum_i^{21} n_i \times c_i + \sum_i^{21} n_i \times e_i \cdot \mathbf{E} + \sum_i^{21} n_i \times s_i \cdot \mathbf{S} + \sum_i^{21} n_i \times a_i \cdot \mathbf{A} + \sum_i^{21} n_i \times b_i \cdot \mathbf{B} + \sum_i^{21} n_i \times l_i \cdot \mathbf{L} \\ \mathrm{Log}P &= \sum_i^{21} n_i \times c_i^{'} + \sum_i^{21} n_i \times e_i^{'} \cdot \mathbf{E} + \sum_i^{21} n_i \times s_i^{'} \cdot \mathbf{S} + \sum_i^{21} n_i \times a_i^{'} \cdot \mathbf{A} + \sum_i^{21} n_i \times b_i^{'} \cdot \mathbf{B} + \sum_i^{21} n_i \times v_i \cdot \mathbf{V} \end{aligned} \qquad (20)$$

Where n_i is the number of group i present in the ionic liquid.

Mutelet et al. [47] proposed to extend the temperature dependent GC-LSER in view of determining the partition coefficient of organic compounds in ionic liquids. The GC-LSER can be rewriting as followed:

$$\mathrm{Log}K_L = \mathrm{const} + \frac{\sum_i^{21} n_i \times c_i + \sum_i^{21} n_i \times e_i \cdot E + \sum_i^{21} n_i \times a_i \cdot A + \sum_i^{21} n_i \times b_i \cdot B + \sum_i^{21} n_i \times l_i \cdot L}{T} \qquad (21)$$

The experimental data used to calculate Abraham's model ion-specific equation coefficients were taken from the collection of [43-46] and were updated with recent data. A total of 1450 gas-liquid partition coefficients and 1410 water-to-liquid partition coefficients were used for the calculation. Solutes were mainly n-alkanes, cycloalkanes, alkenes, alkynes, aromatics, alcohols, ethers, aldehydes, ketones, chloroalkanes. The *E*-scale varies from 0 to 1.5, the *S*-scale from 0 to 1.72, the *A*-scale from 0 to 1.04, the *B*-scale from 0 to 1.28, the *L*-scale from –1.200 to 7.833 and the *V*-scale from 0.109 to 1.799. The dataset is composed of 27 imidazolium based ionic liquids, 3 ammonium, 3 pyridinium and 4 pyrolidinium based ionic liquids. The authors also add sulphonium and phosphonium ionic liquids although only one set of K_L (or P) data may be found for these families. The twenty one groups which are defined in this method are listed in Table 4. The decomposition into groups of the ionic liquids is very easy, that is as simple as possible. No substitution effects are considered. No exceptions are defined. In Figure 3 are represented all ionic liquids studied in this work. Five groups are defined to describe the chains R_1, R_2, R_3 and R_4 grafted on the cation: CH_3, CH_2, -O-, -O-N_{cycl} and –OH. These groups allow the calculation of partition coefficients of alkyl based ionic liquids but also functionalized ionic liquids such as ether, alcohols. The remaining seven groups are: $CH_{2cyclic}$, CH_{cyclic}, C_{cyclic}, N_{cyclic}, N^+ (ammonium cation), P^+ (phosphonium cation) and S^+ (sulphonium cation).

More precisely, N_{cyclic} represents two structures: $-\underset{|}{N}^+=$ and $-\underset{|}{N}-$. Nine groups are used for anions: bis(trifluoromethylsulfonyl)imide $:(TF)_2N^-$, hexafluorophosphate: PF_6^-, tetrafluoroborate: BF_4^-, ethylsulfate: $EtSO_4^-$, octylsulfate: $OcSO_4^-$, thiocyanate: SCN^-, trifluoromethylsulfonate : $CF_3SO_3^-$, trifluoroacetate : ACF_3^- and dicyanamide: $(CN)_2N^-$. As an example, let's have a look at the decomposition of 1-butyl-3-methylimidazolium hexafluorophosphate. In this case, the decomposition of the molecule into elementary groups is: 2 group 1 (-CH_3) + 3 group 2 (-CH_2) + 3 group 7 (C_{cyclic}) + 2 group 9 (N_{cyclic}) + 1 group 14 (PF_6^-).

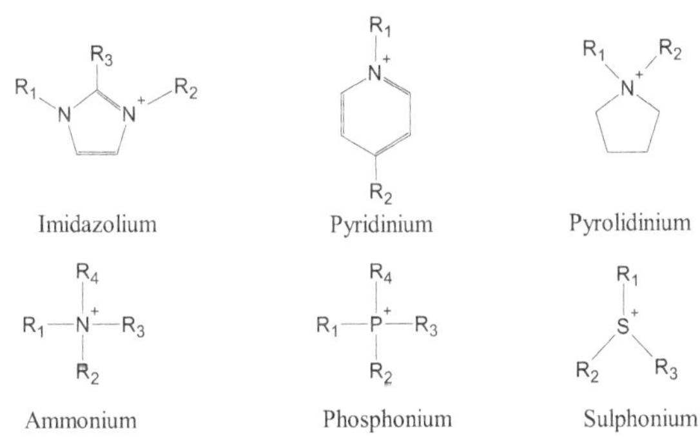

Figure 3. Cation of six families of ionic liquids.

Group contribution model coupled to LSER (GC-LSER) for estimating the gas-to-ionic liquids partition coefficients and water-to-ionic liquids partition coefficients allows to predict with good accuracy Log K_L and Log P at 298 K of not only alkyl based ionic liquids but also functionalized ionic liquids. The parameters of the group contribution methods were determined for imidazolium, pyridinium, pyrrolidinium, phosphonium, ammonium and sulphonium based ionic liquids containing several different anions. A comparison between the experimental and calculated values showed that the proposed models describe the experimental data available with a mean absolute error of about 0.15 log unit. While the model is probably somewhat limited in prediction for pyridinium and pyrrolidinium based ionic liquids because of the poor dataset for these cations, results obtained are satisfactory.

Cation's group		Definition	Anion's group		Definition
Group 1	CH_3-	CH_3 from alkyl chain R_1, R_2,R_3 or R_4	Group 13	$(TF)_2N^-$	bis(trifluoromethyl-sulfonyl)imide
Group 2	-CH_2-	CH_2 from alkyl chain R_1, R_2,R_3 or R_4	Group 14	PF_6^-	Hexafluorophosphate
Group 3	-O-	-O- in alkyl chain R_1, R_2,R_3 or R_4	Group 15	BF_4^-	Tettrafluoroborate
Group 4	-O-N_{cyclic}-	Oxygenated atom connected directly to N_{cyclic}	Group 16	$EtSO_4^-$	Ethylsulfate
Group 5	-OH	-OH from alkyl chain R_1, R_2,R_3 or R_4	Group 17	$OcSO_4^-$	Octylsulfate
Group 6	$CH_{2cyclic}$	CH_2 cyclic in pyrolidinium's cation	Group 18	SCN^-	Thiocyanate
Group 7	CH_{cyclic}	CH cyclic in imidazolium or pyridinium's cation	Group 19	$CF_3SO_3^-$	Trifluoromethyl-sulfonate
Group 8	C_{cyclic}	C cyclic in imidazolium or pyridinium's cation	Group 20	ACF_3^-	Trifluoroacetate
Group 9	N_{cyclic}	Cyclic nitrogen (imdazolium, pyridinium and pyrolidinium)	Group 21	$(CN)_2N^-$	Dicyanamide
Group 10	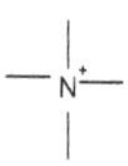	Ammonium's cation			
Group 11	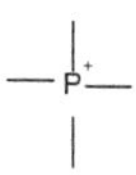	Phosphonium's cation			
Group 12	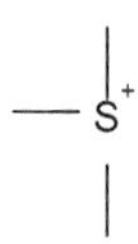	Sulphonium's cation			

Table 4. Description of the 21 groups used for the estimation of $LogK_L$ and LogP.

5. Conclusion

The solvation parameter model is suitable for describing the retention properties of molecules in chromatographic systems. To establish the system properties requires identification of a group of compounds with well known descriptor values. We have shown that all LSER parameters of solutes may be determined using gas chromatography or experimental techniques. The solvation model may be used either for the physico-chemical characterization of the stationary phases or for the establishment of a suitable quantitative structure–property relationship to facilitate the prediction of further system properties for compounds lacking experimental values.

Author details

Fabrice Mutelet
Laboratoire Réactions et Génie des Procédés (CNRS, UPR 3349), Ecole Nationale Supérieure des Industries Chimiques, Nancy CEDEX, France

6. References

[1] Poole, C.F. Chromatographic and spectroscopic methods for the determination of solvent properties of room temperature ionic liquids. Journal of Chromatography A 2004; 1037 49-82.

[2] Mutelet, F. & Rogalski, M. Experimental determination and prediction of the gas-liquid n-hexadecane partition coefficients J. Chromatogr. A 2001; 923 153-163.

[3] Russom, C.L.; Breton, R.L.; Walker, J.D.; Bradury, S.P. An overview of the use of quantitative structure-activity relationships for ranking and prioritizing large chemical inventories for environmental risk assessments Environ. Toxicol. Chem. 2003; 22 1810-1821.

[4] Katritzky, A.R.; Maran, U.; Lobonov, V.S.; Karelson, M. Structurally Diverse Quantitative Structure-Property Relationship Correlations of Technologically Relevant Physical Properties J. Chem. Inf. Comput. Sci. 2000; 40 1-18.

[5] Abraham, M. H.; Grellier, P. L. & Mc Gill R.A. Determination of Olive Oil-Gas and Hexadecane-Gas Partition Coefficients, and calculation of the corresponding Olive Oil-Water and Hexadecane-Water Partition Coefficients. J. Chem. Soc. PERKIN TRANS II 1987; 797-803.

[6] Vitha, M. & Carr, P.W. (2006). The chemical interpretation and practice of linear solvation energy relationships in chromatography. Journal of Chromatography A 2006; 1126 143-194.

[7] Abraham, M.H., Poole, C.F. & Poole, S.K. Classification of stationary phases and other materials by gas chromatography Journal of Chromatography A, 1999; 842 79-114.

[8] Kamlet, M.J.; Abboud, J.L. & Taft, R.W. (1977). The solvatochromic comparison method. 6. The π scale of solvent polarities. Journal of the American Chemical Society 1977; 18 (99) 6027-6038.

[9] Kamlet, M.J.; Abboud, J.-L.M.; Abraham, M.H. & Taft, R.W. Linear solvation energy relationships. 23. A comprehensive collection of the solvatochromic parameters, π^*, α, and β, and some methods for simplifying the generalized solvatochromic equation. Journal of Organic Chemistry 1983; 17 (48) 2877-2887.
[10] Kamlet, M.J. & Taft, R.W. Linear Solvation Energy Relationships. Local Empirical Rules -- or Fundamental Laws of Chemistry? A Reply to the Chemometricians., Acta Chem. Scand., B 1985; 39 611-628.
[11] Kamlet, M.J.; Doherty, R.M.; Abraham, M.H.; Marcus, Y. & Taft, R.W. Linear solvation energy relationships. 46. An improved equation for correlation and prediction of octanol/water partition coefficients of organic nonelectrolytes (Including strong hydrogen bond donor solutes). Journal of Physical Chemistry, (1998); 18 (92) 5244-5255.
[12] Taft, R.W. & Kamlet, M.J. The solvatochromic comparison method. 2. The α-scale of solvent hydrogen-bond donor (HBD) acidities. Journal of the American Chemical Society 1976; 10 (98) 2886-2894.
[13] Abraham, M. H & Whiting, G.S. Hydrogen-bonding. Part 22. Characterization of soybean oil and prediction of activity coefficients in soybean oil from inverse gas chromatographic data. J. Am. Oil Chem. Soc., 1992; 69 1236-1238.
[14] Abraham, M.H.; Andonian-Hanftvan, J.; Osei-Owusu, J.P.; Sakellariou, P.; Urieta, J.S.; López, M.C. & Fuchs, R. Hydrogen bonding. Part 25. The solvation properties of methylene iodide. J. Chem. Soc., Perkin Trans. II 1993; 3 299-304.
[15] Park, J.H. & Carr, P.W. Interpretation of normal-phase solvent strength scales based on linear solvation energy relationships using the solvatochromic parameters π^*, α and β. J. Chromatogr. 1989; 465 (3) 123-136.
[16] Weckwerth, J.D.; Carr, P.W.; Vitha, M.F.; Nasehzadeh, A. A comparison of gas-hexadecane and gas-apolane partition coefficients Anal. Chem. 1998; 70 3712-3716.
[17] Abraham, M. H.; Whiting, G. S. & Doherty R. M. Hydrogen Bonding. Part 13. A New Method for the Characterization of GLC Stationary Phases-The Lafford Data Set. J. Chem. Soc. PERKIN TRANS II 1990; 1451-1460.
[18] Abraham, M. H.; Whiting, G. S.; Doherty, R. M. & Shuely, W. J. Hydrogen bonding XVI. A new solute solvation parameter, π_2^H, from gas chromatographic data. J. Chromatogr. 1991; 587 213-228.
[19] Abraham, M. H. Scales of Solute Hydrogen-bonding: Their construction and Application to Physicochemical and Biochemical Processes. Chem. Soc. Rev. 1993; 22 73-83.
[20] Platts, J.A.; Butina, D.; Abraham, M.H. & Hersey, A. Estimation of molecular linear free energy relation descriptors using a group contribution approach. Journal of Chemical Information and Computer Sciences 1999; 39 835-845.
[21] Abraham, M.H. & Platts, J.A. Hydrogen bond structural group constants. Journal of Organic Chemistry 2001; 66 3484-3491.
[22] Weckwerth, J.D. & Carr, P.W. Study of Interactions in Supercritical Fluids and Supercritical Fluid Chromatography by Solvatochromic Linear Solvation Energy Relationships. Anal. Chem. 1998; 70 (7) 1404-1410.

[23] Zhang, Y.; Dallas, A. J.; Carr, P. W. Critical comparison of gas hexadecane partitioncoefficients as measured with packed and open-tubular capillary columns J. Chromatogr. 1993; 638, 43-56.
[24] Li, Q.; Poole, C. F.; Kiridena, W.; Koziol, W.W. Chromatographic methods for the determination of the logL16 solute descriptorAnalyst, 2000; 125 2180-2188.
[25] Abraham, M.H.; Andonian-Haften, J.; My Du, C.; Osei-Owusu, J. P.; Sakellariou, P. Shuely, W.J.; Poole, C.F.; Poole, S.K. Comparison of uncorrected retention data on a capillary and a packed hexadecane column with corrected retention data on a packed squalane column J. Chromatogr. A 1994; 688 125-134.
[26] Defayes, G.; Fritz, D. F.; Görner, T.; Huber, G.; De Reyff, C.; Kovats, E. sz. J. Chromatogr., Organic solutes in paraffin solvents: Influence of the size of the solvent molecule on solution data. 1990; 500 139-184.
[27] Havelec, P.; Sevcik, J. G. K. Concept of additivity for a nonpolar solute-solvent criterion Log L(16) – Nonaromatic compounds J. Chromatogr. A 1994; 677 319-329.
[28] Havelec, P.; Sevcik, J. G. K. (1996). Extended additivity model of parameter log(L(16)) ,J. Phys. Chem. Ref. Data 1996; 25 1483-1493.
[29] Tranchant, J. Manuel pratique de chromatographie en phase gazeuse, Masson, Paris, 1995.
[30] Dallas, A.J.; Carr, P.W. Direct chromatographic comparison of the relative adsorption activity of various types of capillary transfer tubing Anal. Chim. Acta 1991; 251 83-93.
[31] Kersten, B.R. & Poole C.F. Influence of concurrent retention mechanisms on the determination of stationary phase selectivity in gas chromatography J. Chromatogr. 1987; 399 1-31.
[32] Donovan, S.F. (1996). New method for estimating vapor pressure by the use of gas chromatography J. Chromatogr. A 1996; 749 123-129.
[33] Revelli, A.-L.; Mutelet, F. & Jaubert, J.-N. Partition coefficients of organic compounds in new imidazolium based ionic liquids using inverse gas chromatography. Journal of Chromatography A 2009; 1216 (23) 4775-4786.
[34] Revelli, A.-L.; Mutetet, F.; Turmine, M.; Solimando, R. & Jaubert, J.-N. Activity coefficients at infinite dilution of organic compounds in 1-butyl-3-methylimidazolium tetrafluoroborate using inverse gas chromatography. Journal of Chemical and Engineering Data 2009; 54 (1), 90-101.
[35] Mutelet, F., Jaubert, J.-N.; Rogalski, M.; Harmand, J.; Sindt, M. & Mieloszynski, J.-L. Activity coefficients at infinite dilution of organic compounds in 1-(meth)acryloyloxyalkyl-3-methylimidazolium bromide using inverse gas chromatography. Journal of Physical Chemistry B 2008; 112 (12) 3773-3785.
[36] Mutelet, F.; Jaubert, J.-N.; Rogalski, M.; Boukherissa, M. & Dicko, A. Thermodynamic properties of mixtures containing ionic liquids: Activity coefficients at infinite dilution of organic compounds in 1-propyl boronic acid-3-alkylimidazolium bromide and 1-propenyl-3-alkylimidazolium bromide using inverse gas chromatography. Journal of Chemical and Engineering Data, 2006; 51 (4) 1274-1279.

[37] Mutelet, F. & Jaubert, J.-N. Accurate measurements of thermodynamic properties of solutes in ionic liquids using inverse gas chromatography. Journal of Chromatography A 2006; 1102 (1-2), 256-267.

[38] Domańska, U. & Marciniak, A. Activity coefficients at infinite dilution measurements for organic solutes and water in the ionic liquid triethylsulphonium bis(trifluoromethylsulfonyl)imide. Journal of Chemical Thermodynamics 2009; 41 (6) 754-758.

[39] Krummen, M.; Wasserscheid, P. & Gmehling, J. Measurement of activity coefficients at infinite dilution in ionic liquids using the dilutor technique. J. Chem. Eng. Data 2002; 47 (6) 1411-1417.

[40] Poole, C.F. & Poole, S.K. Extraction of organic compounds with room temperature ionic liquids. Journal of Chromatography A 2010; 1217 (16) 2268-2286.

[41] Abraham, M.H.; Zissimos, A.M.; Huddleston, J.G.; Willauer, H.D.; Rogers, R.D. & Acree, W.E., Jr. Some novel liquid partitioning systems: Water-ionic liquids and aqueous biphasic systems. Ind. Eng. Chem. Res. 2003; 42 413-418.

[42] Acree, W.E., Jr. & Abraham, M.H. The analysis of solvation in ionic liquids and organic solvents using the Abraham model linear free energy relationship. J. Chem. Technol. Biotechnol. 2006; 81 1441-1446.

[43] Sprunger, L.; Clark, M.; Acree, W.E., Jr. & Abraham, M.H. Characterization of room-temperature ionic liquids by the Abraham model with cation-specific and anion-specific equation coefficients. J. Chem. Inf. Model. 2007; 47, 1123-1129.

[44] Sprunger, L.M.; Proctor, A.; Acree, W.E., Jr. & Abraham, M.H. LFER correlations for room temperature ionic liquids: Separation of equation coefficients into individual cation-specific and anion-specific contributions. Fluid Phase Equilib. 2008; 265 104-111.

[45] Sprunger, L.M.; Achi, S.S.; Acree Jr. W.E.; Abraham, M.H.; Leo, A.J. & Hoekman, D. Correlation and prediction of solute transfer to chloroalkanes from both water and the gas phase. Fluid Phase Equilib. 2009; 281 144-162.

[46] Sprunger, L.M.; Gibbs, J.; Proctor, A.; Acree Jr., W.E.; Abraham, M.H.; Meng, Y.; Yao, C. & Anderson, J.L. Linear free energy relationship correlations for room temperature ionic liquids: revised cation-specific and anion-specific equation coefficients for predictive applications covering a much larger area of chemical space. Ind. Eng. Chem. Res. 2009; 48 4145-4154.

[47] Mutelet, F; Ortega-Villa V.; Moise J.-C. Acree Jr., W.E. & Baker, G.A. Prediction of Partition Coefficients of Organic Compounds in Ionic Liquids Using a Temperature-Dependent Linear Solvation Energy Relationship with Parameters Calculated through a Group Contribution Method Journal of Chemical and Engineering Data 2011; 56 3598-3606.

Chapter 16

Chromatographic Retention Parameters as Molecular Descriptors for Lipophilicity in QSA(P)R Studies of Bile Acid

Mihalj Poša

Additional information is available at the end of the chapter

1. Introduction

Beside of separation and identification of chemical compounds, chromatography can be used to obtain molecular parameters that reflect their structural characteristics – molecular descriptors. Most often it is a parameter of hydrophobicity (lipophilicity) of molecules (ions) which is obtained by using retention parameters of reversed phase liquid chromatography or thin layer chromatography of high resolutions (RPHPLC and RPTLC) [1,2]. Hydrophobicity of molecules is an important feature in medical chemistry, and arbitrarily connected with the logarithm of the solute partition (distribution) coefficient log$P$ (P stands for the ratio of the equilibrium concentration of the particle of the same electronic structure in 1-octanol and its equilibrium concentration in water). Partition coefficient is determined by the traditional shake flask method, which has drawbacks: long analysis (for reaching equilibrium) and the results often do not have adequate reproducibility [1,2,3]. While hydrophobicity obtained by chromatographic methods is obtained relatively quickly, it is possible to specify a number of compounds and achieved by high precision and reproducibility of results. Lipophilicity of chemical compounds (which is expressed either as a chromatographically i.e. retention parameters or as an *in silico* molecular descriptors of log$P$) is often included in the regression equations obtained in the QSA(P)R (Quantitative Structure Activity (Property) Relationship) studies [1,2,3].

Bile acids are amphiphilic molecules that have peculiar structure, because molecular descriptors that are obtained on the basis of the molecular graph or fragmentation methods often do not reflect their true structural features [4]. Therefore, the bile acid chromatographic lipophilicity play an important application in obtaining QSA(P)R models that connect biological and pharmacological or other physical-chemical properties (solubility, critical micelle concentration, critical micelle temperature, etc.) for their structure [5]. In the QSA(P)R models chromatographic parameters are independent variables.

In the following section presents the introduction chromatographic parameters (TLC and HPLC) that are used to represent the hydrophobicity of compounds, and presents the main structural features of bile acids.

2. Chromatographic parameters

The basic characteristic of the position of analytes spot on TLC chromatograms is the R_f value (retardaction factor). R_f value is the ratio of the path length that has crossed spot of solute from the start line S_s and path length S_f of the solvent front (Fig. 1).

$$R_f = \frac{S_s}{S_f} \tag{1}$$

If the solute spends more time in the mobile phase, then its chromatographic spot position is closer to the front of solvent and its R_f value is even higher (maximum value of the R_f parameter is 1). Whereas if the solute spends more time in the stationary phase, then the position of his spots closer to the start line, and its R_f value is less (Fig. 1 (a)).

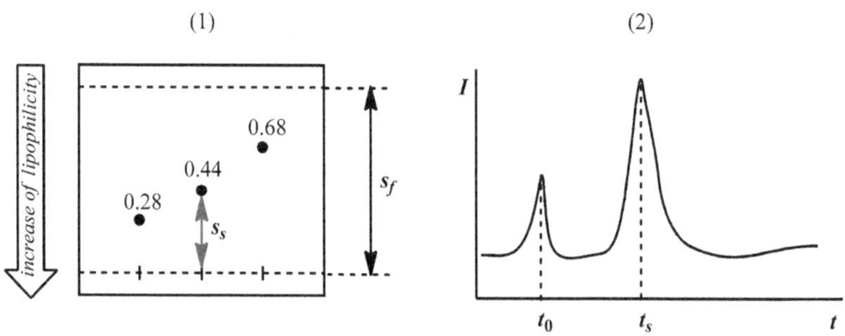

Figure 1. Chromatograms: (1) TLC with R_f values of the solute over their spots, (2) HPLC.

The ratio of time spent by the solute in the stationary phase t_s and the time by spend in the mobile phase t_m is the capacitance factor.

$$k = \frac{t_s}{t_m} \tag{2}$$

The connection between R_f values and capacitive factor has the equation:

$$k = \frac{1}{R_f} - 1 \tag{3}$$

The logarithm of the above expression is:

$$\log k = \log\left(\frac{1}{R_f} - 1\right) \tag{4}$$

and the retention parameter log$k$ usually indicates with R_M. In the reverse phase thin layer chromatography, where the stationary phase is the hydrophobic environment, in the course of the chromatographic process chemical compounds that are more hydrophobic spend more time in the stationary phase than in the polar mobile phase, which resulting in less crossed paths of their chromatographic spots. Therefore, the more hydrophobic is solute its R_f value is more lower or its retention parameter R_M value is more higher. In reverse phase thin-layer chromatography for each compound is determined the dependence of the chromatographic parameters R_M of the volume fraction ϕ of organic modifier in aqueous mobile phase (i.e. for each of the tested compounds from one chromatographic experiment to another experiment varies of mobile phase volume fraction of organic modifier). The most common organic modifier is methanol. If increasing the volume fraction of the organic modifier in aqueous mobile phase results of decrease in hydrophobicity of mobile phase. This is manifested as the reduction of the difference between staying time of solute in the stationary and mobile phase during the chromatographic process – increased R_f values (decreasing value of R_M). Usually between the chromatographic parameters R_M and the volume fraction of organic modifier ϕ is linear relation (usually in the interval: $0.2 \leq \phi \leq 0.8$) [3,6-8].

$$R_M = R_{M0} + S\varphi \tag{5}$$

In the above equation R_{M0} is the extrapolated value of the chromatographic parameters R_M which is governed to the mobile phase contain only water (or buffered aqueous solution), i.e. mobile phase without an organic modifier, while S is the slope of the right (Fig. 2). S is directly related to the specific surface of the stationary phase.

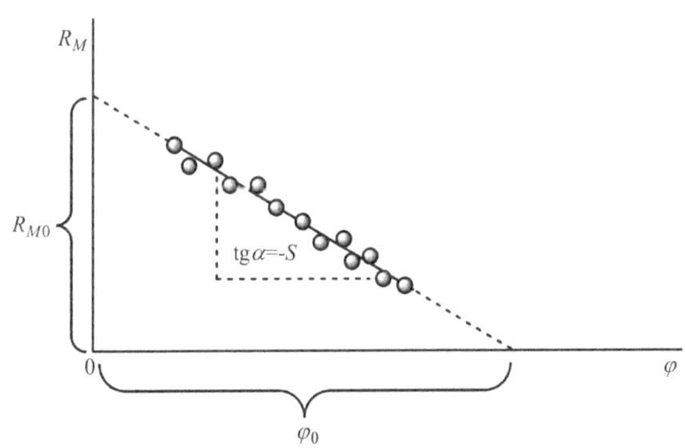

Figure 2. The linear dependence of R_M chromatographic parameters of volume of organic modifier ϕ.

R_{M0} parameters value depends on the type of organic modifier on the basis of which is determined by the function (1.5). R_{M0} is usually in the good correlation of the compounds lipophilicity. In recent times to describe the molecules lipophilicity also used the chromatographic parameters ϕ_0 which is the ratio of R_{M0} and the slope S (Fig. 2) [7,8].

$$\varphi_0 = \frac{R_{M0}}{|-S|} \tag{6}$$

The chromatographic parameters ϕ_0 corresponding volume ratio of organic modifier in aqueous mobile phase in which the same amount of solute in the mobile phase and in the stationary phase. Indeed solute in the above mobile phase composition, during the chromatographic process, spends at the same time in the stationary phase and the mobile phase, therefore capacitive factor (2) is $k = 1$, respectively R_f value (3) is 0.5. This R_f values corresponding to R_M chromatographic parameter (4) whose value is zero. Which means that equation (5) is: $0=R_{M0} + S\phi_0$ from which follows the expression (6). If the solute is more hydrophobic then a larger quantity of organic modifier is needed to equalize the amount of solute in two phases, i.e. the chromatographic parameter ϕ_0 has higher value. With a high efficient reverse phase liquid chromatography, stationary phase is also hydrophobic environment. Solute is characterized by retention time t_s, which represents the elapsed time from injection to the occurrence of the same solute in detectors, i.e. the retention time of solute in the column. Chemical compounds in RP-HPLC analysis is usually characterized by the retention coefficient (capacity factor):

$$k = \frac{t_s - t_0}{t_0} \tag{7}$$

where t_0 is the retention time of solvent from the mobile phase (Fig. 1 (a)). If a chemical compound is more hydrophobic, then more time is spent in the hydrophobic stationary phase, i.e. it takes more time to pass the column and the retention time or retention factor even greater. The logarithm of the retention coefficient $\log k$ is used as a parameter of hydrophobicity of chemical compound. An important chromatographic parameter of lipophilicity is $\log k_w$ which were obtained an extrapolation of the linear equations (8) to the zero volume fraction ϕ of organic modifier in aqueous mobile phase.

$$\log k = \log k_w + S\varphi \tag{8}$$

Similar as in the RP TLC analysis, in RP HPLC also can defined the chromatographic indices ϕ_0 (with the same meaning) as the ratio of $\log k_w$ and the slope from $\log k = f(\phi)$ [9,10].

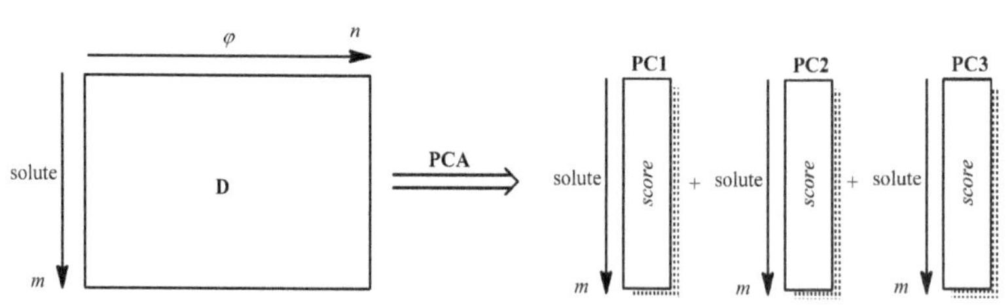

Figure 3. Application of principal component analysis on the retention data matrix D, n = number of different volume fraction of organic modifier, m = number of different compounds, n> 3.

There is a possibility of application principal component analysis (PCA) on chromatographic data to the thin layer and the liquid chromatography. PCA is applied to the data matrix **D** of retention parameters R_M, log*k* (or *k*). The columns of the matrix **D** corresponding retention parameters for different volume fraction ϕ of organic modifier (columns represent the organic modifier) while the matrix rows represent the different types of chemical compounds. PCA is usually applied directly on the covariance matrix $\mathbf{D}^T\mathbf{D}$ retention parameters. As a result of PCAs mathematical procedure (orthogonal diagonalization) are obtained orthonorms PC score vectors of whose number is 2 (PC1, PC2) or 3 (PC1, PC2, PC3), depending on the percentage of variance explained from the data matrix **D**. Accordingly objects (tested molecules) from a multidimensional space matrix of retention parameters **D** mapped to 2d or 3d space of PC. In the graphs of PC scores can be found congeneric group of chemical compounds by their lipophilicity. As a parameter of lipophilicity of chemical compounds in the QSA(P)R studies can be applied also to the scores of PC [3,6].

3. The hydrophobic effect

Generally, the dissolution of an amphiphilic object in water is accompanied by the disruption of the hydrogen bonds between water molecules and formation of a hydration sheath (hydration layer) around the particles of the dissolved substance. If we observe such a system (amphiphilic solution), which consists of two subsystems: hydrophobic part (**a**) and hydrophilic part (**b**), then the thermodynamic functions can be considered separately for each subsystem (Fig. 4). For both subsystems it holds that in the formation of the solvation sheaths (around the amphiphilic objects) approximately the same number of hydrogen bonds are formed as existed between the water molecules in the bulk water, i.e. before introducing the amphiphilic object. Hence the change of the enthalpy for each of the subsystems is equal to zero: $\Delta H(\mathbf{a})\approx\Delta H(\mathbf{b})\approx 0$. Also, in the formation of the solvation sheath, the entropy (translational and rotational) of both subsystems decreases, that is: $\Delta S(\mathbf{a})\approx\Delta S(\mathbf{b})<0$. However, the water molecules from the hydrophilic side (**b**) of the amphiphilic molecule form additional hydrogen bonds; hence for this subsystem there is an additional negative enthalpy ($\Delta H_\varepsilon<0$), which is dissipated as heath in the environment (bulk of the solution), thus giving rise to a positive change of the entropy (of the environment). Therefore, the free enthalpy change for the hydrophilic subsystem is: $\Delta G(\mathbf{b})<0$, on the basis of which the water molecules from the hydrophilic side (**b**) of the amphiphil can be denoted as stabilized water molecules (SWM), while the water molecules from the hydrophobic side (**a**) are nonstabilized water molecules (NSWM) [4,11].

The ratio of the hydrophilic-to-hydrophobic surface area of an amphiphilic molecule determines the overall change of the Gibbs energy of formation of the hydration sheath (ΔG), and, since the hydrophobic surface of the amphiphilic molecule is larger, then $\Delta G>0$. The larger the amount of the amphiphil present in the solution, the more water molecules participate in the formation of the hydration sheath, and the more negative is the overall entropy change. This results in the changes in the system (solution) due to the passing of NSMW from the amphiphil hydration sheath to the bulk of the solution, giving rise to the

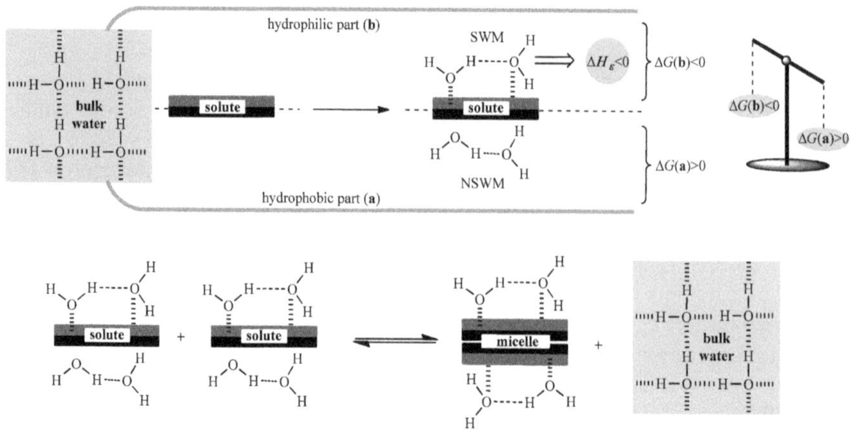

Figure 4. The hydrophobic effect: SWM = stabilised water molecules, NSWM = nonstabilised water molecules.

system entropy. Amphiphilic molecules (ions) are oriented so that their (formally) desolvated hydrophobic sides are to the smallest extent exposed to water molecules. This is realized either in the form of self-association of the amphiphils via their hydrophobic surfaces (Fig. 4) (entropy decrease due to the amphiphil self-association is smaller than the increase in the entropy due to their release from the hydration layer) or by binding of the amphiphil to the hydrophobic surface of the system: hydrophobic stationary phase, proteins, distribution in the organic solvent, etc. [11]. Generally, the hydrophobicity of a particle (molecule, ion) can be expressed via the logarithm of the coefficient of its partition (log*P*) between 1-octanol and water (*P* stands for the ratio of the equilibrium concentration of the particle of the same electronic structure in 1-octanol and its equilibrium concentration in water). If $\log P > 1$, the particle is hydrophobic (lipophilic), and if $\log P < 1$, the particle is hydrophilic [12]. Apart from the partition coefficient, the hydrophobicity of a particle can be also expressed via chromatographic parameters, either of the normal-phase or reversed-phase chromatography.

4. Structure of bile acids

Bile acid enzymatic produced in the liver of human and other mammals are the primary bile acids, and their intestinal microbial transformation to obtain the secondary bile acids. Both groups of bile acids are hydroxy derivatives of 5β-cholanic acid (cholic acid and chenodeoxycholic acid) [13]. The geometry of the 5β-cholanic acid largely determines of bile acids properties. The steroid skeleton of 5β-cholanic acid can be distinguished convex β and concave α surface (Fig. 5) [5].

Hydroxyl groups of bile acids are mostly oriented towards α side of steroid skeleton. X ray diffraction is demonstrated that the carboxyl group of the side chain from C17 carbon is also located on the α side of the steroid ring system. A concave (α) of bile acids steroid

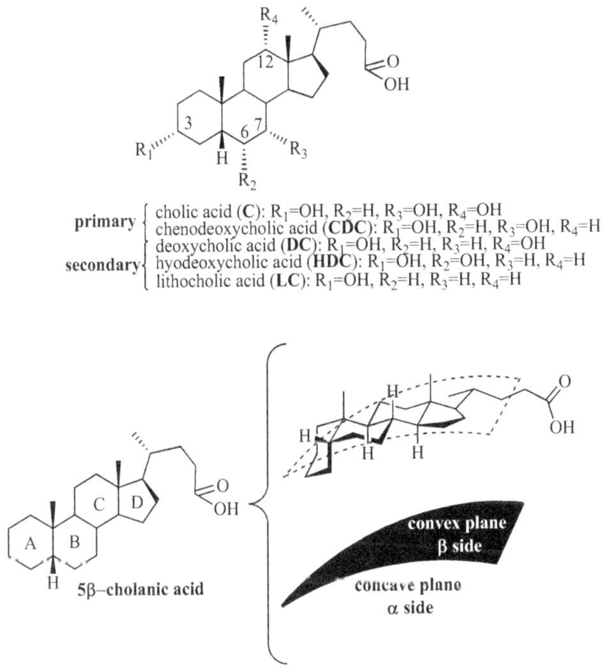

Figure 5. The bile acids general structure.

skeleton is the polar surface – hydrophilic side, while the convex surface (β) is the nonpolar surface – hydrophobic side. Simultaneous presence of hydrophobic and hydrophilic regions in the molecule of bile acids is named as amphiphilic. Since the carboxyl group ionized under physiological conditions, bile acids in biochemical systems belong to the ionic amphiphile. Also, bile acids are a special group of amphiphilic compounds, the group of biplanar molecules [4.5]. This is most apparent in cholic acid molecules (Fig. 6) because the molecules oxygen atoms from the α OH groups are in the same plane, the polar plane. In chenodeoxycholic and deoxycholic acid cannot be talking about the true planar polarity, but the literature often uses the term molecule with a hydrophilic edge.

If it oxidation of C7 or C12 α axial hydroxy group in cholic acid (**C**) (or deoxycolic acid (**DC**) or chenodeocycholic acid (**CDC**)) molecule leads to oxo derivative whose oxigen atom is switched for 60° (Newman projection) in relation to its' starting axial orientation i.e forms an angle of 30° with steroid skeleton mean plain (SSMP) (Newman projection, **Fig. 7**) [11,14]. Oxigen atom from C7 or C12 oxo group has the same steric orientation as has the equatorial C6 OH group from hyodeoxycholic acid (**HDA**).

Convergence of oxigen core (from oxo group) toward steroid sceleton β side means that there is the rise in number of stabilised water molecules (SWM, stabilised by hydrogen bonds) in hydratation layer of bile acids' steroid sceleton while hydrophobic surface of the

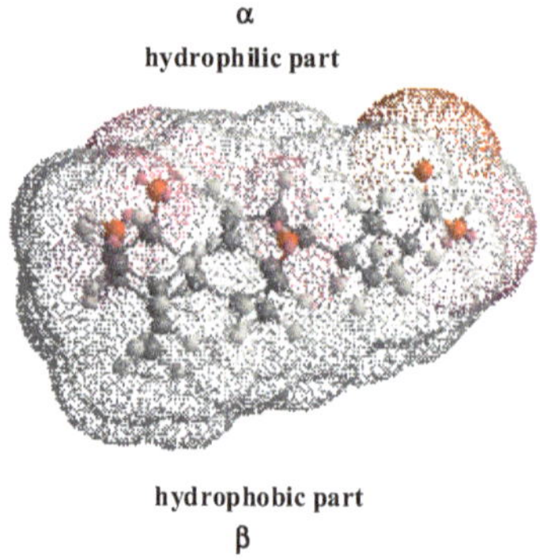

Figure 6. Two different side of cholic acid.

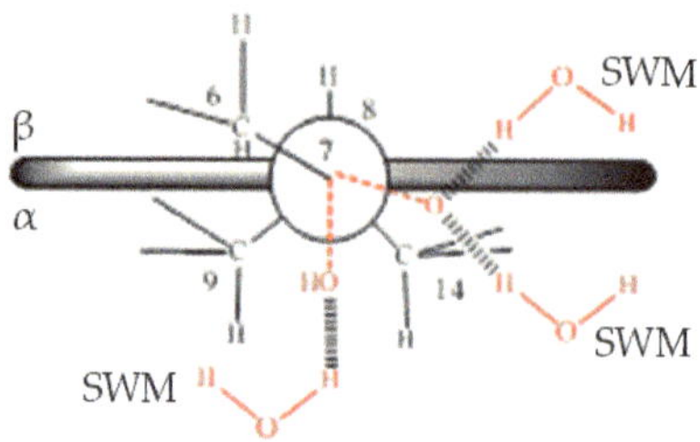

Figure 7. Introduction of oxo group in bile acids' steroid sceleton leads to rise in number of stabilised water molecules (SWM) i.e. the hydrophobic β side of the steroid sceleton decreases.

convex side of the steroid sceleton decreases (Fig. 7). Thus, the number of unstabilised water molecules (NSWM) declines [11].

$$BA\left[NSWM\right]_{n(aq)} \rightleftarrows BA_{(s)} + nNSWM \quad (R1)$$

If bile acids' hydrophobic surface deceases (B), i.e. the number of NSWM is lower, the observed molecule bonds to the hydrophobic stationary phase with lower intensity, i.e. reaction (R1) is moved to the right. This change in hydrophobic surface reflects in the values of retention coefficient k (RPHPLC) and retention parameter R_{M0} (RPTLC).

5. Discussion: bile acids and the hromatographic parametars (indices)

In the RPHPLC(TLC), there is an equilibrium between the bile acid molecules (ions) from the polar mobile phase $BA_{(aq)}$ and the bile acid $BA_{(s)}$ adsorbed on the hydrophobic stationary phase.

$$BA_{(aq)} \rightleftarrows BA_{(s)} \quad (R2)$$

This equilibrium is characterized by the equilibrium constant $K_{ad} = [BA_{(s)}] / [BA_{(aq)}]$, which is connected with the retention factor (capacity factor) (7) via the following relation [15]:

$$k = \frac{K_{ad} V_{hc}}{V_{pf}} = K_{ad} \varphi \tag{9}$$

where V_{hc} represents the volume of hydrocarbon in the stationary phase, whereas V_{pf} is the volume of the polar phase. Their ratio is denoted as ϕ, and it represents a characteristic of the column. The change of the standard Gibbs energy ΔG°_{ad} of the reaction (R2), taking also into account (9), is:

$$\Delta G^{\circ}_{ad} = -RT \ln K_{ad} = -RT \ln k + RT \ln \varphi \tag{10}$$

In both RPHPLC and RPTLC, in the adsorption of bile acids on the hydrophobic stationary phase, the water molecules from the solvation sheath that are not stabilized by hydrogen bonds (NSWM) pass to the hydrophilic mobile phase, giving a positive entropy contribution. Beside that, during the return of these water molecules in the bulk solution, NSWMs surface energy is released, which is then dissipated as a heat in the environment (solvophobic theory: according to this theory change in surface energy is a key factor in the reversed phase chromatographic processes energy). The formally dehydrated hydrophobic surfaces of bile acid molecules (ions) bind to the surface of the adsorbent by hydrophobic interactions (London dispersion forces, and dipol-induced dipole interactions). This makes a negative enthalpic contribution to the adsorption process, which is also dissipated as heat into the mobile phase (environment), giving rise to a positive change of the entropy of the environment.

Hence, the overall entropy change is positive, i.e. the Gibbs energy is lowered. The decrease in the free enthalpy of adsorption ($\Delta_{ad}G^{o}$) of bile acids (molecules or ions) on the hydrophobic stationary phase is more pronounced if the solvation layer of the steroid skeleton contains a greater number of NSWM, since then, there exists a largest increase in the overall entropy. In the binding of bile acids (molecules or ions) to the hydrophobic stationary phase, their steroid skeletons orient in such a way that the formally desolvated surfaces are to the smallest possible extent exposed to the solvent molecules from the hydrophilic mobile phase (methanol and water). Since bile acids are biplanar compounds, their molecules bind to the hydrophobic stationary phase by the side of the steroid skeleton that is most hydrophobic. With natural bile acids and their oxo derivatives, this is usually the convex β side of the steroid skeleton (an exception is the enantiomer of cholic acid with β configurations of all three OH groups, where α side is more hydrophobic than the β side of the steroid skeleton). Thus, the ln$k$ chromatographic parameter has the liner dependence of $\Delta_{ad}G^{o}$, which means that lnk has also linear depending on the hydrophobic surface of the β side of steroid nucleus. Similar is true for R_M chromatographic parameter (RPTLC). Therefore, ln$k$ could serve as a molecular descriptor for the hydrophobicity of bile acids. But the studies dealing with the hydrophobicity of bile acids as expressed via the retention coefficients (k), there is a problem of mutual comparability of the ln$k$ values. Because of that it is usually accustomed to give an order of hydrophobicity of the molecules, while the ln*k* values, though determined under identical experimental conditions, may vary because of the characteristic of the column (RPHPLC), which is in equation (9) denoted with ϕ. Because of that, Heuman introduced for bile acids a relative retention coefficient (k_r) which is defined as the ratio of the retention coefficient of the given bile acid (k(BA)) and the retention coefficient of taurocholic acid (**TC**) (k(TC)) (Fig. 8) [16].

$$k_r(\mathrm{BA}) = \frac{k(\mathrm{BA})}{k(\mathrm{TC})} \tag{11}$$

Taurocholic acid (**37**) was chosen as a reference because its retention coefficient in the pH range of the mobile phase from 2.9 to 9.0 is practically constant (complete ionization) [16]. By combining equations (10) and (11), the free enthalpy of adsorpion of bile acid (BA) on the hydrophobic stationary phase is:

$$\Delta G^{\circ}_{ad}(\mathrm{BA}) = -RT\ln k_r(\mathrm{BA}) + \Delta G^{\circ}_{ad}(\mathrm{TC}) \tag{12}$$

In this way are eliminated the individual characteristics of the column, so that the relative retention coefficient depends only on the composition of the mobile phase and the degree of hydrophobicity of the stationary phase.

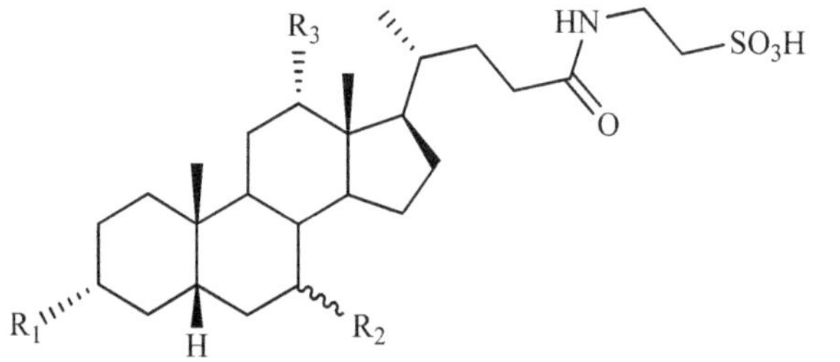

taurocholic acid-**TC** (**37**): R_1=OH, R_2=α–OH, R_3=OH
taurolithocholic acid-**TL** (**38**): R_1=OH, R_2=H, R_3=H
tauroursocholic acid-**TU** (**39**): R_1=OH, R_2=β–OH, R_3=OH
tauroursodeoxycholic acid-**TUD** (**40**): R_1=OH, R_2=β–OH, R_3=H
taurodeoxycholic acid-**TD** (**41**): R_1=OH, R_2=H, R_3=OH

Figure 8. Tauro conjugated bile acids.

Besides, Heuman defined also the hydrophobicity index *HI* as the ratio of the logarithm of the relative retention coefficient of the given bile acid (BA) and of the relative retention coefficient of taurolithocholic acid (**TL**) (lnk_r(TL)) (Fig. 7):

$$HI(\mathrm{BA}) = \frac{\ln k_r(\mathrm{BA})}{\ln k_r(\mathrm{TL})} \tag{13}$$

By introducing equation (13) into (12), one obtains the following equation:

$$\Delta G^{\circ}_{ad}(\mathrm{BA}) = HI(\mathrm{BA}) \times \left(-RT \ln k_r(\mathrm{TL})\right) + \Delta G^{\circ}_{ad}(\mathrm{TC}) \tag{14}$$

from which is evident that there is a linear relation between the hydrophobicity index of a bile acid and the Gibbs energy of adsorption on the hydrophobic stationary phase.

The *HI* value for taurocholic acid (**TC**) is zero since its relative retention coefficient equation (11) is equal to one, and the same value (one) holds also for taurolithocholic acid (**TL**). If a certain bile acid is more hydrophilic than taurocholic acid (**TC**), its retention constant (*k*) is smaller than the *k* value of taurocholic acid (**TC**), so that the relative retention coefficient of that bile acid is smaller than unity (ln*k*<0), which means that, according to equation (13), the *HI*(BA) is negative. In the opposite case, if a certain bile acid is more hydrophobic than taurocholic acid (**TC**), then the value of its retention coefficient is greater than the retention coefficient of taurochollic acd (**37**), and the relative retention coefficient for that bile acid is larger than unity (ln*k*>0), that is *HI*>0 (Fig. 9).

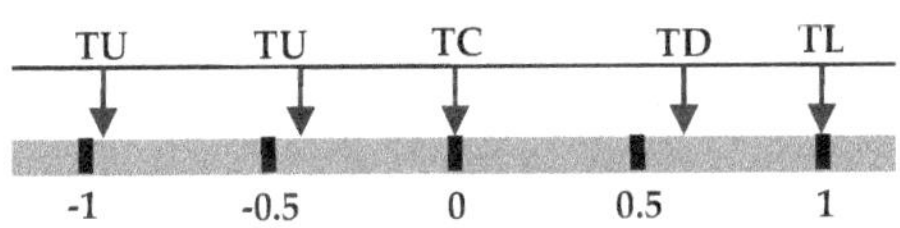

Figure 9. Heumans scale of bile acids hydrophobicity.

Besides of Heumans hydrophobic index (based on data retention) log*k* is used to obtain predictive equations where log*k* is the independent variable while bile acid properties associated with its hydrophobicity is the dependent variable. Armstrong and Carey were given a linear regression equation, where log*k* is associated with the solubilized molar fraction of cholesterol ϑ (solubilized by micellar solutions of bile acid salts in the equilibrium conditions) [17]:

$$\log k = S \log \vartheta - const. \tag{15}$$

In the plane of ϑ and log*k* is obtained of three linear congeneric groups of examined bile acids, a particular group form the non-conjugated bile acids, whereas the glyco- and tauro-conjugated bile acids are also in separate groups. For each of the derived linear congeneric group is determined equation (15) whose regression coefficients (R) is range in 0993-0999. Armstrong and Carey were predicted the capacity of cholesterol solubilization by bile acids oxo derivatives using the equations (15) for each congeneric group of bile acid [17]. It can be shown that log*k* is indeed associated with a number of hydration layers non-stabilized water molecules n_{NSWM} from β side of steroid skeleton:

$$\frac{d \ln k}{d\left(\Phi m_{NaCl}\right)} = n_{NSWM} M_{H_2O} \tag{16}$$

where Φm_{NaCl} is the product of the osmotic coefficient and the molality of NaCl, and M_{H_2O} is the molecular weight of water [11]. Therefore with the increasing of the number of the α

axial OH groups which is replacement with equatorial OHs or oxo groups resulting in a decrease of NSWM number from the hydratation shell of bile acids, which are then reflected in the increase of log*k* values i.e. increasing hydrophobicity of bile acids. However, the cholesterol solubilization depends on the size of hydrophobic domain of micelles, which depends on the hydrophobic surface of the β side of steroid nucleus, that's all mean that the log*k* adequately describes of cholesterol (or any other hydrophobic guest molecules) solubilization. However it should be noted that the use of log*k* as molecular descriptors, log*k* for each molecule, as the molecules from teaching set and as the molecules from the control set must be obtained in an identical column. Contrary to the different values of ϕ (9) increases the error of prediction.

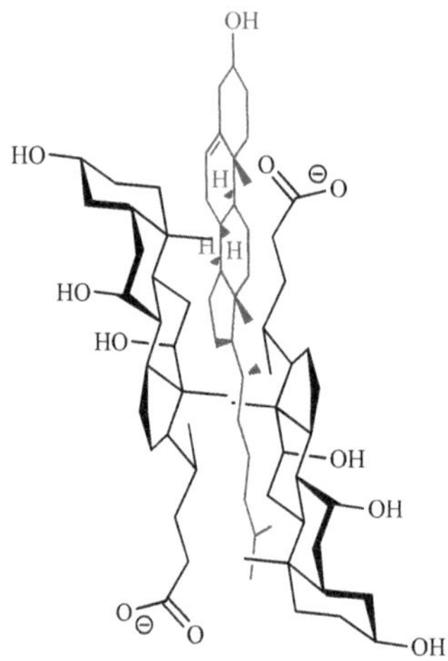

Figure 10. Micellar solubilization of cholesterol: cholesterol is in the contact with the hydrophobic surfaces of β side of bile acids steroid skeleton.

Natalini et al. as an experimental molecular descriptor used chromatographic parameter ϕ_0 (6) for modeling the negative logarithm of the critical micellar concentration ($pCMC$ = -logCMC) [9,10]. The study was included 20 non-conjugated and conjugated bile acids. For the full set of bile acids were obtained following linear regression equation:

$$pCMC = 3.85\varphi_0 - 0.82 \tag{17}$$

$$R = 0.8246$$

Regression coefficient is improved if the equations of linear regression obtained separately for conjugated (18) and for non-conjugated bile acids (19):

$$pCMC = 4.78\varphi_0 - 1.57 \tag{18}$$

$$R = 0.9643$$

$$pCMC = 1.05\varphi_0 + 1.32 \quad (19)$$

$$R = 0.9380$$

Equations (18) and (19) have a good predictiveness in their congeneric groups, also with ϕ_0 is well described hydrophobicity of the bile acids steroid skeleton β side. The mutual association of bile acid salts, especially around the first critical micellar concentration, which is through to their hydrophobic β side of steroid skeleton (Fig. 11) indicated that the chromatographic parameter ϕ_0 adequately be used to predict its critical micellar concentration. Advantages of chromatographic parameters ϕ_0 (6) with respect to the parameters R_{M0} (5) and $\log k_w$ (8) is reflected in the absence of appropriate extrapolation of retention coefficients. Extrapolation error can be large if the linear area of equations (5) and (8) are away from the mobile phase without an organic modifier.

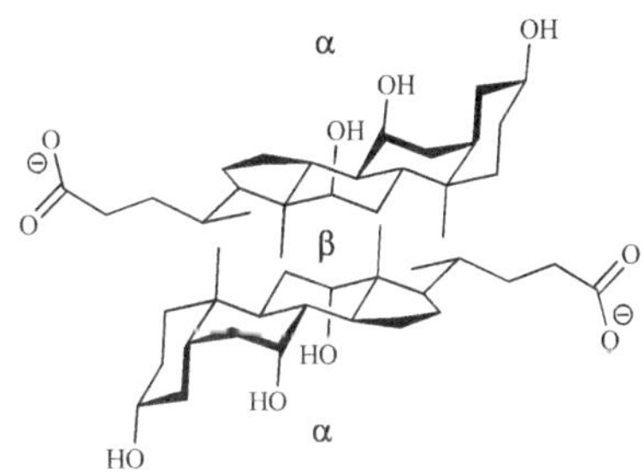

Figure 11. Dimeric micelles of cholic acids anion.

Sarbu et al. in RPTLC experiments examined which chromatographic parameter the best describes the hydrophobicity of bile acids. In the studied bile acids (Table 1) between R_{M0} parameter and the slope S (Equation 5) there is a good correlation, suggesting the presence of congeneric groups among chemical compounds in the tested assembly [3].

Bile acids	Abbreviation	Position and orientation of OHs
Lithocholic acid	LC	3α
Deoxycholic acid	DC	3α, 12α
Chenodeoxycholic acid	CDC	3α, 7α
Ursodeoxycholic acid	UDC	3α, 7β
Hyodeoxycholic acid	HDC	3α, 6α
Hyocholic acid	HC	3α, 6α, 7α
Cholic acid	C	3α, 7α, 12α
Glycolithocholic acid	GLC	3α
Taurolithocholic acid	TLC	3α
Glycochenodeoxycholic acid	GCDC	3α, 7α
Taurochenodeoxycholic acid	TCDC	3α, 7α
Glycodeoxycholic acid	GDC	3α, 12α
Taurodeoxycholic acid	TDC	3α, 12α
Glycocholic acid	GC	3α, 7α, 12α
Taurocholic acid	TC	3α, 7α, 12α

Table 1. Studied bile acids

Application of PCA on the matrix of R_M values (**D**) of the studied bile acids are obtained space of principal components with smaller dimensions then matrix **D**. Screen plot showed that the significant first and second principal components (PC1 and PC2) that explained 99.75% of variance in the starting matrix of retention parameters. Bartlett's statistic shows that PC1, PC2 and PC3 are significant which explained of 99.92% of the total variance. In the set of studied bile acids (Table 1) there is a good correlation between R_{M0} chromatographic parameters and scores of PC1:

$$R_{M0} = 5.379 - 2.624\,\mathrm{PC1} \tag{20}$$

$$\mathrm{R} = -0.9203$$

	Compound	R_{M0}	PC1	PC2	PC3	Rodas data:	
				scores		$\log P_{HA}$	$\log P_{HA}$
1	LC	5.33	0.195	0.122	0.080		
2	DC	4.78	0.346	0.179	0.111	3.50	2.65
3	CDC	5.01	0.340	0.177	0.082	3.28	2.25
4	UDC	3.65	0.545	0.226	0.114	3.00	2.20
5	HDC	3.88	0.488	0.208	0.110	3.08	2.28
6	HC	3.66	0.559	0.229	0.107	2.80	1.84
7	C	3.84	0.533	0.217	0.105	2.02	1.10
8	GLC	3.39	0.674	0.246	0.119		
9	TLC	2.12	1.000	0.249	0.077		
10	GCDC	2.87	0.819	0.237	0.088	2.12	0.45
11	TCDC	2.27	1.249	0.194	0.082		
12	GDC	2.90	0.850	0.247	0.087	2.25	0.80
13	TDC	2.30	1.165	0.191	0.090		
14	GC	2.30	1.102	0.186	0.090	1.65	-0.40
15	TC	2.29	1.524	0.137	0.125		

Table 2. Chromatographic parameters

Sarbu for assessing whether R_{M0} or score of PC adequate to describe lipophilicity of bile acids using the experimental data of particion coefficients between 1-octanol and water of bile acids in molecular (HA) and ionized (A) forms. Between the logarithm partition coefficients and chromatographic parameters on the basis of Table 1 are obtained by the following equation of linear regression:

$$\log P_{HA} = 0.366 + 0.620 R_{M0} \tag{21}$$

$$\mathrm{R} = 0.8559$$

$$\log P_{A} = -2.255 + 1.017 R_{M0} \tag{22}$$

$$R = 0.8751$$

$$\log P_{HA} = 3.995 - 2.195\,PC1 \quad (23)$$

$$R = 0.8675$$

$$\log P_{A} = 3.824 - 3.807\,PC1 \quad (24)$$

$$R = 0.9379$$

$$\log P_{HA} = 4.407 - 2.122\,PC1 - 2.168\,PC2 \quad (25)$$

$$R = 0.8715$$

$$\log P_{A} = 3.387 - 3.885\,PC1 + 2.290\,PC2 \quad (26)$$

$$R = 0.9395$$

Linear regression equations (23 and 24) in which the chromatographic parameter R_{M0} replaced with a score of PC1 has correlation coefficients with larger values compared to the equations (21 and 23). Correlation coefficient is slightly increased in the multiple linear regression equation for log*P* (molecular and ionized form) when the equations involved scores of PC1 and PC2. Increasing the coefficient of correlation in the above equations can be explained with Wolds assumption according to which measurement errors are contained in the principal components with small eigenvalues, so the scores of PC1 is purified from the errors. In Fig. 12 bile acids are shown from the Table 2 in the plane of the scores of principal components. It can be seen to form three linear congeneric groups: non-conjugated bile acids (**BA**), glyco-conjugate (**G-BA**) and tauro-conjugated (**T-BA**). Grouping of bile acids on the basis of the scores of PC as the appropriate regression equation for log*P* (from 23 to 26), indicating that the PC1 scores is adequate replacement for R_{M0} chromatographic parameters to describe the lipophilicity (hydrophobicity) of bile acids, namely the scores of PC1 can be used as molecular descriptors for hydrophobicity of the steroid skeleton β (bile acids).

If increase the set of bile acids from Table 1 by including oxo and acetoxy derivatives, R_{M0} chromatographic parameter also included in the multiple linear regressions for the logP_{HA} and the logP_{A} with molecular descriptors such as polarizability and molar specific sum of absolute values of the charges on each atom of the molecule (electronic parameter), molecular volume, the third order of the connectivity index and Wiener index (shape parameter). Based on the loadings of PC can be concluded that R_{M0} carries different information about the structure of the studied bile acids than the above *in silico* molecular descriptors [6].

Effect of temperature on the chromatographic process in reverse phase chromatography can also be applied to obtain chromatographic parameters that have properties of molecular

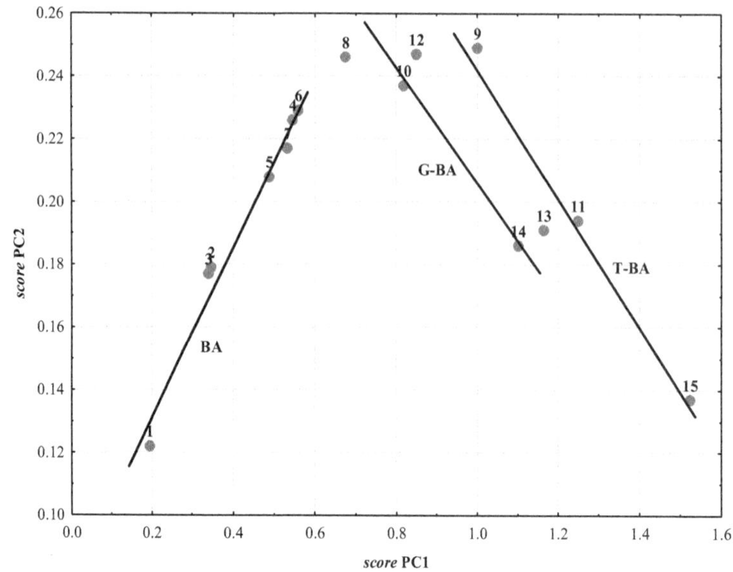

Figure 12. Sore of PC: bile acids from Table 1.

descriptors [15]. Namely, the reaction (R2) is exothermic (H_{ad} <0), then based on the van' t Hoff equation, the equilibrium constant of reaction (R2) decreases

$$\frac{d\ln K_{ad}}{dT} = \frac{dH_{ad}}{RT^2} \tag{27}$$

This means that the retention factor *k* (9) also decreases. Thus, the increase in temperature has the same effect on the value of *k* as there are increasing the volume fraction of organic modifier in aqueous mobile phase (reversed phase chromatography). In the experiment, the influence of temperature with an the retention factor at constant mobile phase included 25 different bile acids (conjugated and non-conjugated) (Table 3) [18].

For each tested bile acids between retention factor *k* and temperature *t* there is a linear relation:

$$k = a + bt \tag{28}$$

The linear model (28) explains from 96% to 99% of the variance at tested bile acids (Table 4). In the set of the studied bile acids (Table 3) between the parameters of equation (28) (*a* and *b*) there is a good correlation, indicating the formation of congeneric groups in the plane of *a* and *b*:

$$a = 0.7683 + 58.8522b \tag{29}$$

R=0.9996

Bile acids	Abbreviation	Position and orientation of OH and oxo groups
Lithocholic acid	**LC**	3α-OH
Deoxycholic acid	**DC**	3α-OH, 12α-OH
Chenodeoxycholic acid	**CDC**	3α-OH, 7α-OH
Cholic acid	**C**	3α-OH, 7α-OH, 12α-OH
Ursodeoxycholic acid	**UDC**	3α-OH, 7β-OH
Hyodeoxycholic acid	**HDC**	3α-OH, 6α-OH
Hyocholic acid	**HC**	3α-OH, 6α-OH, 7α-OH
Glycochenodeoxycholic acid	**GCDC**	3α-OH, 7α-OH
Taurodeoxycholic acid	**TDC**	3α-OH, 12α-OH
Glycocholic acid	**GC**	3α-OH, 7α-OH, 12α-OH
Glycodeoxycholic acid	**GDC**	3α-OH, 12α-OH
Taurolithocholic acid	**TLC**	3α-OH
Taurochenodeoxycholic acid	**TCDC**	3α-OH, 7α-OH
Glycolithocholic acid	**GLC**	3α-OH
12-Monoketocholic acid	**12-MKC**	3α-OH, 7α-OH, 12α-Oxo
7-Monoketocholic acid	**7-MKC**	3α-OH, 7α-Oxo, 12α-OH
7,12-Diketocholic acid	**7,12-DKC**	3α OH, 7α-Oxo, 12α-Oxo
3,7-Diketocholic acid	**3,7-DKC**	3α-Oxo, 7α-Oxo, 12α-OH
3,12-Diketocholic acid	**3,12-DKC**	3α-Oxo, 7α-OH, 12α-Oxo
Dehydrocholic acid	**TKC**	3α-Oxo, 7α-Oxo, 12α-Oxo
12-Monoketodeoxycholic acid	**12-MKD**	3α-OH, 12α-Oxo
Diketodeoxycholic acid	**DKD**	3α-Oxo, 12α-Oxo
7-Monoketochenodeoxycholic acid	**7-MKCD**	3α-OH, 7α-Oxo
Diketochenodeoxycholic acid	**DKCD**	3α-Oxo, 7α-Oxo
6-Monoketohyodeoxycholic acid	**6-MKHD**	3α-OH, 6α-Oxo

Table 3. Tested bile acids in the experiment the influence of temperature on the values of retention factor k

Fig. 13 shows that the tested bile acids in the plane of parameters of equation (23) formed by three congeneric groups. The first group (**I**) formed by bile acids with oxo groups in the steroid nucleus and the bile acids that in addition to C3 have one more equatorial OH groups. The α equatorial OH groups have the same position relative to the steroid skeleton mean plane such have the α oxo groups. However, in this group also includes the cholic acid and the glycocholic acid which is structurally not fit into the above mentioned group. The second group (**II**) formed deoxycholic acid and chenodeoxycholic acid with its conjugates. The third group (**III**) forms lithocholic acid and its conjugates. Question may arise as whether can be improved with grouping of bile acid gained by in the plane of a and b, especially in terms of group **I** which provides the vendor cholic acid? The answer may be the application of methods of principal components of the matrix of the data of retention factor dependence on temperature. The first principal component (PC1) explains 99.99715%

of variance in the starting data matrix, while the remaining variance explained by PC2. In Table 4 presents the scores of principal components of bile acids studied.

Compound		*a*	*b*	R^2	PC1	PC2
					score	
1	LC	43.0501	0.7200	0.9879	5.37637	0.023150
2	DC	21.8893	0.3518	0.9906	1.79015	-0.051081
3	CDC	18.4260	0.3021	0.9807	1.05575	0.020646
4	C	8.4873	0.1319	0.9828	-0.66245	0.016292
5	UDC	4.4884	0.0680	0.9843	-1.40668	0.031446
6	HDC	5.6203	0.0804	0.9538	-1.11512	-0.034642
7	HC	5.5477	0.0864	0.9836	-1.23205	0.041543
8	GCDC	16.6862	0.2651	0.9946	0.85970	-0.024730
9	TDC	19.5503	0.3165	0.9802	1.30724	-0.001736
10	GC	7.8569	0.1213	0.9841	-0.77443	0.014749
11	GDC	20.6286	0.3336	0.9786	1.51098	-0.012582
12	TLC	39.1628	0.6544	0.9728	4.65836	0.029036
13	TCDC	16.3808	0.2671	0.9796	0.70668	0.007651
14	GLC	41.1016	0.6926	0.9750	4.95053	0.026326
15	12-MKC	3.1340	0.0402	0.9704	-1.58139	-0.004996
16	7-MKC	2.8202	0.0353	0.9676	-1.64238	-0.002356
17	7,12-DKC	0.7141	0.0078	0.9769	-2.12906	0.056420
18	3,7-DKC	0.6100	0.0054	0.9974	-2.11715	0.056756
19	3,12-DKC	0.6237	0.0060	0.9752	-2.12981	0.056122
20	TKC	0.5042	0.0056	0.9738	-2.17086	0.066570
21	12-MKD	7.7659	0.1040	0.9652	-0.57871	-0.116428
22	DKD	4.1490	0.0511	0.9681	-1.32315	-0.048688
23	7-MKCD	6.1746	0.0817	0.9665	-0.91824	-0.078586
24	DKCD	3.4944	0.0423	0.9704	-1.46618	-0.031508
25	6-MKHD	6.3614	0.0913	0.9667	-0.96809	-0.039373

Table 4. The parameters of the linear model (28) and scores of principal components

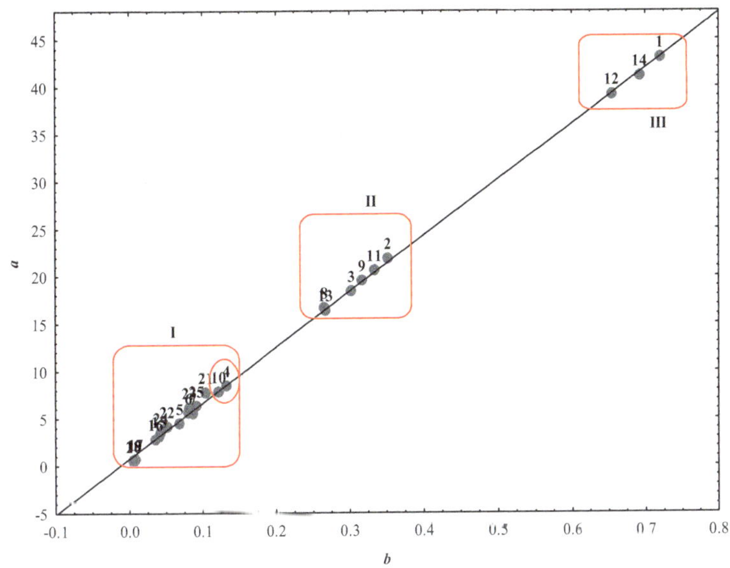

Figure 13. Grouping the bile (Table 4) acid on the basis of linear regression parameters of the retention factor temperatures dependence.

In the plane of principal component scores are obtained by analogous groups such as Fig. 14. However in the group **I** the oxo derivatives and the derivatives with α equatorial OH groups forming linear congeneric group, while based on Cooks distance cholic acid and glycocholic acid may be considered as outliners. With loadings of principal component can be determined by mutual correlation of different *in silico* descriptors (topological and electronic) and obtained experimental descriptors *a*, *b*, and PC1 on the set of 25 bile acids (Table 4). It may be concluded that chromatographic parameters *a* and PC1 are more or less orthogonal to most topological descriptors (Winers index, Balabans index, shape attributes, konectivities, etc.), while PC1 and *a* are mutually parallel so that explain the same structural characteristics of bile acids. Namely, as flat molecules nitrazepan guest molecule is incorporated in the hydrophobic domain of micelles. Therefore, if the β side of bile acids steroid skeleton more hydrophobic then nitrazepans partition coefficient has a higher value. Using the method of partial least square (PLS) regression equation was obtained for the nitrazepans partition coefficients in which experimentally determined variables *a* and PC1 are the most significant with respect to *in silico* descriptors. Which means that PC1 and *a* can be used as molecular descriptors of bile acids lipophilicity, and that the temperature dependence of retention factors may obtain information regarding the structure of bile acids (lipophilicity).

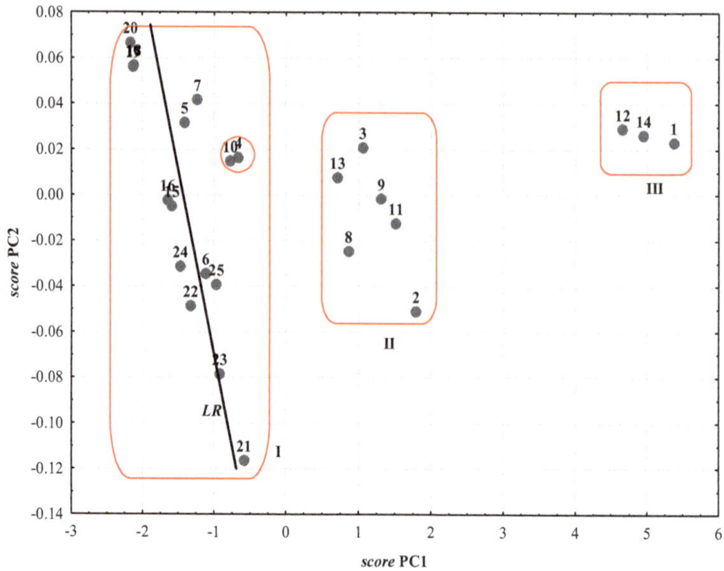

Figure 14. Grouping of bile acids (Table 4) based on the scores of principal component.

6. Conclusions

Chromatographic data (parameters) can be used for prediction of following feaatures of bile acids:

- Critical Micellar Concentration
- Haematolityc potentitial
- Binding for Albumin
- Solubilization of cholesterol, lecitin, vitamin E etc.
- Mixed micelles parameters (for example partition coefficient of nitrazepan in bile acids' micelles)

Application of retention parameters for prediction of bile acids' properties is convenient for finding the size of the starting set of molecules. If the set is too small (less than ten molecules), by using *in silico* descriptors multiple regression equations are obtained which are overfitted, while by using chromatographic parameters that posses appropriate structural informations about bile acids, simple linear equations are derived with acceptable predictive power.

Author details

Mihalj Poša

Department of Pharmacy, Faculty of Medicine, University of Novi Sad, Serbia

Acknowledgement

This study was supported by the Secretariat for Science and Technological Development of AP Vojvodina, Republic of Serbia (Grant No. 114-451-2113/2012).

Abbreviations

QSA(P)R = Quantitative Structure Activity (Property) Relationship
RPHPLC = Reversed Phase High Pressure (Resolutions) Liquid Chromatography
RPTLC = Reversed Phase Thin Layer Chromarography
PCA = Principal Component Analysis
PC = Principal Component
SWM = Stabilised Water Molecules
NSWM = Non Stabilised Water Molecules
SSMP = steroid skeleton mean plane

7. References

[1] Camile W (2003) The Practice of Medicinal Chemistry. Oxford: Academic Press.

[2] King F.D (2002) Medicinal Chemistry Principles and Practice. Cambridge: R.S.C.

[3] Sârbu C, Kuhajda K, Kevrešan S (2001) Evaulation of the lipofilicity of bile acids and their derivatives by thin-layer chromatography and principal component analysis. J. Chromatogr. A 917: 361-366.

[4] Poša M (2011) QSPR study of the effect of steroidal hydroxy and oxo substituents on the critical micellar concentrations of bile acids. Steroids 76(1-2): 85-93.

[5] Poša M, Kuhajda K (2010) Hydrophobiciti and haemolytic potential of oxo derivatives of cholic, deoxycholic and chenodeoxycholic acids. Steroids 75(6): 424-431.

[6] Sârbu C, Onişor C, Poša M, Kuhajda K, Kevrešan S (2008) Modeling and prediction (correction) of partition coefficients of bile acids and their derivatives by multivariate regression methods. Talanta 75: 651-657.

[7] Onişor C, Poša M, Sârbu C (2011) Modeling of chromatographic lipophilicity indices of bile acids and their derivatives. Rev. Roum. Chim. 56(3): 217-229.

[8] Onişor C, Poša M, Kevrešan S, Kuhajda K, Sârbu C (2010) Modeling chromatographic lipophilicity of bile acids and their derivatives estimated by HP-RPTLC. J. Sep. Sci. 33(20): 3110-3118.

[9] Natalini B, Sardella R, Camaioni E, Gioiello A, Pellicciari R (2007) Correlation between CMC and chromatographic index: simple and effective evaluation of the hydrophobic/hydrophilic balance of bile acids. Anal. Bioanal. Chem. 388: 1681-1688.

[10] Natalini B, Sardella R, Camaioni E, Macchiarulo A, Gioiello A, Carbone G, Pellicciari R (2009) Derived chromatographic indices as effective tools to study the self-aggregation process of bile acids. J. Pharm. Biomed. Anal. 50: 613-621.

[11] Poša M, Pilipović A, Lalić M (2010) The influence of NaCl on hydrophobicity of selected, pharmacologically active bile acids expressed with chromatographic retention index and critical micellar concentration. Colloids Surf. B: Biointerfaces 81: 336-343.

[12] Roda A, Minutello A, Angellotti M.A, Fini A (1990) Bile acid structure-activity relationships: evaluation of bile acids lipophilicity using 1-octanol/water partition coefficient and reverse phase HPLC. J. Lipid Res. 31: 1433-1443.

[13] Mikov M, Fawcett J.P (2007) Bile Acids. Geneva: Medishet Publisher.

[14] Poša M, Farkaš Z (2010) Cholesterol solubilization by oxo derivatives of selected bile acids and their osmotic resistance. Collect. Czech. Chem. Commun. 75(8): 767-784.

[15] Poša M, Pilipović A, Lalić M, Popović J (2010) Hydrophobicity and retention coefficient of selected bile acid oxo derivatives. Acta Chim. Slov. 57(4): 828-835.

[16] Heuman M.D (1989) Quantitative estimation of the hydrophilic-hydrophobic balance of mixed bile salt solutions. J. Lipid Res. 30: 719-730.

[17] Armstrong J.M, Carey C.M (1982) The hydrophobic-hydrophilic balance of bile salts. Inverse correlation between reverse-phase high performance liquid chromarographic mobilities and micellar cholesterol-solubilizing capacities. J. Lipid Res. 23: 70-80.

[18] Poša M, Pilipović A, Lalić M, Popović J (2011) Determination and importance of temperature dependence of retention coefficient (RPHPLC) in QSAR model of nitrazepams' partition coefficient in bile acid micelles. Talanta 83(5): 1634-1642.

Chapter 17

A Review of Current Trends and Advances in Analytical Methods for Determination of Statins: Chromatography and Capillary Electrophoresis

Biljana Nigović, Ana Mornar and Miranda Sertić

Additional information is available at the end of the chapter

1. Introduction

Statins are now among the most frequently prescribed agents for reducing morbidity and mortality related to cardiovascular diseases (Figure 1) and analysis of these drugs is a current problem. The major therapeutic action of statin drugs is reduction of circulating atherogenic lipoproteins as a result of inhibition of 3-hydroxy-3-methylglutaryl coenzyme A (HMG-CoA) reductase [1]. The key enzyme catalyzes the conversion of HMG-CoA to mevalonate, a critical intermediary in the cholesterol biosynthesis. This mechanism was discovered in 1976, when Endo and co-workers isolated a compound mevastatin from *Penicillium citrinum* that exhibited cholesterol-lowering effects [2]. Clinical studies have shown that statins significantly reduce the risk of heart attack and death in patients with proven coronary artery disease, and can also reduce cardiac events in patients with high cholesterol levels [3]. Beside lipid-lowering activity, statins improve endothelial function, maintain plaque stability and prevent thrombus formation. There is also an increased interest in statins non-lipid activities such as an anti-inflammatory action [4].

Ischemic heart disease is the leading cause of death in middle- and high-income countries, killing over 7 million people each year. Cardiovascular disease has no geographic, gender or socio-economic boundaries, and will remain the leading cause of death globally in the future. Therefore, the development of new analytical methods for statin drugs is of great importance. Analytical methods are employed through entire life cycle of a drug, from design and manufacture, elucidating the mechanism of biotransformation, clinical trials, dosage scheme adjustment, its introduction into the marketplace, quality control and pharmacovigilance to drug recycling and disposal with emphasis on environmental protection.

Statins can be grouped into fermentation-derived and chemically synthesized. Lovastatin, also called mevinolin, was isolated as secondary metabolite of fermentation process of various fungi such as *Aspergillus terreus, Monascus ruber* and *Penicillium* species [5, 6]. Lovastatin was the first commercially available compound for treatment of hypercholesterolemia, approved for use in 1987. It is produced biosynthetically from the fungus *Aspergillus terreus*. Whereas lovastatin is a natural product, simvastatin and pravastatin are semi-synthetic. Simvastatin is obtained by synthesis from lovastatin by replacement of 2-methylbutyryl side chain with 2,2-dimethylbutyryl group, while pravastatin is produced by microbial hydroxylation of mevastatin by *Strepromyces carbophilus*. Fluvastatin, atorvastatin, pitavastatin and rosuvastatin are completely synthetic compounds. Although all statins share a common mechanism of action and structural component that is very similar to the HMG portion of HMG-CoA reductase, they differ in terms of their chemical structures (Figure 1). The statins differ from each other in the rigid, hydrophobic structures covalently linked to the HMG-like moiety. The naturally derived statins contain a substituted decalin ring structure. Only pravastatin has a hydroxyl substituent on the hexahydronaphthalene nucleus which causes higher hydrophilicity. Fully synthetic statins have fluorophenyl groups linked to the HMG-like moiety. Depending upon chemical structure, statins have different affinities for HMG-CoA reductase and different pharmacokinetic properties [7]. Clinical trials have demonstrated rosuvastatin to be the most effective in reducing LDL cholesterol. In addition to the standard statin pharmacophore, rosuvastatin molecule contains a polar methyl sulfonamide group that forms a unique interaction with the catalytic site of HMG-CoA reductase. Cerivastatin was a synthetic statin drug, approved in 1997. Unfortunately, due to its fatal rhabdomyolysis, as a sever side effect, it was voluntarily withdrawn from the market in 2001.

Statins exist in two forms, lactone and open-ring hydroxy acid forms. Lovastatin and simvastatin are administered as lactone prodrugs and subsequently transformed to active metabolites in contrast to other statins, which are formulated in the pharmacologically active β-hydroxy acid form. *In vivo*, lactone prodrugs are enzymatically hydrolyzed to their hydroxy acid pharmacophores in the liver to achieve pharmacological activity [8]. The lactone forms can be converted in aqueous solutions to their corresponding hydroxy acid equilibrium products. Such interconversion may occur even in the biological matrix before collecting aliquots of the sample, during sample preparation and analysis of the drug. Therefore, it is crucial to optimize the multiple steps of the analytical method in order to minimize the interconversion during the analysis. On the other hand, statins in β-hydroxy acid form possess two hydroxyl groups in an alkyl chain at the β and δ positions with respect to the carboxylic acid group. The carboxylic acid group and the hydroxyl group at the δ position are prone to lactonize. Therefore, all statins may exist in solutions in the free acid form or the lactone form or as an equilibrium mixture of both forms in a pH-dependent manner [9]. For samples of hydroxy acid and lactone forms, maintaining the pH of solution around 4-5 minimizes interconversion. Increasing the pH above 6 facilitates the conversion of lactone to acid, whereas lowering pH enables the conversion from acid to lactone or lactone to acid in the non-ionized form. Consequently, great care must be exercised when handling these compounds in order to isolate them in high yields and the analytical methods should be designed for the simultaneous quantification of two analytes that can potentially undergo interconversion during analysis.

Statins are considered for long-term therapy and thus the purity assessment of these drugs is of great significance. Development of selective methods for monitoring their potential impurities and degradation products is highly required. Identification and determination of drug-related substances is an important aspect because impurities and degradation products of drugs are often responsible for some side-effects. The estimation of the impurity profiles of bulk drugs or dosage formulations requires methods involving high sensitivity and resolution as well as acceptable analysis time. The hyphenated technique that incorporates the efficient separation using liquid chromatography and specific and sensitive detection by mass spectrometry has become indispensable tool for identification and structure elucidation of unknown impurities in statin drugs as well as quantification of trace impurity levels.

Various chromatographic methods for determination of statins and their related impurities in the bulk drug forms and pharmaceutical formulations were developed. Almost all methods used for the separation of statins are based on high-performance liquid chromatography. In pharmaceutical applications UV detection was most commonly used. Analytical methods for determination of statins were developed individually as expected from their different structural and chemical properties. This approach to the analysis was chosen most probably because statins are not used in combination with other statin molecules during therapy. However, the development of a rapid analytical procedure that is not limited to the analysis of only one statin can be considered as a very useful assessment in quality control. Numerous chromatographic methods for quantification of statins in different biological fluids were developed. The levels of statins in biological fluids are very low because only about 5% of dosed statin reaches the systematic circulation. The liquid chromatography coupled to tandem mass spectrometry has become the method of choice for therapeutic plasma level monitoring of statins and their metabolites in pharmacokinetic investigations [10]. Generally, hyperlipidemic patients are treated with multiple-drug regime which commonly leads to drug interaction. The simultaneous determination of statins and drugs usually combined in cardiovasculary therapy in human plasma is important to get more insight in their possible interactions with a consequent increased risk to toxic effects. Due to different physical and chemical properties of co-administrated drugs development of methods for their simultaneous analysis is an over going challenge.

This chapter will present recent advances in chromatographic and capillary electrophoretic methods for the determination of statin drugs in various fields of application. Current trends in developing new methods for analysis of the most frequently used drugs will be discussed.

2. Pharmaceutical application

Pharmaceutical analysis provides information on the identity, purity, content and stability of starting material, excipients and active pharmaceutical ingredients (APIs). A distinction is made between analysis of the pure active ingredients and pharmaceutical formulations. Specification and test methods for the commonly used API and excipients are described in detail in pharmacopoeias.

Impurity profiling is of great importance in new drug substance and new drug product because of their potential unwanted pharmacological effects, possible toxicity, side effects, and their eventual impact on the activity, efficacy and the stability of the drug, its bioavailability and the results of the drug analysis. International Conference on Harmonization (ICH) gives strict regulatory guidelines for identification and quantification of trace impurities in drugs. Any compound that does not have the same chemical entity as the active substance, present at levels higher than 0.1% or 0.05% (depending on the daily dose), needs to be identified. Therefore there is a permanent need for developing new accurate, selective, and sensitive methods for the determination of drug impurities. Impurities can come from starting materials, they can be intermediars and by-products from the synthesis of the API (process related impurities), degradation products formed during manufacturing process and long-term storage, interaction products between API and other active ingredients and excipients or primary container.

Stability indicating methods are quantitative test methods that can detect changes of API and drug products during time and under certain conditions. Information on type and amount of degradation products over time is important for quality, safety and efficacy of the drug. Therefore, Food and Drug Administration (FDA), European Medicines Agency and other regulatory agencies, along ICH and good manufacturing practise require development and validation of stability indicating methods. General purpose of stability testing is to provide evidence on how the quality of an API or a finished pharmaceutical product changes during time under the influence of different environmental factors such as temperature, humidity and light. After these tests have been performed, recommendation on storage conditions and shelf life of the product can be given. ICH guidelines give detailed description of forced decomposition studies (stress testing). Stress testing of the API can help identify possible degradation products. It should include the effect of temperature (in 10 °C increments), humidity (≥ 75% relative humidity), oxidation, photolysis and hydrolysis of the API at a wide range of pH (acidic, neutral and alkali conditions).

In this section a review of chromatographic methods applied for identification and quantification of statins in bulk drug and pharmaceutical dosage forms will be given (Table 1). Each statin commercially available on the market will be covered in this review. Special emphasis will be given to stability indicating methods and papers describing impurity profiling.

Statins are often manufactured in combined pharmaceutical formulations together with ramipril, acetylsalicylic acid, amlodipine etc., and especially ezetimibe, a novel lipid-lowering agent that inhibits the absorption of cholesterol in the intestine by blocking Niemann-Pick C1-like protein cholesterol transporter. A synergic effect in reducing plasma concentrations of LDL cholesterol is achieved, mainly by the combination of statin and ezetimibe. Since statins are often co-administered with other drugs in therapy of cardiovascular disease, i.e. acetylsalicylic acid, antihypertensive medicines (ACE inhibitors, calcium channel blockers), but also in combined therapy of multiple disorders, e.g. antidiabetics, diuretics, nonsteroidal anti-inflammatory drugs and other analgetics, antibiotics etc. In order to avoid problems with patient compliance when a combination of acetylsalicylic acid, antihypertensives, lipid-lowering drugs and etc. is required, a polypill, a fixed-dose combination containing three or more drugs in a single pill, would be the solution. Methods describing simultaneous analysis of these combined pharmaceutical products will also be mentioned.

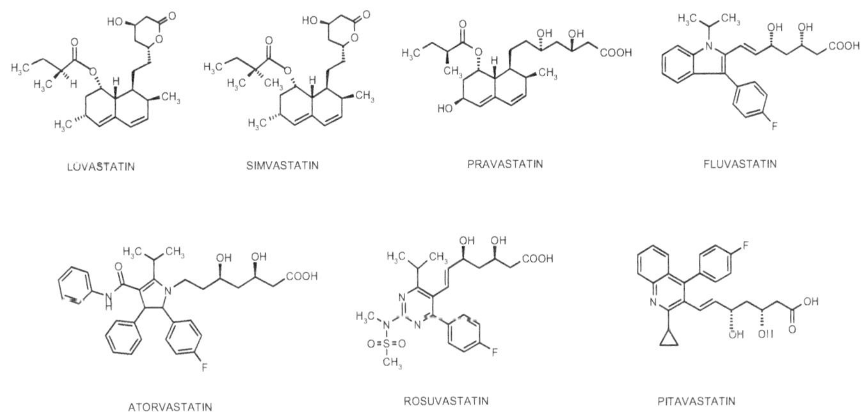

Figure 1. Chemical structures of statins

2.1. Lovastatin

The first statin registered as a drug was lovastatin. Nowadays, in therapy it is greatly replaced by new synthetic products, mainly atorvastatin and simvastatin. Therefore there are not many new methods for determination and quantification of lovastatin in bulk drug and pharmaceutical formulations.

There are scarce reports investigating the conversion of statins from lactone to their corresponding hydroxy acid forms. Yang and Hwang studied the conversion of lovastatin and simvastatin from lactone to corresponding hydroxy acid forms [11]. They concluded that the conversion of lactone forms to corresponding hydroxy acid forms would occur in water or 70% acetonitrile. However, this conversion could be retarded by addition of acetic acid to the solution. Hence a mobile phase with acetic acid added to the composition is recommended for HPLC analysis. Furthermore, lactone forms could only be transformed to their corresponding hydroxy acid forms in 0.1 M NaOH or 0.05 M KOH prepared in 25% or 50% acetonitrile in water. When alkaline methanolic solutions were used further transformation to methyl ester of hydroxy acid form would take place. Recently another paper was published investigating conversion of lovastatin [12]. The identity of all three forms, lovastatin, lovastatin hydroxy acid and its methyl ester was confirmed by electrospray ionization (ESI) mass spectrometry (MS). Their results imply that also under acidic conditions, with increase of storage time, lactone is converted to hydroxy acid form and further transformed to methyl ester form.

Bearing in mind the interconversion problem, special attention should be given to the choice of a mobile phase for HPLC analysis, the extraction procedure and sample storage time. Methanol in acidic conditions should be avoided because it induces the conversion and transformation of lovastatin forms. Hence, most recently developed LC methods utilize pH around 4.5.

Lovastatin is an active pharmaceutical ingredient in red yeast rice products, used as a dietary supplement. In such products lovastatin is mostly refered to as monakolin K, and is accompanied by 13 more monacolins naturally occurring in red yeast rice. These products are frequently used by millions of people as a complementary and alternative therapy for lowering total lipid and LDL cholesterol levels. Unfortunately dietary supplements do not follow strict quality control as medicines do, active ingredients are not standardized and published on labels, and considerable variations can be found among different manufactures even between lots of the same manufacture. Therefore there is a growing need for specific and precise methods for determination of lovastatin in red yeast rice dietary supplements in order to ensure standardization, efficacy and safety of these products.

Identification and chemical profiling of all 14 monacolins in red yeast rice and its formulated products was conducted using HPLC with photodiode array detecore (PDA) and MS [13]. Because red yeast rice has a complex matrix, sample extraction procedure was carried out with 75% ethanol. Chemical profiling was performed using electrospray ionization and ion trap mass analyzer. Since lovastatin content depends on the fermentation process of the rice by *Monascus purpureus*, an LC-PDA-ESI-ion trap method was published investigating differences in raw material powder and finished products [14].

A stability-indicating method for the stress test of red yeast rice was also performed [15]. An assay of seven main monacolins, monacolin K (lovastatin), monacolin J, monacolin L and their corresponding hydroxy acid forms and dehydromonacolin K, representing 97% of total monacolins, was determined. In order to shorten the analysis time Song et al. proposed a fast screening method of lovastatin in red yeast rice products by flow injection tandem mass spectrometry without LC separation [16].

2.2. Simvastatin

Simvastatin is along atorvastatin the most often used statin drug and there is a great number of analytical methods developed. Novakova et al. published a review paper on HPLC methods for the determination of simvastatin and atorvastatin [17]. An oversight on different areas of application, pharmaceutical formulations, clinical medicine (human plasma) and environmental (aqueous samples) was given. A more detailed overview will be given on papers not covered by this review.

A simple HPLC-UV method was optimized according to the USP chromatographic method for simvastatin [18]. By changing the column length from 30 cm to a Chromolith RP18 monolithic column, 10 cm in length and reducing the pH to 3.0, a reduction in elution time was about 60%, resulting in analysis time less than 4 min. Method was applied to determine the quality of 60 compounding simvastatin 40 mg capsules. The mean content and weight variation evaluation, content uniformity, determination of simvastatin concentration, determination of lovastatin as an impurity and the dissolution test were performed. Results were devastating. The mean content of the capsules varied from 70 mg to 316 mg. In ten Brazilian pharmacies more than one tested capsule was outside the range from 85-115%. Only three pharmacies presented content uniformity with values complying to reference ones. Capsules from all the pharmacies resulted in simvastatin content less than 100% of the declared value. In 6 of them the content ranged from 4-87% of the declared amount. These results do not meet the requirements for simvastatin contents, resulting in underdosing. These appalling results emphasize the need for the control of raw material, compounding process and finished products quality, efficacy and safety.

Tablet splitting is a somewhat controversial topic among pharmacy practitioners, patients, managed care organizations and many other associations involved in health care. However it has become increasingly common, especially within geriatric and psychiatry communities. There are many concerns surrounding tablet splitting program, mainly if there will be considerable weight fluctuations, will the daily dose be the same in two half's, and will tablet splitting deliver same clinical outcomes at a lower cost. Hill et al. presented an HPLC-UV method, taken from the USP monograph and adapted to half-tablets, for drug content and weight uniformity for half-tablets of six commonly split medications, including simvastatin [19]. There analysis found 38.80 mg as target drug content, while the measured drug content mean was 40.06 mg, with a RSD 4.29%. Target drug content ranges from 95.21% to 111.35%. These small changes in daily dose should have no significant impact on long-term clinical end points.

RP-HPLC method was developed and validated for simultaneous analysis of simvastatin and tocotrienol and tocopherols isoforms in simvastatin-tocotrienol nanoparticles manufactured as potential targeted therapy of breast cancer [20]. In order to obtain good resolution in short analysis time the separation was carried out on a Phenomenex Onyx C18 monolithic column (100 mm x 4.6 mm) with a gradient elution.

Preparation and evaluation of a high-dose nicotinic acid loaded sustained-release pellets coated with double polymer and immediate release simvastatin was introduced by Zhao and co-workers [21]. After the preparation of drug-loaded pellets, drug content analysis was performed by HPLC for both nicotinic acid and simvastatin. However, unnecessary, different methods, using similar columns and mobile phases, were employed.

There are a number of methods describing simultaneous determination of simvastatin and ezetimibe from their combination drug products [22-25]. Stability indicating studies on combined pharmaceutical products of simvastatin and ezetimibe have also been published [24, 25]. Different approaches to forced degradation study, chromatographic conditions and determination of degradation products were performed. Hefnawy and co-workers proposed a very fast and sensitive stability indicating method for simultaneous determination of ezetimibe and simvastatin in tablet dosage form [25]. Instead of traditional chromatographic columns packed with porous particles, they used a monolithic stationary phases, i.e. RP Merck Chromolith Performance column (RP-18e, 100 mm x 4.6 mm). Due to monolithic stationary phase, an elevated flow rate is possible, resulting in a run-time five-fold reduced (analysis time under 2 min), consumption of mobile phase about two-fold decreased, while the resolution between peaks remained unaffected.

Several methods have been developed for identification and quantification of known impurities, but many also studied fragmentation and structural determination of unknown simvastatin impurities [26-29]. Structural characterization and identification of a new compound, an unknown simvastatin by-product generated during the industrial synthesis starting from lovastatin was published [26]. After HPLC-diode array detector (DAD) analysis, ESI-ion trap mass analyzer was employed to obtain MS/MS spectra, followed by Fourier transform-infrared spectroscopy (FT-IR) and nuclear magnetic resonance (NMR) analysis.

Plumb et al. [27] proposed a method using high resolution sub 2 μm particle LC column together with hybrid quadrupole orthogonal time-of-flight (TOF) mass spectrometer used to profile and identify simvastatin impurities. All common impurites were identified in a single 10 min run. A new impurity of simvastatin was detected and identified as the saturated ring form of simvastatin. The same group published a paper on screening pharmaceutical products by ultra performance liquid chromatography (UPLC) coupled to TOF-MS [28]. Principal components statistical analysis was used for rapid classification of batches of simvastatin tablets according to their impurity profile.

Reddy et al. [29] performed HPLC separation of simvastatin and its two main impurities, anhydro-simvastatin and simvastatin dimmer. An unknown impurity was detected. MS/MS spectrum was obtained by ESI^{+} and ion-trap mass analyzer and the structure of the so far unknown simvastatin impurity was proposed. Recently, a paper on synthesis, characterization and quantification of simvastatin's metabolites and impurities was published [30]. This method emphasizes use of non-compendial reference standards for quantification, with shorter analysis time and improved sensitivity. β-hydroxy acid and methyl ester of simvastatin were synthesized as non-compendial reference standards. After complete and detailed characterization by MS, FT-IR and NMR, they were used as reference standards in quantification of simvastatin impurities.

2.3. Pravastatin

An HPLC method for quantification of pravastatin in tablets was published [31]. However, an unnecessary complicated extraction procedure and linearity investigation was performed. Identification of an impurity in pravastatin was performed by application of collision-activated decomposition mass spectra both in positive and negative ionization mode [32]. The impurity is an analogue of pravastatin with an additional methyl group on ester side chain.

Two stability indicating studies of pravastatin under different forced degradation conditions were conducted [33, 34]. Forced degradation study was performed for neutral, acid and basic hydrolysis, chemical oxidation, photochemical degradation and thermal stress using HPLC-UV [33]. Under neutral hydrolysis a decrease in the peak area of pravastatin was observed accompanied by two additional peaks. In basic hydrolysis a 90% decrease of pravastatin peak was noted and an additional peak is obtained, while in acidic conditions pravastatin peak completely disappeared and two new signals appeared. Degradation of pravastatin was also observed under oxidative conditions, while under thermal stress no change was percived.

Results obtained by Brain-Isasi et al. [34] are somewhat different then those previously published [33]. They argue that the previously described method was to short (7 min) to observe all degradation peaks obtained by acid hydrolysis while all of them are eluting after pravastatin. This indicates they are more liphophyllic than the parent drug, probably formed after esterification and lactonization of pravastatin. By the use of MS/MS spectra obtained in the positive mode, one of the peaks was identified as pravastatin lactone form. In alkaline medium only one product was observed and after preparative TLC it was isolated and identified by ^{1}H-NMR and ^{13}C-NMR as the 8-hydroxy derivate of pravastatin.

2.4. Fluvastatin

Photodegradation study of fluvastatin has been studied and examined by high-performance-thin-layer chromatography (HPTLC) and spectrophotometry [35]. Photoproducts were separated by HPTLC on a nonpolar C18 stationary phase with a mixture of phosphate buffer and methanol as a mobile phase. Both in water and methanol solutions, photochemical decomposition led to the formation of three major products.

2.5. Atorvastatin

Of all seven statins, atorvastatin is the most often administered statin drug. In fact, it is one of the most often prescribed prescription drugs overall. Therefore many methods are developed for determination and quantification of atorvastatin in bulk drug and pharmaceuticals. Since Novakova et al. in 2008 [17] gave a review of HPLC methods for the determination of atorvastatin in pharmaceutical assays, only papers published afterwards will be presented.

There are several stability indicating methods for determination of atorvastatin using different techniques and detectors. A RP-HPTLC method using aluminium sheets precoated with silica gel 60 RP18F(254) as the mobile phase consisted of methanol-water was used for determination of atorvastatin in bulk drug and pharmaceutical formulation [36]. Quantification was conducted densitometrically at 246 nm. Under acidic conditions drug underwent significant hydrolysis, while it was stable under alkali, oxidation, dry heat and photodegradation conditions. HPLC method using fluorescence detector (282 nm excitation, 400 nm emission) was introduced for analysis of atorvastatin and its degradation products in bulk drug and tablet form [37]. HPLC method with UV detection at 247 nm was developed for determination of atorvastatin and its degradation products in bulk drug, marketed tablet and in-house prepared nanoemulsion formulation [38].

Another stability indicating method was proposed for simultaneous determination of atorvastatin and amlodipine alongside with their degradation products in commercial combined tablets [39]. An UPLC method using ethylene bridged hybrid C18 column (50 mm x 2.1 mm, 1.7 μm) was used for simultaneous determination and quantitation of atorvastatin, acetylsalicylic acid and their four known and six unknown degradation products in combined dosage forms [40].

Two LC-MS method were reported for structure determination and identification of atorvastatin degradation products. An LC method employing an atmospheric pressure chemical ionization (APCI) source in positive mode with TOF mass spectrometer for acquiring accurate mass and an ion trap analyzer for complete fragmentation pattern was introduced [41]. Six unknown atorvastatin degradation products formed under stress conditions of hydrolysis, oxidation and photolysis were identified. Preparative HPLC method with Luna prep C18(2) column (200 mm x 50 mm, 10 μm) was used for isolation of four oxidative degradation products [42]. HPLC coupled to MS, high resolution MS and NMR spectroscopy were applied for the structure elucidation. Quantitative NMR spectroscopy was used for assay determination of isolated oxidative atorvastatin degradation products. A fast UPLC method with analysis time of 3 min was employed for determination of atorvastatin, fenofibrate and their degradation products in combined dosage forms [43].

We have developed HPLC/DAD/ESI/MS^n method for separation and identification of atorvastatin and its four related impurities [44]. To avoid hydrolysis of the atorvastatin lactone and the lactonization of acid form, ammonium buffer pH 4.0 was used. In order to achieve separation between atorvastatin and its diastereomer, several mobile phases were examined. Finally, a gradient elution mode was chosen to achieve good separation between peaks adjacent to the drug components, as well as to keep short analysis time of lipophilic impurities (Fig. 2.). Mass spectra were obtained by ESI source in the positive ion mode and ion trap analyzer. For quantitative analysis of atorvastatin and its four known impurities multiple reaction monitoring (MRM) mode was employed. Several unknown impurities were identified through MS/MS fragmentation analysis, i.e. diamino-atorvastatin, photolytic oxo-product, photolytic degradation product and diastereomer of atorvastatin lactone. Method was successfully applied to bulk drug and pharmaceutical dosage forms provided by different manufactures (Figure 2).

HPLC-UV method was developed for simultaneous determination of atorvastatin and seven related compounds specified as process-related impurities and possible degradation impurities. Experimental design was used during method optimization and robustness testing [45]. Artificial Neural Networks were used for the modelling and prediction of chromatographic retention of atorvastatin and its impurities in micellar liquid chromatography [46].

Atorvastatin in combined dosage forms, e.g. with ezetimibe, fenofibrate, ramiprile was determined by HPTLC methods [47, 48]. HPLC methods were published for simultaneous determination of atorvastatin in combination with amlodipin [49], fenofibrate [50], ezetimibe [51] and ramiprile [47, 52]. An improved HPLC method, with higher sensitivity and shorter analysis time using a chemometric protocol (statistical experimental design and Derringer's desirability function) was developed for simultaneous analysis of amlodipine and atorvastatin in pharmaceutical formulations [53]. Three HPLC methods have been published for analysis of atorvastatin and acetylsalicylic acid in combination with clopidogrel [54] and ramipril [55].

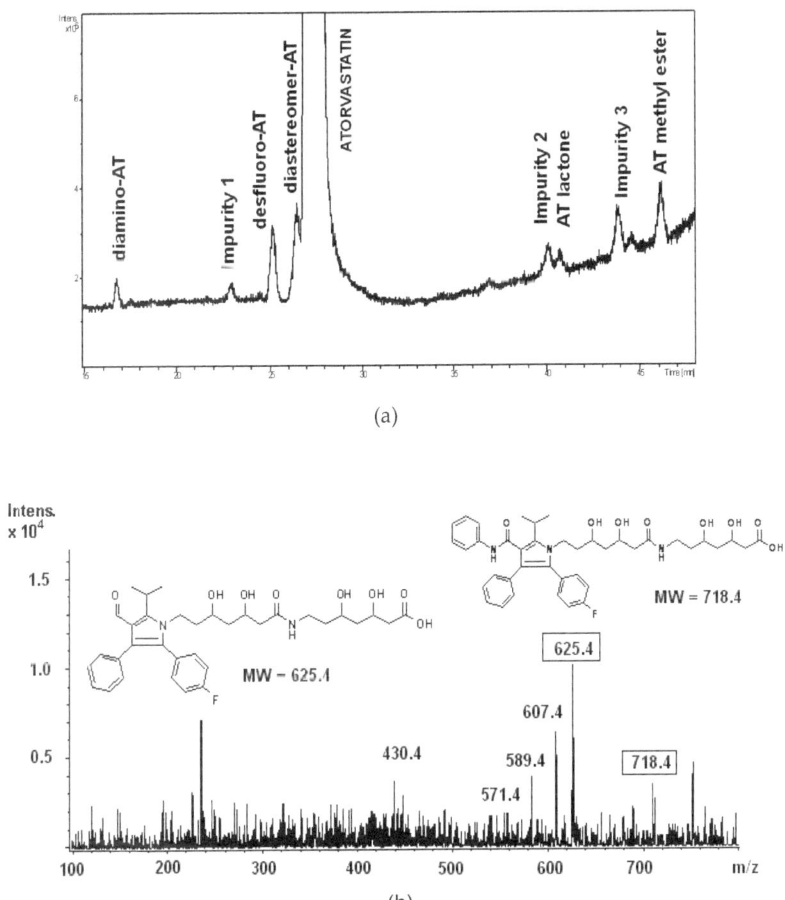

Figure 2. Total ion current chromatogram of atorvastatin pharmaceutical dosage form (A) and MS spectra of its process related impurity diamino-atorvastatin (B)

HPLC method was used for investigation of polypills for the treatment of cardiovascular diseases [56]. Seven drugs, i.e. lisinopril, aspirin, atenolol, hydrochlorothiazide and simvastatin/pravastatin/atorvastatin in the presence of their major interaction and degradation products were separated on a C8 column. In order to obtain mass spectra of the interaction and degradation products, ESI-MicroTOFQ mass spectrometer was employed. Atenolol, lisinopril, simvastatin and atorvastatin mass spectra were acquired in positive ESI mode, while hydrochlorotiazide and aspirin were ionized better in negative mode. Pravastatin gave good molecular ions in both modes. All the interaction and degradation products gave satisfactory mass spectra in positive ESI modes, except for two pravastatin related products which showed better molecular ions in negative mode. Results suggested that use of pravastatin in relate to other statins resulted in more interaction and degradation products, as well did the combination with atenolol by comparison with hydrochlorotiazide. This is a very nice approach that can be utilized for drug-drug interactions and stability studies of the polypill. Drawbacks of the proposed method are long analysis time of 90 min, replacement of the phosphate buffer with water for MS analysis and three different gradient methods for each of the statins.

2.6. Rosuvastatin

Far to our knowledge first HPLC method for the determination of rosuvastatin in bulk drug and in its dosage form was published by Mehta et al [57]. A forced degradation study was done at various pH values, under hydrolytic, oxidative, photolytic and thermal stress conditions. Developed method was able to resolve the degradation products formed during the stress study.

Not so commonly used in quality control analysis of pharmaceuticals, HPTLC method was proposed [58] for determination of rosuvastatin in its bulk drug and pharmaceutical preparations. Analysis was performed in a Camag twin-trough chamber on silica gel 60F(254) HPTLC plates. Aceclofenac was used as internal standard. Optimized mobile phase consisted of toluene-methanol-ethyl acetate-formic acid. Quantitation was performed densitometrically at 265 nm.

A paper employing both HPTLC and HPLC for determination of rosuvastatin and ezetimibe in combined tablet dosage forms was published [59]. HPLC analysis was performed on a Chromolith C18 column (100 mm x 4.6 mm) with PDA detector set at 245 nm. HPTLC separation was carried out on an aluminum-backed sheet of silica gel 60F (254) layers using n-butyl acetate-chloroform-glacial acetic acid as the mobile phase. Quantification of analites was performed with UV densitometry at 245 nm. A stability indicating method for simultaneous estimation of rosuvastatin and ezetimibe in their combination drug product was introduced [60]. Under oxidation, thermal and photodegradation conditions, both drugs were relatively stable. For rosuvastatin a high degree of degradation was observed in acidic hydrolytic conditions (0.1 M HCl at 80 °C for 1h), while ezetimibe was stable. On the contrary, ezetimibe was completely degradated with 0.1 M NaOH at 80 °C in 30 min, while rosuvastatin remained stable.

Simultaneous determination and quantification of atenolol, rosuvastatin, spironolactone, glibenclamide and naproxen sodium in bulk drugs, pharmaceutical formulations and in spiked human plasma was performed by HPLC [61].

2.7. Pitavastatin

Pitavastatin is the newest statin on the market available in Japan since 2003, and approved for use in US in 2009. Currently pitavastatin is under evaluation in Europe (in UK it was approved in 2010). Hence, not many methods have been reported for determination of pitavastatin in bulk drug and pharmaceutical formulations.

Two HPTLC methods were reported for the determination of pitavastatin in commercial pharmaceutical dosage forms [62, 63]. Validation was performed and both methods were shown to be selective, sensitive and accurate.

A HPLC method was proposed for determination of pitavastatin in pharmaceutical dosage forms by Kumar et al [64]. Separation was achieved on a Phenomenex C18 column (250 mm x 4.6 mm, 5 μm) in isocratic mode. Different mobile phases were tested and based on the best separation, analysis time, cost-effectives, sensitivity and suitability for the stability studies, a mobile phase consisted of 0.5% acetic acid:acetonitrile (35:65, *v/v*) was chosen. Four different drugs were tried out as the internal standard, and based on peak shape, resolution and elution time, paracetamol was chosen.

Several stability indicating methods have been published [65-68]. Panchal and co-workers proposed two different methods, using liquid chromatography and ultraviolet spectrophotometry for determination of pitavastatin in tablet dosage forms [66]. Additionally forced degradation study was conducted under acidic, basic, oxidative, thermal and photolytic conditions. No change in the area of pitavastatin peak and no additional peaks were detected under photodegradation conditions. Both acidic and basic hydrolysis and thermal conditions generated additional peaks. After oxidative degradation a significant decrease of pitavastatin peak and additional peaks were observed. Linearity range of the LC method was 0.1-2.5 μg/mL, while for the UV method it ranged from 2-20 μg/mL. The limit of detection (LOD) of the LC method was 0.0055 μg/mL, whereas for the UV method it was much higher, 0.4062 μg/mL. Statistical comparison between two methods by applying the paired t-test was performed and no statistically significant difference was observed.

UPLC stability indicating method was developed for degradation study of pitavastatin [67]. Separation of pitavastatin and its degradation products and impurities was performed in less then 5 min. More detailed photodegradation study of pitavastatin was conducted by Grobelny et al. [68]. Pitavastatin solution was exposed to UV-A radiation. HPLC analysis was performed to monitor the changes of pitavastatin. Identification of four photoproducts was conducted by MS analysis.

A single method is reported for simultaneous determination of pitavastatin and ezetimibe [69]. After optimization and validation, the proposed method was successfully applied for determination of pitavastatin and ezetimibe in a prepared binary mixture. However, no real sample was tested.

Analyt	Application	Separation technique and detector	Stationary phase	Mobile phase	Ref.
LOV	stability indicating study	HPLC PDA 237 nm	Symmetry C18 (150 x 3.9 mm, 5 μm)	Gradient elution A: ACN B: 0.1% TFA	15
SIM, EZE	combined dosage form	HPLC UV 240 nm	Chromolith Performance monolitich column RP-18e (250 x 4.6 mm)	ACN:50 mM ammonium acetate (65:35)	25
SIM	impurity profiling	HPLC DAD 240 nm, ESI-ion trap	Symmetry Shield RP 18 (250 x 4.6 mm, 5 μm)	ACN:water (85:15)	26
SIM	combined dosage form	UPLC Q-TOF MS	Acquity BEH C18 (100 x 2.1 mm, 1.7 μm)	gradient elution ACN:ammonium acetate pH 6	28
PRA	impurity profiling	HPLC APCI-CAD MS	Betasil C18 (250 x 4.6 mm)	gradient elution A: 30%methanol+10 mM ammonium acetate B: 100% methanol + 10 mM ammonium acetate	32
PRA	stability indicating study	HPLC UV 238 nm	Alltima C18 (150 x 4.6 mm, 5 μm)	methanol:0.02 M phosphate buffer pH 7	33
ATO, AML	stability indicating study	HPLC UV 237 nm	Perfectsil Target ODS-3 (250 x 4.6 mm, 5 μm)	ACN:0.025 M sodium dihydrogen phosphate pH 4.5 (55:45)	39
ATO, ASA	stability indicating study	UPLC UV 247 nm	BEH C18 (50 x 2.1 mm, 1.7 μm)	ACN:0.1 M phosphate buffer	40
ATO, FEN	stability indicating study	UPLC UV 247 nm	BEH C18 (50 x 2.1 mm, 1.7 μm)	ACN:0.01 M ammonium acetate pH 4.7	43
ATO	related compounds	HPLC UV 248 nm	Zorbax XDB C18 Rapid Resolution HT (50 x 4.6 mm, 1.8 μm)	gradient elution A: Tetrahydrofuran:ACN (90:10) B: 0.025M phosphate buffer pH 3.5	45
ATO, RAM	combined dosage form	HPLC HPTLC	Phenomenex Luna C18 (250 x 4.6 mm, 5 μm) Silica gel 60F254	0.1% phosphoric acid:ACN (38:62) methanol-benzene-glacial acid (19.6:80.0:0.4)	47
ATO, FEN	combined dosage form	TLC UV 258 nm	Aluminum foil silica gel 60 F-254	toluene-methanol-triethylamine (7:3:0.2)	48
ATO, SIM, PRA	pharmaceutical dosage form	HPLC UV 225 nm ESI-QTOF MS	Supelco C8 (250 x 4.6 mm, 5 μm)	gradient elution A: ACN B: phosphate buffer pH 2.3	56

Analyt	Application	Separation technique and detector	Stationary phase	Mobile phase	Ref.
ROS, EZE	combined dosage form	HPLC HPTLC UV 245 nm	Chromolith C 18 (100 x 6 mm) Aluminium-backed silica gel 60F(254)	0.1% orthophosphoric acid pH 3.5:ACN (63:37)	59
ROS, EZE	stability indicating study	HPLC UV 242 nm	Hypersil C18 (150 x 4.6 mm, 5 μm)	0.05 M phosphate buffer pH 2.5:methanol (45:55)	60
PIT	pharmaceutical dosage form	HPTLC UV 245 nm	Aluminum backed Silica gel 60F(254)	ethyl acetate-methanol-ammonia+1drop formic acid (7:2:0.8)	62
PIT	photostability study	HPLC MS	LiChrospher RP-18 (250 x 4.6 mm, 5 μm)	gradient elution A:ACN B:10 mM phosphate buffer	66
ROS, ATO, SIM, LOV, PRA	pharmaceutical dosage form	HPLC GC UV 246 nm	Symmetry C18 (250 x 4.6 mm, 5 μm) HP-1 (30 m x 0.25 mm x 0.25 μm)	ACN:water (70:30) pH 2.5 1.2 mL/min 2.9 mL/min (GC)	123
PRA, FLU, ATO, ROS	stability indicating study	HPLC	RP C18	methanol-water (60:40)-PRA,ROS Methanol-water (70:30)-FLU,ATO	124

LOV-lovastatin, SIM-simvastatin, PRA-pravastatin, FLU-fluvastatin, ATO-atorvastatin, ROS-rosuvastatin, PIT-pitavastatin, EZE-ezetimibe, AML-amlodipine, ASA-acetylsalicylic acid, FEN-fenofibrate, RAM-ramiprile, ACN-acetonitrile

Table 1. Chromatographic methods for analysis of statin drugs in pharmaceuticals

3. Bioanalytical methods

There have been three reviews on analytical methods for the determination of HMG-CoA reductase inhibitors in biological samples. The first one, published by Ertürk and co-workers in 2003 [70], reviews bio-analytical methods for lovastatin, simvastatin, pravastatin, fluvastatin and atorvastatin. The second one, published in 2007, is focused only on chromatography-mass spectrometry methods for the quantification of statins in biological samples [10]. In 2008 Nováková and co-workers [17] have published a review on HPLC methods for the determination of simvastatin and atorvastatin in various fields of application, including bioanalytical assays. Since these reviews have been published, a number of bioanalytical methods have been developed for all HMG-CoA reductase inhibitors. Most of the methods published since 2007 were applied for investigation of HMG-CoA reductase inhibitors in human plasma or serum. Far to our knowledge since 2007 only two LC/MS/MS methods for determination of statins in human urine have been developed. The sample preparation procedures and analytical assays for quantification of statins in biological samples are listed in Tables 2 and 3.

3.1. Sample preparation

Sample preparation is a quite tedious but still unavoidable procedure in bioanalitical methods. The objective of this delicate and challenging step is to transfer analyte of interest into a form that is purified, concentrated and compatible with the analytical system. The extraction and enrichment of analytes from the sample matrix are often realized by procedures such as, protein precipitation, liquid-liquid extraction (LLE) and solid-phase extraction (SPE). These conventional sample preparation procedures are still dominating in the preparation of biological samples for determination of statin drugs as well as their metabolites.

In most of the methods protein precipitation reagent is used as a dilution solvent for internal standard in order to reduce the number of reagent additions [71-74]. Still, Apostolou and co-workers [75] suggested addition of protein precipitation reagent after the internal standard in order to ensure a more satisfying binding of internal standard molecules with plasma proteins, simulating the binding of proteins with analytes in real human plasma. A number of different protein precipitation reagents were tested [76]. Despite the good recoveries obtained with phosphoric acid, the authors recommended to avoid acidic precipitants due to degradation of fluvastatin in acidic conditions. The highest recoveries were obtained with organic solvents. Although no significant differences were observed between methanol and acetonitrile, the second one was used as it offered a more compact precipitate minimizing the risk of SPE cartridge obstruction.

The simplest way to concentrate the analyte is certainly LLE. Hence, Hamidi and co-workers [73] tested a wide spectrum of organic solvents from various physicochemical categories with different volume fractions as well as combinations for extraction of lovastatin from human plasma. The best extraction efficacy was obtained using diethyl ether as extraction solvent. The same solvent was used for extraction of pitavastatin from human plasma [77]. An addition of hidrocloric acid to the plasma samples afore the extraction procedure was shown to be necessary in order to obtain the non-ionized form of analyte which considerably improved extraction efficacy. Assays employing LLE with ethyl acetate [78] and ethyl ether [79] as extraction solvents for determination of rosuvastatin in plasma samples were already published with extraction recoveries of 74 and 69%, respectively. However, in the preliminary study by Lan and co-workers [80], it was found that the extraction recovery of rosuvastatin from plasma in most common organic solvents, such as above mentioned ethyl acetate and ethyl ether, was less than 20%, resulting in an insufficient, imprecise and inaccurate extraction procedure. The authors presumed that low extraction efficacy of rosuvastatin was due to its extremely low water solubility. However, a carboxyl group in its structure forms a salt with calcium ion which indicates that rosuvastatin was apt to ionization. The application of ion paring with tetrabutyl ammonium hydroxide was suggested for improvement of rosuvastatin solubility and subsequently extraction efficacy. Finally, using ion paring LLE, extraction efficacy of rosuvastatin in ethyl acetate was improved from around 10% to more than 50%. A somewhat unusual LLE method for determination of timolol maleate, rosuvastatin and diclofenac in human plasma and aqueous humor from the bovine eyes was proposed [81]. The mobile phase, consisted of acetonitrile and 0.2% triethylamine, was used as extraction solvent. The quite high extraction efficacy of all investigated compounds was obtained using this uncommon extraction solvent.

Although LLE is generally considered to be providing cleaner extracts and lower matrix effect than the SPE, lower recovery due to the transfer of a fraction of the organic extract after the extraction may be the main disadvantage of the LLE technique. Moreover, when low concentrations have to be detected it is necessary to use a large solvent volumes and sample preparation becomes time consuming and labor invasive. In order to reduce organic solvent consumption, sample volume and sample preparation time, Apostolou and co-workers [75] have presented a fully automated high-throughput two-step LLE-LC/MS/MS method for the quantification of simvastatin and its acid form using a robotic liquid handling workstation with 96-deepwell plates. Another fully automated high-throughput salting-out (SA) assisted LLE-LC/MS/MS method was introduced by Zhang and co-workers [82]. Due to the compatibility between SALLE and LC/MS/MS, the extracts of simvastatin and its acid from human plasma were injected directly into LC system immediately after sample extraction. In this way extract solvent evaporation was eliminated and consequently sample preparation procedure was simplified. Also, the exposure of the extracts to the room temperature was minimized and hence minimal interconversion between simvastatin and its acid was achieved.

Among the SPE methods, the reverse phase cartridges have been extensively used for extraction of statins from biological samples. Gonzalez and co-workers [76] have presented a nice work regarding the traditional one-variable-at a time optimization for SPE extraction of fluvastatin together with other drugs from human plasma. The optimization of conditioning and washing solution composition, pH for conditioning and washing step and elution solvent selection were described in details. The SPE procedure has also been used as sample preparation step for quantification of atorvastatin and simvastatin as well as their metabolites in serum from patients with end stage renal disease [83]. In order to obtain the satisfactory and repeatable extraction efficacy and to remove matrix effects, several different reversed-phase SPE sorbents have been tested. The best results were obtained using ZORBAX SPE C-18 (Agilent Technologies) and Discovery DSC-18 SPE (Supelco) cartridges. As ZORBAX SPE C-18 columns were withdrawn from commercial market circulation during optimization of method, further investigations were performed using Discovery DSC-18 SPE cartridges. For the purpose of minimization of the interconversion between lacton and open-ring hydroxy acid forms of simvastatin and atorvastatin, SPE sorbents were conditioned and analytes were eluated with solvents containing 0.1 M acetate ammonium buffer pH 4.5. In the work of Di and co-workers [84) SPE sample preparation procedure was used for determination of pitavastatin with rosuvastatin as internal standard in human plasma. The influence of pH on extraction efficacy of statin drugs was investigated in detail. The authors have pointed out importance of 0.5 M potassium dihydrogenphosphate buffer (pH 4.0) as conditioning reagent for cartridges. At pH lower than 4 both molecules were protonated, leading to a decrease in its partitioning in reversed-phase SPE and recovery. At pH higher than 4, the carboxylic group in both pitavastatin and rosuvastatin undergo ionization, which also resulted in a decrease in the recovery for the same reason. Furthermore, it was found that pitavastatin degradation was much faster at lower than at high pHs. Also, it was found that pitavastatin was sensitive to sunlight. It was recommended to minimize the exposure of samples to sunlight as well as to dissolve the dried extract rather in methanol and water than in mobile phase containing formic acid.

To reduce the time of sample preparation, Mertens and co-workers [85] have used an automated SPE on disposable extraction cartridges to isolate pravastatin and its metabolites together with fenofibric acid, another lipid-regulating agent, from the human plasma and to prepare cleaner samples before injection and analysis in the LC/DAD/MS/MS system. Different kinds of disposable extraction cartridges containing bonded silicas of different polarities (ethyl, endcapped ethyl, octyl, endcapped octyl, octadecyl, endcapped octadecyl and cyanopropyl) were tested. The best recoveries for all investigated compounds were reported when disposable extraction cartridges filled with octyl functionalized silica sorbent were used.

Unfortunately, conventional SPE and LLE approaches are multi-step, time-consuming and the sample required for analyses as well as the consumption of organic solvent are quite high, particularly in case of LLE. A solvent-minimized sample preparation approach has been popular in last decades, therefore Farahani and co-workers [71] have published liquid-liquid-liquid microextraction procedure (LLLME), a miniaturized format of LLE, for determination of atorvastatin in human plasma. A number of factors affecting the microextraction efficiency were studied in detailed and the optimized conditions were established. They have obtained quite high extraction efficacy of atorvastatin from human plasma using proposed sample preparation procedure. Vlčková and co-workers [86] have developed fast and simple extraction procedure using microextraction by packed sorbent (MEPS) for sample purification and concentration of atorvastatin and its metabolites from human serum. Briefly, MEPS is a miniaturization of conventional SPE, but it differs from commercial SPE by fact that packing is inserted directly into the syringe, not into a separate column. In addition, they have compared a previously described [83] SPE procedure for extraction of atorvastatin and its metabolites from human serum with newly developed MEPS approach. The results of samples treated by SPE and MEPS were compared by means of Student *t*-test. The difference between obtained concentrations was statistically not significant. Hence, MEPS procedure was found to be simpler and faster sample preparation technique using smaller volume of sample, which is regardful to the patients and smaller volume of solvents, which is environmentally friendly.

3.2. Liquid chromatography

3.2.1. High performance liquid chromatography

The high performance liquid chromatography has become the method of choice for bioanalytical methods. Generally, in the HPLC methods reversed-phase C18 chromatographic columns were used for analysis of statin drugs in biological fluids. The recently developed columns based on BEH particles technology were employed in several methods [83, 86, 87]. Only in one assay reversed-phase C8 chromatographic column was used [88]. Unusually, reversed-phase narrow bore phenyl column was employed for investigation of atorvastatin, rosuvastatin and their metabolites [74, 89]. The length and diameter of columns differed fairly from 50 to 250 mm and from 2.0 to 4.6 mm, respectively. Although in most of the cases columns with particle size 5 μm were used, several authors preferred columns with smaller particles in order to obtain better peak shapes, resolution and thus shorter analysis time [72, 82, 83, 86, 87]. Analytical run times have been very variable, the shortest 2 min, the longest about 20 min.

The selection of mobile phase was quite a challenging task in all investigations. In most of the methods acetonitrile or methanol were present in the mobile phase as organic solvent. The percentage of organic solvents was optimized such that the retention times of analytes were kept as short as possible. In most assays percentage of organic solvent was quite high, usually more than 70%. The majority of publications emphasize the pH as the most critical variable for separation of the statin drugs [76, 82, 84]. In order to minimize the interconversion, it is critical to maintain pH of mobile phase between 4 and 5.

The influence of mobile phase pH on retention of atorvastatin and rosuvastatin has been investigated [90]. Since both of the analytes are acidic compounds, their retention on the reversed-phase column was expected to be pH dependant. When pH of the mobile phase was decreased from 4.0 to 3.0, the retention times of the analytes decreased unexpectedly and with further decreases in the pH to 2.0 the retention times increased once again. This behavior was explained by a change in binding of the analytes to the stationary phase and also changes in the solubility of the analytes in the mobile phase. The pH 3.0 was chosen as optimum pH because of the reasonable retention times while the resolution between peaks, as well as peak shapes, were satisfactory.

The pH of mobile phase was also a critical variable for the separation of the fluvastatin from valsartan and its metabolite during the optimization of LC/PDA/FLD method [76]. The pH of the mobile phase was limited by the native fluorescence of valsartan and its metabolite, which disappears in the basic form (pK_a = 3.7). On the other hand, spectrophotometric studies showed that fluvastatin degradation was accelerated in acidic conditions. Mobile phases with different formic acid/formate proportions were tested in order to establish the range where fluvastatin was stable and valsartan and its metabolites kept their fluorescence. 0.01% formic acid/10 mM ammonium formate (pH 4.1) was finally chosen as appropriate buffer. Uncommon pH was used for quantification of lovastatin in human plasma [73]. Mobile phase consisted of acetonitrile and 0.05 M phosphate buffer with pH 7, adjusted with phosphoric acid.

The flow rate of the mobile phase was in range from 0.2 up to 1.5 mL/min. In all of the assays the flow rate did not change during the chromatographic analysis except in the reference [76] where the flow rate was gradually changed after three minutes.

The chromatographic separation of most of the methods was performed at room temperature. In order to shorten analysis time, in the several cases the column temperature

Extracted analytes	Matrix	Sample preparation procedure	Stationary phase	PP reagent / LLE reagent / SPE eluent	Recovery (%)	Ref.
ATO	plasma	PP, LLLME	-	methanol , HCl, trichloroacetic acid /1-octanol	91	71
SIM, MET IS=propranolol hydrochloride	plasma	PP	-	methanol:water (1:1)	83-91	72
ROS + metabolites IS=deuterium labeled	plasma	PP	-	0.1% acetic acid in methanol	88–106	74
FLU, VAL + metabolite, CLT IS=candesartan cilexetil	plasma	PP, SPE	Phenomenex Strata-X polymeric C18	ACN/methanol	78–91	76
ROS IS=hydrochlorothiazide	plasma	LLE	-	ethyl ether	69–72	79
ROS IS=estrone	plasma	ion pair LLE	-	ethyl acetate	47–63	80
ROS, TIM, DIC IS=naproxen	plasma, bovine aqueous humor	PP, LLE	-	methanol/mobile phase	95–99	81
SIM, SIM-acid IS=deuterium labeled	plasma	SALLE	-	ACN, 5 M ammonium formate buffer (pH 4.5)	71–79	82
SIM, ATO + metabolites IS= deuterium labeled	serum	SPE	Supelco Discovery DSC-18	ACN:0.1 M ammonium acetate buffer pH 4.5 (95:5)	65-100	83
PIT IS=ROS	plasma, urine	SPE	Supelco Superclean™ LC-18 SPE Tubed	methanol	plasma 84–88 urine 86–96	84
PRA + metabolites, FFA IS=triamcinolone	plasma	at-SPE	Disposable extraction cartridges C8 silica sorbent	methanol	50–77	85
ATO + metabolites IS=deuterium labeled	serum	MEPS	C8	ACN:0.1 M ammonium acetate pH 4.5 (95:5)	89–116	86
PIT, PIT-lactone IS=racemic *i*-prolact	plasma, urine	LLE	-	methyl-terc-butyl ether	plasma 70–75 urine 74–83	88
PRA, ASA IS=furosemide	plasma	LLE	-	tertiary butyl methyl ether	51–66	94

SIM-simvastatin, PRA-pravastatin, FLU-fluvastatin, ATO-atorvastatin, ROS-rosuvastatin, PIT-pitavastatin, FFA-fenofibric acid, TIM-timolol maleate, MET-metoprolol, DIC-diclofenac, VAL-valsartan, ASA-acetylsalycilic acid, CLT-chlorthalidone, ACN-acetonitrile, IS-internal standard

Table 2. Sample preparation procedures utilized for the determination of statins in biological samples was maintained above 30 °C [76, 77, 81, 83, 86, 87]. The effect of column oven temperatures on the analysis of atorvastatin and rosuvastatin in the range 25 to 35 °C was investigated and best results were observed at 25 °C in terms of retention factor and resolution [90]. Increasing temperature above 25 °C resulted in the rapid elution of rosuvastatin close to the solvent front.

LC/DAD methods are rarely sensitive enough for quantification of statins as well as their metabolites in human plasma samples due to the poor UV-absorption properties of statin molecules. Furthermore, the levels of statins and their metabolites in biological fluids are very low due to low amount of drug reaching the systemic circulation. Their typical plasma concentrations are in ng/mL levels. However, several sensitive LC/DAD methods for determination of pravastatin [31], atorvastatin [71], lovastatin [73], rosuvastatin [81], and atorvastatin with rosuvastatin [90] have been developed with limit of quantification (LOQ) in range of 1 - 10 ng/mL. Less sensitive LC/DAD method for quantification of lovastatin in human plasma was developed [91]. The LOQ value for lovastatin was relatively high, 400 ng/mL. Another even less sensitive LC/DAD method for quantification of several HMG-CoA reductase inhibitors in human plasma was developed by Sultana and co-workers [92]. The LOQ values were between 376 and 1006 ng/mL. In fact, both of these methods were not used on real plasma samples.

Fluorescence detection has not been widely employed in the determination of HMG-CoA reductase inhibitors, as most of statins do not possess a natural native fluorescence. Still, Gonzalez and co-workers [76] have developed a SPE-HPLC/PDA/FLD method for determination of fluvastatin and valsartan in human plasma. Comparing results obtained with spectrophotometic and fluorimetric detector superior selectivity and sensitivity by fluorescence detection of fluvastatin could be perceived.

3.2.2. *Ultra performance liquid chromatography*

Recently UPLC is becoming a leading chromatographic technique in modern bio-analytical methods. Nováková and co-workers [83] have investigated its potential in combination with MS/MS detection for the fast, sensitive, reliable and selective detection of atorvastatin and simvastatin together with their main metabolites and interconversion products in human serum. Iriarte and co-workers [87] have investigated UPLC technique as a faster alternative to HPLC for simultaneous analysis of fluvastatin and other drugs usually prescribed in cardiovascular therapy. Acquity UPLC Columns Calculator software was used for transfer of previously developed HPLC method [76].

The UPLC technology has significantly improved the method optimization process since shorter analysis and re-equilibration times allowed a greater number of experimental testing conditions than with a conventional HPLC. The sample volume required was much lower than in HPLC method. Furthermore, shorter analysis time together with slower flow rates reduced the organic solvent consumption. The sharper and higher chromatographic peaks, thereby improved peak capacity, was obtained using UPLC technology. Still, the sensitivity of UPLC method was found to be analyte dependent as the improvement was not achieved for all analytes.

3.2.3. Liquid chromatography coupled to tandem mass spectrometry

In pharmacokinetic investigations of statins LC/MS/MS technique is unequivocally the method of choice. Recently, several procedures were described in the literature taking the advantages of the benefits of mass spectrometry. Both ESI and APCI sources as well as triple quadrupole analyzator were applied in most LC/MS/MS sample analysis.

As it was mentioned above the selection of appropriate mobile phase composition for determination of statins in biological fluids is quite challenging task which is even more complicated when detection and quantification of statins is performed using MS. Only few additives could enable good stability at pH range 4 to 5 as well as volatility and sensitive mass spectrometric response. Therefore, Di and co-workers [84] have pointed out the importance of the formic acid in lowering the pH of mobile phase. In this way pitavastatin was obtained in non-ionized form and a symmetrical peak shape was observed. The concentration of formic acid was optimized not only to maintain a symmetrical peak shape in the chromatographic system but also to render good ionization and fragmentation of pitavastatin in the MS/MS detector. An addition of 0.025% formic acid to the aqueous phase was found to be an important factor for acquiring the high sensitivity of another LC/MS/MS method for determination of pitavastatin in human plasma [77].

Nováková and co-workers [83] have presented a nice example of optimization of the buffer pH and concentration in order to get the best signal to noise ratio of MS detector. Ammonium formate and ammonium acetate at pH 4.0 and 4.5 were tested at the concentration range 0.01 to 10 mM. The best response of atorvastatin and simvastatin was observed at 0.5 mM buffers. The concentrations higher than 5 mM significantly decreased the response of mass spectrometer. On the other hand, the concentrations lower than 0.5 mM were not sufficient to keep buffering capacity and thus had negative influence to the response of mass spectrometer. Ammonium acetate was preferred before ammonium formate because of better peak shapes. Finally, the optimized mobile phase composition was 70% of acetonitrile and 30% of ammonium acetate buffer 0.5 mM (pH 4.0). In most of bioanalytical methods isocratic elution has been utilized, still when more analytes with different polarities were separated, gradient elution had to be applied.

Tandem mass spectrometry detection for identification and quantification of simvastatin and atorvastatin together with their metabolites and lacton/hidroxy acid interconversion forms was employed [83, 86]. All analytes were monitored using electrospray positive ionization (ESI^+) mode and for all analytes protonated molecule $[M+H]^+$ was the most intensive ion in mass spectra. Quantification of all analytes was performed using selected reaction monitoring (SRM) and two specific transitions were optimized for each molecule in order to increase selectivity and sensitivity of the method. In the paper published afterwards simvastatin in its lactone form was determined in ESI^+ mode, while its hydroxy acid form was determined in ESI^- mode due to poor sensitivity of hydroxy acid form in positive ion mode [82].

LC/MS/MS method developed by Apostolou and co-workers [75] consisted also of two periods combining both negative and positive ionization modes. The mass spectrometer operated in the negative detection mode for 1.21 min until simvastatin and lovastatin hydroxy acid forms were eluted from chromatographic column. Afterwards a period of 0.69 min followed in the positive mode during witch simvastatin and lovastatin lacton forms were eluted. Comparing LOQ values for simvastatin acid obtained by these three methods it can be seen that lower LOQ values and thus better sensitivity were obtained in the last two methods. Unfortunately, simvastatin forms various adducts influenced by mobile-phase and matrix composition and such adducts sometimes give higher intensity than protonated molecule $[M+H]^+$, which is an ideal precursor ion for SRM transition and quantification studies. However, Senthamil Selvan and co-workers [72] have observed very high signal of $[M+Na]^+$ in the spectra of simvastatin next to the $[M+H]^+$. Consequently, it was used as precursor ion for quantitation of simvastatin. Also, Zhang and co-workers [82] have used the methylammonium aduct $[M+CH_3NH_4]^+$ as a parent ion for simvastatin because the adduct ion showed the best signal to noise ratio.

Rosuvastatin has a pyrimidine ring and a carboxylic group in its structure, hence it could be detected either in positive or negative ionization mode. However, the quantification of rosuvastatin in positive ionization mode is more common and was used for determination of rosuvastatin [80] and rosuvastatin together with its metabolites [74], respectively. In the both assays the major ion was protonated molecule $[M+H]^+$ in full-scan mode and principal product ion was at *m/z* 482. Macwan and co workers [74] have also reported two minor fragments at *m/z* 300 and *m/z* 272.

During the method development, Gao and co-workers [79] also attempted to optimize ESI conditions under positive ionization mode. However, the observed signal intensity was not sensitive enough for determination of expected rosuvastatin's concentrations, especially for low dosage administration. Low sensitivity of positive ionization mode could be explained by a number of fragment ions produced in the product ion spectrum of $[M+H]^+$. In order to improve the sensitivity of the method, the negative ESI detection was taken into consideration. Under negative ESI mode, rosuvastatin produced abundant deprotonated molecule $[M-H]^-$ at m/z 480. In the product ion mass spectrum of $[M-H]^-$, fewer fragment ions were formed compared with that of $[M+H]^+$. Also, it was pointed out that negative ESI mode produced lower chemical background noise than positive. Comparing LOQ values obtained by these three methods, it can be observed that almost five times lower LOQ value for rosuvastatin was obtained using negative ESI detection.

Pitavastatin has similar structure to rosuvastatin. It contains alkaline nitrogen ion on the quinoline ring and a carboxylic group, therefore positive and negative ionization mode could be also employed. Both of ionization modes for determination of pitavastatin in human plasma and urine by LC/MS/MS method were applied [84]. The results showed that the response intensity of pitavastatin in negative mode was lower and furthermore the response was quite unstable. Pitavastatin was scanned under Q1 MS full-scan mode to determine the parent ion and under Q1/Q3 product ion scan mode to locate the most abundant production. The protonated molecular ion, $[M+H]^+$, was the predominant ion in the Q1 spectrum and was used as the parent ion to obtain the product ion spectra. The most sensitive mass transition was from *m/z* 422.0 to 290.1, which was similar to the MS/MS spectrum of pitavastatin reported in reference [88] and [93], while in the previously reported LC/MS/MS method the highest collision energy gave the most abundant product ion at *m/z* 318.0 [77].

Recently, two LC/MS/MS methods have been developed for determination of pravastatin in human plasma [85, 94]. Both methods utilized ESI but in different modes. In the method developed by Martens and co-workers [85] the mass spectrometer was operated in the positive mode. The MS/MS detection was set up in MRM mode. The full scan mass spectra of pravastatin and its metabolites were scanned. The collision energy in Q2 produced different significant fragment ions. The MS/MS ion transitions selected for quantification purpose were *m/z* 442.2 to 269.1, 442.2 to 269.1 and *m/z* 424.3 to 183.0 for pravastatin, 3-OH metabolite and its lacton form. On the contrary, Polagani and co-workers [94] have found high sensitivity and stability using negative ionization mode. Deprotonated form of pravastatin, $[M\text{-}H]^-$ ion was the parent ion in the Q1 spectrum and was used as the precursor ion to obtain Q3 product ion spectra. The most sensitive mass transition was monitored from *m/z* 423.3 to 100.8.

Internal standards have been used in most of the assays leading to more corrected results. In some cases one of the statins has been used as internal standard [73, 75, 77, 84], while other works utilized internal standards of various structure, including hydrochlorothiazide [79], estrone [80], naproxen [90], gemfibrozil [91], pioglitazone [95] etc. The best internal standards for precise and accurate quantification in MS or tandem MS are stable-isotope-labeled standards. Only a few works employed deuterium labeled standards [74, 82, 83, 86]. In the case of atorvastatin, [d5] labeling usually occurs on the phenyl ring, which does not contain fluorine. [d3] labeling of simvastatin occurs on the side chain, while [d6] labeling of rosuvastatin occurs on isopropyl group attached to pyrimidin ring. In most of investigations only one compound was used as internal standard.

However, Mertens and co-workers [85] have used two different internal standards for quantification of fenofibric acid, pravastatin and its metabolites in human plasma by automated SPE-LC/DAD/MS/MS technique. To avoid the need for plasma dilution and two time-consuming analytical runs, the use of two internal standards was necessary as the concentration of fenofibric acid was too high and MS signal appeared saturated. Hence, the sulindac was selected for the quantification of fenofibric acid by UV-detector, while the triamcinolone was used for MS/MS quantification of pravastatin and its metabolites. As it was mentioned above, in the method developed by Zhang and co-workers [82], the LC/MS/MS data acquisition for simvastatin was conducted in positive ionization mode, whereas the data acquisition for simvastatin acid was conducted in negative ionization mode. Therefore, it was inevitable to use two internal standards deuterium labeled simvastatin and deuterium labeled simvastatin acid, respectively.

3.3. Gas chromatography

Several GC/MS methods for determination of statins in biological samples have been reported [10]. Unfortunately, these methods are limited and not recommended for routine applications as they include analyte derivatization step prior to anaysis in order to obtain volatile derivatives of the drug molecule and therefore a complicate sample preparation procedures.

Simultaneous determination of lovastatin, simvastatin and pravastatin in plasma using GC with chemical ionization mass spectrometry has been described [70]. The analytes were isolated from plasma by SPE procedure which separated the lactone and acid forms of the drugs. The lactone forms were converted to the corresponding acid forms, which were subsequently derivatized by pentafluorobenzylation of the carboxyl group, and trimethylsilylation of the hydroxyl functions. The method has sufficient sensitivity for the analysis of clinical samples containing the drugs administered at therapeutic doses with recoveries between 79 and 90%. In another method, simvastatine and its acid form were derivatized with ferroceneboranic acid.

Far to our knowledge since 2001 no method for determination of statin drugs in biological samples using gas chromatography has been published due to imprecise and time consuming derivatization procedures which is an unavoidable step in analysis of statin molecules and the biggest disadvantage of using this technique.

3.4. Pharmacokinetic studies

Since all HMG-CoA reductase inhibitors are given to the patients once daily, monitoring plasma concentrations over a period of 24 hours is necessary. In all published papers monitoring plasma concentration levels were performed at least over 24 hours, except in references [74, 89] were the blood samples were collected at various time points during a period of 12 hours after a single oral dose of rosuvastatin and atorvastatin, respectively. Also, in pharmacokinetic and bioavailability study of simvastatin in healthy volunteers and moderately hyperlipemic patients' drug plasma concentrations were monitored during 12 hours [96]. In the most of investigations pharmacokinetic parameters of statins were investigated after only one pharmaceutical tablet dosage.

Pharmacokinetic parameters of rosuvastatin have been investigated after single doses of 5, 10 and 20 mg [79]. The peak plasma levels obtained from this study were 8.32, 14.8 and 20.1 ng/mL, respectively. It was found that plasma exposure to rosuvastatin appeared increasing dose-proportionally and the plasma elimination half-lives were prolonged with increased doses. Not so many methods for determination of statins in human urine have been developed. The SPE-LC/MS/MS method was successfully applied to quantify the pitavastatin concentration in plasma and urine which were collected from Chinese volunteers [84]. The urinary excretion ratio of pitavastatin accounted for less than 0.6%, which suggested that pitavastatin was not excreted primarily by kidney. Quite similar data were obtained using LLE-LC/MS/MS method [88].

Several above described bioanalytical assays have been used in bioequivalence studies of statin drugs. The pharmacokinetic parameters derived from drug plasma concentrations, including maximum plasma concentration, area under the plasma concentration-time curve from 0 h to the last measured data, area under the plasma concentration-time curve from 0 h

Analytes	Matrix / Sample preparation	Stationary phase	Mobile phase	Separation technique and Detection	LOQ	Ref.
ROS + metabolites, IS=deuterium labeled	plasma	Agilent Zorbax-SB Phenyl, Rapid Resolution HT (100 x 2.1 mm, 3.5 μm)	gradient elution A: 0.1% glacial acetic acid in 10 % methanol in water B: 40% methanol in ACN	HPLC ESI^+ MS/MS MRM	0.1– 0.5 ng/mL	74
SIM, SIM acid IS= LOV, LOV acid	plasma	YMC ODS-A (50 x 4.0 mm)	ACN:5 mM ammonium acetate pH 4.5 (82:18)	HPLC APCI/ESI^-/ESI^+ MS/MS MRM	0.1 ng/mL	75
FLU, VAL + metabolite, CLT IS=candesartan cilexetil	plasma	Waters Atlantis dC18 (100 x 3.9, 3 μm)	gradient elution A: ACN, 0.01% formic acid, 10 mM ammonium formate B: 0.01% formic acid, 10 mM ammonium formate pH 4.1	HPLC UV 229, 254, 236 nm FD 254, 378 nm	UV: 31-85 μg/mL FD: 10-20 μg/mL	76
ROS IS=hydrochlorothiazide	plasma	Agilent Zorbax XDB-C18 (150 x 4.6 mm, 5 μm)	methanol:water (75:25), pH 6.0 with ammonia	HPLC ESI^- MS/MS MRM	0.02 ng/mL	79
SIM, SIM-acid, ATO + metabolites IS=deuterium labeled	serum	Waters Acquity UPLC™ BEH C18 (100 x 2.1 mm, 1.7 μm)	gradient elution A: ACN B: 0.5 mM ammonium acetate buffer pH 4.0	UPLC ESI^+ MS/MS SRM	0.09–4.38 nM	83
PIT IS=ROS	plasma, urine	Shimadzu Shim-pak VP-ODS (150 x 4.6 mm, 5 μm)	methanol:water:formic acid (75:25:0.05)	HPLC ESI^+ MS/MS SRM	0.08 ng/mL	84
PRA + metabolites, FFA IS=triamcinolone	plasma	Phenomenex Synergi Max-RP (150 x 2 mm, 4 μm)	ACN:methanol:5 mM ammonium acetate buffer pH 4.5 (30:30:40)	HPLC ESI^+ MS/MS MRM	0.05–0.5 ng/mL	85
ATO + metabolites IS=deuterium labeled	serum	Waters BEH C_{18} (100 x 2.1 mm, 1.7 μm)	gradient elution A: ACN B: 0.5 mM ammonium acetate pH 4.0	UPLC ESI^+ MS/MS SRM	0.08-0.66 nM	86
FLU, VAL + metabolite, CLT IS=candesartan cilexetil	plasma	Waters Acquity UPLC™ BEH C18 (50 x 2.1 mm, 1.7 μm)	gradient elution A: 10 mM ammonium formate, 0.01% formic acid B: ACN, 10 mM ammonium formate, 0.01% formic acid, pH 4.1	UPLC UV 220 nm	20-110 μg/mL	87
PIT, PIT-lacton IS=racemic i-prolact	plasma, urine	Thermo BDS Hypersil C8 (50 x 2.1 mm, 3 μm)	methanol:0.2% acetic acid in water (70:30)	HPLC ESI^+ MS/MS MRM	1 ng/mL	88

LOV-lovastatin, SIM-simvastatin, PRA-pravastatin, FLU-fluvastatin, ATO-atorvastatin, ROS-rosuvastatin, PIT-pitavastatin, VAL-valsartan, FFA-fenofibric acid, CLT-chlorthalidone, ACN-acetonitrile, IS-internal standard

Table 3. Analytical methods for the determination of statins in biological samples to the infinity, the time to reach peak concentration, the apparent elimination rate constant, showed that there was no statistically significant difference between two investigated pharmaceutical formulations [72, 73].

Not so many chromatographic methods have been developed for the quantification of HMG-CoA reductase inhibitors in combination with their metabolites. They undergo quite extensive first-pass metabolism during witch active and inactive metabolites are produced. The actual plasma concentrations of both parent compounds and metabolites are of major interest in pharmacokinetics studies. Therefore, analytical methods for simultaneous determination of statins and their metabolites are quite valuable. Although simultaneous determination of statins and their metabolites was considered being difficult owing to the different polarities of the analytes, several methods have been published.

Recently, Apostolou and co-workers [75] published fast and fully automated LLE-LC/MS/MS method, while Zhang and co-workers [82] presented a high-throughput salting-out assisted LLE-LC/MS/MS method for simvastatin in lactone and acid form. Both of methods were very fast with analytical runs less than two min and fairly sensitive with LOQ values around 0.1 ng/mL. Nováková and co-workers [83] have developed fast selective and reliable SPE-UPLC/MS/MS method for simultaneous determination of simvastatin and atorvastatin as well as their active and inactive metabolites. The main advantage of the method was applicability of the method for determination of two clinically widely used statins using one sample preparation procedure and one chromatographic run, while the main limitation of study was slightly higher LOQ value obtained for simvastatin in open-ring hydroxy acid form.

More recently Vlčková and co-workers [86] have presented a new MEPS-UPLC method for determination of atorvastatin and its metabolites, faster and more sensitive comparing to previously published ones. A simple, fast and reproducible method for determination of rosuvastatin and metabolites in human plasma has been described [74]. The major advantages of the method were the requirement for small plasma volume and simple sample preparation procedure, protein precipitation. The major limitation of method was its inability to determine *N*-desmethyl rosuvastatin in the patient samples although its LOQ was quite low, 0.5 ng/mL. The patients included in the study took a single dose of rosuvastatin at 20 mg. *N*-desmethyl rosuvastatin is a minor metabolite that is present in much lower concentrations than rosuvastatin. Therefore, the authors anticipate that the methods should be sensitive enough to measure its concentration in patients receiving rosuvastatin on a routine basis.

A sensitive and accurate procedure based on solid-phase extraction coupled at-line to a LC/MS/MS for determination of pravastatin and its two metabolites in human plasma has been presented [85]. Optimized and validated LLE-LC/MS/MS method for determination of pitavastatin and its lacton form in human plasma as well as in urine is described [88].

Furthermore, a LC/MS/MS method for separation of fluvastatin from its *threo* isomers metabolites to support a bioequivalence study has been developed [97].

The advantage of the methods for simultaneous determination of several co-administered drugs is that the one sample preparation and one chromatographic run are required for monitoring therapeutic levels of several drugs. Therefore, these methods could be useful in daily routine sample handling, when many samples from patients taking different drugs together with HMG-CoA reductase inhibitors are analyzed in clinical laboratories. Recently, several chromatographic methods have been developed for the quantification of statin drugs in combination with other drugs, most of them are commonly used in treatment of cardiovascular disease: atenolol, spironolactone, glibenclamide [61], metoprolol succinate [72], valsartan and chlorthalidone [76, 87], timolol maleate, diclofenac sodium [81], fenofibric acid [85], ezetimibe [91], ceftriaxone [92], acetylsalicylic acid [94], amlodipine [98] and losartan, atenolol, acetylsalicylic acid [99].

Recently, several papers were published regarding prediction of statins' pharmacokinetics. In our work the usefulness of reversed-phase high performance chromatography in building models that would allow the prediction of pharmacokinetics parameters of statins was evaluated [100]. In order to get better insight into the nature of their chromatographic behavior, the retention times were measured using octyl and octadecyl chromatographic columns. Obtained chromatographic data were compared with pharmacokinetic parameters predicted by use of 17 different computer programs. Significant correlations were found between chromatographic data and lipophilicity of statins. In addition, with the combine set of descriptors (chromatographic data, solubility, quantum chemical and topological indices) the highly significant correlations with pharmacokinetic parameters have been found, which confirms the utility of HPLC technique for prediction of pharmacokinetic behavior of statin drugs.

In order to predict the bioavailability of statins, the association mechanism with phosphatidylcholine using immobilized artificial membrane high performance liquid chromatography technique was studied. Moreover, the thermodinamic driving forces for the statin molecules with phosphatidylcholine monolayers were analyzed in detail [101].

4. Capillary electrophoresis

Capillary electrophoresis (CE) is an alternative separation technique which is designed to separate species based on their size to charge ratio in an electric field in the interior of a small capillary filled with background electrolyte. Driving forces in CE are electrophoretic migration and the electro-osmotic flow (EOF). CE has become a useful tool in pharmaceutical analysis because of its advantages over other separation techniques, such as high resolution, high selectivity, simplicity, short analysis time, cost efficiency and low consumption of solvents and reagents [102]. Mainly employed CE modes for drug analysis are capillary zone electrophoresis (CZE) based on charge-to-mass ratio and micellar electrokinetic chromatography (MEKC) based on chromatographic partition of analytes between micelles and background electrolyte. MEKC is the most appropriate electrophoretic technique for impurity profiling because the neutral compounds and charged components that have similar electrophoretic mobilities can be separated simultaneously [103]. CE is currently recommended in several pharmacopoeias. Principal advantage of CE over well-established and widely used HPLC technique is its ability to deliver high efficiency in short analysis times [104]. However, CE methods proposed for the determination of statin drugs are scarce.

CE has been applied for determination of pravastatin in fermentation broth in order to optimize its production in bioreactors [105]. Pravastatin is produced in two-step fermentation. In the first step, mevastatin is produced by *P. citrinum*, and in the second step, bioconverted to pravastatin by *S. carbophylus*. The method successfully separated pravastatin from interfering matrix, mevastatin and 6-*epi* pravastatin. Its determination in production media was also performed using two HPLC methods. All three proposed methods had runtimes under 1 min. However two HPLC methods, performed on a particle and a monolithic LC column had superior sensitivity compared to MEKC, with LOD around 0.01 ng/mL, 0.2 ng/mL and 20 ng/mL, respectively.

We have developed CZE method for determination of pravastatin in pharmaceutical dosage form [106]. Rapid migration of negatively charged pravastatin molecule was obtained in alkaline buffer by the application of electric field of 30 kV. The alkaline buffer generated strong EOF that enabled determination of a fully charged drug molecule within 2.5 min. Pravastatin retention time is about 21 min in the assay procedure listed in European Pharmacopoeia (Ph. Eur.) using the HPLC with UV detection. Relatively short analysis time is the main advantage of the CZE method developed. Pravastatin is administered to patients in its active form as the hydroxy acid sodium salt. However, the drug exists in solution with its lactone equilibrium product reversibly formed at acidic pH. Pravastatin is also susceptible to an isomerization reaction which is relatively rapid [107]. The MEKC method was established to separate the drug and its degradation products in acidic media. Introduction of sodium dodecyl sulphate (SDS) in the background electrolyte solution plays a key role in the separation of negatively charged and neutral species. The proposed method allows baseline separation of pravastatin, C-6 epimer of pravastatin and their corresponding lactone forms that appear as interconversion products depending on the pH value. The migration times of degradation compounds ranged from 2.8 to 6.2 min The above mentioned interconversion compounds of pravastatin represent its related impurities defined in Ph. Eur. and are also potential biotransformation products. CE has also been applied to the screening of anionic impurities in bulk drug [108].

The application of CE to rapidly quantitate lovastatin production levels by *Aspergillus terreus* mutants has been described [109]. The fermentation broths of thousands of mutated strains were efficiently and inexpensively screened for increased lovastatin production by the developed high-throughput method. Determination of lovastatin in the presence of its oxidation products after exposure to an oxidative atmosphere has been carried out using CE technique [110]. The method developed is suitable for the routine analysis of lovastatin.

The quantitative analysis of lovastatin in urine samples based on CE has significance for the control of clinical therapy [111]. The concentration sensitivity is poor in CE because of the short optical path length limited by the inner diameter of the capillary and small volume of sample injected. Such low sensitivity has hampered the use of this method in clinical drug monitoring. However, the sensitivity was enhanced by using a simple stacking method for the determination of trace lovastatin in biologic fluids.

The CZE method was developed for the separation and determination of lovastatin as active ingredient in the red yeast rice product [112]. Prior to determination, lovastatin was extracted from capsule by ethanol. In this study, high pH (10.5) was selected in order to convert lovastatin to its acidic form completely. However, earlier reported studies revealed that lovastatin and lovastatin hydroxy acid are the two main components which contribute to up to 90% of the total quantity of monacolins in the read yeast rice [13]. Hence, the main disadvantage of the proposed CE method is that the content of the main components contributing to the pharmacology effect in red yeast rice supplement was not determined individually.

Only one CE method for the analysis of simvastatin is available till date [113]. This method was developed for the quanitication of both lovastatin and simvastatin in pharmaceutical dosage forms.

In the literature, CZE method has been reported for determination of atorvastatin [114]. The separation was optimized on capillary, but it was further miniaturized to a microchip platform with linear imaging UV detection. Even though CE is a rather good alternative for evaluation of impurity profile and enantiomeric purity of a drug, it is not enough applied. Therefore, we have developed a new MEKC method for separation and simultaneous quantitation of atorvastatin and its related substances diastereomer-atorvastatin, desfluoro atorvastatin, atorvastatin methyl ester and atorvastatin lactone [115]. The separation was carried out in an extended light path capillary in order to improve sensitivity at applied voltage of 30 kV using a background electrolyte consisting of 10 mM sodium tetraborate buffer pH 9.5, 50 mM SDS and 20% (*v/v*) methanol. Separation of neutral compounds from each other requires partitioning into charged micelles that migrate at a different rate from the EOF. The addition of methanol to the running buffer resulted in a very effective choice to achieve resolution between the peaks of charged substances adjacent to atorvastatin as well as the peaks of neutral drug-related substances. Linear calibration curves were established over the concentration range 100-1200 μg/mL for atorvastatin and 1.0-12.5 μg/mL for related substances. The applicability of the proposed MEKC method to the assay of atorvastatin in the presence of its related substances was investigated by analyzing the bulk drug provided by different manufacturers and various commercial formulations. The use of very low volumes of electrolyte (μL) and samples (nL) make the new MEKC procedures very interesting for determination of atorvastatin, purity evaluation and quantification of drug-related substances in a single analysis. The drawback of the proposed MEKC method is lower sensitivity compared to one obtained by RP-LC method for the same related substances [45]. The published UPLC method for simultaneous determination of atorvastatin and fenofibrate has better sensitivity and runtime of 3 min but the linearity, LOQ and LOD was established only for atorvastatin lactone [43].

CE method was also developed for the separation and simultaneous determination of atorvastatin and amlodipine in their combination formulations [116]. Degradation products produced as a result of stress studies did not interfere with the detection of both drugs and the assay can thus be considered stability indicating.

The CE method was developed for the enantiomeric purity determination of fluvastatin enantiomers [117]. Its principle involves the formation of diastereoisomer complexes after addition of neutral cyclodextrin to the running buffer. Fluvastatin enantiomers were separated on an uncoated fused silica with 100 mM borate solution containing 30 mg/mL of (2-hydroxypropyl)-β-cyclodextrin as running buffer and fenoprofen as an internal standard. The limit of detection and quantification for (+)-3R, 5S and (-)-3S, 5R-fluvastatin were 1.5 μg/mL and 2.5 μg/mL, respectively. Compared to chiral LC separations, CE analyses are cheaper (no chiral column, no solvent, low consumption of chiral selector) and peak efficiencies are higher by one order of magnitude.

There is only one CE method for quantification of rosuvastatin [118]. Currently, for rosuvastatin only a limited number of analytical methods are reported in literature. This is due to the fact that rosuvastatin is a new statin introduced in the EU in 2002 and approved in the US in 2003.

Using neutral β-cyclodextrin as chiral selector, the CZE method has been established for the chiral separation of pitavastatin calcium enantiomers [119]. Pitavastatin is a novel statin that potentially represents an important addition to the cardiovasculary therapy. In view of this, simple and efficient capillary electrophoretic methods for the determination of rosuvastatin and pitavastatin are highly required.

5. Simultaneous analysis of statin drugs

Since statins differ in their structure, analytical methods for their determination are developed individually. In fact, since statin drugs are never co-administered together during treatment of hyperlipimidemia, some authors even argued that there is no need for their simultaneous analysis. However, lately papers have been published that propose analytical methods that enable separation, identification and quantitative determination for two and even all six statins simultaneously in a single run. This kind of method would allow determination of any statin available on the market without the need of developing a new, separate, individual method for each statin, and could by used for simultaneous analysis of pharmaceutical dosage forms or in routine clinical monitoring.

A HPTLC method was published using precoated silica gel 60F 254 aluminum sheets and detection carried out at 239, 238 and 310 nm for determination of simvastatin, pravastatin and rosuvastatin in tablet dosage forms, respectively [120]. Far to our knowledge first HPLC-PDA method for simultaneous analysis of atorvastatin, lovastatin, pravastatin, rosuvastatin and simvastatin was reported for determination in pharmaceutical formulations and *in vitro* metabolism studies [121]. An uncommon gradient method using 3 mobile phase reservoirs was employed. Downside of the method is relatively long analysis time of 40 minutes and fluvastatin not being included in the simultaneous analysis.

An interesting method for pharmaceutical analysis of atorvastatin, simvastatin and lovastatin using a charged aerosol detector (CAD) was published [122]. CAD is a universal detector for HPLC that operates regardless of the physiochemical and spectral properties of non-volatile analytes. It can provide data complementary to UV or MS detectors. The eluent from the HPLC column is first nebulized and then charged. A highly sensitive electrometer generates a signal proportional to the analyte quantity. Although CAD is considered as a non-linear detector, the authors found a perfectly linear response ($R > 0.9995$). Sensitivity of the CAD detector was two folds greater then the UV detector; LOD of atorvastatin measured with UV and CAD detectors was 0.17 μg/mL and 0.08 μg/mL, respectively.

A HPLC-UV method for quantification of rosuvastatin, atorvastatin, fluvastatin, lovastatin and simvastatin and four fibrates in pharmaceutical dosage forms was developed [123]. In this paper a simple GC-FID method was also proposed for identification of atorvastatin, lovastatin and simvastatin along four fibrates.

Two stability-indicating HPLC methods for quantitative determination of pravastatin, fluvastatin, atorvastatin and rosuvastatin in pharmaceuticals were developed [124].

Methods for their simultaneous determination in biological samples could provide easy quantification of drug level in human plasma without changes in the chromatographic procedures for individual statin. Despite the fact that these drugs seem to be structurally similar, development of the method for their simultaneous determination in complex biological samples is quite challenging task as they differ significantly in terms of solubility, polarity, stability as well as optic characteristics. Until now several analytical methods have been developed for the determination of statins in biological samples simultaneously, simvastatin and atorvastatin [83], rosuvastatin and atorvastatin [90], simvastatin, pravastatin, rosuvastatin and atorvastatin [92].

Investigation of statins in the environment has become an important issue in the last years due to their large worldwide consumption and their potential adverse effects on animal and human health. Three different preconcentration techniques including solid phase extraction, dispersive liquid–liquid microextraction and stir-bar sorptive extraction have been optimized and compared for the simultaneous analysis of statin drugs in wastewater and river water samples by HPLC coupled to quadrupole-time-of-flight mass spectrometry [125].

Due to low sensitivity of CE, three on-line preconcentration strategies were investigated for the analysis of charged and neutral statins by MEKC [126]. A background electrolyte consisting of 20 mM ammonium bicarbonate buffer (pH 8.50) and 50 mM SDS was used for the separation of all statin molecules including mevastatin. The methods were applied for the analysis of statin analytes in wastewater samples. The more frequently prescribed statins are of environmental concern. Consequently, sensitive methods for investigation of distribution of statin drugs in the environment are very valuable.

We have introduced a universal MEKC method with diode-array detection for the simultaneous and short-time analysis of lovastatin, simvastatin, pravastatin, fluvastatin, atorvastatin and rosuvastatin in a single run [127]. Base hydrolysis was used to open lactone ring of lovastatin and simvastatin, administered as lactone prodrugs, in order to transform these compounds to the corresponding acid forms before analysis. This approach offered shorter analysis time due to a decrease of the migration times of negatively charged statin drugs in comparison to neutral lactone forms. The first step in CE method development for optimizing the separation of ionisable statin molecules was the selection of the buffer pH, which determined the extent of ionization and mobility of each drug. As reported in the literature, statins with β-hydroxy acid forms have pKa values between 4.1 and 4.6 [128] and statin molecules are completely in anionic forms above pH 6. Surfactant was added to the electrolyte to improve the selectivity of the separation. With SDS, negatively charged statins molecules were not strongly attracted to the micelles, and drug molecules were separated as a result of differences in their electrophoretic mobilities and lipophilicity. The addition of an organic modifier in the presence of SDS in the electrolyte solution played a key role in the separation of statin molecules. The addition of an organic modifier changes the selectivity and migration times due to the change in electrolyte viscosity, dielectric constant and the zeta potential. The SDS micelles and methanol in the concentration of 10% *v/v* added to the borate buffer (pH 9.5) were employed in order to reduce the analysis time while maintaining good resolution between all six statins. The new developed MEKC method enabled a powerful separation and simultaneous, simple and rapid determination of six statins in 5 min (Figure 3). The method developed was successfully applied to analysis of six different pharmaceutical dosage forms of statin drugs.

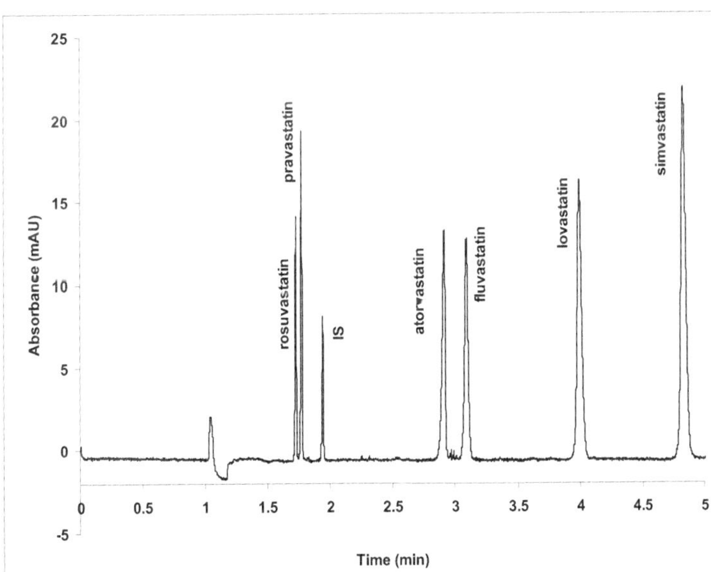

Figure 3. Electropherogram of a simultaneous analysis of statins

6. Conclusion

Development of the analytical methods for identification, purity evaluation and quantification of statin drugs has received a great deal of attention in the field of pharmaceutical analysis in recent years. This review includes trends and advances in separation methods developed for the analysis of statin molecules with different physical and chemical properties. The chapter surveys the application of chromatographic techniques for the determination of statins in pharmaceutical dosage forms and biological samples.

Stability indicating methods and papers describing impurity profiling are discussed in this review. Special emphasis is given to sample preparation as unavoidable and delicate step in bioanalitical methods for quantification of statins and their metabolites. The hyphenated technique that incorporates the efficient separation using liquid chromatography and sensitive detection by mass spectrometry has become an indispensable tool for quantification of statins in biological fluids and pharmacokinetic studies. Methods describing simultaneous analysis of different statins as well as drugs in combined pharmaceutical products and other co-administered drugs in therapy of cardiovascular disease are also described. The application of capillary electrophoresis as alternative separation technique for statins is considered and compared with chromatographic methods.

The use of statin drugs has augmented in recent years and is expected to increase further in the years ahead because high cholesterol and cardiovascular diseases are being diagnosed more frequently. Therefore, the development of new analytical methods for commercially available statins as well as novel upcoming statin drugs will be a future challenging task for many analysts.

Author details

Biljana Nigović, Ana Mornar and Miranda Sertić
University of Zagreb, Faculty of Pharmacy and Biochemistry, Zagreb, Croatia

Acknowledgement

This work was supported through Grant No. 0061117-1240 (Investigation of new methods in analysis of drugs and bioactive substances) from the Ministry of Science, Education and Sports of the Republic of Croatia.

7. References

[1] Mc Taggart F, Jones P (2008) Effects of Statins on High-Density Lipoproteins: A Potential Contribution to Cardiovascular Benefit. Cardiovasc. Drugs Ther. 22: 321-338.

[2] Endo A, Kuroda M, Tsujita Y (1976) ML-236A, ML-236B, and ML-236C, new inhibitors of cholesterogenesis produced by Penicillium citrinium. J. Antibiot. (Tokyo) 29: 1346-1348.

[3] Balk EM, Karas RH, Jordan HS, Kupelnick B, Chew P, Lau J (2004) Effects of statins on vascular structure and function: a systematic review. Am. J. Med. 117: 775-790.

[4] Bu D, Griffin G, Lichtman AH (2011) Mechanisms for the anti-inflammatory effects of statins. Curr. Opin. Lipidol. 22: 165-170.

[5] A.W. Alberts, J. Chen, G. Kuron, V. Hunt, C. Hoffman, J. Rothrock, M. Lopez, H. Joshua, E. Harris, A. Patchett, R. Monaghan, S. Currie, E. Stapley, G. Alberts-Schonberg, O. Hensens, J. Hirshchfield, K. Hoogsteen, J. Liesch, J. Springer (1980) Mevinolin: A highly potent competitive inhibitor of hydroxymethyl glutaryl-coenzyme A reductase and a cholesterol-lowering agent. Proc. Natl. Acad. Sci. USA 77: 3957–3961.

[6] Endo A (1979) Monacolin K, a new hypocholesterolemic agent produced by a *Monascus* sp. J. Antibiot. 32: 852–854.

[7] Schachter M (2005) Chemical, pharmacokinetic and pharmacodynamic properties of statins: an update. Fundam. Clin. Pharmacol. 19: 117-125.

[8] Corsini A, Maggi FM and Catapano AL (1995) Pharmacology of competitive inhibitors of HMG-CoA reductase. Pharmacol. Res. 34: 9-27.

[9] Jemal M, Xia YQ (2000) Bioanalytical method validation design for the simultaneous quantitation of analytes that may undergo interconversion during analysis. J. Pharm. Biomed. Anal. 23: 813-827.

[10] Nirogi R, Mudigonda K, Kandikere V (2007) Chromatography-Mass Spectrometry Methods for the Quantitation of Statins in Biological Samples. J. Pharm. Biomed. Anal. 44: 379-387.

[11] Yang DY, Hwang LS (2006) Study on the Conversion of Three Natural Statins from Lactone Forms to their Corresponding Hydroxy Acid Forms and their Determination in Pu-Erh Tea. J. Chromatogr. A. 119: 277-284.

[12] Huang Z, Xu Y, Li Y, Wang Y (2010) Conversion Investigation for Lovastatin and Its Derivates by HPLC. J. Chromatogr. Sci. 48: 631-636.

[13] Li Y, Zhang F, Wang Z, Hu Z (2004) Identification and Chemical Profiling of Monacolins in Red Yeast Rice Using High-Performance Liquid Chromatography with Photodiode Array Detector and Mass Spectrometry. J. Pharm. Biomed. Anal. 3: 1101-1112.

[14] Huang HN, Hua YY, Bao GR, Xie LH (2006) The Quantification of Monacolin K in Some Red Yeast Rice from Fujian Province and the Comparison of the Other Product, Chem Pharm Bull 54: 687-689.

[15] Li YG, Liu H, Wang ZT (2005) A Validated Stability Indicating HPLC with Photodiode Array Detector (PDA) Method for the Stress Tests of *Monascus purpureus* Fermented Rice, Red Yeast Rice, J. Pharm. Biomed. Anal. 39: 82-90.

[16] Song F, El-Demerdash, Lee SSH, Smith RE (2012) Fast Screening of Lovastatin in Red Yeast Rice Products by Flow Injection Tandem Mass Spectrometry. J. Pharm. Biomed. Anal. 57: 76-81.

[17] Novakova L, Šatinsky D, Solich P (2008) HPLC Methods for the Determination of Simvastatin and Atorvastatin, Trends Anal. Chem., 27(4):352-367.

[18] Ortega Markman BE, Pires Rosa PC, Walter Koschtchak MR (2010) Assessment of the Quality of Simvastatin Capsules from Compounding Pharmacies. Rev. Saude Publica 44: 1055-1062.

[19] Hill SW, Varker AS, Karlage K, Myrdal PB (2009) Analysis of Drug Content and Weight Uniformity for Half-Tablets of 6 Commonly Split Medications. J. Manag. Care Pharm. 15: 253-261.

[20] Ali H, Nazzal S (2009) Development and Validation of a Reversed-Phase HPLC Method for the Simultaneous Analysis of Simvastatin and Tocotrienols in Combined Dosage Forms. J. Pharm. Biomed. Anal. 49: 950-956.

[21] Zhao X, Guofei L, Zhang L, Tao X, Guan T, Hong M, Tang X (2010) Preparation and evaluation of nicotinic acid sustained-release pellets combined with immediate release simvastatin. Int. J. Pharm. 400: 42-48.

[22] Ozaltin N, Ucakturk E (2007) Simultaneous Determination of Ezetimibe and Simvastatin in Pharmaceutical Formulation by Dual-Mode Gradient LC. Chromatographia 66: 87-91.

[23] Oliveira PR, Barth T, Todeschni V, Dalmora SL (2007) Simultaneous Liquid Chromatographic Determination of Ezetimibe and Simvastatin in Pharmaceutical Products. J. AOAC Int. 90: 1566-1572.

[24] Chaudhari BG, Patel NM, Shah PB (2007) Stability-Indicating Reversed-Phase Liquid Chromatographic Method for Simultaneous Determination of Simvastatin and Ezetimibe from their Combination Drug Products. J. AOAC Int. 90: 1242-1249.

[25] Hefnawy M, Al-Omar M, Julkhuf S (2009) Rapid and Sensitive Simultaneous Determination of Ezetimibe and Simvastatin from their Combination Drug Products by Monolithic Silica High-Performance Liquid Chromatographic Column. J. Pharm. Biomed. Anal. 50: 527-534.

[26] Bertacche V, Milanese A, Nava D, Pini E, Stradi R (2007) Structural Elucidation of an Unknown Simvastatin By-product in Industrial Synthesis Starting from Lovastatin. J. Pharm. Biomed. Anal. 45: 642-647.

[27] Plumb RS, Jones MD, Rainville P, Castro-Perez JM (2007) The Rapid Detection and Identification of the Impurities of Simvastatin Using High Resolution sub 2 microm Particle LC Coupled to Hybrid Quadrupole Time of Flight MS Operating with Alternating High-low Collision Energy. J. Sep. Sci. 30: 2666-2675.

[28] Plumb RS, Jones MD, Rainville PD, Nicholson JK (2008) A Rapid Simple Approach to Screening Pharmaceutical Products Using Ultra-Performance LC Coupled to Time-of-Flight Mass Spectrometry and Pattern Recognition. J. Chromatogr. Sci. 46: 193-198.

[29] Reddy GV, Kumar AP, Reddy BV, Sreeramulu J (2009) Application of Ion-Trap Mass Spectrometry for Identification and Structural Determination of an Unknown Impurity in Simvastatin. Pharmazie 64: 638-641.

[30] Bhatia MS, Jadhav SD, Bhatia NM, Choudhari PB, Ingale KB (2011) Synthesis, Characterization and Quantification of Simvastatin Metabolites and Impurities. Sci. Pharm. 79: 601-614.

[31] Campos-Lara M, Mendoza-Espinoza J (2008) Development of a Selective Extraction Method for Pravastatin Qantification in Tablets using HPLC with Ultraviolet Detection. J. Liq. Chromatogr. R. T. 31: 619-623.

[32] Kocijan A, Grahek R, Zupančić-Kralj L (2006) Identification of an Impurity in Pravastatin by Application of Collision-Activated Decomposition Mass Spectra. Acta Chim. Slov. 53: 464-468.
[33] Onal A, Sagirli O (2006) Development of a Selective LC Method for the Determination of Pravastatin Sodium. Chromatographia, 64: 157-162.
[34] Brain-Isasi, Requena C, Alvarez-Lueje A (2008) Stability study of Pravastatin Under Hydrolity Conditions Assessed by HPLC. J. Chil. Chem. Soc. 52: 1684-1688.
[35] Mielcarek J, Grobelny P, Osmalek T (2009) Identification of Photoproducts of Fluvastatin in Solutions. J. Plan. Chromatogr. 22: 137-140.
[36] Shirkhedkar AA, Surana SJ (2010) Development and Validation of a Reversed-Phase High-Performance Thin-Layer Chromatography-Densitometric Method for Determination of Atorvastatin Calcium in Bulk Drug and Tablets. J. AOAC, Int. 93: 798-803.
[37] Khedr A (2008) Stability-Indicating High-Performance Liquid Chromatographic Assay of Atorvastatin with Fluorescence Detection. J AOAC 90: 1547-1553.
[38] Mustafa G, Azeem A, Ahmad FJ, Khan ZI, Shakeel F, Talegaonkar S (2010) Stability-Indicating RP-HPLC Method for Analysis of Atorvastatin in Bulk Drug, Marketed Tablet and Nanoemulsion Formulation. J. Chil. Chem. Soc. 55: 184-188.
[39] Mohammadi A, Rezanour N, Ansari Dogahch M, Ghorbani Bidkorbeh F, Hashem M, Walker RB (2007) A Stabiliy-Indicating High Performance Liquid Chromatographic (JPLC) Assay for the Simultaneous Determination of Atorvastatin and Amlodipine in Commercial Tablets. J. Chromatogr B. 846: 215-221.
[40] Vora DN, Kadav AA (2008) Validated Ultra HPLC Method for the Simultaneous Determination of Atorvastatin, Acetylsalicylic Acid and their Degradation Products in Capsules. J. L. Chromatogr. R. T. 31: 2821-2837.
[41] Shah RP, Kumar V, Singh S (2008) Liquid Chromatography/Mass Spectrometric Studies on Atorvastatin and its Stress Degradation Products. Rapid Commun. Mass Spectrom. 22: 613-622.
[42] Kračun M, Kocijan A, Bastarda A, Grahek R, Plavec J, Kocjan D (2009) Isolation and Structure Determination of Oxidative Degradation Products of Atorvastatin. J. Pharm. Biomed. Anal. 50: 729-736.
[43] Kadav AA, Vora DN (2008) Stability Indicating UPLC Method for Simultaneous Determination of Atorvastatin, Fenofibrate and their Degradation Products in Tablets. J. Pharm. Biomed. Anal. 48: 120-126.
[44] Mornar A, Damic M, Nigovic B (2010) Separation, Characterization, and Quantification of Atorvastatin and Related Impurities by Liquid Chromatography-Electrospray Ionization Mass Spectrometry. Anal. Lett. 43: 2859-2871.
[45] Petkovska R, Cornett C, Dimitrovska A (2008) Development and Validation of Rapid Resolution RP-HPLC Method for Simultaneous Determination of Atorvastatin and Related Compounds by Use of Chemometrics. Anal. Lett. 41: 992-1009.
[46] Malenovic A, Jancic-Stojanovic B, Kostic N, Ivanovic D, Medenica M (2011) Impurities Retention in Micelaar Liquid Chromatography. Chromatography. 73: 993-998.

[47] Panchal HJ, Suhagia BN (2010) Simultaneous Determination of Atorvastatin Calcium and Ramipril in Capsule Dosage Forms by High-Performance Liquid Chomatography and High-Performance Thin Layer Chromatography. J. AOAC Int. 93: 1450-1457.

[48] Skirkhedkar AA, Surana SJ (2009) Simultaneous Densitometric TLC Analysis of Atorvastatin Calcium and Fenofibrate in the Bulk Drug and in Pharmaceutical Formulations. J. Planar Chromatogr. 22: 355-358.

[49] Raja Rajeswai K, Sankar GG, Rao AL, Seshagiriraio JVLN (2006) RP-HPLC method for the Simultaneous Determination of Atorvastatin and Amlodipine in Tablet Dosage Form. Indian J. Pharm. Sci. 68: 275-277.

[50] Nakarani NV, Bhatt KK, Patel RD, Bhatt HS (2007) Estimation of Atorvastatin Calcium and Fenofibrate in Tablets by Derivate Spectrophotometry and Liquid Chromatography. J. AOAC Int. 90: 700-705.

[51] Sechachalam U, Kothapally CB (2008) HPLC analysis for simultaneous determination of atorvastatin and ezetimibe in pharmaceutical formulations. J. Liq. Chromatogr. R. T. 31: 714-721.

[52] Lincy J, Mathew G, Venkata RR (2008) Simultaneous Estimation of Atorvastatin and Ramipril by RP-HPLC and Spectroscopy. Pak. J. Pharm. Sci. 21: 282-284.

[53] Sivakumar T, Manavalan R, Muralidharan C, Valliappan K (2007) An Improved HPLC Method with the Aid of a Chemometric Protocol: Simultaneous Analysis of Amlodipine and Atorvastatin in Pharmaceutical Formulations. J. Sep. Sci. 30: 3143-3153.

[54] Londhe SV, Deshmukh RS, Mulgund SV, Jain KS (2011) Development and Validation of a Reversed-Phase HPLC Method for Simultaneous Determination of Aspirin, Atorvastatin Calcium and Clopidogrel Bisulphate in Capsules. Indian J. Pharm. Sci. 73: 23-29.

[55] Damle MS, Patole SM, Potale LV, Khodke AS (2010) A validated HPLC method for Analysis of Atorvastatin Calcium, Ramipril and Aspirin as the Bulk Drug and in Combined Capsule Dosage Forms. Pharmacie Globale 1: 1-5.

[56] Kumar V, Shah RP, Singh S (2008) LC and LC-MS Methods for the Investigation of Polypills for the Treatment of Cardiovascular Diseases Part 1. Separation of Active Components and Classification of their Interaction/Degradation Products. J. Pharm. Biomed. Anal. 47: 508-515.

[57] Mehta TN, Patel AK, Kulkarni GM,, Suubbaiah G (2005) Determination of Rosuvastatin in the Presence of its Degradation Products by a Stability-Indicating LC Method. J. AOAC Int, 88: 1142-1147.

[58] Sane RT, Kamat SS, Menon SN, Inamdar SR, Mote MR (2005) Determination of Rosuvastatin Calcium in its Bulk Drug and Pharmaceutical Preparations by High-Performance Thin-Layer Chromatography. J. Planar Chromatogr. 18: 194-198.

[59] Varghese SJ, Ravi TK (2010) Determination of Rosuvastatin and Ezetimibe in a Combined Tablet Dosage form using High-Performance Column Liquid Chromatography and High-Performance Thin-Layer Chromatography. J. AOAC Int. 93: 1222-1227.

[60] Gajjar AK, Shah VD (2010) Development and Validation of a Stability-Indicating Reversed-Phase HPLC Method for Simultaneous Estimation of Rosuvastatin and Ezetimibe from their Combination Dosage Forms. Eurasian J. Anal. Chem.. 5:.265-283.

[61] Sultana N, Arayne MS, Iftikhar B (2008) Simultaneous Determination of Atenolol, Rosuvastatin, Spironolactone, Glibenclamide and Naproxen Sodium in Pharmaceutical Formulations and Human Plasma by RP-HPLC. J. of the Chin. Chem. Soc.. 55:1022-1029.

[62] Kumar/Satheesh NK, Baghyalakshmi J (2007) Determination and Quantification of Pitavastatin Calcium in Tablet Dosage Formulation by HPTLC Method. Anal. Lett. 40: 2625-2632.

[63] Panchal HJ, Suhagia BN, Patel NJ, Patel BH (2008) A Simple and Sensitive HPTLC Method for Quantitative Analysis of Pitavastatin Calcium in Tablets. J. Planar Chromatogr. 21: 267-270.

[64] Kumar NS, Nisha N, Nirmal J, Sonali N, Bagyalakshmi J (2011) HPLC determination of pitavastatin calcium in pharmaceutical dosage forms, Pharm Anal Acta 2: 119 doi:10.4172/2153-2435.1000119

[65] Panchal HJ, Suhagia BN (2011) Stability-Indicating Liquid Chromatographic Method for Analysis of Pitavastatin Calcium in Tablet Dosage Forms. Acta Chromatogr. 23: 81-64.

[66] Grobleny P, Viola G, Vedaldi D, Dall'Acqua F, Gliszczynska-Swiglo A, Mielcarek J (2009) Photostability of Pitavastatin – A Novel HMG-CoA Reductase Inhibitor. J. Pharm. Biomed. Anal. 50: 597-601.

[67] Panchal H, Suhagia BN (2011) Simultaenous Determination and Validation of Pitavastatin Calcium and Ezetimibe in Binary Mixture by Liquid Chromatography. Int. J. Pharm. Tech. Research. 3: 2155-2161.

[68] Gomas AR, Ram PR, Srinivas N, Sriramulu J (2010) Degradation Pathway for Pitavastatin Calcium by Validated Stability Indicating UPLC Method. Am. J. Anal. Chem. 2: 83-90.

[69] Panchal HJ, Suhagia BN, Patel MM, Patel BH (2009) Estimation of Pitavastatin cCalcium in Tablet Dosage Forms by Column Liquid Chromatography and Ultraviolet Spectrophotometry. J. AOAC Int. 92: 158-165.

[70] Ertürk S, Önal A, Çetin SM (2003) Analytical Methods for the Quantitative Determination of 3-hydroxy-3-methylglutaril Coenzyme A Reductase Inhibitors in Biological Samples. J. Chromatogr. B. 793: 193-205.

[71] Farahani H, Norouzi P, Beheshti A, Sobhi HR, Dinarvand R, Ganjali MR (2009) Quantitation of Atorvastatin in Human Plasma using Directly Suspended Acceptor Droplet in Liquid–liquid–liquid Microextraction and High-performance Liquid Chromatography-ultraviolet Detection. Talanta. 80: 1001-1006.

[72] Senthamil Selvan P, Pal TK (2009) Chromatography–tandem Mass Spectrometry Method for the Simultaneous Quantitation of Metoprolol Succinate and Simvastatin in Human Plasma. J. Pharm. Biomed. Anal. 49: 780-785.

[73] Hamidi M, Zarei N, Shahbazi M-A (2009) A Simple and Sensitive HPLC-UV Method for Quantitation of Lovastatin in Human Plasma: Application to a Bioequivalence Study. Biol. Pharm. Bull. 32: 1600-1603.

[74] Macwan JS, Ionita IA, Akhlaghi F (2012) A Simple Assay for the Simultaneous Determination of Rosuvastatin Acid, Rosuvastatin-5S-lactone, N-Desmethyl rosuvastatin in Human Plasma using Liquid chromatography-tandem Mass Spectrometry (LC-MS/MS). Anal. Bioanal.Chem. 402: 1217-1227.

[75] Apostolou C, Kousoulos C, Dotsikas Y, Soumelas G-S, Kolocouri F, Ziaka A, Loukas YL (2008) An Improved and Fully Validated LC–MS/MS Method for the Simultaneous Quantification of Simvastatin and Simvastatin Acid in Human Plasma. J. Pharm. Biomed. Anal. 46: 771–779.

[76] Gonzalez O, Iriarte G, Ferreirós N, Maguregui MI, Alonso RM, Jiménez RM (2009) Optimization and Validation of a SPE-HPLC-PDA-fluorescence Method for the Simultaneous Determination of Drugs Used in Combined Cardiovascular Therapy in Human Plasma. J. Pharm. Biomed. Anal. 50: 630-639.

[77] Lv H, Sun J-G, Wang G-J, Zhu X-Y, Zhang Yi, Gu S-H, Liang Y, Sun J (2007) Determination of Pitavastatin in Human Plasma via HPLC-ESI-MS/MS and Subsequent Application to a Clinical Study in Healthy Chinese Volunteers. Clin. Chim. Acta. 386: 25-30.

[78] Trivedi RK, Kallem RR, Mullangi R, Srinivas NR (2005) Simultaneous determination of rosuvastatin and fenofibric acid in human plasma by LC–MS/MS with electrospray ionization: Assay development, validation and application to a clinical study. J. Pharm. Biomed. Anal. 39: 661-669.

[79] Gao J, Zhong D, Duan X, Chen X (2007) Liquid Chromatography/negative Ion Electrospray Tandem Mass Spectrometry Method for the Quantification of Rosuvastatin in Human Plasma: Application to a Pharmacokinetic Study. J. Chromatogr. B. 856: 35-40.

[80] Lan K, Jiang X, Li Y, Wang L, Zhou J, Jiang Q, Ye L (2007) Quantitative Determination of Rosuvastatin in Human Plasma by Ion Ppair Liquid–liquid Extraction Using Liquid Chromatography with Electrospray Ionization Tandem Mass Spectrometry. J. Pharm. Biomed. Anal. 44: 540-546.

[81] Nasir F, Iqbal Z, Khan A, Ahmad L, Shah Y, Khan AZ, Khan JA, Khan S (2011) Simultaneous Determination of Timolol Maleate, Rosuvastatin Calcium and Diclofenac Sodium in Pharmaceuticals and Physiological Fluids Using HPLC-UV. J. Chromatogr. B. 879: 3434- 3443.

[82] Zhang J, Rodila R, Gage E, Hautman M, Fan L, King LL, Wu H, El-Shourbagy TA (2010) High-throughput Salting-out Assisted Liquid/liquid Extraction with Acetonitrile for the Simultaneous Determination of Simvastatin and Simvastatin Acid in Human Plasma with Liquid Chromatography. Anal. Chim. Acta. 661: 167-172.

[83] Nováková L, Vlčková H, Šatínský D, Sadílek P, Solichová D, Bláha M, Bláha V, Solich P (2009) Ultra High Performance Liquid Chromatography Tandem Mass Spectrometric Detection in Clinical Analysis of Simvastatin and Atorvastatin. J. Chromatogr. B. 877: 2093-2103.

[84] Di B, Su M-X, Yu F, Qu L-J, Zhao L-P, Cheng M-C, He L-P (2008) Solid-phase Extraction and Liquid Chromatography/tandem Mass Spectrometry Assay for the Determination of Pitavastatin in Human Plasma and Urine for Application to Phase I Clinical Pharmacokinetic Studies. J. Chromatogr. B. 868: 95-101.

[85] Mertens B, Cahay B, Klinkenberg R, Streel B (2008) An Automated Method for the Simultaneous Determination of Pravastatin, 3-hydroxy Isomeric Metabolite, Pravalactone and Fenofibric Acid in Human Plasma by Sensitive Liquid Chromatography Combined with Diode Array and Tandem Mass Spectrometry Detection. J. Chromatogr. A. 1189: 493-502.

[86] Vlčková H, Solichová D, Bláha M, Solich P, Nováková L (2011) Microextraction by Packed Sorbent as Sample Preparation Step for Atorvastatin and its Metabolites in Biological Samples-Critical evaluation. J. Pharm. Biomed. Anal. 55: 301-308.

[87] Iriarte G, Gonzalez O, Ferreirós N, Maguregui MI, Alonso RM, Jiménez RM (2009) Validation of a Fast Liquid Chromatography–UV Method for the Analysis of Drugs Used in Combined Cardiovascular Therapy in Human Plasma. J. Chromatogr. B. 877: 3045-3053.

[88] Tian L, Huang Y, Jia Y, Hua L, Li Y (2008) Development and Validation of a Liquid Chromatography–tandem Mass Spectrometric Assay for Pitavastatin and its Lactone in Human Plasma and Urine. J. Chromatogr. B. 865: 127-132

[89] Macwan JS, Ionita IA, Dostalek M, Akhlaghi F (2011) Development and Validation of a Sensitive, Simple, and Rapid Method for Simultaneous Quantification of Atorvastatin and its Acid and Lactone Metabolites by Liquid Chromatography-tandem Mass Spectrometry (LC-MS/MS). Anal. Bioanal. Chem. 400: 423-433.

[90] Shah Y, Iqbal Z, Ahmad L, Khan A, Khan MI, Nazir S, Nasir F (2011) Simultaneous Determination of Rosuvastatin and Atorvastatin in Human Serum using RP-HPLC/UV Detection: Method Development, Validation and Optimization of Various Experimental Parameters. J. Chromatogr. B. 879: 557-563.

[91] Khan IU, Kausar T, Ashfaq M, Sharif S (2010) Development and Validation of Liquid Chromatographic Method for the Simultaneous Estimation of Ezetimibe and Lovastatin in Human Plasma. J. Chil. Chem. Soc. 55: 461-464.

[92] *Sultana N, Arayne MS, Shahzad W* (2010) Simultaneous Determination of Ceftriaxone Sodium and Statin Drugs in Pharmaceutical Formulations and Human Serum by RP-HPLC. J. Chil. Chem. Soc. 55: 193-198.

[93] Shen-Tu J, Xu X, Liu J, Hu X, Chen J, Wu L, Huang M, Zhou H (2009) Determination of Pitavastatin in Human Plasma by LC-MS-MS. Chromatographia. 69: 1041-1047.

[94] Polagani SR, Pilli NR, Gandu V (2012) High Performance Liquid Chromatography Mass Spectrometric Method for the Simultaneous Quantification of Pravastatin and Aspirin in Human Plasma: Pharmacokinetic Application. J. Pharm. Anal. In press

[95] Kotagiri H, Gannu R, Palem CR, Yamsani SK, Yamsani VV, Yamsani MR (2011) Simultaneous Determination of Glimepiride and Atorvastatin in Human Serum by High-performance Liquid Chromatography: Application to Pharmacokinetic Study. J. Liqu. Chromatogr. R. T. 34: 2420-2432.

[96] Ahmad M, Qamar-Uz-Zaman M, Madni A, Usman M, Atif M, Akhtar N, Murtaza G (2011) Pharmacokinetic and Bioavailability Studies of Commercially Available Simvastatin Tablets in Healthy and Moderately Hyperlipidemic Human Subjects. J. Chem. Soc. Pak. 33: 49-54.

[97] Taillon MP, Côté C, Furtado N, Garofolo F (2011) Potentially New Isobaric Metabolite of Fluvastatin Observed by LC-MS/MS During Incurred Sample Analysis. Bioanalysis. 3: 1827-1835.

[98] Yu Q, Hu ZY, Zhu FY, Zhu JH, Wan LL, Li Y, Guo C (2011) HPLC-MS-MS for the Simultaneous Determination of Atorvastatin and Amlodipine in Plasma of Hypertensive Patients. Chromatographia. 73: 257-262.

[99] Bhatia NM, Gurav SB, Jadhav SD, Bhatia MS (2012) RP-HPLC Method for Simultaneous Estimation of Atorvastatin Calcium, Losartan Potassium, Atenolol, and Aspirin from Tablet Dosage form and Plasma. J. Liqu. Chromatogr. R. T. 35: 428-443

[100] Mornar A, Damić M, Nigović B (2011) Pharmacokinetic Parameters of Statin Drugs Characterized by Reversed Phase High-performance Liquid Chromatography. Anal. Lett. 44: 1009-1020.

[101] Sarr FS, André C, Guillaume YC (2008) Statins (HMG-coenzyme A Reductase Inhibitors)-biomimetic Membrane Binding Mechanism Investigated by Molecular Chromatography J. Chromatogr. B 868: 20-27.

[102] Altria K, Marsh A, Sanger-van de Griend C (2006) Capillary Electrophoresis for the Analysis of Small-Molecule Pharmaceuticals. Electrophoresis 27: 2263-2282.

[103] Terabe S (2009) Capillary Separation: Micellar Electrokinetic Chromatography. Annu. Rev. Anal. Chem. 2: 99-120. [104] Suntornsuk L (2007) Recent Advances of Capillary Electrophoresis in Pharmaceutical Analysis. J. Chromatogr. Sci. 45: 559-577.

[105] Kocijan A, Grahek R, Bastrada A, Zupancic-Kralj L (2005) Fast Analysis of Pravastatin in Production Media. J. Chromatogr. B 822: 311-315.

[106] Nigović B, Vegar I (2008) Capillary Electrophoresis Determination of Pravastatin and Separation of Its Degradation Products. Croat. Chem. Acta. 81: 615-622.

[107] Mulvana D, Jemal M, Pulver SC (2000) Quantitative Determination of Pravastatin and its Biotransformation Products in Human Serum by Turbo Ion Spray LC/MS/MS. J. Pharm. Biomed. Anal. 23: 851-866.

[108] Nair JB, Izzo CG (1933) Anion Screening for Drugs and Intermediates by Capillary Ion Electrophoresis. J. Chromatogr. A 640: 445-461.

[109] Kittell J, Borup B, Voladari R, Zahn K (2005) Parallel Capillary Electrophoresis for the Quantitative Screening of Fermentation Broths Containing Natural Products. Metab. Eng. 7: 53-58.

[110] Javernik Rajh S, Kreft S, Štrukelj B, Vrečer F (2003) Comparison of CE and HPLC Methods for Determining Lovastatin and Its Oxidation Products after Exposure to an Oxidative Atmosphere. Croat. Chem. Acta 76:263-268.

[111] Li M, Fan LY, Zhang W, Cao CX (2007) Stacking and Quantitative Analysis of Lovastatin in Urine Samples by the Transient Moving Chemical Reaction Boundary Method in Capillary Electrophoresis. Anal. Bioanal. Chem. 387: 2719-2725.

[112] Li M, Fan LY, Zhang W, Sun J, Cao CX (2007) Quantitative Analysis of Lovastatin in Capsule of Chinese Medicine Monascus by Capillary Zone Electrophoresis with UV–vis Detector. J. Pharm. Biomed. Anal. 43: 387-392.

[113] Srinivasu MK, Raju AN, Reddy GO (2002) Determination of Lovastatin and Simvastatin in Pharmaceutical Dosage Forms by MEKC. J. Pharm. Biomed. Anal. 29: 715-721.

[114] Guihen E, Sisk GD, Scully NM, Glennon JD (2006) Rapid Analysis of Atorvastatin Calcium Using Capillary Electrophoresis and Microchip Electrophoresis. Electrophoresis 27: 2338-2347.

[115] Nigović B, Damić M, Injac R, Kočevar Glavač N, Štrukelj B (2009) Analysis of Atorvastatin and Related Substances by MEKC. Chromatographia 69: 1299-1305.

[116] Hefnawy MM, Sultan M, Al-Johar H (2009) Development of Capillary Electrophoresis Technique for Simultaneous Measurement of Amlodipine and Atorvastatin from Their Combination Drug Formulations. J. Liq. Chromatogr. Relat. Technol. 32: 2923-2942.

[117] Trung TQ, Dung PT, Hoan NN, Kim DJ, Lee JH and Kim KH (2008) Chiral Separation of Fluvastatin Enantiomers by Capillary Electrophoresis. Arch. Pharm. Res. 31: 1066-1072.

[118] Süslü İ, Çelebier M, Altinöz S (2007) Determination of Rosuvastatin in Pharmaceutical Formulations by Capillary Zone Electrophoresis. Chromatographia 66: S65-S72.

[119] Cheng X, Wang L, Yang G, Cheng J, Zhang Y (2010) Chiral Separation of Pitavastatin Calcium Enantiomers by Capillary Zone Electrophoresis. Chin. J. Chrom. 28: 1089-1093.

[120] Chaduhari BG, Patel NM, Shah PB (2007) Determination of Simvastatin, Pravastatin Sodium and Rosuvastatin Calcium in Tablet Dosage Forms by HPTLC. Indian J. Pharm. Sci 69: 130-132.

[121] Pasha K, Muzeeb S, Basha SJS, Shashikumar D, Mullangi R, Srinivas N (2006) Analysis of Five HMG-CoA Reductase Inhibitors- Atorvastatin, Lovastatin, Pravastatin, Rosuvastatin and Simvastatin: Pharmacological, Pharmacokinetic and Analytical Overview and Development of a New Method for use in Pharmaceutical Formulations Analysis and *in vitro* Metabolism Studies.Biomed. Chromatogr. 20: 282-293.

[122] Novakova L, Lopez Sofia Arnal, Solichova D, Šatinsky D, Kulichova B, Horna A, Solich P (2009) Comparison of UV and Charged Aerosol Detection approach in Pharmaceutical Analysis of Statins. Talanta 78: 834-839

[123] Kublin E, Malanowicz E, Kaczmarska-Graczyk B, Mazurek AP (2012) Development of Xhromatographic Method for Determination of Drugs Reducing Cholesterol Level. Acta Pol. Pharm. 69: 139-143.

[124] Gomes FP, Garcia PL, Alves JMP, Singh AK, Kedor-Hackmamm ERM, Santoro MIRM (2009) Development and Validation of Stability-Indicating HPLC Methods for Quantitaive Determination of Pravastatin, Fluvastatin, Atorvastatin and Rosuvastatin in Pharmaceuticals. Anal. Lett. 42: 1784-1804.

[125] Martín J, Buchberger W, Alonso E, Himmelsbach M, Aparicio I (2011) Comparison of Different Extraction Methods for the Determination of Statin Drugs in Wastewater and River Qater by HPLC/Q-TOF-MS. Talanta. 85: 607-615.

[126] Dawod M, Breadmore MC, Guijt RM, Haddad PR (2010) Strategies for the On-line Preconcentration and Separation of Hypolipidaemic Drugs Using Micellar Electrokinetic Chromatography. J. Chromatogr. A 1217: 386-393.
[127] Damić M, Nigović B (2010) Fast Analysis of Statins in Pharmaceuticals by MEKC. Chromatographia 71: 233-240.
[128] Ishihama Y, Nakamura M, Miwa T, Kajima T, Asakawa N (2002) A Rapid Method for pKa Determination of Drugs Using Pressure-Assisted Capillary Electrophoresis with Photodiode Array Detection in Drug Discovery. J. Pharm. Sci. 91: 933-942.

Permissions

The contributors of this book come from diverse backgrounds, making this book a truly international effort. This book will bring forth new frontiers with its revolutionizing research information and detailed analysis of the nascent developments around the world.

We would like to thank Leonardo de Azevedo Calderon, for lending his expertise to make the book truly unique. He has played a crucial role in the development of this book. Without his invaluable contribution this book wouldn't have been possible. He has made vital efforts to compile up to date information on the varied aspects of this subject to make this book a valuable addition to the collection of many professionals and students.

This book was conceptualized with the vision of imparting up-to-date information and advanced data in this field. To ensure the same, a matchless editorial board was set up. Every individual on the board went through rigorous rounds of assessment to prove their worth. After which they invested a large part of their time researching and compiling the most relevant data for our readers. Conferences and sessions were held from time to time between the editorial board and the contributing authors to present the data in the most comprehensible form. The editorial team has worked tirelessly to provide valuable and valid information to help people across the globe.

Every chapter published in this book has been scrutinized by our experts. Their significance has been extensively debated. The topics covered herein carry significant findings which will fuel the growth of the discipline. They may even be implemented as practical applications or may be referred to as a beginning point for another development. Chapters in this book were first published by InTech; hereby published with permission under the Creative Commons Attribution License or equivalent.

The editorial board has been involved in producing this book since its inception. They have spent rigorous hours researching and exploring the diverse topics which have resulted in the successful publishing of this book. They have passed on their knowledge of decades through this book. To expedite this challenging task, the publisher supported the team at every step. A small team of assistant editors was also appointed to further simplify the editing procedure and attain best results for the readers.

Our editorial team has been hand-picked from every corner of the world. Their multi-ethnicity adds dynamic inputs to the discussions which result in innovative outcomes. These outcomes are then further discussed with the researchers and contributors who give their valuable feedback and opinion regarding the same. The feedback is then collaborated with the researches and they are edited in a comprehensive manner to aid the understanding of the subject.

Apart from the editorial board, the designing team has also invested a significant amount of their time in understanding the subject and creating the most relevant covers. They scrutinized every image to scout for the most suitable representation of the subject and create an appropriate cover for the book.

The publishing team has been involved in this book since its early stages. They were actively engaged in every process, be it collecting the data, connecting with the contributors or procuring relevant information. The team has been an ardent support to the editorial, designing and production team. Their endless efforts to recruit the best for this project, has resulted in the accomplishment of this book. They are a veteran in the field of academics and their pool of knowledge is as vast as their experience in printing. Their expertise and guidance has proved useful at every step. Their uncompromising quality standards have made this book an exceptional effort. Their encouragement from time to time has been an inspiration for everyone.

The publisher and the editorial board hope that this book will prove to be a valuable piece of knowledge for researchers, students, practitioners and scholars across the globe.

List of Contributors

Emma B. Casanave
Cátedra de Fisiología Animal, Dpto. de Biología, Bioquímica y Farmacia, Universidad Nacional del Sur, Bahía Blanca, Buenos Aires, Argentina
Consejo Nacional de Investigaciones Científicas y Técnicas, (CONICET) Argentina

M. Soledad Araujo
Cátedra de Fisiología Animal, Dpto. de Biología, Bioquímica y Farmacia, Universidad Nacional del Sur, Bahía Blanca, Buenos Aires, Argentina

Gustavo H. López
Cátedra de Bioanalítica II, Dpto. de Biología, Bioquímica y Farmacia, Universidad Nacional del Sur, Bahía Blanca, Buenos Aires, Argentina

Leonardo de A. Calderon, Juliana P. Zuliani and Rodrigo G. Stábeli
Center of Applied Biomolecular Studies in Health (CEBio), Fiocruz Rondônia, Oswaldo Cruz Foundation - FIOCRUZ and Medicine Department, Federal University of Rondônia - UNIR, Porto Velho, RO, Brazil

Andreimar M. Soares
Center of Applied Biomolecular Studies in Health (CEBio), Fiocruz Rondônia, Oswaldo Cruz Foundation - FIOCRUZ, Porto Velho, RO, Brazil

Anderson M. Kayano, Antonio C. Neto, Andrea A. de Moura and Gizeli S. Gimenez
Experimental Biology, Federal University of Rondônia - UNIR, Porto Velho, RO, Brazil

Kayena D. Zaqueo and Rodrigo S. Silva
Biodiversity and Biotechnology, Bionorte Web, Porto Velho, RO, Brazil Experimental Biology, Federal University of Rondônia - UNIR, Porto Velho, RO, Brazil

Cleópatra A. S. Caldeira
Biological Sciences, Federal University of Rondônia - UNIR, Porto Velho, RO, Brazil

Phan Van Chi and Nguyen Tien Dung
Department of Protein Biochemistry, Institute of Biotechnology (IBT), Vietnam Academy of Science & Technology (VAST), Hanoi, Vietnam

Paolo Lucci
Department of Nutrition and Biochemistry, Faculty of Sciences, Pontificia Universidad Javeriana, Bogotà D.C., Colombia

Deborah Pacetti and Natale G. Frega
Department of Agricultural, Food and Environmental Sciences, Marche Polytechnic University, Ancona,
Italy

Oscar Núñez
Department of Analytical Chemistry, University of Barcelona, Barcelona, Spain

Robert Roškar and Tina Trdan Lušin
University of Ljubljana, Faculty of Pharmacy, Slovenia

JolantaRubaj, Waldemar Korol and Grażyna Bielecka
National Research Institute of Animal Production, National Feed Laboratory, Poland

Juan M. Traverso-Soto, Eduardo González-Mazo and Pablo A. Lara-Martín
Departamento de Química Física, Facultad de Ciencias del Mar y Ambientales, Campus de Excelencia
Internacional del Mar (CEI·MAR), Universidad de Cádiz, Puerto Real, Spain

Saksit Chanthai
Department of Chemistry, Faculty of Science, Khon Kaen University, Khon Kaen, Thailand

Thanee Tessiri
Academic and Research Services Unit, Faculty of Pharmaceutical Sciences, Khon Kaen University, Khon Kaen, Thailand

Manabu Asakawa and Keisuke Miyazawa
Department of Bioresource Science and Technology, Department of Bioresource Science and Technology, Graduate School of Biosphere Science, Hiroshima University, Kagamiyama, Higashi-Hiroshima, Hiroshima, Japan

Yasuo Shida
Clean Energy Research Center, University of Yamanashi, Takeda, Kofu, Yamanashi, Japan

Tamao Noguchi
Tokyo Healthcare University, Setagaya, Tokyo, Japan

Jin HaiRu and Jiang Xiangyan
College of Chemistry and Life Sciences, Zhejiang Normal University, Jinhua, China

Wangsa T. Ismaya
Department of Chemistry, Faculty of Mathematics and Natural Sciences, Universitas Padjadjaran,

Jatinangor, Sumedang, Indonesia
Institute for Bioengineering and Bioscience, School of Chemistry and Biochemistry, Georgia Institute of Technology, IBB Petit Building, Atlanta, Georgia, USA

Khomaini Hasan
Department of Chemistry, Faculty of Mathematics and Natural Sciences, Universitas Padjadjaran, Jatinangor, Sumedang, Indonesia
Loschmidt Laboratories, Department of Experimental Biology and Research Centre for Toxic Compound in the Environment, Faculty of Science, Masaryk University, Brno, Czech Republic

Toto Subroto and Soetijoso Soemitro
Department of Chemistry, Faculty of Mathematics and Natural Sciences, Universitas Padjadjaran,
Jatinangor, Sumedang, Indonesia

Dessy Natalia
Biochemistry Research Division, Faculty of Mathematics and Natural Sciences, Institut Teknologi
Bandung, Bandung, Indonesia

Mahitosh Mandal
School of Medical Science and Technology, Indian Institute of Technology, Kharagpur, West Bengal, India

Moganavelli Singh
Department of Biochemistry, School of Life sciences, University of KwaZulu-Natal, Westville, Durban, South Africa

Sankar Ramachandran
School of Medical Science and Technology, Indian Institute of Technology, Kharagpur, West Bengal, India
Department of Biochemistry, School of Life sciences, University of KwaZulu-Natal, Westville, Durban, South Africa

Vijay Prabha and Siftjit Kaur
Department of Microbiology, Punjab University, Chandigarh, India

Maria Helene Giovanetti Canteri
Federal Technological University of Paraná, Ponta Grossa, Paraná, Brazil

Alessandro Nogueira and Gilvan Wosiacki
State University of Ponta Grossa, Av. General Carlos Cavalcanti, Ponta Grossa, Paraná, Brazil

Carmen Lúcia de Oliveira Petkowicz
Federal University of Paraná, Curitiba, Paraná, Brazil

Fabrice Mutelet
Laboratoire Réactions et Génie des Procédés (CNRS, UPR 3349), Ecole Nationale Supérieure des Industries Chimiques, Nancy CEDEX, France

Mihalj Poša
Department of Pharmacy, Faculty of Medicine, University of Novi Sad, Serbia

Biljana Nigović, Ana Mornar and Miranda Sertić
University of Zagreb, Faculty of Pharmacy and Biochemistry, Zagreb, Croatia